Outlines of Paint Technology

Third Edition

W M MORGANS

Consultant Paint Technologist

CBS PUBLISHERS & DISTRIBUTORS PVT. LTD.

NEW DELHI • BENGALURU • PUNE • KOCHI • CHENNAI

CBS Pubs. ISBN : 81-239-0430-4
Ed. Arnold ISBN : 0-85264-308-X

First Indian Reprint : 1996
Reprint : 2000

This edition has been reprinted in India by arrangement with Edward Arnold (Publisher) Limited, UK

Published by Satish Kumar Jain and produced by V.K. Jain for
CBS Publishers & Distributors Pvt. Ltd.,
CBS Plaza, 4819/XI Prahlad Street, 24 Ansari Road, Daryaganj,
New Delhi - 110002, India. • Website: www.cbspd.com
e-mail: delhi@cbspd.com, cbspubs@airtelmail.in
Ph.: 23289259, 23266861, 23266867 • Fax: 011-23243014

Branches:

- ***Bengaluru:*** Seema House, 2975, 17th Cross, K.R. Road, Bansankari 2nd Stage, Bengaluru - 560070
  • Ph.: +91-80-26771678/79 • Fax: +91-80-26771680
  • E-mail: cbsbng@gmail.com, bangalore@cbspd.com
- ***Pune:*** Bhuruk Prestige, Sr. No. 52/12/2+1+3/2, Narhe, Haveli (Near Katraj-Dehu Road By-pass), Pune - 411041
  • Ph.: +91-20-64704058/59, 32342277 • E-mail: pune@cbspd.com
- ***Kochi:*** 36/14, Kalluvilakam, Lissie Hospital Road, Kochi - 682018, Kerala • Ph.: +91-484-4059061-65
  • Fax: +91-484-4059065 • E-mail: cochin@cbspd.com
- ***Chennai:*** 20, West Park Road, Shenoy Nagar, Chennai - 600030
  Ph.: +91-44-26260666, 26208620 • Fax: +91-44-42032115
  • E-mail: chennai@cbspd.com

Printed at :
J.S. Offset Printers, Delhi

Contents

Preface to the second edition

VOLUME 1

The object of this work remains essentially as in the first edition, namely, to provide an 'Outline' of the subject and so provide the reader with a general framework which can be filled in, if desired, by a study of more specialized works and original papers. The book makes no claim to be a comprehensive treatise, and the number of examples quoted as illustrations in each class is, in general, but a fraction of the total.

When considering this revision, it became apparent that several advantages would accrue from a division of the work into two volumes, Volume One to cover Materials and Volume Two, Finished Products. Volume One is likely to be of interest to technologists in other industries such as printing inks and artists' colours to whom Volume Two is unlikely to make a great appeal. Other factors in favour of two volumes are ease of handling and division of expenditure.

Advances made in paint technology since the date of publication of the first edition have necessitated a number of changes to the text. In addition, the increased awareness of the health hazards associated with the use of certain paint ingredients has resulted in several Acts and Regulations to which reference is made in the text.

The manufacture of pigments is now a highly specialized industry and so the description of manufacturing operations and plant has been reduced, more attention being paid to properties and performance. Colour science has become more important and this is reflected in the text. Some pigments (and resins) which are no longer used but have some historic interest are listed in Appendix E.

Oleoresinous varnishes based on phenolic or modified phenolic resins continue to be used on a limited scale, but those based on fossil resins are virtually obsolete in this country. The treatment of these, therefore, has been curtailed and more attention paid to the wide range of alkyds and polymeric binders generally. Increased space has also been given to water-based media which are growing in importance for reasons of low toxicity, non-pollution of the atmosphere, and absence of fire risk. Oil-bound distempers, which have been replaced by latex emulsions, have been deleted. The section on water-soluble media has been extended to include media for both anodic and cathodic methods of electrodeposition.

The author extends his thanks to many friends and colleagues for helpful discussions and suggestions and to the following companies for their ready assistance: BASF (UK) Ltd, Coulter Electronics Ltd, Degussa, ECC International Ltd, Pye-Unicam Ltd, Tintometer (Sales) Ltd, and Tioxide International Ltd. To those who kindly allowed reproduction of illustrations, acknowledgement is made with each caption.

A special tribute is due to my wife for her patience and understanding and to Mrs Audrey Walford who so efficiently deciphered my notes and typed the manuscript.

Finally, the author's thanks are accorded to the publishers for their patience and help at all times.

W. M. MORGANS

Walberswick, May 1981

VOLUME 2

Since publication of the first edition considerable changes have taken place in the technology of surface coatings. A great number of new materials have been introduced (many of which are discussed in Volume 1), resulting in the availability of coatings suited to practically every type of surface and exposure condition. To discuss every type in a work of moderate size is clearly impossible, and it is hoped that the examples chosen and the depth of treatment will provide at least a suitable foundation for further study.

A deeper understanding of the basic principles of pigment dispersion, coupled with the use of new types of polymer and dispersion aids, has brought about a radical change in the methods of paint production. Roll mills are used only for certain types of printing ink and other high consistency materials. Emphasis is now on high-speed dispersion, and the ball mill, which for many years formed the backbone of paint production units, is being displaced by high-speed open dispersers and bead mills.

Among the requirements of modern building construction is one for 'long-life' or 'high-durability' paint which will provide satisfactory protection, especially in areas difficult of access, for periods of 15 to 20 years or longer. In order to achieve their potential, such paints must be applied to a clean, properly prepared surface. The methods of surface preparation of most common surfaces are discussed, together with the present-day methods of application and cure.

Discussion of the composition of many paints is illustrated by type formulations. These are presented as typical examples only and may not necessarily be the best for a particular application; paint formulation allows considerable latitude in the choice and manipulation of raw materials. They are put forward in good faith but without liability, since conditions of manufacture and use are not within the author's control.

The production of this volume would not have been possible without the invaluable help and suggestions from a number of friends to whom the author extends his sincere thanks. He offers his thanks also to the following companies who provided literature, discussion and illustrations; the latter are acknowledged also in the text.

The DeVilbiss Company Ltd
Jenag Equipment Ltd
J. B. Marr & Co. Ltd
The Mastermix Engineering Co. Ltd
The Mickle Laboratory Engineering Co. Ltd
Ransburg UK Ltd
Research Equipment (London) Ltd
Sheen Instruments (Sales) Ltd
Silverson Machines Ltd
Tioxide UK Ltd

The typing of the manuscript (which involved deciphering the author's atrocious handwriting) was most efficiently carried out by his daughter, Helen, to whom he is deeply indebted.

The author is, as ever, deeply grateful to his wife for her inexhaustible patience and understanding and to the publishers for their help and co-operation at all times.

W. M. MORGANS

Walberswick
April 1983

Preface to the third edition

At the time of publication of the Second Edition there were valid reasons for presenting the work in two volumes but in preparing this Third Edition serious consideration has, of necessity, been given to changes in the structure of the industry as well as to escalating costs of book production. The outcome was a decision of revert to a single volume which, it is hoped, readers will find more convenient.

The object of the work remains essentially the same as that of earlier editions, namely, to present a general framework which can be filled in, if desired, by a study of more specialised works or original papers. The text is illustrated, wherever possible, with typical examples of coatings for various end uses but the book is not claimed to be a comprehensive treatise nor are the examples quoted necessarily the best for any particular application. Paint formulation allows considerable latitude in the selection and proportioning of ingredients and the formulations are put forward in good faith but without liability since conditions of manufacture and use are not within the author's control.

The general structure of the work follows that of earlier editions. Chapters 1–13 discuss materials and 14–27 are concerned with finished products, application, substrates and uses. The following comments are of general interest.

In the field of pigments, particle size distribution and surface characteristics are more closely controlled resulting in easier dispersion and improved dispersion stability. Refinements in the measurement of colour, particularly parameters associated with colour space, are reflected in a selection of instruments capable not only of measuring very small colour differences but also of providing correction figures.

The decline in use and therefore of interest in lead and chromate pigments on toxicity grounds together with the increased use of organics has reduced the space devoted to coloured inorganic pigments. These are discussed in Chapter 5.

Among the organic pigments there are many improved versions of pigments based on established chromophores and new developments in polycyclic and heterocyclic types have extended the range of pigments possessing outstanding lightfastness and durability.

Changes in the pattern of use of certain types of solvents have resulted from studies of health hazards. Threshold Limit Values (TLV) have been replaced by the similarly time-weighted Occupational Exposure Limits (OEL). New and very low OEL values have been assigned to certain solvents including ethers and ether-esters of ethylene glycol. These have been replaced in paint formulations by the relatively less toxic propylene glycol derivatives. OEL values are published annually by the Health and Safety Executive in Guidance Note No. 40 so that users can be kept informed annually of all alterations and additions to the list. For this reason OEL values are not quoted in the text.

The reduction of atmospheric pollution caused by organic solvents, mainly in the paint-using industries, has stimulated interest in water-borne and other types of coating. Water-borne coatings are now of sufficient importance to justify a separate chapter (19). Other types are the high-solids and solvent-free coatings. Powder coatings are, of course, solvent free and have made serious inroads in the field of industrial finishes.

Although natural vegetable oils are the main source of fatty acids for resin manufacture, very little oil is used in paint manufacture. Both linseed and tung oils continue to be used with hard resins in varnish manufacture and so are treated in this context.

In paint media, alkyd resins continue to hold pride of place in terms of quantity used. Improvements in drying characteristics have resulted from the introduction of 'chain stopped' alkyds whilst new long-oil types compete with the widely-used silicone/alkyd co-polymers for use in 'long life' coatings. A wide variety of stoving finishes is available by cross linking reactions involving modifications of basic types such as epoxies, hydroxylated acrylics and polyesters with amino resins and isocyanate derivatives.

A notable change in the industrial application field has been the replacement of anodic electrodeposition by the cathodic system. The latter offers many advantages in terms of film integrity and protection against corrosion.

Many other interesting developments, not the least being the use of 'anti-carbonation' coatings for the protection of reinforced concrete, are described in the text and it is hoped that the work will present a realistic outline of present day paint technology.

The author is very greateful to many friends and colleagues for helpful discussions and suggestions. He also extends his thanks to the following companies for their ready and willing co-operation.

Akzo Chemical Division
Analytical Measuring Systems Ltd
BIP Chemicals Ltd
Buckman Laboratories Ltd
Ciba-Geigy Pigments & Plastics Divisions
Colourgen Ltd
Columbian (UK) Ltd
Coulter Electronics Ltd
Degussa Ltd
Diffusion Systems Ltd
Kerstol Ltd
Kremlin Spray Painting Equipment Ltd
Malvern Instruments Ltd
Minolta (UK) Ltd
OBS Machines Ltd
Perchem Ltd
SCM Chemicals Ltd
Sheen Instruments Ltd
Techmation Ltd
Tioxide UK Ltd

ICI Chemicals & Polymers Group
Industrial Colour Systems

The author is deeply grateful to his wife for her constant support and patience and to the publishers for their ready advice and co-operation.

Walberswick 1988

W. M. Morgans

1

Introduction

The nature of paint

Paint can be defined as a fluid material which, when spread over a surface in a thin layer, will form a solid, cohesive and adherent film.

Paints have been used for decorative purposes for many centuries. The paintings produced in ancient Egypt have been shown to incorporate gum arabic, gelatin, eggwhite and beeswax. Gum arabic was used also by the Persians. In classical Greece extensive use was made of paint in sculpture (for the hair, lips and eyes of statues), architecture and in painting ships. It was used also in interior decoration and, of course, by artists.

The evolution and use of paints in Europe, prior to the Industrial Revolution, was mainly in the hands of artists although some was used in ship painting. The Industrial Revolution created a demand for paint for a number of purposes and this can be regarded as the beginning of the modern paint industry.

Ingredients of paint

A liquid paint contains three major ingredients together with smaller quantities of 'additives' which play an essential rôle. The major ingredients are pigments (including extenders), binder (or film former) and solvent or thinner. A dispersion of the pigment in the binder constitutes the paint film the properties of which depend, to a large degree, on the nature of the binder, but the nature and quantity of pigment are also important. The solvent or thinner is used to render the pigment/binder mixture sufficiently fluid for application as a thin film after which it is lost by evaporation and plays no part in the performance of the dry paint film. The additives employed comprise materials such as driers, anti-skinning and anti-settling agents, fungicides or bactericides and surface active agents used to assist pigment dispersion.

The following is a brief summary of the general nature and functions of the major paint ingredients, all of which are discussed in some detail later in the book.

Pigments

These are finely divided solids, the average particle size of which can vary from 0.2 to 10 μm (micrometres). They may be inorganic or organic in constitution. The general classification of pigments can be illustrated by the following scheme.

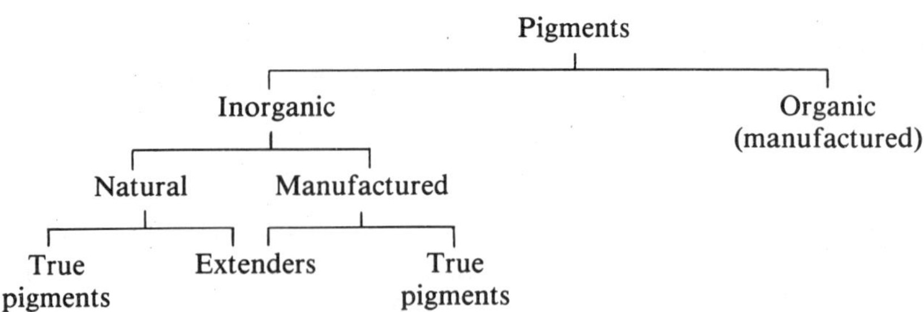

Extenders have been included with the inorganic pigments since they are all inorganic solids. They differ from 'true' pigments in their behaviour when dispersed in organic media. True pigments exhibit opacity or hiding power in varying degree, whereas extenders are practically transparent. Extenders are used in certain types of paints (notably undercoats, primers and some low-gloss finishes) to modify or control physical properties. They make no contribution to colour (unless they are very impure) or to opacity.

True pigments are used to provide colour and opacity or hiding power. In finishes they contribute to durability. A pigmented film is more weather-resistant than an unpigmented film of the same binder. In primers for metals, specific pigments are used to check or inhibit corrosion of the metal.

The majority of natural pigments are oxides or hydroxides of iron but may contain appreciable quantities of clay or siliceous matter. The colours are less bright than the corresponding manufactured oxides and hydroxides.

The manufactured inorganic pigments contain the whites and a wide range of colours, including yellows, reds, oranges, greens and blues. Carbon black, consisting essentially of elementary carbon, is usually included in the inorganic pigments.

The organic pigments cover the entire spectrum range, but the brilliance and opacity vary considerably. There are no white organic pigments and organic blacks find only limited use; carbon black satisfies most paint requirements.

In general, organic pigments are brighter than the inorganic counterparts but show much greater variation in opacity and in lightfastness, the latter particularly when mixed with white. Organic pigments derived from plant and animal sources are no longer employed in paint manufacture. Although some possess bright self-colours, they lack permanence. A number which are of historic interest are listed in Appendix C.

The chemical constitution of pigments (and dyes) is set out in the Colour Index [1] and in the following pages the Colour Index description (CI –) follows the pigment name.

The testing of pigments is usually carried out by the generally accepted methods set out in British Standard 3483 [2] to which frequent reference is made in the text.

Following the introduction of the Health and Safety at Work Act, 1974, greater attention is now paid to the avoidance of any toxic or irritant effects arising from dusts created during the use of pigments and extenders (as well as solvent vapours). This matter is discussed under Toxicity in Chapter 2.

Binder (or film-former)

This is the continuous phase in a paint film and is largely responsible for the protective and general mechanical properties of the film. Film properties depend also on the nature of the pigment, the degree to which the pigment is dispersed in the binder, and the volume of film occupied by the pigment (the 'pigment–volume concentration').

The majority of binders are organic materials – oleoresinous varnishes, resins containing fatty acids from natural oils (alkyds, epoxy esters, urethane oils), treated natural products (cellulose nitrate, chlorinated rubber), and completely synthetic polymers.

A few inorganic binders are used, notably pre-hydrolysed ethyl silicate, quaternary ammonium silicate and alkali silicates (sodium and lithium) which are pigmented with zinc dust to give primers for steelwork.

Organic binders can be divided into two general classes – convertible and non-convertible.

Convertible binders These undergo a chemical reaction in the film. In the conventional oxidizing binders, oxygen is absorbed and the film slowly sets and then dries to a product which is no longer soluble in the solvents used in the liquid paint. The drying and hardening of the film is catalysed by 'driers' and since oxygen absorption is involved, there is a limit to the thickness of film which will dry through.

The two-pack materials, notably epoxies and polyurethanes, cure by chemical reaction between two components in the film. Absorption of oxygen is not involved and thicker coatings can be applied. By the use of low molecular-weight resins and hardeners, 'solventless' epoxy systems can be produced and very thick films can be applied and cured.

Thermosetting binders These are also known as 'thermohardening' and as 'stoving' types. As in other convertible coatings, film hardening is the result of three-dimensional linking of the polymer molecules, but in this case it is brought about by the action of heat. The film is cured much more rapidly and attains the full resistance to reagents and solvents in a fraction of the time required by the air-drying types.

Stoving of the films is carried out in either hot-air convection ovens or in infra-red ovens heated by either electricity or gas. Certain types of coating, especially those which cure by a free radical mechanism, can be cured in seconds only, by either an electron beam or ultraviolet radiation. Since no heat is involved, this process is suited to finishes on sensitive substrates such as wood or card.

Non-convertible binders These do not depend on chemical reaction of any sort for film formation, the process consisting solely of evaporation of solvents. The film remains soluble in the parent solvent blend and is therefore classed as 'non-convertible'. Such materials are often designated 'lacquers' and are usually applied by spray (especially a second coat) since the solvents present tend to dissolve the previously applied coat. This would be disturbed or 'lifted' if brushing were employed. Typical examples of non-convertible binders are chlorinated rubber and cellulose nitrate.

The familiar vinyl emulsion paints contain a non-convertible polymer dispersed in water in which the pigment also is dispersed. Film formation results from evaporation of the water followed by coalescence of the binder in which the pigment becomes embodied.

Solvents (or thinners)

These are usually volatile organic liquids in which the binder or film former is soluble. When the paint film is deposited on a surface the solvent should evaporate completely. It plays no part in film formation and is used solely as a means of conveying the pigment/binder mixture to the surface as a thin uniform film.

Classes of materials used as solvents include aliphatic hydrocarbons (white spirit and the SBP solvents), aromatic hydrocarbons (toluene, xylene and the trimethyl benzenes), alcohols, esters, ketones, esters and ether-esters of propylene glycol. Water is the solvent in water-borne and emulsion paints.

The fact that these materials are volatile, flammable and possess a degree of toxicity has raised a number of serious problems. Atmospheric pollution by solvents can be appreciable in areas containing a number of paint-consuming industries. The problem was particularly acute in Los Angeles and resulted in the 'Rule 66' legislation [3] designed to control the emission of solvent vapours into the air.

Organic solvents can be taken into the human body by swallowing (this is uncommon), by the respiratory system, or by absorption through the skin. The effects can vary from skin irritation to damage to internal organs. A great deal of legislation, including the Health and Safety at Work Act, 1974, now exists setting out the precautions to be taken in handling solvents and paints. This matter is discussed in more detail in Chapters 2 and 9.

A second hazard associated with most organic solvents is that of fire and explosion and this also is discussed in Chapter 9.

CLASSIFICATION OF PAINTS

As mentioned above, paints can be divided into two general types, convertible and non-convertible. Further subdivisions are possible – particularly with convertible coatings – into types of curing mechanism (air-dry, stoving, etc.) and performance characteristics (heat-resisting, corrosion-resisting, etc.). Both these subdivisions will be employed in subsequent chapters.

A further, very general classification is based on the order of application, a method familiar to professional decorators, industrial finishers and the general public. In brief, the types are the following.

Primers are, with a few exceptions, the first coats applied and are formulated to meet the requirements of the surface. Primers for steel, for instance, are required to meet an entirely different set of conditions from primers for wood, and the formulations differ considerably.

Sealers Where surfaces are porous, absorption of paint medium can upset the pigment/binder ratio of the paints applied and sealers are therefore used to eliminate the suction. In some cases the formulation of the primer can be modified to enable it to function as a sealer also.

Stoppers and fillers When surfaces contain holes or deep indentations they are treated with paste stopper after application and drying of the primer. Stoppers are very highly pigmented so that they dry hard without shrinking and can be rubbed down to a smooth surface before application of further coats of paint.

If the surface contains merely shallow indentations or irregularities, these are corrected by the application of filler. This is liquid and can be applied by spray or brush. Like the stopper, it dries hard and can be rubbed down to a smooth finish. Both stoppers and fillers are porous when dry and require treatment with sealer before further painting.

Undercoats are applied over the primer or filler. They are fairly highly pigmented to give good obliterating power and to give *build* or thickness to the paint film. They usually dry to an *eggshell* finish which is suitable for good adhesion by the finishing coat.

Finishing coats can be designed to dry to films which exhibit a high gloss, semi-gloss, eggshell or matt finish.

High build, high solids and solventless paints

The protection afforded by a paint film is, in general, proportional to the thickness and the purpose of these paints is to provide a means of achieving a desired film build or thickness in the minimum number of coats.

High build paints contain a structuring or gelling agent which allows thick films to be applied without risk of sagging or forming 'curtains' on vertical surfaces.

High solids paints contain only a small quantity of solvent so that shrinkage of the film on drying is minimal. These paints are environmentally attractive as the low solvent content means less atmospheric pollution than is caused by traditional types of paint.

Solventless paints are based on liquid resins of low viscosity and are used where the presence of solvent is undesirable as, for example, in tank interiors and other enclosed spaces. They are used also in the underwater painting of steel and concrete structures.

These high build paints are formulated to cure either by chemical reaction (two pack types such as epoxies or polyurethanes) or are non-convertible types (high build vinyls or chlorinated rubber) which dry solely by solvent evaporation.

With oxidizable media, such as alkyds or oleoresinous types, there is a limit to the film thickness which can be applied in one operation. Oxygen is required for cure and must be available throughout the film until it is dry and hard. With thick films oxidation of the surface layers produces a 'skin' which reduces the rate of oxygen penetration to the lower layers. These can remain soft for a considerable time

Very thick films should contain a minimum of solvent. Films of normal thickness can retain small amounts of solvent for a considerable time [4] and this is more pronounced with thick films since migration of solvent molecules to the surface is retarded by increasing rigidity of the film. This must be borne in mind when multicoat systems are applied; recoating intervals must be adequate.

APPLICATION OF PAINTS

Whilst decorative paints continue to be applied by brush or roller, newer methods have been introduced in the maintenance and industrial fields, and these are described in Chapter 20. Among these are (a) electro-deposition, both anodic and cathodic, used mainly for primers for metals and made possible by the introduction of water-soluble resins, and (b) airless and air-assisted spray which permits the rapid coating of large areas and is capable of applying 'high-build' or thixotropic paints to give greater film build than is possible with conventional paints.

The demand from a number of industries for shorter drying or curing times has been met by the use of new and modified resins; industrial air-drying primers and undercoats with drying times of 10–20 minutes and over-coating times of 1–2 hours are now common.

Films which cure by a free radical mechanism are cured very rapidly by irradiation with either electron beams or ultraviolet. The electron beam method has been applied to metal strip (coil coating) whilst ultraviolet irradiation is suited to wood finishes.

The paint film

The loss of solvent by evaporation from a freshly applied paint film results in a shrinkage or decrease in thickness of the film. The ratio of the thickness of the dry film to that of the wet film is controlled by the volume of thinner used or, conversely, by the volume of solid material in the paint. This is known as the 'volume solids' or 'solids by volume' and is expressed as a percentage:

$$\%\ \text{Volume solids} = \frac{\text{Dry film thickness} \times 100}{\text{Wet film thickness}}$$

The figure is generally quoted on manufacturers' data sheets. It can also be calculated from the paint formula (using specific gravities) or can be determined practically by the method of BS 3900, Part A 10. Painting specifications usually require a minimum dry film thickness for each coat of paint and for the thickness of the complete system. Film thickness is usually expressed in micrometres (symbol μm) but the terms 'thou' and the US 'mil' (both of which equal 25.4 μm) are often encountered.

In order to apply a given dry film thickness an operater requires to know the corresponding wet film thickness which can be readily checked on the wet film (see p. 428). The required wet film thickness is calculated by the equation

$$\text{Wet film thickness} = \frac{\text{Dry film thickness} \times 100}{\text{\% Volume solids}}$$

Coverage rate

This can be defined as the area covered by unit volume of paint when applied at a specified film thickness. It is usually expressed in square metres per litre and can be calculated from the dry film thickness and the volume solids using the equation

$$\text{Coverage (m}^2\text{/litre)} = \frac{\text{\% Volume solids} \times 10}{\text{Dry film thickness } (\mu\text{m})}$$

The figure derived is the ideal or theoretical coverage. Practical coverage rates are always lower since they are influenced by design of structure, method and technique of application and general losses. The site engineer or painting contractor is best qualified to assess practical coverage rates which can vary from 50 to 80 percent of the theoretical value.

Pigment Volume Concentration

A number of the properties of a paint film are determined by the volume of the film occupied by the pigment – the 'Pigment Volume Concentration' (PVC). This is defined by the expression

$$\text{Pigment Volume Concentration} = \frac{\text{Volume of pigment} \times 100}{\text{Volume of pigment} + \text{volume of non-volatile binder}}$$

Glossy films are characterized by a low PVC, but with increasing PVC a point is reached where the gloss falls rapidly and radical changes occur in the film properties. This is illustrated in Fig. 1.1, in which film properties (in arbitrary units) are plotted against PVC. Other film properties such as moisture permeability and tendency to blister also change radically at about the same point. This point of rapid change in film properties is known as the 'Critical Pigment Volume Concentration' (CPVC) [5]. It represents the situation where

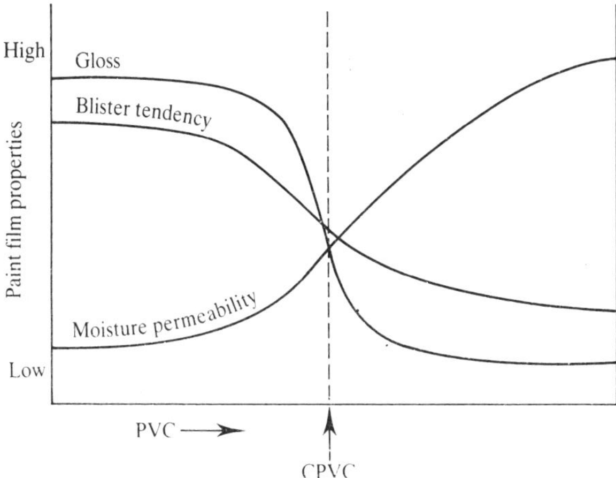

Fig. 1.1

just sufficient binder is present to coat the particles of pigment and to fill the interstices between them. An increase in pigment content will result in a film containing voids (an underbound film), whereas an increase in binder will become evident by an increase in gloss. The value of the CPVC is characteristic of a given system and varies from one system to another.

Stieg [6] has traced the historical development of the concepts of both Pigment Volume Concentration and Critical Pigment Volume Concentration and has discussed their application in paint formulation. PVC figures are quoted for many of the paint formulations discussed in subsequent chapters.

References to Chapter 1

[1] Colour Index. Published by the Society of Dyers & Colourists, Bradford.

[2] BS 3483. Methods for testing pigments for paints. British Standards Institution.

[3] Larson, E. C. and Supple, H. E., Los Angeles Rule 66 and exempt solvents, *J. Paint. Tech.*, 1967, **39**, No. 508, 258.

[4] Ramsbotham, J., Solvent loss from paint films, *JOCCA*, 1979, **62**, 359

[5] Asbeck, W. K. and Van Loo, M., *Ind. Eng. Chem.*, 1949, **41**, 1470.

[6] Stieg, F. B., Influence of pigment volume concentration on paint properties, *Progr. Org. Coatings*, 1973, **1**, No. 4, 351.

2

Pigments: general physical properties

A pigment can be defined as a finely divided powder which can be dispersed in media of various types to produce paints or inks. It is insoluble in the medium and confers the following properties on the mixture.

1 **Colour**, for decorative effect or aesthetic appeal.
2 **Obliteration or hiding power** It should obliterate the surface, whether it is metal, wood, stone, or an earlier coat of paint. As will appear later, pigments differ considerably in their hiding power and, in some instances, several coats of paint may be necessary to effect complete obliteration of the surface.
3 **Protective properties** These are important in paints on exposed wood and certain types of metals, particularly iron and steel. A suitably pigmented varnish is more weather-resistant and therefore more protective than the unpigmented material.
4 **Corrosion repression** Primers for iron and steelwork contain special pigments chosen for their corrosion-repressing properties.

A special class of pigment, usually described as 'extender', generally possesses little or no colour or hiding power. Such materials are often included in paints to modify physical characteristics such as gloss and flow properties.

Classification of pigments

Pigments may be inorganic or organic in composition. The inorganic types may be processed ores or may be manufactured, but the organic pigments in present-day use are all manufactured. Many earlier organic types were obtained from vegetable or animal sources, but these have been replaced by the manufactured pigments which are more consistent and possess greater all-round stability. Since many of the natural colouring-matters are of historic interest, a selection is listed in Appendix C.

Physical properties

A number of physical properties are common to all pigments and are described in the following pages. Chemical properties are more specific and will be considered under the pigment classes or individual pigments.

COLOUR

Colour can be defined as a sensation resulting from activation of the retina of the eye by electromagnetic vibrations which we describe as light-waves. These light-waves differ in wavelength and each wavelength produces a specific sensation. Thus the longest waves produce the sensation which we describe as red and the shortest give the sensation of blue. The range of sensations between red and blue, which we call colours, is known as the *visible spectrum*. The fact that daylight contains the whole visible spectrum is shown by the simple experiment of passing a narrow beam of white light through a glass prism. The beam is opened out and a spectrum is produced.

The relationship between colour and wavelength is illustrated in the following diagram.

Relationship between colour and wavelength

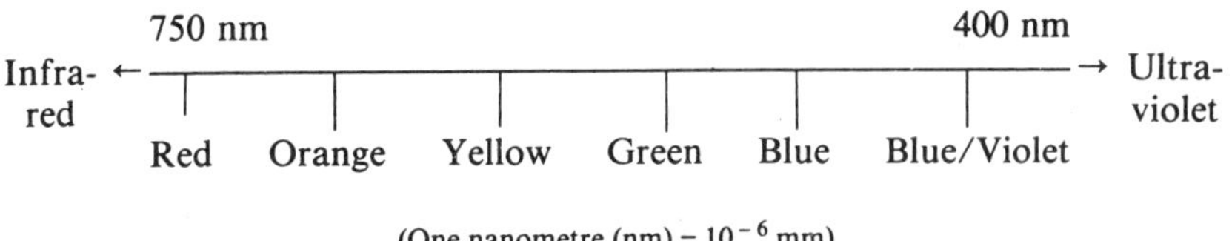

(One nanometre (nm) = 10^{-6} mm)

The visible spectrum consists of only a part of the whole electromagnetic range of wavelengths or vibrations. Beyond the red are the infra-red (heat) rays and the much longer radio waves. Wavelengths shorter than the blue/violet are the invisible but intensely reactive ultraviolet, X-rays, and gamma rays.

Primary colours

By the use of mixtures of narrow wavebands of three spectrum colours – red, green and blue – it is possible to produce white light as well as a very close approximation to the range of spectrum colours. The three spectrum colours are known as the 'light primaries' (or, sometimes, 'additive primaries') and correspond approximately to wavelengths of 650 nm, 530 nm and 460 nm.

Additive colour mixing

The light primaries can be combined by directing the light from three lanterns, each fitted with the appropriate filter to transmit the desired wavelength, on to a white surface. The reflected colour is a mixture of the three incident beams and no appreciable energy is lost; the mixing is therefore said to be 'additive'. The matching or measurement of colours by mixtures of light primaries is the basis of colorimetry (p. 14).

Complementary colours

When the three primary colours are mixed in pairs, red and blue give magenta, red and green give yellow, green and blue give cyan. If each of these resulting colours is mixed with the remaining primary, the result is white light, i.e. magenta + green, yellow + blue, red + cyan are pairs of colours which can produce white light on mixing. Such pairs are known as complementary colours or, sometimes, as complementaries.

Another, more practical, way of describing complementary colours is by considering a coloured material such as a pigment. When illuminated by white light, the reflected wavelengths (the observed colour) are complementary to those absorbed.

Magenta, yellow and cyan are sometimes known as 'subtractive primaries'.

Colour of objects

When white light falls on a solid object it can be affected in one or more of the following ways.

1 It can be completely reflected or scattered from the surface. The object will then appear white.
2 It can be completely absorbed and the object will appear black. Complete reflectance or absorption are not achieved in practice, although certain surfaces such as magnesium oxide and high-quality carbon black, respectively, are close approximations.
3 Some of the wavelengths can be absorbed and some reflected. The object will then appear coloured, the colour being determined by the nature of the reflected wavelengths.

In practice, **3** is often accompanied by a certain amount of **1** or **2**. If accompanied by **1**, that is, by some reflectance over the whole spectrum (white), the colour will appear lighter. When accompanied by **2** or absorption over the whole spectrum (black), the colour will appear darker or often dirtier.

Subtractive colour mixing

When a mixture of coloured pigments is exposed to white light, the colour observed depends on the absorption and reflectance characteristics of the individual pigments. Consider, for example, a mixture of yellow and blue pigments. The blue absorbs the red, orange and yellow wavelengths whilst the yellow absorbs the blue and violet. The mixture absorbs all the spectrum colours except green, which is the resulting colour. The green is produced by subtractive colour-mixing. (This is an ideal condition – in practice the absorption bands are not clearly differentiated and overlap occurs. There are, for instance, 'reddish' and 'greenish' blues, 'reddish' and 'greenish' yellows; and a range of greens will be produced from the combinations.)

Again, a mixture of red and green pigments, unlike the light primaries, will not produce yellow on account of absorption. Yellow is therefore introduced as a primary in pigment mixing, and the pigment primaries are red, yellow and blue. However, it is not possible to produce white by any combination of these, but, as a result of absorption effects, a range of greys is obtained.

Colour perception

On entering the eye, rays of light pass through the transparent lens and are focused on the interior surface at the back of the eye. This surface comprises the sensitive ends of the optic nerve and is known as the 'retina'. The wavelengths of the incident light produce characteristic impulses which pass to the brain by the optic nerve and produce the sensations associated with colour.

The ends of the optic nerve form a large number of *rods* and *cones* which have distinct functions. The rods do not react to colour but are sensitive to differences in intensity. They are responsible for night vision, when objects appear as different intensities of dark grey or black.

The cones are sensitive to colour but do not function efficiently in poor light. There is an old theory, put forward in the nineteenth century by Young and Helmholtz, in which cones are classified in three types, i.e. (i) those sensitive to long wavelengths which produce the sensation of red, (ii) those sensitive to medium wavelengths which produce the sensation of green, and (iii) those sensitive to short wavelengths which produce the sensation of violet. This theory has been difficult to prove but fits in with modern ideas of colour-blindness.

The sensitivity of the average human eye varies with the wavelength of the light. It is greatest in the green waveband (about 560 nm) and decreases rapidly on each side (Fig. 2.1).

Colour-blindness

A person with normal colour vision can match any colour of the visible spectrum by using red, green, and blue lights in the correct proportions. Such a normal person is known as a *trichromat*.

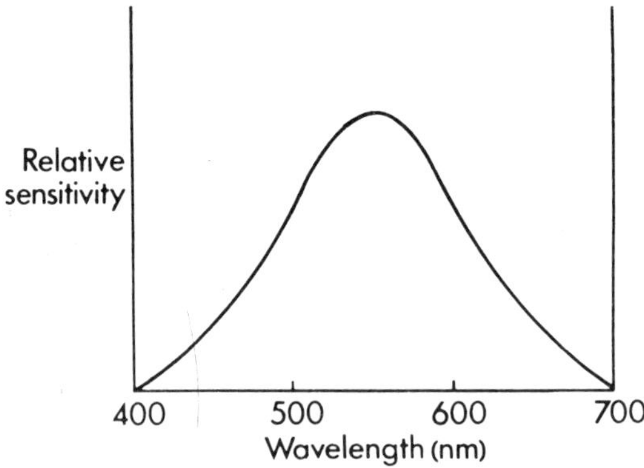

Fig. 2.1 Sensitivity of the human eye

Colour-blindness in females is comparatively rare, but with men a common type involves difficulty with one of the primaries, so that the colour match produced will be different from that of a trichromat. About 1 percent of males have difficulty with red and about 4·6 percent with green. Difficulty with blue is uncommon. Persons suffering from this defect are known as *anomalous trichromats*.

Dichromats are completely blind to one of the primaries and would use only two for a colour matching. The most common forms are blindness to red or green. Blindness to blue is very rare.

Totally colour-blind people, *monochromats*, are extremely rare. They can distinguish only light and dark, and their world, therefore, is one of varying shades of grey.

It is possible to diagnose deficiencies in colour vision by the use of the Ishihara Colour Chart [1], or the Colour Matching Aptitude Test [2].

Colour matching

This involves the mixing of suitable pigments (or paints) to match a given standard and requires (i) an operator who has normal colour vision, and (ii) a standard source of illumination. The latter is important for comparison of the standard and the matching; and whenever possible, in the Northern hemisphere, a north light is used. In the absence of a north light it is possible to use a source of illumination in which the wavelengths in white light have been simulated by blending the light of fluorescent and tungsten lamps. These are often called *daylight lamps*.

The colour of an object illuminated with white light can be different when viewed in, for example, tungsten light since the latter is deficient in the blue/violet wavelengths and richer in the red wavelengths, compared with white light. Two colours can be matched in daylight but may not match under other types of illumination unless identical pigments have been used. This phenomenon is known as 'metamerism'.

A colour which matches a standard under all types of illumination is known as a 'spectral' match and can be obtained only if identical pigments are used in both.

Illuminant standards

Standard light-sources are specified in the CIE system (see below) and called 'Standard Illuminants'. They refer to 'colour temperature' which is the temperature to which a black-body radiator is heated to emit light which matches the light in question. The standards generally adopted are the following.

Illuminant A colour temperature 2850°K, corresponds to the light from a tungsten filament lamp.

Illuminant B colour temperature 4800°K, corresponds to noon sunlight.

Illuminant C colour temperature 6740°K, corresponds to north sky light or average daylight. This is the most frequently used.

Illuminant D_{65} colour temperature 6500°K, is a more recent addition to the range. It includes more ultra-violet radiation than does Illuminant C and covers all wavelengths between 300 and 830 nm. Like Illuminant C, it is a close approximation to average daylight.

COLOUR MEASUREMENT

Instruments for the measurement of colour fall into two general classes.

1 Colorimeters, in which the colour is measured in terms of three primaries.
2 Spectrophotometers, which measure the reflectance at a number of wavelengths over the entire spectrum.

Colorimeters

The basic principle of these instruments is evaluation of a coloured surface either by matching with an additive mixture of the three primaries or by measuring the reflectance of the three primaries from the surface by photoelectric methods. The former method is embodied in visual colorimeters which, although the least expensive, have been largely replaced by photoelectric instruments in which the human factor is eliminated.

Photo-electric colorimeters

These operate on one of two basic principles.

1 The coloured surface is illuminated by each of the three primaries in turn, the intensity of the reflected beam being measured by photoelectric cell and galvanometer. A matt white surface is used as a reference standard.
2 The coloured surface is illuminated by a standard source, and the reflected beam is passed through red, green and blue filters in turn. The intensities of the transmitted beams are measured as in **1** and the results recorded as tristimulus values.

Although the optical and electrical systems in **2** are more complex than in **1**, the majority of commercial instruments operate on this principle. It is not possible, within the confines of this text, to describe the many tristimulus colorimeters available today. They vary from relatively simple instruments which measure tristimulus values only to sophisticated types which either embody a computer or to which a computer can be attached. These are capable of calculating CIE x,y,Y and CIELAB co-ordinates (p. 18) as well as small colour differences ΔE.

Colorimeters which perform the functions described are available from a number of manufacturers some of whom are listed [3] in the References at the end of this chapter.

Some colorimeters employ a portable sensing head which enables colour measurements to be made on surfaces which cannot be placed in or brought to the instrument. The principle was first employed by the Paint Research Association in the Fibre Optics Colorimeter [4]. The sensing head and detector are connected by a flexible lead consisting of a bundle of glass fibres. Light is conducted along the fibres which, in order to prevent loss of light energy, are surrounded by a protective material of lower refractive index.

Reflectance spectrophotometers

If, instead of measuring reflectance values at three narrow wave-bands as in colorimetry, a large number of reflectance values can be determined covering the entire spectrum, then a more complete picture of the reflectance characteristics of a colour can be obtained. If these reflectance values are plotted against wavelength a 'spectral reflectance curve' (Fig. 2.2) is obtained which is a permanent record of the reflectance characteristics of a coloured surface or pigment.

The general principle of a spectrophotometer is illustrated in Fig. 2.3, but commercial instruments can vary appreciably in design. A narrow beam of white light is split into a spectrum by a prism or diffraction grating and the spectrum thrown on to a screen. The screen embodies a movable slit by means of which a very narrow band of wavelengths is allowed to illuminate the standard white surface and the colour to be measured. The intensity of the reflected beams is measured by photocell and a special type galvanometer.

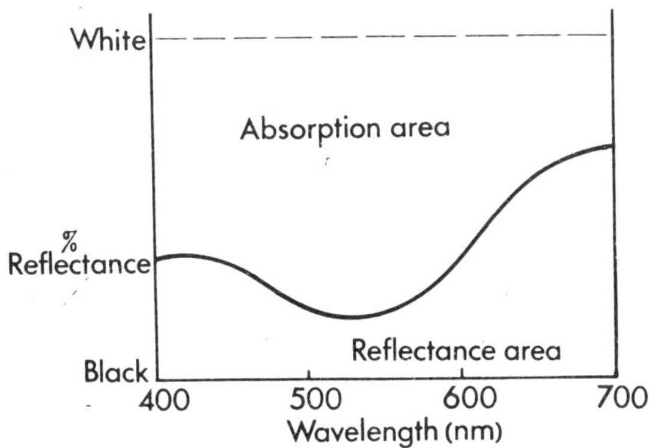

Fig 2.2 Spectral reflectance curve

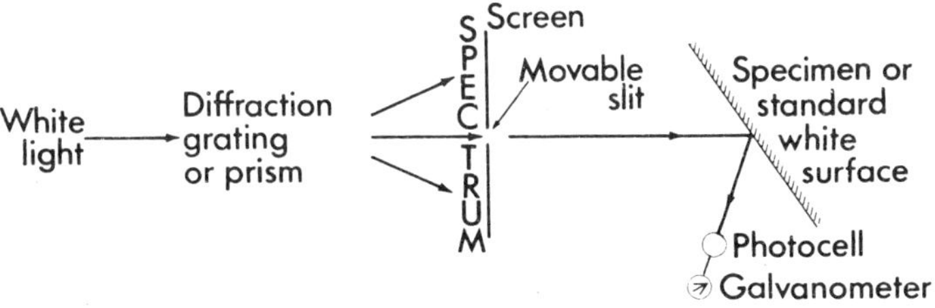

Fig 2.3 Principles of the spectrophotometer

The figure obtained is the reflectance of the particular waveband from the coloured surface in relation to the reflectance from the white surface. Any number of measurements over the spectrum can be made in this way, and modern spectrophotometers [5] can present the results in tabulated form or as a spectral reflectance curve. With computer attachment they will also furnish tristimulus values as well as CIE x,y,Y and CIELAB co-ordinates.

Abridged spectrophotometers are simpler and less sophisticated instruments and produce a spectral reflectance curve for a limited number of measurements. Although less accurate than the full spectrophotometer, they are often adequate for a number of control and routine measurements. They incorporate nine to sixteen narrow-band filters mounted round a wheel through which the sample and the standard magnesium carbonate can be illuminated. The filters are used in turn to illuminate the magnesium carbonate and then the specimen. The reflected beams are directed to a photoelectric cell and galvanometer. The figures obtained give the percentage of the particular wavelength reflected from the specimen relative to that reflected from the white surface. Filters can be fitted also to enable X, Y and Z values to be determined. Instruments of this type are the EEL Reflectance Spectrophotometer and the Sheen Miniature Colorimeter [6].

Spectral reflectance curves

These provide much more information on the reflectance characteristics of a coloured surface than do tristimulus values. They are particularly useful in colour matching in the avoidance of metamerism. Spectral colour matches will give identical spectral reflectance curves but if metamerism is present the curves usually cross each other, often more than once. Another important use of these curves is in calculation of absorption and scatter coefficients (*below*) used in colour-mixture formulation and control.

Use of spectral reflectance curves in colour-matching

Kubelka and Munk [7], as a result of their studies on the optical properties of dispersed powders, derived a relationship between the reflectance of a pigmented film and the absorption and scatter coefficients of the pigments. This is now referred to as the Kubelka–Munk (or KB) equation of which the following is a simplified form:

$$\frac{K}{S} = \frac{(1-R)^2}{2R}$$

where K = coefficient of absorption,
S = coefficient of scatter,
R = reflectance of film at maximum opacity, i.e. when the opacity no longer increases with the thickness.

For opaque pigmented films the expression is valid at any wavelength in the visible spectrum.

Values of R can be taken out at intervals from a spectral reflectance curve and from these the corresponding K/S values can be calculated. With mixtures of pigments the *K/S* values are additive and for pale shades, where absorption is small in relation to scatter, the values can be programmed for an analogue computer. By this means, and utilising the previously determined characteristics of the stock tinting pigments, it is possible to match or correct a shade. The calculations are more involved with deep or saturated colours and the original Kubelka–Munk equation may be required.

Assessment of whiteness

There is no agreed standard of whiteness but several formulae have been suggested using red, green and blue reflectance values. Of these the following form is widely used:

$$\begin{matrix}\text{Colour index} \\ \text{(of whiteness)}\end{matrix} = \frac{100(R-B)}{G}$$

where R, G and B are percentage reflectances compared with magnesium oxide in red, green, and blue light [8]. The formula is useful in assessing the degree of yellowing of white paint films as well as for comparisons. Positive values indicate yellowness and negative values blueness. A value of zero indicates neutrality.

COLOUR SYSTEMS AND ATLASES

The CIE System [9]

This is an internationally agreed system of colour specification which was agreed in 1931 by the Commission Internationale de l'Eclairage, a body consisting of a number of international organizations. The essence of the system is the definition of colour in terms which are independent of a particular individual. It assigns numerical values to the sensations experienced by an average individual in looking at colour.

According to the CIE system, colour possesses three subjective attributes – hue, saturation, and lightness or luminosity. These correspond to the objective variables dominant wavelength, purity, and luminance; they correspond also to the three subjective variables of the Munsell system, namely hue, chroma and value (p. 20). The latter represents the intensity of light reflected from a coloured surface relative to that reflected from a white surface.

In observing a coloured object, three factors are involved. These are **1** the nature of the light-source or illuminant, **2** the visual response of the observer, and **3** the reflectance properties of the object. These are specified as follows.

1 The light source or illuminant Three sources were specified orginally and called 'Standard Illuminants'. A further source (D_{65}) was introduced more recently. These have been described on pp. 13–14.

2 The visual response of the observer The system defines a 'standard observer' as one whose response to the three primaries is the average of a large number of observers examined.

3 Reflectance properties of the object.

In the practical matching of certain spectral colours using the three additive primaries, difficulties have arisen due to the fact that mixtures of the primaries are less saturated than the spectral colour to be matched. The latter requires 'desaturating' and this introduces difficulties. To overcome this problem the CIE system has postulated three 'unreal' primaries designated X, Y and Z which can be calculated from the tristimulus values determined colorimetrically. They are also known as CIE tristimulus values and are indicated directly on modern colorimeters.

In order to represent a colour on a two-dimensional chart, the CIE tristimulus values are converted to chromaticity co-ordinates in the following way:

$$x = \frac{X}{X+Y+Z}, \quad y = \frac{Y}{X+Y+Z}, \quad z = \frac{Z}{X+Y+Z}$$

Since $x + y + z = 1$, only two of these are required to locate a colour on a two-dimensional graph and are designated x and y. A third co-ordinate is the 'luminance factor', designated Y, the value of which is related to the spectral response of the human eye when observing the colour.

The chromaticity diagram

Guild and Wright determined the tristimulus values at a large number of wavelengths over the entire spectrum and converted the figures to CIE chromaticity co-ordinates. When these were plotted the curve illustrated in Fig. 2.4 was obtained. The curved line is known as the 'spectrum locus' and extends from blue (400 nm) to red (700 nm). The straight line joining the red and violet represents the non-spectral purples. All real colours are contained within the closed area.

Saturation or purity of a colour

Point C indicates the position of standard illuminant C (or white); suppose D is the position of a colour as given by the chromaticity co-ordinates. If the line CD is extended to meet the spectrum locus at E, then this will be the dominant wavelength (or hue) of the colour. The saturation or purity of the colour is given by CD divided by CE. The complete CIE specification for a colour is then given by the chromaticity co-ordinates (x,y), the luminance factor Y (which is not shown in the diagram but which would be perpendicular to the surface), and the nature of the illuminant (A,B,C or D_{65}) – C and D_{65} being most commonly used.

It is possible for two samples to possess identical chromaticity co-ordinates (x and y) but to differ in brightness i.e. in the values of Y. The complete specification would then be represented by two points in colour space.

Colour difference measurements

In batch manufacture of paints which are tinted to match a given standard the permissible colour variations from batch to batch have to be established. These

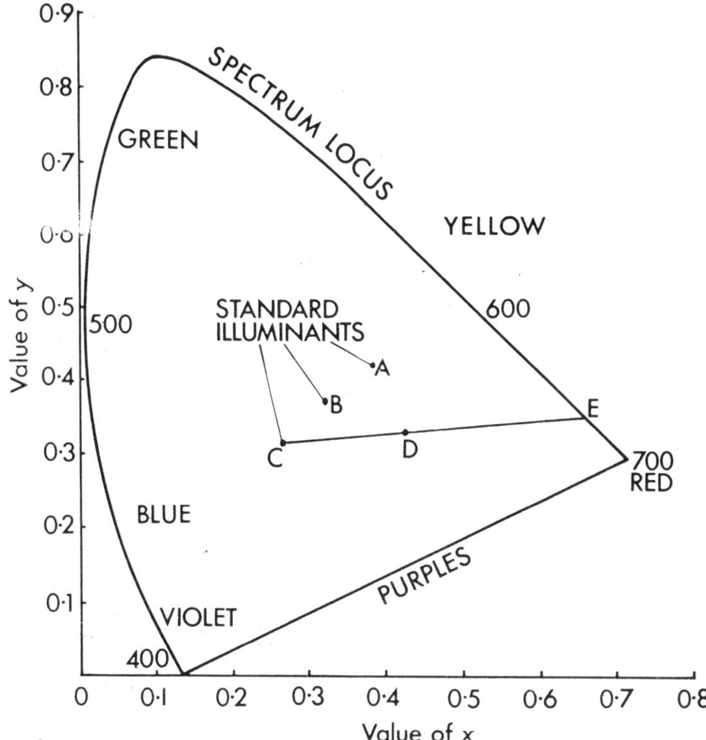

Fig 2.4

tolerances, which represent just noticeable differences in colour, should be represented as the distance between two points in the CIE chromaticity diagram. However, it has been found that, with different colours, the distances between these points can vary considerably [10]. This is attributed to the non-uniformity of colour space. Instead of the x, y, Y space, CIE suggested a form of colour space (designated CIELAB or CIE $L^+a^+b^+$) which bears a close resemblance to that devised by Munsell (below) and uses the same concepts of *hue* (or basic colour), *chroma* (purity or saturation) and *lightness* (or greyness).

The CIELAB system is illustrated in Fig. 2.5. The hues are arranged round the circumference of a horizontal plane in which the two axes, a and b, lie at right angles to each other and to the vertical axis L. Axis a represents redness/greenness (a^+ redness, a^- greenness); axis b represents yellowness/blueness (b^+ yellowness, b^- blueness). Axis L represents lightness (L^+ whiteness, L^- blackness).

The colour difference between two samples, represented by the distance between two points in LAB colour space is given by the equation

$$\Delta E = \sqrt{(\Delta L)^2 + (\Delta a)^2 + (\Delta b)^2}$$

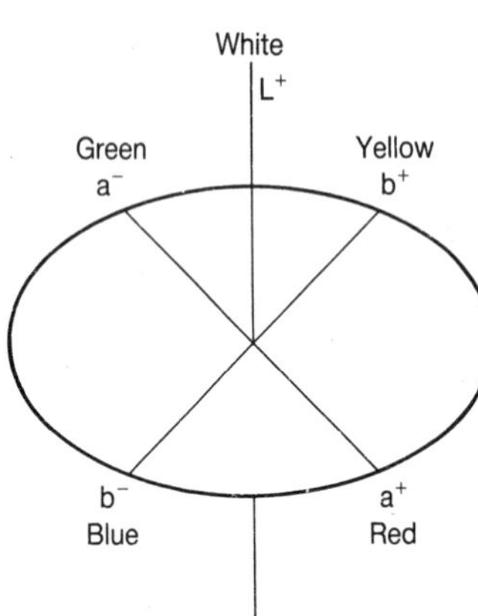

Fig 2.5 CIELAB System

where ΔE = colour difference

$\left.\begin{array}{r}\Delta L\\ \Delta a\\ \Delta b\end{array}\right\} =$ differences between the respective values of L, a and b for the two samples.

Values for L, a and b can be calculated from the tristimulus values. In spectrophotometer/computer work, these values as well as ΔE can be produced.

The Munsell system [11]

This comprises a colour atlas which was devised by Munsell, who assigned three subjective variables to each colour, namely *hue* (basic colour corresponding to dominant wavelength), *value* (lightness), and *chroma* (purity or saturation). He envisaged (see Fig. 2.6) all colours as contained in a solid having a vertical axis with white at the top and black at the bottom. This axis or trunk is divided into nine blocks which represent values (grey content).

On a horizontal circumference surrounding the axis are placed ten principal colours or hues which are designated by their initial letters. These hues are Red, Red/Purple, Purple, Purple/Blue, Blue, Blue/Green, Green, Green/Yellow, Yellow, and Yellow/Red. Each hue is divided into ten, with the pure colour at 5 and the positions of the hues as illustrated in Fig. 2.7. Points on lines connecting the hues and the axis will represent purity or chroma. These are shown as blocks and are numbered in six steps of 2 (2 to 12) from the axis.

If the solid is cut vertically to the axis in very thin sections, each section will represent all possible mixtures of a particular hue with black and white. These sections comprise the pages in the Munsell book, a page being represented by Fig. 2.8.

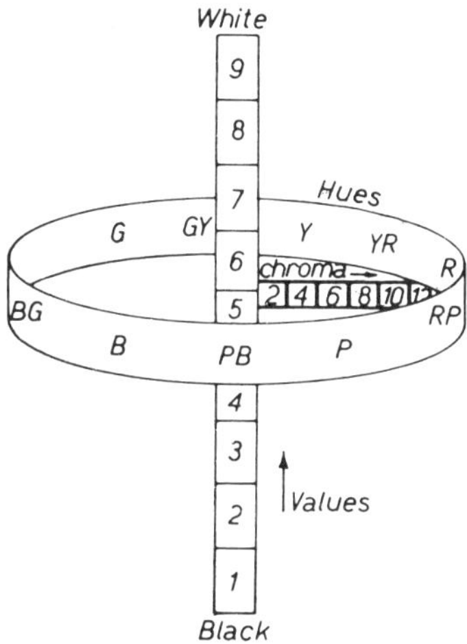

Fig 2.6 The munsell 'Tree'

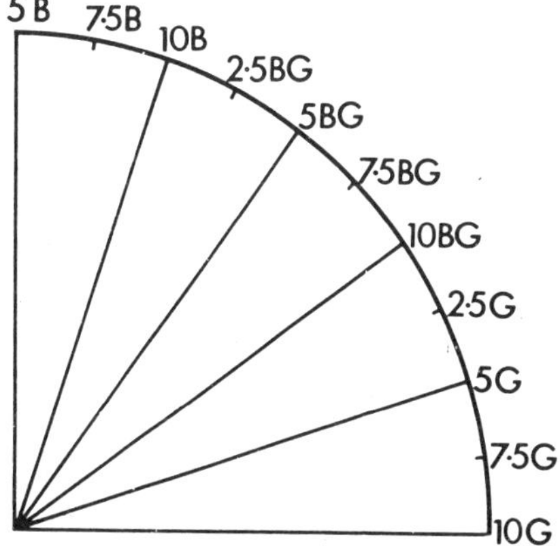

Fig 2.7

The complete Munsell designation of a colour is given by the sequence hue, value and chroma. For example, a colour designated 7·5 G 4/6 would be a yellowish green (hue 7·5 G) with a lightness value of 4 and a chroma of 6.

It is essential in using the Munsell colour atlas that all colour comparisons be made under standard illumination.

The Munsell description of a colour corresponds to the CIE attributes of dominant wavelength (hue), luminance (lightness), and saturation (purity). All Munsell colours have been assigned the corresponding CIE values.

This system possesses an advantage over that of Ostwald (*below*) in that

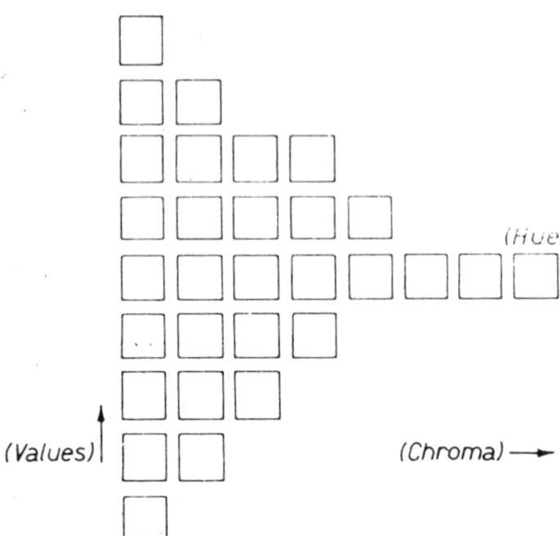

Fig 2.8 Form of page from Munsell colour atlas

additional hues and chroma can be added when necessary. It is now used internationally and forms the basis of British Standard 4800 (*below*). The American Standards Association also uses the system for colour specification.

The Ostwald system [12]

Ostwald was particularly interested in colour harmony and this largely inspired his colour system. He arranged his *full* colours round the circumference of a circle in such a way that *complementaries* were diametrically opposite. To produce other colours his *full* colours were mixed with black and white. In making the mixtures the quantities of black and white were increased in geometric progression to give, in his view, *subjectively equal* steps. His mixings were done with sectors of colour, black, and white on a rotating disc, and as the proportions then added up to the same total, he evolved the relationship:

$$\text{Full colour} + \text{Black} + \text{White} = 1.$$

By taking a full colour and adding black and white he was able to set out the results in the form of an equilateral triangle (Fig. 2.9) with the full colour at one corner, black at the second and white at the third. This actually represents a section through a colour solid since, if the black–white axis is arranged vertically, the series of full colours will form a circumference round the axis and the whole will be a symmetrical double cone, base to base.

The vertical axis White–Black contains the greys in blocks of definite composition. Lines drawn parallel to the White–Colour side give mixtures with equal percentages of black. Those parallel to the Black–Colour side have equal percentages of white.

In the Ostwald colour designation, a number is assigned to each full colour or hue. The white content is indicated by the letters a, c, e, g, i, l, n, and p, decreasing in this order. Black content is indicated by the same letters in the reverse order. The Ostwald designation for a colour can then be, for example,

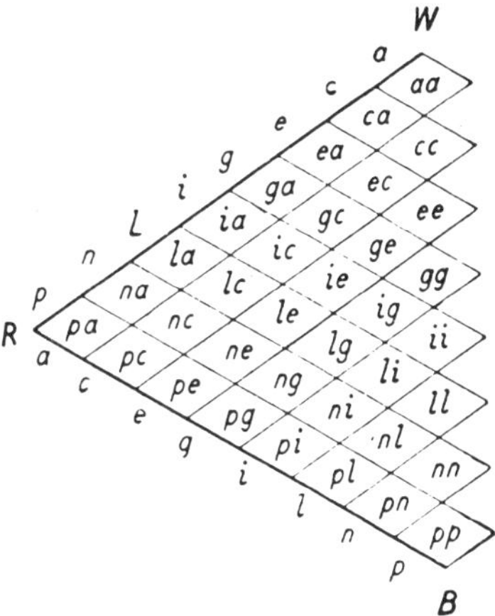

Fig 2.9 The ostwald triangle

8ge, indicating hue No. 8, with a white content g, and black content e.

Possibly as a result of its emphasis on colour harmony, the Ostwald system is popular with architects and in schools of art.

Colour cards and patterns

The need for a system of colour patterns based on scientific principles has been recognized for many years, and it is now usual for paint manufacturers in this country to use the patterns set out in BS 4800 (Paint Colours for Building Purposes) [13].

BS 4800 contains 86 colours together with black and white. The colours are based on 12 hue positions in the Munsell scale covering the important regions and chosen so that the identity of each hue appears constant at all levels of lightness and saturation. All colours are given a Munsell reference and are described in terms of hue, greyness, and weight which correspond to hue, value, and chroma of the Munsell system.

On the card the colour patterns are arranged in blocks according to hue and are given the following numbering system: an even number indicating hue, a single letter indicating greyness, and an odd number indicating weight or intensity of colour. Examples are 08 B 15 (Magenta) and 20 D 45 (Royal Blue). The British Standards Institution also publishes BS 381C (Colours for identification, coding and special purposes) as well as colour cards with more restricted uses.

The *Colour Index*, published by the Society of Dyers and Colourists [14] is a comprehensive work setting out the generic name, chemical constitution, etc. of pigments and dyestuffs. It is used internationally.

Colour comparisons of pigments

A comparison of pigment colours in the dry state is of little value as appearances often change when they are dispersed in paint media. For comparison purposes pigments are usually dispersed in acid-refined linseed oil, but other media may be used. The following is a brief summary of the method described in BS 3483, Part A1 (Comparison of colour) [15] in which either a palette knife or an automatic muller (*below*) can be used to effect the dispersion.

The dry pigment is placed on a glass palette and the oil added dropwise from a burette. The pigment is dispersed well by thorough rubbing with a stiff palette knife and the addition of oil continued until a thin paste is obtained. The comparison standard is treated in the same way. The two pastes are then applied in parallel and touching strips on a glass slide and examined under north daylight or a daylight lamp.

The colour of the undiluted pigment is often referred to as the *self colour* or *mass tone*.

The Automatic Muller

In comparing self colours and tinting strengths it is essential that the sample and standard receive exactly the same treatment so that the degree of dispersion is the same in each case. This is not easy with a palette knife but is possible by the use of the A & W Automatic Muller. In this machine the pigment is dispersed between two ground-glass plates, one of which rotates, the other being fixed. The top (fixed) plate can be loaded so that various pressures can be applied. A mechanism is provided for giving the material a prescribed number of rotations or continuous running, and an automatic cut-out is incorporated. By this means it is possible to give the same treatment to a number of samples and to obtain reproducible results.

Tinting strength (or staining power)

Coloured pigments are frequently added to whites to produce pastel shades or to other pigments to modify the colour. The weight of coloured pigment required by a given weight of white to produce a given depth of tint is an indication of the tinting strength. Tinting strengths differ widely from one pigment to another and generally increase with decreasing particle size. They are measured against a standard pigment of the same colour and composition so that tinting strengths are relative.

There is no general relationship between tinting strength and opacity. For example, carbon black has high opacity and high tinting strength, but Prussian blue, which has high tinting strength, has very poor opacity.

The tinting strength of a pigment is determined by the method described in BS 3483, Part A4 (Comparison of relative tinting strength (or equivalent colouring value) and colour on reduction in linseed stand oil using the automatic muller) [15]. The following is a brief summary of the method.

Equal weights of the standard pigment and of the sample under test are dispersed at the same concentration in the stand oil using the automatic muller. A dispersion of titanium dioxide, together with a proportion of calcium stearate and synthetic silica in a urethane-modified linseed oil, is prepared on a triple roll mill.

White paste and colour dispersions in the desired ratios (1/10, 1/20, etc.) are mixed thoroughly on the automatic muller and the colours compared on glass slides. Differences in colour intensity or hue are noted. A quantitative measure of difference in tinting strength is obtained by adjusting the ratio of white paste to colour dispersion until the same intensity is reached. Suppose x parts of test sample produce the same intensity as y parts standard pigment; then relative tinting strength $= y/x \times 100$.

The term 'tinting strength' is not generally applied to white pigments which are seldom used as stainers. In this case the terms 'tint resistance', 'stain resistance', or 'lightening power' are used. Measurement is carried out by a method similar to that for tinting strength but using a dispersed coloured pigment (generally ultramarine blue). The method is described in detail in BS 3483, Part A5 (Comparison of lightening power of white pigments) [15].

Opacity

This is also known as 'hiding' or 'obliterating power' and can be defined as the ability of a pigment to hide or obliterate a surface when dispersed in a medium and spread in a uniform film over the surface.

Pigments and extenders are opaque as dry powders but show differences in opacity when dispersed in media. Pigments are opaque in both organic and aqueous media, although the degree of opacity, particularly of organic pigments, can vary considerably; extenders are practically transparent in organic media but opaque in aqueous systems.

The opacity of a pigment or extender is a function of the degree to which the incident light is scattered from the surface, and this depends on the difference in refractive index between pigment and dispersion medium. The refractive index of air is taken as 1·00. The figures for drying oils average 1·48 and for resins 1·55. Table 2.1 shows the refractive indices of some of the common white pigments and extenders.

Table 2.1

		Refractive index
White pigments	Titanium dioxide (rutile)	2·71
	Titanium dioxide (anatase)	2·55
	Zinc sulphide	2·37
	Zinc oxide	2·08
	White lead	2·09
Extenders	Whiting	1·58
	Silica	1·55
	Talc	1·49
	Magnesium carbonate	1·57

The refractive indices of the white pigments all differ appreciably from those of water (1·33) and organic media and they are therefore opaque in all media. The figures for extenders, however, are extremely close to those of oils and resins and it is well known that they are virtually transparent in organic systems. They are, however, sufficiently far removed from that of water to account for their opacity in aqueous media. White pigments are generally highly reflective and scatter the incident light, but with black and coloured pigments absorption can play an important part.

A rough figure for the opacifying power of a pigment can be obtained by using the Fresnel equation

$$R = \frac{(\eta_1 - \eta_2)^2}{(\eta_1 + \eta_2)^2}$$

where R = reflectivity of film
η_1 = refractive index of pigment
η_2 = refractive index of binder

The relationship between the coefficients of absorption (K) and of scatter (S) is given by the Kubelka–Munk equation (p. 16). The value of S is dependent on the wavelength of the incident light, the ratio of refractive indices of pigment and medium, the particle size and particle size distribution, the degree of dispersion, and the pigment-volume concentration. It increases with decreasing particle size until the latter approaches 750 nm after which the value of S decreases and the opacity of the pigment also decreases. An example is afforded by red iron oxide, a pigment with good opacity in the particle size range normally employed in paint. The particle size can be reduced to a degree which results in a transparent film when the pigment is dispersed in a paint binder.

Measurement of opacity [16]

The optimum scattering efficiency of the particles in a pigment dispersion is attained when the particles are separated by a distance equal to or greater than their diameter. The pigment volume concentration (PVC) is, therefore, of paramount importance and there is a maximum beyond which pigment packing reduces the scattering efficiency of the individual particles. Hence, in a paint, there is a maximum pigment concentration for opacity. No increase is attained with higher pigment loadings.

From the foregoing it follows that, in making comparisons of pigment opacities, the pigments must be at the same PVC and dispersed to the same degree. Comparisons can then be made using fluid dispersions of the same thickness (as with the Pfund Cryptometer, below) or by using dry films of the same thickness. The latter gives results which are more significant for practical purposes since pigment flocculation (p. 259) can occur during drying and this generally reduces the opacity. There are, however, occasions when it is desired to measure the opacity of a fluid dispersion and so the well known method is included.

Opacity of liquid paints or dispersions

This is generally determined by the Pfund cryptometer (Fig. 2.10), although

Fig 2.10 The Pfund cryptometer (diagrammatic section)

rapid visual comparisons can be made using Morest charts (*below*). The base A consists of blocks of white and black glass fused together and has a linear scale engraved on the surface. The liquid paint is placed in the space between the movable plate C, of clear glass, and the metal distance-piece B. The paint is viewed from above and the wedge of paint is moved until the join in the base is just obliterated. A check reading is obtained by reversing the wedge and repeating the procedure. The thickness of paint over the line is known as the 'critical thickness' and can be calculated from the scale reading and a knowledge of the wedge angle.

Opacity of dry paint films

Use of Morest hiding-power charts

These are sheets of stiff glazed card, one square foot in area, printed in contrasting patterns of black and white, or grey and white. The following are two methods by which the charts can be used for opacity assessment.

1 Equal weights (or volumes) of the two dispersions under test are applied uniformly to the charts and a visual comparison made when the films are dry.
2 The pigment dispersion is applied uniformly by brush until the contrast pattern is just obliterated. The amount of dispersion used is determined by weighing, and the opacity or hiding power of the pigment can then be expressed as sq. feet per lb, or as sq. metres per kg (multiplying by the factor 10·76).

Contrast ratio method

In this method the pigment dispersion is applied to transparent cellulose acetate sheet in a film of known thickness. This is allowed to dry and the reflectance is measured photoelectrically over white and over black substrates. The difference is the amount of light which has penetrated the film and been absorbed by the black substrate.

The method is described in detail in BS 3900, Part D4 [17], of which the following is a brief summary.

The film on cellulose acetate sheet is produced by using steel shims 0·05 mm (0·002 in.) thick and a doctor blade. When dry the film is cut to the required size for the substrate block which comprises two blocks of glass, one black and one white, fused together. The reflectance characteristics of the blocks are standardized by comparison with magnesium carbonate.

The film is first placed on the white substrate, the opacity head placed on it, and the galvanometer set to 100. The opacity head contains a standard lamp for illumination and a photoelectric cell. The film and opacity head are then placed on the black substrate and the galvanometer reading taken. After correcting for the reflectance characteristics of the substrates, the contrast ratio is given by the equation

$$\text{Contrast ratio}(\%) = \frac{\text{Reflectance over black}}{\text{Reflectance over white}} \times 100$$

CHARACTERISTICS OF PIGMENT PARTICLES

Particle size

The particle size of a pigment exerts an influence on one or more of a number of paint properties such as gloss, opacity, freedom from settlement, and consistency.

The size, or average diameter of pigment particles, is usually expressed in micrometers (symbol μm, where μm = 0·001 mm). Pigments are never uniform in particle size but this varies over a range, with the majority of the particles approximating to one particular value and the numbers falling off rapidly on each side. Modern pigments are generally free from both coarse particles and ultra-fines. The former entail extra processing for their separation whilst the latter, if their average diameter approaches half the wavelength of light, have little or no opacity. The ultra-fines are also thought to be the cause of the *bronzing* of certain pigments, e.g. Prussian blue.

In many natural pigments and extenders the particle sizes are greater than those of prepared pigments. They are reduced by grinding in mills such as the Micronizer which is described in Chapter 3.

Determination of particle size: sieving method

Sieves are used only to separate or classify relatively coarse particles. The limits of these methods are evident from the following figures which give the aperture sizes for some British Standard sieves:

100 mesh BS sieve	Aperture = 152 μm
240 mesh BS sieve	Aperture = 64 μm
350 mesh BS sieve	Aperture = 44 μm

Sieving methods are little used for pigments at the present time as these are supplied with a particle size range of less than 10 μm. They therefore fall into the sub-sieve range.

Particle size distribution

In view of the fact that a pigment contains a range of particle sizes it is usual to express this as a particle size distribution, that is, the percentage (by number or weight) occurring between certain limits, e.g. below 0·1 μm, 0·1 to 0·2 μm, 0·2 to 0·3 μm, and so on. These results are expressed graphically in the form of histograms which take the general form illustrated in Fig. 2.11.

A very large number of methods have been suggested for the determination of particle size distribution, but comparatively few have been adopted. Those adopted can be classified as follows.

1 Sedimentation methods (based on the classical Stokes equation – *below*) in which the pigment is separated into a number of fractions under gravity or,

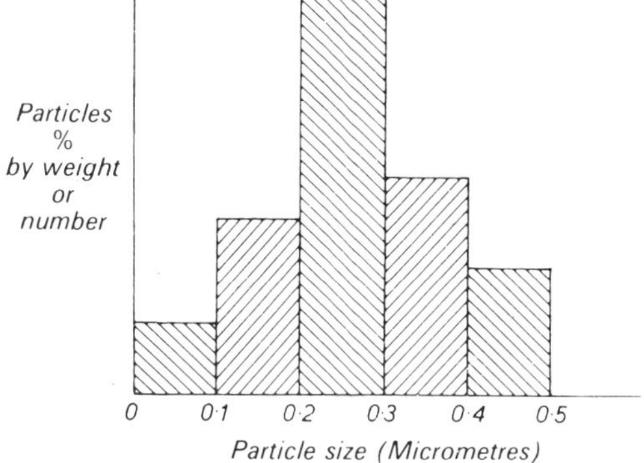

Fig 2.11 General form of histogram

more rapidly, by the use of a centrifuge. Alternatively, measurements can be made directly on the pigment dispersion (see *below*). Results are usually expressed as a mass distribution.

2 Air elutriation methods which also give a mass distribution.

3 Counting methods using the optical or electron microscope or the Coulter counter. In these methods results are expressed as a number distribution.

Sedimentation methods

These are based on the principles first explained by Stokes who derived an expression relating the velocity at which a particle falls through a liquid to the size of the particle, the viscosity of the liquid, and the difference in densities of particle and liquid. The best-known form of the Stokes equation is the following:

$$v = \frac{2r^2(\rho_1 - \rho_2)g}{9\eta}$$

where v = velocity of fall of the particle
r = radius of the particle
ρ_1 = density of the particle
ρ_2 = density of the liquid
η = coefficient of viscosity of the liquid
g = gravitational constant.

It is usual to refer to pigment diameter rather than radius, so that r in the equation can be written as $d/2$. Also if v is represented as distance fallen (h) in time t, the equation becomes

$$\frac{h}{t} = \frac{2\left(\frac{d}{2}\right)^2(\rho_1 - \pi_2)g}{9\eta}$$

and this can be written in the form

$$d = \left(\frac{18\eta h}{(\rho_1 - \rho_2)gt}\right)^{\frac{1}{2}}$$

Now if h is measured in centimetres, t in seconds, η in poises, ρ_1 and ρ_2 in grams per cm^3 and g in cm/sec^2, the value of d obtained will be in centimetres. It is usual to express pigment diameter in micrometres ($\mu m = 10^{-4} cm$), and to obtain these units the right-hand side of the above expression must be multiplied by 10^4. This is the general form of the equation set out in BS 3406 [18].

The specification describes in detail four methods of sedimentation analysis. These are of two general types.

1 The suspended pigment is allowed to settle in a cylinder into which is fitted a pipette of special type connected to a stopcock and tap. The latter allows samples to be withdrawn at given time-intervals from a constant level indicated by the bottom of the pipette. The samples are evaporated to dryness and the weights determined. From the weights determined over a range of time-intervals the rates of sedimentation and particle sizes can be calculated, using the Stokes equation above. The well-known Andreason pipette [19] is of this type.

2 The suspended pigment is allowed to settle in a cylinder, as before, but the weight of settlement is determined at time-intervals, either by drawing off from the bottom, drying and weighing, or by using a sedimentation balance [20]. In the latter instrument a horizontal plate, which forms one pan of a torsion balance, is placed very close to the bottom in the settlement cylinder. The weight of sediment can then be determined at suitable time-intervals. These methods do not give the particle size distribution curve directly but involve a somewhat complex calculation.

All sedimentation methods are slow and also have a lower limit of about 0·2 μm, below which settlement is too slow. The time required for separation can be reduced very considerably by the use of a centrifuge.

The ICI–Joyce Loebl Disc centrifuge is now widely used. A suitable amount of the pigment dispersion is placed at the centre of the rotating disc and samples are removed from the periphery at known time-intervals and weighed. The method was used by Beresford [21] to determine the particle size distribution of organic pigments which, on account of their low specific gravities, settle very slowly under gravity. He found the method satisfactory for particles in the range 0·04 to 1·0 μm.

The separation of samples is avoided by methods which determine particle size concentration at various depths in a sedimentation cell. In the Microscan Particle Size Analizer [22] a narrow beam of low energy X-rays scans the cell from bottom to top. The X-ray beam is attenuated in proportion to the mass of particles in the beam and, by incorporating a microcomputer to control the analysis, continuous plots of particle size distribution by mass are provided. The instrument has a range of 0.1 to 300 micrometers.

A method which does not involve the Stokes equation is based on laser diffraction of suspended particles and is embodied in a range of instruments marketed by Malvern Instruments Ltd [23]. The suspension of particles is illuminated by a low power laser. The particles scatter some of this light at angles which are characteristic of their sizes forming a series of diffraction patterns. The scattered light is collected by a Fourier optical system, focussed on to a detector and then digitised for processing by a computer. In this method the diffraction pattern is not influenced by the position and movement of the particle. It can be used also to study the state of dispersion of particles, i.e. the degree of flocculation of a pigment dispersion.

Air elutriation methods

These methods are described in BS 3406, Part 3 [24] and the specification should be consulted for details. The general principle underlying the methods consists in the determination of the proportion of a pigment which is removed by an upward flow of a gas at a given velocity in a vertical column. The Stokes equation is used to calculate the largest size of particle removed by a gas at the given velocity.

In the determination, the gas velocity is first adjusted so that the smallest particles are carried off and collected. The velocity is then increased to remove the next fraction and the process repeated until the entire range has been separated.

Counting methods

The optical microscope method [25] This method consists of direct observation of a representative sample of the pigment, using a microscope fitted with a graticule. The graticule is inscribed with circles of known sizes as standards of comparison for the magnified images of the pigment particles. The number of particles in each size-class is counted and gives a size distribution by number. If the particles are of the same shape, the relative volume distribution can be calculated, and this will give a size distribution by weight if it is a simple pigment of known specific gravity.

The optical microscope has a limit of resolution of about 0·25 μm, but accurate measurement at this size is difficult. Below about 0·4 μm it is necessary to use the electron microscope.

An interesting development from the optical microscope method is the Optomax V System [26], an automatic 'image analyzer' capable of carrying out particle size and shape analysis on microscope images. The system comprises a high performance TV camera, central processing unit with built-in monitor and key-pad, high resolution colour display monitor and data processing software for a compatible computer. Optomax V contains provision for interfacing to scanning and transmission electron microscopes.

Electron microscope method The construction of the electron microscope and its application to the study of powders have been amply described in the literature [27], and the following is a brief summary of the salient features of its application to the examination of pigments and paint films.

The great advantage of the electron microscope over the optical instrument lies in the greater resolution obtainable. The use of an electron beam combined with electromagnetic lenses having objectives of small aperture gives a great depth of focus.

Very small samples are used for examination, and the mounting of these on suitable substrates requires very special techniques.

In the examination of pigments, information on the *surface relief* is usually required and this can be obtained either by *shadowing* the particles with a heavy metal such as gold, or, if the particles are embedded in a matrix such as a paint film, surface replicas can be made.

Shadowing The heavy metal is deposited under high vacuum and is directed at the specimen from a low angle. Each relief feature facing the source is coated with metal but casts a shadow on the supporting substrate. No metal is deposited in these shadows. When examined in the electron beam the deposited metal appears dark as a result of electron scattering, while the shadows appear light. The photomicrographs are often printed as negatives instead of positives to obtain a realistic picture. From the length of the shadows and the angle of shadowing, the heights of the particles can be calculated.

Surface replicas Surface replicas are prepared when the surface to be examined is in the form of a thick film or is part of a solid body.

The specimen is coated with a dilute solution of a suitable polymer and this is stripped off after drying, a negative replica of the surface being obtained. If necessary, a positive replica can then be cast from the negative and this can be shadowed and the contours examined.

Some electron photomicrographs of a few common pigments are illustrated in Figs 4.1, 4.2, 5.1, 5.2, 7.1, 8.1.

The Coulter counter [28] This instrument embodies an ingenious method in which size and numbers of particles are determined automatically; it is applicable to powders with a mean particle diameter of 0·6 μm to 800 μm. A very dilute dispersion of the pigment in an electrolyte is drawn through a fine orifice across which an electric current path has been established by means of two electrodes. As an individual pigment particle passes through, it displaces electrolyte in the aperture, producing a pulse proportional to its displaced volume. The pulses are amplified and appear as vertical spikes on an oscilloscope trace. They are also recorded on a digital register.

The method can be used with both aqueous and non-aqueous electrolytes, is very rapid, and can be applied to on-line quality control as well as laboratory measurements. The method is described in detail in BS 3406, Part 5 [24].

Surface area of pigment particles

Many paint properties, including viscosity, flow characteristics and dispersion stability, can be influenced by interaction between pigment and medium. Adsorption of certain polymer fractions and/or driers on to the pigment surface takes place frequently, and the total amount of adsorption will depend on the nature and surface area of the pigment particles. A knowledge of the latter is, therefore, of importance and the figure is commonly expressed as square metres per gram. The figure is related to the average particle size of the powder, and a rough figure for surface area can be obtained using the expression

$$\text{Surface area}(\text{m}^2/\text{gram}) = \frac{6}{D\rho}$$

where D = volume/surface mean diameter, i.e. the diameter of a particle possessing the same volume-to-surface ratio as the whole sample,

ρ = specific gravity of the pigment,

6 = factor applicable only to spherical particles; with particles of irregular shape, such as pigments, the expression can give only approximate figures.

Surface area and oil absorption

The oil absorption value of a pigment (p. 35) (which represents the amount of oil required to coat the surface and to fill the voids) increases with decreasing particle size. The smaller voids resulting from closer packing of smaller particles are more than offset by the increased surface area.

Pigment particles may be smooth or irregular individuals or they may be aggregates of various sizes. These will all behave as individual particles towards oil and other media and, in the case of aggregates, the 'practical' surface area will be appreciably less than the 'total' surface area of the component particles. The difference depends on the type of pigment and the degree of grinding. For many manufacturing operations a knowledge of the 'practical' surface area is sufficient, but in any study involving pigment surface a knowledge of 'total' surface area is required.

Determination of surface area

There are two general methods in use known as the 'gas adsorption' and 'gas permeability' methods. A number of variations of each of these methods have been described in the literature [29], [30].

Gas adsorption methods

The best known is the BET method [31] which consists in determining the quantity of nitrogen adsorbed by unit weight of pigment to coat the entire available surface with a layer one molecule thick. From the weight (and therefore the number of molecules) of nitrogen adsorbed and a knowledge of the cross-sectional area of the nitrogen molecule, the surface area can be calculated.

In the determination a known weight of the dried pigment is placed in an adsorption bulb, degassed under high vacuum, and the bulb and contents cooled in liquid nitrogen (– 195°C). A known volume of nitrogen gas is admitted and the pressure noted when equilibrium is reached. The procedure is repeated with further known volumes of nitrogen. For the calculation of results, use is made of the equation of Brunauer, Emmett and Teller [31]

$$\frac{P}{V(P_0 - P)} = \frac{1}{V_m C} + \frac{(C-1)P}{V_m C P_0}$$

where V = volume of gas adsorbed (in cm^3 per gram pigment) at equilibrium pressure P,

P_0 = saturation vapour pressure of the adsorbate,

V_m = volume of gas (per gram pigment) when surface is coated with a monolayer,

C = a constant.

If $P/\{V(P_0 - P)\}$ is plotted against P/P_0 a straight line is obtained. The values of C and V_m can be determined from the slope and intercept. From the value of V_m and the known cross-sectional area of the nitrogen molecule the total surface area of the pigment can be calculated.

A range of instruments for measuring BET surface area as well as total pore volume and absorption/desorption isotherms is marketed by Quantachrome [22].

The shapes of pigment particles

It is well known that solid materials are either amorphous (possessing no definite or characteristic structure) or crystalline. Crystalline materials possess a characteristic structure and shape which is retained if the particles are reduced in size by crushing. The shapes of the crystal faces and the angles between them can sometimes be used as a means of identification. If, however, the material has been reduced in size by the Micronizer, identification in this way is difficult since this machine tends to remove corners and sharp edges. Figure 2.12 illustrates four of the basic types of crystal form.

A considerable number of pigments are crystalline, but in many cases the crystalline form is apparent only under high magnification as is obtained with the ultra or electron microscope. X-ray diffraction is used for fine powders. The lead chromes afford some of the best examples of crystal structure and these can usually be observed under an ordinary microscope. Thus, Primrose Chrome is rhombic, Lemon and Mid-Chromes are mono-clinic, whilst Scarlet Chrome is tetragonal. Other examples could be quoted, but these are mentioned under the individual pigments.

The shape of pigment particles influences many paint properties, including flow characteristics, tendency of the pigment to settle, the nature of settlement, and the durability of the paint film.

Specific gravity of pigments

According to Stokes' Law, the rate at which a solid particle settles in a liquid is proportional to the difference between the specific gravities of the solid and the

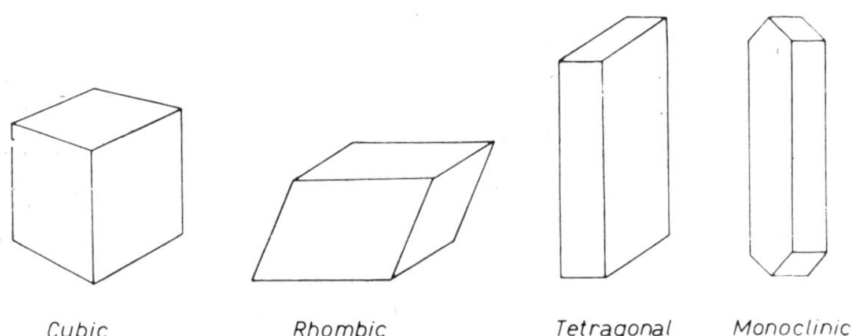

Fig 2.12 Basic types of crystal form

liquid. This is borne out by pigments in paints, where those with high specific gravity such as red lead (sp. gr. 9·0) settle rapidly, whereas carbon black (sp. gr. 1·8) shows little or no tendency to do so. (In the case of carbon black other factors such as particle shape and nature of the surface also play an appreciable part.) A degree of thixotropy or structure is generally built into paints containing heavy pigments in order to reduce settlement.

Specific gravities of pigments are best determined by the displacement method as set out in BS 3483 [15], using either vacuum (Part B8) or centrifuge (Part B9) to remove entrained air.

Oil absorption of pigments

Pigments differ widely in respect of the amount of oil (or other medium) required to convert them into a workable paste. The 'oil absorption' of a pigment is defined as the weight of refined linseed oil required to convert 100 parts by weight of the pigment into a smooth paste. It is determined by the method described in BS 3483, Part B7 [15] of which the following is a summary.

> A known weight of the pigment is placed on a palette and acid-refined linseed oil is added dropwise from a 25 cm^3 burette. After each addition of oil the pigment and oil are well mixed with a stiff palette knife. The additions of oil and the rubbing are continued until a stiff paste of smooth consistency is obtained. The volume of oil used is converted to weight and the oil absorption calculated on 100 parts by weight of the pigment.

The oil absorption figure represents the minimum weight of oil required to coat each pigment particle and to fill the voids between them; it depends on the surface area of the pigment (which is related to particle size) and also on the nature of the pigment surface which influences the wettability.

Ease of wetting of pigments

This is of importance in the dispersion of pigments and depends on both the nature of the pigment surface and the nature of the wetting liquid.

When pigment powders are exposed to the air the particles adsorb gas and/or moisture on the surface. The extent to which this takes place and the type of material adsorbed varies from one pigment to another, and pigments can be classified as hydrophobic (water repelling) or hydrophilic (water attracting). Before a pigment can be wetted properly by an organic liquid the adsorbed gas and/or moisture must be displaced, otherwise an unstable dispersion results.

A method for comparison of the ease of dispersion of pigments is contained in BS 3483, Part B5 [15].

Surface-treated pigments

As a result of the demand for increased rates of paint production, pigment manufacturers have given considerable attention to modifying the pigment surface in order to improve wettability. This applies particularly to hydrophilic pigments which are to be dispersed in organic media. The method used consists in coating the particles with an organic compound – generally described as a surface-active agent – so that the particles assume a hydrophobic condition and are then more readily wetted by organic media.

Many organic pigments are coated with resin to facilitate dispersion. The pigment aggregates consist of an open lattice formed by a small quantity of resin. This is readily penetrated by paint media and dissolved, leaving the pigment particles in a dispersed condition. Such pigments can be dispersed readily by a high-speed stirrer.

Surface treatment is also applied to certain types of pigment or extender to check or prevent hard settlement, and an example is afforded by the deposition of stearic acid on whiting or on precipitated calcium carbonate. The adsorbed acid increases the volume of the particles considerably and leads to an appreciably lower overall specific gravity. In paints these materials are usually mixed with heavier pigments so as to check the rate of settlement and, in addition, the presence of the bulky particles in the sediment prevents the heavier pigments from packing into a hard cake. The sediment remains soft and is readily reincorporated on stirring.

The quantity of surface-coated calcium carbonate employed in a paint formulation should not exceed 10 percent by weight of the total pigment. Larger amounts are likely to interfere with the flow properties of the paint.

Certain grades of titanium dioxide, mainly the rutile type, are surface treated with oxides of zinc, aluminium and silicon to improve their dispersibility and weathering properties (see p. 55). Organic compounds also are incorporated.

Fastness to light

The ability to retain its colour when exposed to light is an essential property of a pigment. The wavelengths responsible for the photochemical breakdown lie in the ultraviolet waveband.

In addition to the effect on certain pigments, ultraviolet radiation can cause colour changes, most commonly yellowing or darkening, in some types of oxidizing media. When assessing lightfastness properties of a pigment it is essential, therefore, to ensure that the organic binder is unaffected, and for this reason acrylic media are commonly employed.

The lightfastness of a pigment depends on a number of factors, the most important of which are chemical constitution, purity, physical condition, and conditions of exposure.

Chemical constitution

Most inorganic pigments, in the pure state, are fast to light and, with a few exceptions, are fast in admixture with other pigments. Organic pigments, however, show a wide variation in lightfastness both in the pure form and when mixed with white ('reduced'). It is an interesting fact that the fastness to light (and the general stability) of organic pigments improves with increasing molecular size and complexity.

Purity

The term 'purity' in this instance means freedom from admixture with other pigments. A pigment may be fast to light in the pure condition but may lose this fastness when diluted with white (or other pigment). The extent of the loss depends on the degree of dilution, and consequently lightfastness figures are determined at internationally agreed ratios of coloured pigment to white (usually titanium dioxide). They constitute the ISO (International Standards Organization) Standard Depths and are described in BS 2661 [32]. Ratios commonly employed vary between 1/10 and 1/200.

Physical condition

Lightfastness is little affected by particle size, but crystal structure and lightfastness are sometimes related, as in the lead chromes. Again, although it is not an example of fading, the anatase form of titanium dioxide is affected by light to a greater degree than the rutile form.

Conditions of exposure

The lightfastness of a pigment is important only when the pigment is dispersed in a medium, and the nature of this medium can affect the performance of the pigment. When lightfastness alone is measured (described below) the film of pigment dispersion is exposed to ultraviolet radiation under controlled conditions. In practice, however, films may be exposed to the combined effects of sun and rain. The latter penetrates the film to a greater or lesser degree and can affect the performance of the pigment. The term *durability* is generally used to denote the performance of the paint film on exterior exposure.

Measurement of lightfastness

The basis of the method is the assessment of the degree of fading which takes place when films of the pigment, and the standard, in the same medium are exposed simultaneously to the same conditions. This can be sunlight but is more usually the radiation from a carbon arc lamp housed in a Fugitometer (Fig. 2.13).

For many years Madder Lake was employed as a standard for lightfastness and was assigned an arbitrary value of 10. The degree of fading of other pigments was then assessed visually against this standard and they were assigned values of 1 (very poor) to 10 (good).

A more scientific method is now employed whereby a closer and more accurate assessment can be made. The method is described in BS 1006 [33] and involves the use of a range of eight blue wool specimens which have been prepared with dyes of differing lightfastness. Thus, No. 1 has very low, and No. 8 very high, lightfastness. Each standard is approximately twice as fast to light as the preceding one.

The pigment is dispersed in a suitable medium and the dispersion applied to a card and allowed to dry. This is exposed either to daylight under glass, or to a carbon-arc lamp, at the same time as the series of blue wool specimens. At the end of the exposure period the sample is compared with the standards, and the degree of fastness quoted is the number of the wool specimen which has faded to the same degree. When changes in colour are involved, a grey wool scale is employed. This is described in BS 2662 [34].

Fig 2.13 Fugitometer

In the fugitometer the specimens are exposed round the periphery of the drum and arranged so that part of the specimen is shielded from direct exposure. A carbon-arc rated at approximately 1500 to 1800 watts provides the ultraviolet light. Temperature is controlled by means of a fan in the base of the instrument.

Carbon-arc lamps produce radiation which is much richer in the active ultraviolet wavelengths than normal daylight, and results can sometimes be misleading. The spectral energy distribution of the radiation from a xenon-arc lamp is closer to that of normal daylight and these lamps are replacing the carbon-arc type.

Lightfastness figures can be obtained from the pigment manufacturers' literature, but a simple comparison of a pigment with one of known lightfastness (and the same type) may be needed. In this case the expensive blue wool patterns are not required, and use can be made of the method described in BS 3483, Part A3 [15] which also involves the use of the fugitometer.

Resistance to heat

Many paint films encounter heat either during a stoving process, which may be in the region of 120°C to 200°C (248° to 392°F) for periods of a few minutes to an hour, or they may be applied to surfaces which remain at high temperatures for long periods.

It is essential that the pigments used in such paints should be resistant to the temperatures encountered and they have, therefore, to be chosen with care. Pigments vary considerably in their resistance to heat and this property depends largely on the chemical composition. Most inorganic pigments – and many organics – will withstand 120°C (248°F) for short periods without decomposition. Above this temperature, however, the number of resistant pigments decreases progressively. An indication of heat resistance is given under the pigment classes. A method for the comparison of the heat stability of pigments is described in BS 3483, Part C8 [15].

Solubility ('bleeding') of pigments

'Bleeding' is the term used to describe the discoloration or staining which sometimes occurs when white or light coloured paint is applied over a deeper colour. It is generally (but not invariably) due to the solubilizing effect of solvents in the second coat on pigments in the first coat. The organic Paranitraniline Red is well known for this tendency. A pigment which is very slightly soluble in the paint medium can function quite satisfactorily, provided the paint is not overcoated as mentioned above.

Solubility in water

With one or two exceptions, notably strontium and zinc chromes, inorganic pigments are insoluble in water. Among organic pigments the majority are insoluble in cold water, but a few tend to bleed in hot water.

Solubility in solvents, resins and oils

Inorganic pigments are insoluble. The solubility of organic pigments varies with the pigment type as well as with the nature and degree of substitution. Generally toners and lakes are less soluble than the pigment dyestuffs, but strong solvents of the ketone and ester types give greater bleeding than the more widely used hydrocarbon solvents.

A method for the comparison of pigments for their resistance to bleeding is described in BS 3483, Part C7 [15].

Water-soluble matter in pigments

In the manufacture of many pigments, water-soluble chlorides and sulphates are produced as by-products. These are sometimes difficult to remove completely and for many uses their presence is not detrimental. If, however, the pigments are used in primers for metals, traces of these salts can actually stimulate corrosion, and for this reason it is advisable to determine the soluble salt content of the pigment before use. This is best carried out by the method described in BS 3483, Parts C2 and C6 [15].

Toxicity of pigments

The pigments which can exert toxic effects in the human body are compounds of antimony, barium, cadmium, chromium (in the form of chromates) and lead (metallic lead and compounds). In order that a material shall be an active poison it must be soluble in the fluids of the human body, and an insoluble material, although it may be a compound of a metal whose soluble salts are poisonous, will pass through the body without causing injury. For example, the soluble salts of barium, as well as the carbonate, are active poisons, but barium sulphate is insoluble and is, in fact, administered as a barium meal before X-ray photographs of the alimentary tract are taken. The most common of the toxic pigments are those containing lead, which is a cumulative poison; adsorption of small amounts over a period can result in chronic poisoning.

The Lead Paints (Protection against Poisoning) Act, 1926, now superceded by Control of Lead at Work Regulations, 1980, S. I. 1980 No. 1248 [35] was designed to protect operatives who come into contact with lead pigments or lead paints. The pigments can enter the body through the respiratory or digestive systems during the handling of the pigments themselves or during the application (particularly spraying) of paints containing them. Protection during the handling of the pigments can be provided by suitable clothing, respirators, and efficient fume extractors.

For lead paints to be applied by spraying, the Act lays down a maximum figure (5 percent) for the lead compounds soluble in 0·25 percent hydrochloric acid, which is the concentration in normal digestive fluids. This soluble-lead figure is calculated as lead monoxide (PbO) and is expressed as a percentage on the pigment. It is determined by the method described in BS 282:1963 [36].

Under the conditions of this method white lead and litharge are completely soluble, but the soluble lead contents of lead chromes and chrome greens vary according to the composition. Figures for the soluble lead content of these pigments are usually available from the suppliers, but if they are used in admixture with other pigments it is necessary to determine the figure for the mixture.

Another possible source of poisoning from pigments is the paint applied to children's toys, and limits to the content of certain metals are specified in the Toys (Safety) Regulations, 1974 [37]. The maximum content of 'total' lead is 2500 parts per million of the dry paint film; limits are also applied to 'soluble' antimony, barium, cadmium and chromium.

Lead pigments have been eliminated from paints for decorative purposes. Where they continue to be used for special purposes the can labels bear the appropriate warning.

Factory regulations

The Health and Safety at Work Act, 1974, is concerned, *inter alia*, with the purity or degree of contamination of the atmosphere in places of work, and strict controls are in force.

Occupational Exposure Limits (OEL) (formerly TLV) have been assigned to most paint raw materials. These figures represent concentrations of air-borne substances below which the majority of workers can be exposed repeatedly without ill effect. They are time-weighted average concentrations which allow brief exposure to higher concentrations, provided it is followed by exposure to equivalent lower concentrations. 'Ceiling values' are assigned to certain materials; these must not be exceeded.

Occupational Exposure Limit figures are published annually by the Health and Safety Executive [38]. This publication should be consulted for up-to-date figures as data on many materials are changed from time to time in the light of experience.

OELs are expressed either as parts per million (of air) or as milligrams per cubic metre. Some solids, generally in the extender class, are classed as 'nuisance particulates', and their OELs are expressed as millions of particles per cubic foot or per cubic metre.

Toxicity ratings

Toxic materials, in general, are given a 'toxicity rating' designated LD_{50} which is the dose which will kill 50 percent of the particular class of animal to which it is administered. It is expressed as milligrams or grams per kilogram body weight. Paint components can be classified roughly as follows [39].

LD_{50} value	Toxicity
1 mg (or less)	High
100 mg	Moderate
1 gram	Slight
15 grams (or more)	Relatively innocuous

The majority of organic pigments are considered to be reasonably safe. One or two, e.g. chromium compounds of certain dyestuffs, can produce slight skin irritation.

References to Chapter 2

[1] Ishihara, S., *Tests for Colour Blindness*. Lewis, London, 1959.

[2] Colour Matching Aptitude Test. Federation of Societies for Coatings Technology, Philadelphia, USA.

[3] BYK Color-Gloss (Westlairds Ltd, North Green, Datchet, Slough)
Erichsen Colorimeter 511 (Pearson Panke Equipment Ltd, London)
Hunterlab (BOC Automation, Daventry)
Minolta Tristimulus Colorimeter (Minolta (UK) Ltd, Milton Keynes).

[4] The Paint Research Association, Teddington, Middx. *see also* Ishak, I. G. H., The fibre optics colorimeter and its application in paint manufacture. *JOCCA*, 1971, **54**, 129.

[5] The following companies produce reflectance spectrophotometers which may be used either as individual instruments or in a system with computer and printer.
Applied Colour Systems Inc. (Kerstol Ltd, Stalybridge, Cheshire)
Beckman Instruments Inc., Fullerton, California
Colourgen Ltd, Birchwood, Warrington
Hunter Associates Inc. (Kerstol Ltd, Stalybridge, Cheshire)
Industrial Colour Systems, Newbury, Berks
Pye Unicam Ltd, Cambridge.
Note: this list is not exhaustive.

[6] Diffusion Systems Ltd, London, W7 2EW
Sheen Instruments Ltd, Teddington, Middx.
[7] Kubelka, P. and Munk, F. *Zeit. Tech. Physik.*, 1931, **12**, 593.
[8] Tioxide Brochure, No. 201A, Measuring Colour, p. 22.
[9] Clulow, F. W., *Colour: its Principles and their Application*. Fountain Press, London, 1972.
[10] McDonald, R., A review of the relationship between visual and instrumental assessment of colour difference. *JOCCA*, 1982, **65**, 43.
[11] Clulow, F. W., *ibid.*
[12] Clulow, F. W., *ibid.*
[13] BS 4800:1981. Paint colours for building purposes. British Standards Institution.
[14] Society of Dyers & Colourists, Bradford, Yorkshire.
[15] BS 3483. Methods of testing pigments for paints. British Standards Institution.
[16] *see also* Simpson, L. A., *Measuring Opacity*. Tioxide Group PLC.
[17] BS 3900. Methods of test for paints. British Standards Institution.
[18] BS 3406. Methods for determination of particle size distribution. British Standards Institution.
[19] Anon. *J. Appl. Chem. Abs.*, 1962, i–612; 1963, ii–238.
[20] Bostock, W. A., A sedimentation balance for particle size analysis in the sub-sieve range. *J. Sci. Instr.*, 1952, **29**, 209.
[21] Beresford, J., Size analysis of organic pigments using the ICI-Joyce Loebl disc centrifuge. *JOCCA*, 1967, **50**, 594.
[22] Quantachrome Corpn New York. (Techmation Ltd, Edgeware, Middx).
[23] Malvern Instruments Ltd, Malvern, Worcestershire.
[24] BS 3406. Part 3. Air elutriation methods. British Standards Institution.
[25] BS 3406. Part 4. Optical microscope methods. British Standards Institution.
[26] Analytical Measuring Systems Ltd, Saffron Walden, Essex.
[27] Watt, I. M., *The Principles and Practice of Industrial Microscopy*. Cambridge University press. 1985.
[28] Coulter Electronics Ltd, Harpenden, Herts. *see also* BS 3406, Part 5.
[29] Lowell, S. and Shields, J. E., *Powder Surface Area and Porosity*, 2nd Edn. Chapman & Hall, London, 1984.
[30] Grigg, S. J. and Sing, K. S. W., *Adsorption, Surface Area and Porosity*, 2nd Edn. Academic Press, 1982.
[31] Brunauer, S., Emmett, P. H. and Teller, E. J. *J.Am.Chem.Soc.*, 1938, **60**, 309.
[32] BS 2661. Methods for determination of colour fastness of textiles. British Standards Institution.
[33] BS 1006. Methods of test for colour fastness of textiles and leather. British Standards Institution.
[34] BS 2662. Grey scale for assessing change in colour. British Standards Institution.
[35] *see also* Cowley, A. C. D., Living with lead legislation. *JOCCA*, 1982, **65**, 357.
[36] BS 282:1963. Lead chromes and zinc chromes for paints. British Standards Institution.
[37] The Toys (Safety) Regulations. 1974. H. M. Stationery Office, London.
[38] Guidance Note EH 40. Health and Safety Executive.
[39] Brunner, H. in Hess, M., *Paint Film Defects*, Ch. 8, 3rd Edn. Chapman & Hall Ltd, London, 1979.

3

The processing of pigments

The maintenance of high quality in paints is possible only if the ingredients themselves are of high quality. Thus the pigments used should possess as narrow a range of particle size as possible and should be free from coarse particles, ultra-fines, and harmful impurities.

It is convenient, for the present discussion, to divide pigments and extenders into two classes: (i) those derived from natural sources, and (ii) those produced by chemical reactions. The processing of materials in class (i) is, broadly speaking, size reduction followed by grading by levigation or air-flotation. A large proportion of class (ii) materials is produced by precipitation which, in some cases, is followed by calcination. Other processes involve calcination or vapour-phase oxidation.

MINERAL PIGMENTS AND EXTENDERS

These materials are either mined or quarried and are obtained in the form of coarse lumps. The lumps are broken down to a coarse powder which is subjected to further grinding in one of the mills described below.

Grinding

Ball mills

Ball mills used for pigment grinding are similar in construction to those used in paint production and consist of a horizontal steel cylinder which is rotated on its axis. The mill contains a charge of steel balls, and the grinding action arises from the cascading of these balls when the mill is rotated (Fig. 3.1). For each size of mill there is a critical speed which ensures optimum cascading. Very low speeds give little cascading, whilst high speeds will result in the balls and charge being carried round by centrifugal action.

The gross volume occupied by the balls should be about 40 to 45 percent of

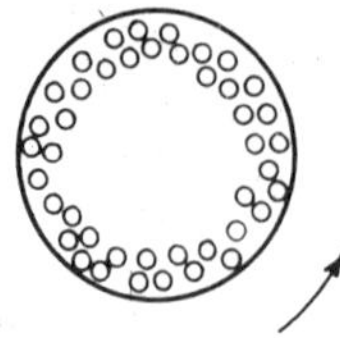

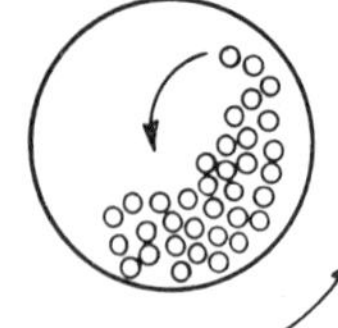

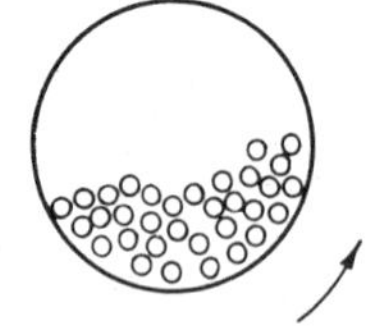

Fig 3.1 Action of ball mills

the total volume of the mill, and for maximum efficiency the material to be ground should just fill the voids in the ball charge.

These mills are used for both dry and wet grinding and for the former purpose are often fitted with an exterior dust casing. They are batch mills and in dry grinding suffer considerable wear on the balls and steel lining. When white pigments are ground the mill is lined with silica or porcelain blocks, and the steel balls are replaced by steatite or high-density alumina balls. However, the lower specific gravity of these materials results in much longer grinding times than when steel balls are used.

The micronizer

Size reduction in this mill takes place as a result of inter-particle attrition. The mill contains no moving parts and the energy required is derived from high-velocity compressed fluids entering the grinding chamber through tangential jets. For this reason it is classed as a fluid energy mill.

The mill, shown in Fig. 3.2, consists of a flat cylindrical grinding chamber fitted with inlet and outlet nozzles. The former are arranged tangentially round the outside and the compressed fluid – compressed air or superheated steam – creates a violent turbulence in the mill. The material to be ground is introduced by a venturi through which the compressed air or superheated steam is passed.

The violent turbulence and high velocities result in rapid size reduction by attrition. The fine particles are taken off in a stream of exhaust fluid and enter a collector. Uniformity of particle size is controlled by regulating the rate of feed and/or the pressure of the grinding fluid. The micronizer is used mainly to produce powders with an average particle size of about five microns or less, but it can be used, if necessary, to produce coarser products.

Many natural pigments such as oxides of iron, as well as extenders such as barytes, talc, and dolomite are processed in these mills.

The pin disc mill

This mill consists, basically, of two circular metal discs to which are attached steel pins arranged in concentric circles. The upper disc, with pins on the lower

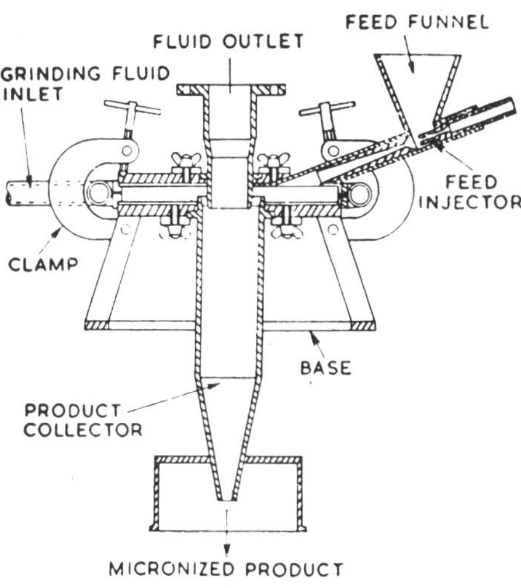

Courtesy: F. W. Berk

Fig 3.2 The micronizer

face, is fixed and has a central hole through which the pigment is introduced. The pins in the lower disc are on its upper face and the circles of pins pass between those of the upper disc. The lower disc is rotated, causing the material to be broken down between the pins, and centrifugal action throws the ground material off the edges of the discs where it is collected. Pin disc mills are suited to the softer types of pigment as they are pulverizing rather than grinding mills.

Levigation

If a suspension of a pigment or extender in water is allowed to overflow from one to another of a series of vats of increasing diameter, the particles which settle out will be progressively smaller. The rate of flow of the liquid is reduced by the increasing diameter of the vats, so that the smaller particles are given the longer time required to settle out. In this way a material can be separated into a series of fractions of decreasing mean particle size, but there is no sharp line of demarcation between the fractions. Each will comprise a particle size range which will be determined by the size of the vats and the initial rate of flow of the liquid. There will be a degree of overlap with fractions above and below.

This process, known as *levigation*, is employed to remove oversize particles and grit from several natural extenders where there is a plentiful supply of water near the source of the powder. It is generally preceded by a grinding operation. The deposit in each vat is filtered, dried, and finally reground.

MANUFACTURED PIGMENTS

The processes employed in the manufacture of pigments can be divided into the following types: (i) precipitation, (ii) precipitation followed by calcination, (iii) calcination, (iv) vapour-phase oxidation. A few pigments are prepared by methods not within this classification and these will be described under the pigments concerned.

Precipitation

General principles

When two solutions, each containing a water-soluble salt, are mixed, a solid product will be precipitated if two of the radicals or ions present are capable of forming such a material. For example, if solutions of lead nitrate and sodium chromate are mixed, a yellow precipitate of lead chromate is produced, according to the equation:

$$\underset{331}{Pb(NO_3)_2} + \underset{162}{Na_2CrO_4} \rightarrow \underset{323}{PbCrO_4} + \underset{2 \times 85}{2NaNO_3}$$

The figures beneath the formulae are the molecular masses of the compounds involved. In practice it often happens that a product is improved if a slight excess of one of the reactants is used. In the above instance a slight excess of lead gives a brighter and more satisfactory product (Mid Chrome).

Another interesting example of precipitation is afforded by the manufacture of lithopone. The reaction in this case is somewhat unique in that both products of the reaction are insoluble. The process consists of mixing solutions of barium sulphide and zinc sulphate, when zinc sulphide and barium sulphate are precipitated:

$$BaS + ZnSO_4 \rightarrow ZnS + BaSO_4$$

Again, a slight excess of zinc produces a more stable product. The precipitated mixture is filtered, dried, and calcined.

Reduced pigments

If an insoluble base such as barytes is present when precipitation occurs, the precipitate will form on the particles of the base. Such pigments are known as *reduced*, and their tinting strength will vary with the amount of base present. Reduced pigments can also be made by dry-grinding the components.

Co-precipitated pigments

Some pigments contain more than one salt of a metal. For example, Lemon Chrome contains both lead chromate and lead sulphate. It is possible to make a pigment of this nature by mixing the two salts, but a far superior product is obtained if the two compounds are precipitated simultaneously. In this case a mixed solution of sodium chromate and sodium sulphate can be added to a solution of lead nitrate:

$$2Pb(NO_3)_2 + Na_2CrO_4 + Na_2SO_4 \rightarrow PbCrO_4PbSO_4 + 4NaNO_3$$

This produces a very intimate mixture of the two products and the process is known as co-precipitation. It is used to produce a number of pigments.

Conditions of precipitation

It is a well-known fact that the physical characteristics of a precipitate depend largely on conditions such as the concentration and temperature of the reacting solutions. This applies particularly to pigment manufacture, many of the properties of the pigment being influenced by the conditions of precipitation, of which the following are important.

Concentration of solutions

In general, fine precipitates are formed from dilute solutions, and coarser precipitates from more concentrated solutions. In practice, a compromise is often used. Whilst a fine particle size may be desirable for the end use, it can be difficult to filter. In a few cases the latter difficulty can be overcome by boiling the pigment suspension and so causing the particles to flocculate. Ten percent solutions have been found to be a very suitable compromise for a number of pigments, but there are several exceptions.

Temperature

High temperatures usually result in coarse precipitates, whilst finer precipitates are formed at lower temperatures. In most cases precipitation is carried out at 10°C to 20°C (50° to 68°F) unless the precipitate formed is too fine to filter. Prussian Blue is a pigment of this type and is therefore precipitated from boiling or near-boiling solutions.

Uniformity of stirring

The rate of stirring should be directly related to the rate at which the solutions are mixed. Considerable variation in particle size can result if these are not uniform.

pH value

For many pigments this is critical and the optimum value has, in most cases, been derived by experience. Variations from the critical value can affect colour and crystal form, as, for example, in the scarlet chromes (Chapter 5).

Precipitation plant

This is comparatively simple, the basic unit consisting of a precipitation vat holding up to 5000 gallons (22 700 litres), flanked by smaller dissolving vats.

The vats are constructed of oak or pitch pine and, when necessary, are lined with lead, rubber, or other acid-resisting material. In special cases, stainless-steel vats are used, but these are very expensive. Stirrers are of wood or coated metal and the vats are also fitted with open or closed steam pipes for heating, a cold water supply, and a bottom outlet to the filtration plant.

Filtration

The filtration of the pigment suspension from the precipitation vat can be carried out by either a filter press or a vacuum filter.

Filter presses

In these the suspension is pumped into chambers lined with filter cloths. The aqueous solution passes through the cloths to waste, and the pigment, in the form of a paste, remains in the chambers. Clean water is then passed through the paste to remove soluble by-products and the paste is 'dewatered' by blowing air through the press. Some water remains and the product is then ready for drying or flushing.

Vacuum filters

Vacuum filters are of two general types which can be designated as the *rotary vacuum filter* and the *leaf type filter.*

Rotary vacuum filter

The general principle of the machine is illustrated by Fig. 3.3. A large drum is rotated on a horizontal axis and is partly submerged in a trough of the pigment slurry. The ends of the drum are solid, but the curved surface is constructed of perforated plate to which filter cloth is attached. Vacuum is applied to the interior of the drum. As the latter rotates, the water is drawn into it and the pigment collects as a paste on the outside, from which it is removed by a scraper. In some cases the drum is designed so that the adherent pigment is washed and partially dried by air-blowing before reaching the scraper.

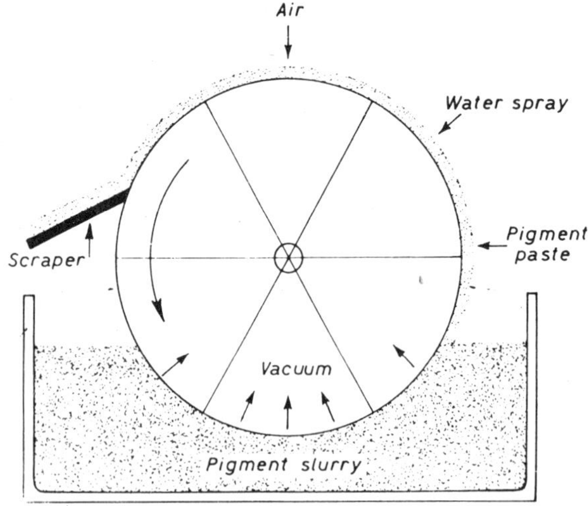

Fig 3.3 Principle of the rotary vacuum filter

It is necessary for the level of the pigment slurry in the trough to be kept constant, and the rotary vacuum filter is therefore admirably suited to continuous process filtration.

Leaf type vacuum filter

In this type of filter rectangular frames are made up of a number of leaves and attached to a vacuum line. The leaves are then covered with filter cloths. When the frames are lowered into a pigment slurry and vacuum applied, the liquid passes through the cloths and a layer of pigment paste builds up on the outside. When the layers are sufficiently thick the frames are lifted from the slurry and lowered into tanks of water for washing and the filtration repeated. After washing, the pigment cakes are removed and transferred to the drying ovens. This type of filter can be used for batch or continuous production.

Drying

The conditions of drying can influence the texture and colour of many pigments. It is essential that the temperature be kept as low as possible otherwise hard clusters or *aggregates* of pigment particles can be formed and these can prove difficult to break down in subsequent operations. Ideally, the pigment when dry should be in the same physical condition as when filtered, but this is difficult to attain. Again, some organic pigments are very heat-sensitive when in the water-wet state and these must be dried at a low temperature.

The filter-press cakes are usually passed through an extruder in order to expose as large a surface as possible and spread in trays. These are either passed through a conveyor oven, where they meet a counter current of warm air, or placed in a low-temperature vacuum oven.

After drying, the pigment aggregates are reduced to their final size by grinding (for the harder aggregates) in mills of the type described earlier. For the softer aggregates, such as the organic and some inorganic pigments, the pin disc mill (p. 44) is used.

Flushing [1]

Pigments can be divided into hydrophilic (water attracting) and hydrophobic (water repelling) types. If the wet filter-press cake of a hydrophobic pigment such as Hansa Yellow G (Chapter 7) is mixed with linseed oil or other organic medium in a suitable mixer, the pigment will pass into the organic medium and the water will separate out. This process is known as *flushing* and can be represented as follows:

$$\text{pigment/water} + \text{oil} \rightarrow \text{pigment/oil} + \text{water}$$

The water can be run off leaving the pigment dispersed in the medium. In addition to being used for hydrophobic pigments the process is also used with white lead. In this case the basic nature of the pigment assists the transfer to the organic medium.

Many of the inorganic pigments are hydrophilic and are therefore difficult to flush. The addition of suitable surface-active agents can convert them to the hydrophobic condition.

In spite of the obvious advantages in flushing, such as good pigment dispersion and avoidance of the costs of drying and subsequent dispersion, the process is used to a limited extent. This is probably due to the wide range of media and pigments used in modern paint manufacture.

Calcination

This term is applied to the process of roasting or heating to a high temperature in a suitable furnace. It is used as a process in itself or as further treatment for materials which have been precipitated, such as titanium dioxide.

The rotary furnace is the most widely used type of calciner for bulk production. It consists of a long steel cylinder mounted on trunnions. It is inclined at a slight angle to the horizontal so that material introduced at the top rolls round slowly and is discharged at the bottom. Heating is provided by combustion of a mixture of gas and air injected at the lower end. This type of furnace is employed in the manufacture of titanium dioxide.

Other types of calciners include the muffle furnace, the reverberatory furnace, and earthenware or silica crucibles.

Subsequent processing

During the process of calcination a certain amount of sintering of the pigment often takes place, resulting in the formation of clinker. To assist in breaking this up, the hot mass is discharged into cold water, when the clinker disintegrates. The pigment is then filtered, washed, dried, ground and graded.

Examples of pigments produced by direct calcination are red lead and Ultramarine Blue. Lithopone and titanium dioxide are produced by precipitation followed by calcination. Among the natural pigments, calcination is employed to convert Raw Sienna and Umber to the Burnt varieties (Chapter 5).

Vapour-phase oxidation

The metals zinc and antimony can be volatilized if heated to a sufficiently high temperature, and the vapours combine with oxygen by admixture with air to produce pigments of fine particle size.

In the case of zinc, the volatilized metal is mixed with air and the non-volatile zinc oxide precipitates. The zinc oxide is carried by the airstream to a classifier. (If the metallic zinc is volatilized in absence of air and then cooled rapidly, zinc dust is produced (Chapter 6).)

Antimony oxide is produced by vapour-phase oxidation of the metal, but the oxide is volatile at the temperature of reaction. It is removed to cooling chambers and cooled as rapidly as possible to produce a fine particle-size pigment.

There is also a vapour-phase oxidation method for the production of basic lead sulphate from galena.

The formation of carbon black (Chapter 6) by the controlled combustion of petroleum gases is also an example of this type of process.

Reference to Chapter 3

[1] Gomm, A. S., Hull, G. and Moilett, J. L., Mechanism of the pigment flushing process. *JOCCA*, 1968, **51**, 143.

4

White pigments and extenders

TITANIUM DIOXIDE (CI Pigment White 6)

Among the white pigments titanium dioxide is pre-eminent. It is unrivalled in respect of colour, opacity, stain resistance and durability; it also is non-toxic. The main consuming industries are paint, printing inks, plastics and ceramics which together account for between 60 and 70 percent of the total demand.

Sources

The two principal sources of titanium are the ores ilmenite and rutile. Ilmenite is a black ferrous titanate which occurs in many parts of the world. Rutile is a natural titanium dioxide which is not as widely distributed as ilmenite but has been found to be more suitable for the manufacture of titanium dioxide by the chloride process (below). However, difficulties in supply of rutile have necessitated modifications to this process to enable ilmenite to be used.

Compounds of titanium are very widely distributed among a number of minerals but in small quantities. Titanium is also present in many plants and can be detected in the residue when these are ashed.

Manufacture of titanium dioxide

The sulphate process

This uses ilmenite as the raw material. The ore is crushed, separated from associated materials, ground and dried. It is then dissolved in oleum. This reaction is smooth, whereas with concentrated sulphuric acid a violent reaction takes place. The solution in oleum is diluted gradually, first with sulphuric acid and then with water. This yields a solution of the sulphates of titanium and iron (ferrous and ferric) together with insoluble siliceous matter. The latter is filtered off. Scrap iron is then added to the solution to reduce all iron to the ferrous condition. The solution is separated, filtered, and transferred to a vacuum evaporator where most of the ferrous sulphate crystallizes out.

The next operation is a crucial one since it controls both the particle size and the nature of the resulting pigment. The solution is boiled to hydrolyse the titanium sulphate to titanium hydroxide, and the conditions of hydrolysis must be rigidly controlled. The precipitated hydroxide is washed thoroughly to remove acid and traces of iron, filtered, and then calcined in a rotary furnace at about 800°C (1472°F).

Finally the titanium dioxide is given the appropriate surface treatment. The sequence of operations is as follows:

$$\left.\begin{array}{c}\text{Ilmenite}\\ +\\ \text{Oleum}\end{array}\right\}\xrightarrow{\text{Dilute}}\begin{array}{l}\text{Titanium}\\ \text{sulphate}\end{array} + FeSO_4 + Fe_2(SO_4)_3$$

↓ (Scrap iron)

Titanium sulphate + $FeSO_4$

↓ Evaporation

Titanium sulphate solution — $FeSO_4\ 7H_2O$ separates

↓ Hydrolyse

$$Ti(OH)_4 \xrightarrow[\text{calcine}]{\text{Filter, dry}} TiO_2 \xrightarrow[\text{treatment}]{\text{Surface}} TiO_2 \text{ pigment}$$

The process outlined above yields the anatase type. The rutile type is produced by 'seeding' with rutile particles before the hydrolysis stage. The pigment then goes forward to the surface treatment stage.

The chloride process [1]

This is of more recent introduction and offers several advantages over the sulphate process in respect of accurate control of purity, particle size and amount of waste requiring disposal.

The process involves two stages. In the first, the titanium ore is mixed with coke and treated with chlorine at about 950°C. Titanium tetrachloride is produced together with chlorides of impurities, largely iron.

$$2TiO_2 + 3C + 4Cl_2 \rightarrow 2TiCl_4 + 2CO + CO_2$$

On cooling the gases, the bulk of the less volatile chlorides of the impurities condense out at a temperature above the boiling point of the titanium tetrachloride. The latter is then condensed to a liquid by further cooling and is purified by distillation.

In the second stage, the titanium tetrachloride is oxidised to titanium dioxide by reaction with oxygen:

$$TiCl_4 + O_2 \rightarrow TiO_2 + 2Cl_2$$

The reaction must be carried out at a temperature above 1000°C to ensure the correct particle size distribution. The liberated chlorine is re-cycled. The ease with which the titanium tetrachloride can be purified coupled with the close control of the oxidation process results in a pigment of outstanding whiteness and consistent particle size range.

Both chloride and sulphate type pigments are next subjected to 'surface treatment'.

Surface treatment

It has been found that the general working properties (e.g. wettability, ease of dispersion) and some aspects of the performance (e.g. resistance to chalking) of titanium dioxide can be improved by surface treatment with oxides or hydroxides of aluminium, silica or zirconium. Certain organic compounds also are used. Surface treatment reduces the photo-catalytic activity of the pigment surface (see Chalking p. 55).

The relative proportions of the oxides and the amount of coating depend on the purpose for which the particular grade of titanium dioxide is intended, e.g. decorative gloss finishes, industrial stoving enamels, emulsion paints, etc. The TiO_2 content of the treated pigment can vary between 90 and 95 percent.

Properties of titanium dioxide

Titanium dioxide has greater opacity and tint resistance than any other white pigment and consequently has largely replaced other white pigments in a number of applications. It is non-toxic, resistant to heat and to the media and solvents used in paint manufacture. It also possesses great chemical stability and is dissolved only by hot concentrated sulphuric acid – and this process is slow.

The pigment exists in two forms known as anatase and rutile. Both forms fall into the tetragonal system, but anatase has the lower density arising from a greater distance between the titanium and oxygen atoms. When exposed to ultraviolet radiation both forms absorb strongly. Anatase becomes highly 'excited' and exhibits intense surface catalytic activity. Excitation and catalytic activation of rutile is very much less, most of the absorbed energy being dissipated as heat (see Chalking, *below*).

The ultraviolet absorption band of rutile overlaps the far blue end of the visible spectrum (up to about 410 nm) with the result that in daylight the reflected light is deficient in blue and gives rutile the characteristic yellowish cast.

The physical characteristics of the two grades are as follows.

	Rutile	Anatase
Colour	Slight yellowish cast (chloride types are cleaner and brighter than sulphate process types)	Cold bluish-white
Specific gravity	4·0–4·1	3·7–3·8
Refractive index	2·71	2·55
Oil absorption	17–24	19–22
Relative tint resistance	1·2–1·4	1·0
Resistance to 'chalking'	Good	Poor
Mean particle size range	0·2–0·3 μm	0·18–0·25 μm

Chalking

This is a condition of a paint surface when free or 'unbound' pigment is present as a result of decomposition of the surface layer of binder. Gentle rubbing of the finger-tip across the surface removes the free pigment as a white powder or 'chalk'.

Paint films pigmented with both rutile and anatase types of titanium dioxide will 'chalk' after periods of exposure to weather. With anatase pigmentation the onset of chalking is much more rapid, and the photochemical breakdown of the film more severe than with rutile. In fact the rutile grades, particularly the surface-treated types commonly used, cause only a slight tendency to chalking of the film after exposure for a number of years. The rutile grades are in general use in exterior finishes, but it should be mentioned that the nature of the binder can influence the chalking tendency of a paint film.

The breakdown of the surface layer of binder which leads to chalking is due to accelerated oxidation in the presence of titanium dioxide. The latter acts as a catalyst, the anatase grade being much more active than rutile. The fact that titanium dioxide will catalyse certain oxidation reactions has long been known, and Murley [2] as well as Kennedy *et al.* [3] have demonstrated that, under ultraviolet radiation, anatase absorbs oxygen to a greater degree than does rutile. In the paint film both oxygen and certain components of the binder are thought to be adsorbed on the activated titanium dioxide surface. Adsorption of reactants at the surface of a solid generally results in a lowering of the activation energy of the reaction (in this case oxidation) which then proceeds more rapidly. Simpson [4] considers that when oxygen and moisture are present, free radicals, $OH^{\bullet}$ and $HO_2^{\bullet}$, are formed and these initiate breakdown of the binder. The function of the surface treatment is reaction with or prevention of formation of these free radicals.

In addition to photochemical breakdown of the medium which is initiated at the pigment surface, there is general oxidation of the binder using atmospheric oxygen and catalysed by UV radiation. Breakdown of surface layers of binder leads to release of pigment, a reaction which accounts for chalking with pigments which are not excited by UV radiation in the same manner as titanium dioxide. Some pigments, of which red oxide of iron and carbon black are notable examples, are absorbers of UV radiation and so 'protect' the binder in which they are dispersed.

Uses of titanium dioxide

It is the most widely used white pigment. The two grades, with suitable surface treatment, are used in the following types of paint.

Rutile decorative, maintenance and industrial gloss finishes; decorative semi-gloss, eggshell and matt finishes; marine paints; undercoats; emulsion paints for interior and exterior use.

Anatase used in industrial finishes where good whiteness is important but where weather resistance is not required, e.g. domestic and hospital equipment.

Identification of titanium dioxide

The outstanding chemical resistance of the pigment enables it to be separated from most other pigments (but not from some extenders which, however, do not interfere with the tests). Titanium dioxide may be recognized by the following simple test.

A small quantity of the pigment is placed in a depression on a spotting tile and moistened with concentrated sulphuric acid. Addition of a few drops of hydrogen peroxide solution results in an orange or orange/yellow coloration.

Spindrift [5] is an interesting product which utilizes the greater scattering power of titanium dioxide when in contact with air compared with that when in contact with an organic medium. The greater scattering power results in improved opacity. In Spindrift, particles of titanium dioxide, surrounded by air, are enclosed in spherical shells of a polymer – usually a co-polymer of styrene and an unsaturated polyester. The spheres range in diameter from 3 to 50 micrometres. The use of Spindrift is confined to semi-matt and matt finishes; it has been found to function successfully in emulsion paints.

Toxicity of titanium dioxide

The pigment is regarded as non-toxic, but dust arising during handling is classed as a 'nuisance particulate'.

ZINC OXIDE (CI Pigment White 4)

This pigment has been used in paints for very many years and is known to artists as *Chinese White*.

With the advent of titanium dioxide and the use of alkyd and other synthetic resins the use of zinc oxide as a pigment in paints has declined very considerably, and today little is used for this purpose in this country. A certain amount of colloidal zinc oxide (*below*) is thought to assist gloss retention in some alkyd gloss finishes and is used for this purpose.

Manufacture of zinc oxide

This is a vapour-phase oxidation and is based on the fact that metallic zinc begins to volatilize when heated above its melting point (419°C, 786°F), and the vapour combines rapidly with oxygen to form the non-volatile zinc oxide. Two methods are in use, known as the direct and indirect processes.

Direct process

In this process the oxide is produced directly from the ore in one operation. The ore – preferably a carbonate – is mixed with coal or coke and heated in a furnace. The ore is reduced to metal which volatilizes and is then mixed with a current of air. Zinc oxide is formed and is carried by the stream of air into suitable collecting chambers.

The product obtained by this process contains impurities, the chief of which is basic lead sulphate (up to 5 percent).

Indirect process

Metallic zinc, known industrially as *spelter*, is first produced and in this operation ores such as the sulphide (zinc blende) can be used. The spelter is heated in crucibles to volatilize the metal, which passes out through holes in the lids. The metal vapour is then mixed with a stream of air and the zinc oxide produced is carried into the collecting chamber. This oxide usually contains less than 0·2 percent of impurities.

Properties of zinc oxide

The bulk of the pigment produced is the amorphous type, but by adjustment of the manufacturing conditions it is possible to produce a crystalline ('acicular') or a very fine ('colloidal') type. Amorphous zinc oxide (indirect type) possesses the following characteristics:

Specific gravity 5·65	Oil absorption 16–23
Refractive index 2·0	Mean particle size range 0·2–0·35 μm

Zinc oxide is a reactive pigment and forms zinc compounds or 'soaps' when mixed with drying oils or varnishes, the extent of reaction depending on the acid value and the degree of polymerization of the medium. Films deposited from zinc oxide dispersed in such media exhibit a high gloss but undergo progressive embrittlement. Such paints were formerly used for interior decorative work. The acicular grade, however, is less reactive.

Uses of zinc oxide

In present-day paints the use is limited to certain specific rôles.

1 It has pronounced fungistatic properties and is incorporated into emulsion paints for use in warm and damp climates where fungus attack is a major problem.
2 As mentioned earlier, a small quantity of the colloidal grade is sometimes included in certain alkyd-type decorative finishes to assist gloss retention. The amount employed is usually critical and depends on the nature of the medium.

Identification tests

Zinc oxide has the well-known property of turning yellow on heating and reverting to white on cooling. If it is moistened with cobalt chloride solution and re-heated a green coloration is produced.

The pigment is soluble in dilute mineral and acetic acids.

Toxicity of zinc oxide

The pigment does not present any significant hazard. 'Fresh' zinc oxide fume, as produced in welding or flame-cutting of films containing metallic zinc, can cause 'zinc fume fever' but the effects are temporary.

ZINC PHOSPHATE (CI Pigment White 32)

A white, non-toxic pigment which is prepared by precipitation, it is assigned the formula $Zn_3(PO_4)_2 2H_2O$.

Properties of zinc phosphate

Zinc phosphate possesses the following physical characteristics:

Specific gravity 3·3	Oil absorption 18–20
pH of aqueous extract 6·8–7·0	Mean particle size range 2–4 μm

It has poor opacity, but this is not a serious drawback as the main use of the pigment is in corrosion-resisting primers for steel and in undercoats. Zinc phosphate is compatible with and performs satisfactorily in a number of media including alkyds, chlorinated rubber, epoxy esters, two-pack epoxies, and polyurethanes.

The corrosion-resisting properties have attracted a number of investigators, but the mechanism of protection is not completely understood. The general opinion is in favour of a physical barrier effect rather than an electro-chemical mechanism. The pigment appears to pack in the film in a unique manner which presents a high resistance to the passage of water molecules and ions. It is possible, however, that some electro-chemical reaction leading to suppression of corrosion cell reactions is also present.

Identification test

Zinc phosphate is soluble on warming with 4M mineral acid. The solution gives the reactions of zinc and phosphate ions.

Toxicity of zinc phosphate

The pigment is regarded as free from toxic hazard.

Zinc phospho-oxide

A white non-toxic pigment marketed under the trade name Nalzin SC–1 [6] and which is claimed to possess excellent corrosion-inhibiting properties when used as a replacement for red lead. It is also claimed to be specially suited to electrodeposition.

Properties of zinc phospho-oxide

Zinc phospho-oxide possesses the following physical characteristics:

Specific gravity 4·06	Mean particle size range <1 μm
Oil absorption 50–70	

A further interesting use of this pigment has been described by Davidson [7] who found it to be an excellent control agent for cedar stain.

Lithopone

This pigment which is sometimes known as *zinc white* has been used for many years and contains zinc sulphide and barium sulphate.

Manufacture of lithopone

The basic reaction involved is double decomposition using solutions of barium sulphide and zinc sulphate. The reaction is somewhat unusual in that both products are insoluble in water:

$$BaS + ZnSO_4 \rightarrow ZnS + BaSO_4$$

The precipitate is a mixture of zinc sulphide and barium sulphate and of little interest as a pigment. It is filtered, washed, dried and then calcined at 700°C (1292°F) in an inert atmosphere to develop the pigmentary properties. The hot mass from the calciner is discharged into cold water to break up the *frit*, and then filtered, dried and ground.

Properties and use of lithopone

Lithopone possesses the following physical characteristics:

Specific gravity 4·2	Oil absorption 12–15
Refractive index 1·90	Mean particle size range 1–3 μm

It is clean white and does not react with acid media, although it is attacked by mineral acids. Lithopone is very little used in solvent-based paints, but it is occasionally encountered in emulsion paints in which it can form part of the pigment mixture.

Toxicity of lithopone

Lithopone is considered to be free from toxic hazard.

ANTIMONY OXIDE (CI Pigment White 11)

Antimony oxide (Sb_2O_3) is a pigment whose usage in paint has declined in recent years. It is now used mainly for certain specialized purposes.

Manufacture of antimony oxide

The chief source of the metal is the ore stibnite (Sb_2S_3) from which the metal is extracted by roasting with iron:

$$Sb_2S_3 + 3Fe \rightarrow 2Sb + 3FeS$$

Antimony oxide is manufactured from the metal and the process bears a certain resemblance to the indirect process for zinc oxide. The metal is volatilized by heat and the vaporized metal is oxidized to oxide in a current of air. Antimony oxide, however, differs from zinc oxide in that it is volatile at the temperature of formation and is carried by the airstream into condensing chambers where the solid pigment is deposited on cooling. The rate of cooling determines the particle size and crystalline form. Rapid cooling produces an amorphous pigment of fine particle size, whereas slow cooling gives a less desirable crystalline product. The oxidation reaction can be represented by the equation

$$4Sb + 3O_2 \rightarrow 2Sb_2O_3$$

Properties and uses of antimony oxide

Antimony oxide possesses the following physical characteristics:

Specific gravity 5·5	Refractive index 2·20
Oil absorption 11–13	Mean particle size range 0·5–2·0 μm

It possesses very good opacity and high stain resistance. It is non-reactive toward organic media and for this reason it was at one time used with zinc oxide to counteract the film embrittlement caused by the latter. It also works well in cellulose nitrate lacquers.

When exposed to sunlight, paints based on antimony oxide do not chalk and for some years the pigment was used in conjunction with anatase titanium dioxide to suppress chalking. With the introduction of the chalk-resistant rutile titanium dioxide the use of antimony oxide declined. The principal use of the pigment today is in fire-retardant paints where it is used in conjunction with a chlorinated medium.

Identification tests

On strong heating, antimony oxide melts and then volatilizes. The vapour will form needle-like crystals on cooling.

It is insoluble in dilute mineral acids but dissolves in concentrated hydrochloric acid. The solution gives an orange precipitate of antimony sulphide with hydrogen sulphide. If the solution in concentrated hydrochloric acid is diluted considerably with water, a white precipitate of oxychloride is produced:

$$SbCl_3 + H_2O \rightarrow SbOCl + 2HCl$$

WHITE LEAD (CI Pigment White 1)

This material has been known for many centuries and to artists was familiar as *flake white*. It consists of a basic carbonate of lead and the composition approximates to the formula $Pb(OH)_2.2PbCO_3$.

Manufacture of white lead

Over the years a number of methods have been employed for the manufacture of white lead but most are now of historic interest only [8]. The present day method involves precipitation and consists in treatment of a solution of basic lead acetate with carbon dioxide under finely regulated conditions.

$$3Pb(CH_3COO)_2Pb(OH)_2 + 2CO_2 \rightarrow Pb(OH)_2.2PbCO_3 + 3Pb(CH_3COO)_2 + 3H_2O$$

The pigment is separated by filtration and the lead acetate solution recirculated for conversion to the basic acetate. Following filtration the white lead is either dried or flushed into organic medium.

Properties of white lead

White lead possesses the following physical characteristics:

Refractive index 2·0   Oil absorption 7·5–10·0
Specific gravity 6·7

It is chemically reactive and forms lead 'soaps' when mixed with linseed oil or varnish. These soaps are largely responsible for the elasticity of the films of white lead paints and for their protective qualities.

For many years white lead paint was the standard type of finish and was satisfactory under rural and marine conditions. In urban atmospheres the hydrogen sulphide present caused discoloration, the extent of which varied with locality.

The introduction of alkyd resins and rutile titanium dioxide which give durable gloss finishes with good flow caused a decline in the use of white lead in finishing paints. Although the protective qualities of white lead oil paint were excellent, the flow and gloss retention were inferior to those of modern alkyds.

White lead is a highly toxic pigment and, as a lead compound, acts in the human system as a cumulative poison. It is completely soluble in the stomach acid (0·25 percent hydrochloric acid) and so is readily ingested.

Both the manufacture and use of white lead (in common with other lead pigments) are subject to the regulations of the Lead Paints (Protection against Poisoning) Act, 1926 and subsequent amendments.

On account of its high toxicity and the widespread use of titanium dioxide white lead has been phased out of the majority of paints, but it continues to be specified in some undercoats for certain types of exterior maintenance work.

Identification tests

On heating, white lead turns orange and finally yellow, due to the formation of litharge.

It is completely soluble in dilute acetic and nitric acids with evolution of carbon dioxide. Reaction takes place with hydrochloric acid, but the lead chloride produced is soluble only to a limited extent in the cold but dissolves on boiling. Treatment of the solution with hydrogen sulphide gives a black precipitate of lead sulphide. Addition of potassium iodide solution to the hot solution of lead chloride leads to the formation of golden-yellow spangles on cooling.

Treatment of white lead with dilute sulphuric acid results in an initial reaction which is rapidly stifled by the film of insoluble lead sulphate which forms on the surface.

BASIC LEAD SULPHATE (CI Pigment White 2)

The composition of this pigment approximates to the formula $PbO.2PbSO_4$, but variations occur due to fluctuations in quality of the raw material used.

Manufacture of basic lead sulphate

Basic lead sulphate can be manufactured by roasting natural lead sulphide (galena) so that it volatilizes and is led into a second chamber where it is oxidized by a cold air blast. The pigment is then collected.

Precipitation methods are also employed and consist in treating an aqueous suspension of litharge or litharge mixed with finely divided lead with carefully regulated quantities of sulphuric acid.

Properties and uses of basic lead sulphate

Basic lead sulphate is less reactive than white lead towards oils and varnishes and has not been used to the same extent in gloss paints and undercoats. It is toxic and the soluble-lead content is in excess of that permitted for use in spray paints. At the present time the main use is in certain types of primer for the underwater areas of ships' hulls. The pigment possesses the following physical characteristics:

Oil absorption 12–15	Refractive index 1·93
Specific gravity 6·4	

Identification tests

Basic lead sulphate dissolves completely on boiling with medium strength hydrochloric acid, and the solution gives the usual qualitative tests for lead and sulphate.

It is, of course, subject to the Lead Paint Regulations.

EXTENDERS

These are, in general, chemically inert materials some of which are processed minerals while others are prepared chemically. They are characterized by the fact that their refractive indices fall within a narrow range and this is close to that of oils and resins. Consequently extenders are practically transparent in paint media and make no contribution to colour or opacity. They are opaque in water and can therefore be used as pigments in certain types of water paint.

When pure, extenders are colourless, but many natural products are coloured by impurities which are often difficult to remove.

The term *extender* suggests a material used to cheapen a product, and whilst it is true that on occasions extenders are used for this purpose they are more often used to confer desirable properties on a paint. For example, they can be used to control consistency, as flatting agents, and to prevent hard settlement of heavy pigment. The term 'auxiliary pigment' might therefore be more appropriate.

Classification of extenders

Extenders are best classified by chemical composition, and the majority can be classified as compounds of the metals barium, calcium, aluminium and magnesium. The exception is silica, and the natural and prepared silicas are considered as a separate class.

BARIUM COMPOUNDS

These comprise the sulphate which occurs naturally as barytes or heavy spar and the precipitated form known as blanc fixe.

Barytes

This mineral occurs extensively in many parts of the world, but the colour and quality vary considerably. Some of the purest grades are obtained in Germany and there are extensive deposits of high-grade material in the United States; other sources are Algeria and Greece. Home-produced barytes is generally of poor colour and the quality varies considerably.

In view of the variation in colour, barytes is classified as *white* and *off-colour*, the latter term covering a very wide range of materials.

Processing of barytes

If the mineral is a pure crystalline grade of barytes it is crushed, ground and graded by air-flotation. Off-colour grades are often treated with sulphuric or hydrochloric acid to improve the colour by removing impurities, which may be iron oxide or compounds of lead. Calcium sulphate or carbonate may also be present. As barytes is a heavy pigment and settles rapidly the acid can readily be washed out by decantation. The barytes is then filtered, dried, ground (much is now micronized) and graded by air-flotation.

Grades

Barytes is available in ordinary grades in which the particle size can range up to 20 to 25 μm, or in micronized form. In the latter it is possible to obtain a number of size ranges and these are marketed with an average particle size of 2 μm, 5 μm, 10 μm, etc. These micronized grades are now widely used and are displacing the ordinary grades. In addition they have a lower oil absorption than blanc fixe and for many purposes are displacing this extender.

Properties and uses of barytes

Barytes possesses the following physical characteristics:

Specific gravity 4·5	Refractive index 1·67
Oil absorption 10–12	Mean particle size range Varies according to grade

When pure it is practically transparent in organic media, but most commercial grades show some slight opacity due to impurities.

It is inert toward all paint media and chemically is very stable. It will dissolve only in concentrated sulphuric acid on heating but is reprecipitated on cooling or dilution.

Barytes is widely used in undercoats and primers in which it is thought to give physical reinforcement to the film. It is also used in fillers, stoppers, and some types of masonry finish.

Identification tests

Natural barytes is distinctly crystalline under the microscope, but the crystalline nature is less obvious in the micronized grades. This process tends to round off the corners and sharp edges.

In the platinum wire flame test, barytes gives the characteristic green-coloured flame of barium.

The material is extremely inert, and in order to confirm the presence of barium and sulphate it is necessary to fuse it with anhydrous sodium carbonate. Decomposition takes place according to the equation

$$BaSO_4 + Na_2CO_3 \rightarrow BaCO_3 + Na_2SO_4$$

The fused mass is extracted with water and filtered. The filtrate is acidified with dilute hydrochloric acid and barium chloride solution is added. A white precipitate indicates sulphate. The residue is dissolved in dilute hydrochloric acid, and then dilute sulphuric acid is added. Formation of a white precipitate indicates barium.

Toxicity of barium sulphate

Barium sulphate is insoluble in the digestive fluids and is considered to be non-toxic. It is used in medicine as the basis of the 'barium meal' given to patients before intestinal X-ray examination.

Blanc fixe

This is a precipitated form of barium sulphate and generally contains about 98 percent $BaSO_4$. It can be produced by treatment of barium sulphide solution with a soluble sulphate such as sodium sulphate, according to the equation

$$BaS + Na_2SO_4 \rightarrow BaSO_4 + Na_2S$$

The product is washed well, filtered and dried.

An alternative method of production utilizes natural barium carbonate, witherite. This is dissolved in hydrochloric acid to produce barium chloride which is then treated with sulphuric acid:

$$BaCl_2 + H_2SO_4 \rightarrow BaSO_4 + 2HCl$$

This is followed by filtering, washing and drying.

Properties and uses of blanc fixe

Blanc fixe possesses the following physical characteristics:

Specific gravity 4·3	Refractive index 1·67
Oil absorption 14–16	Mean particle size range 1–5 μm

As a precipitated product, blanc fixe has a finer texture and higher oil absorption than barytes. It dissolves more readily than barytes in concentrated sulphuric acid but otherwise shows the same chemical inertness.

It has been used as an extender in undercoats but has tended to be replaced by micronized barytes which gives better flow. In addition, the use of high percentages of blanc fixe in undercoats can lead to *sinkage* and loss of gloss in finishing coats.

Blance fixe alone and as a co-precipitated base with alumina is fairly widely used as a base for organic pigments (Chapter 7).

In water paints, the finer texture of blanc fixe gives it greater opacity than ordinary barytes and confers smoother working properties on the paints.

Toxicity of blanc fixe

See Barytes (*above*).

CALCIUM COMPOUNDS

The calcium-based extenders of interest in paint manufacture comprise the carbonate, as natural whiting or as the precipitated form; the double carbonate with magnesium as the mineral dolomite; and, to a lesser extent, the sulphate.

Whiting

The purer grades of this material are sometimes known as Paris White. Whiting is obtained when natural chalk (mainly $CaCO_3$) is reduced to a powder. Chalk consists of the remains of tiny creatures (coccoliths) laid down in ocean beds over very long periods of time. In electron micrographs of whiting, the ring-like structure of the coccoliths is often quite apparent.

Chalk comprises the basic structure of much of south-east and north-east England and of some hill ranges such as the Yorkshire Wolds and the Chilterns.

Processing of whiting

There are two general methods in use for the treatment of the chalk and the basic sequence of operations is the same, namely, grinding, drying and grading. The different processes arise from the fact that whitings differ in hardness and density. The softer types are treated by a wet process and the harder types by a dry process.

The wet process

The soft type of whiting is broken up by stirring in water. Any nodules of flint separate out, and the slurry is passed through hydrocyclones which separate the coarse material. The slurry is next placed in shallow tanks for settlement. The paste obtained after decanting off the upper liquor is filtered by press or rotary vacuum filter and dried. It is finally disintegrated and packed.

The dry process

The raw material is crushed and screened to remove the coarse particles and grit. It is then passed through a pin disc mill such as the Kek where it is reduced to a fine powder. Hot air can be blown through at the same time and serves both to dry the powder and to carry it to a cyclone for grading.

Properties of whiting

In this country whiting is marketed in two forms, known as paint grades and putty grades. The former is obtained from the softer chalks of south-east England, whereas the harder putty grades are derived from the north-east of the country. Both grades contain about 96 to 97 percent $CaCO_3$, the remainder consisting of fine clay. The pinkish colour is due to traces of iron oxides.

The physical characteristics of the two grades are as follows:

	Paint grade	Putty grade
Specific gravity	2·5–2·8	2·5–2·8
Oil absorption	17–18	16–17
Refractive index	1·58	1·58
Mean particle-size range	2–5 μm	2–5 μm but with a number of particles over 10 μm

Whiting shows an alkaline reaction in water and is readily soluble in dilute mineral and acetic acids.

Uses of whiting

Glazing putty for wooden window frames consists of whiting dispersed in linseed oil. Whiting is used in many fillers and stoppers and, to varying extents, in certain undercoats and primers. The presence of a proportion of whiting in primers for steelwork is thought to be an advantage in providing an alkaline environment. Whiting is often incorporated in vinyl emulsion paints where it acts as a 'buffer' against the possibility of a fall in pH value.

Identification tests

Whiting shows an alkaline reaction when mixed with distilled water. It is decomposed on strong heating leaving a residue of lime, and the platinum wire flame test gives the characteristic crimson colour of calcium. It is soluble in dilute mineral and acetic acids giving carbon dioxide and a solution which shows the usual reaction for calcium ions, i.e. it gives a white precipitate with ammonium oxalate under slightly alkaline conditions.

Coated whitings

The chemical reactivity of whiting enables it to react with some of the long-chain fatty acids such as stearic acid, whereby a coating of the stearate is formed over the surface. The amount of acid can be controlled up to 2 percent and the products then exhibit hydrophobic properties in contrast to the hydrophilic properties of the original whiting, i.e. they are more readily wetted by organic media.

Precipitated calcium carbonate ($CaCO_3$)

This material is obtained in large quantities as a by-product in water-softening plants when sodium carbonate has been used. It is also produced from the purer grades of limestone (which is essentially $CaCO_3$) by calcining to calcium oxide followed by slaking with water to give calcium hydroxide. Coarse particles are separated and the calcium hydroxide in suspension converted to carbonate by passing carbon dioxide or addition of sodium carbonate solution. The product is filtered, washed and dried.

Properties and uses of precipitated calcium carbonate

Like whiting, it is hydrophilic and is similar to whiting in reactivity; it yields an alkaline aqueous extract and dissolves in dilute hydrochloric, nitric and some organic acids. The physical characteristics are as follows:

Specific gravity 2·7	Oil absorption 24–26
Refractive index 1·6	Mean particle size range 2–4 μm

The main use is for gloss control in low sheen paints in which it has a stabilizing effect on pigment dispersions. In some paint systems it can be used to control floating and flooding.

Identification tests

As for whiting but no residue is obtained when dissolved in dilute acid.

Coated calcium carbonate

Precipitated calcium carbonate will combine with stearic acid to give a product which is hydrophobic in character. It usually contains about 3 percent stearic acid, the hydrocarbon chains of which are directed away from the particle surface. The result is a particle of lower overall density than the original and of effectively greater diameter. When mixed with pigments which tend to settle to a compact mass the latter is kept softer and more voluminous. It is, therefore, more readily re-dispersed. The amount used should not exceed 10 percent on the total weight of pigment; more than this is likely to impair the flow properties of the paint.

Coated calcium carbonate is marketed under a number of proprietary names [9].

Toxicity of calcium carbonate

The compound is considered to be non-toxic but dusts can be regarded as 'nuisance particulates'.

Calcite

This is a crystalline form of calcium carbonate and is marketed under the trade name 'Millicarb' [10]. It is harder than the chalks and is useful where a hard extender with low specific gravity is required.

Specific gravity 2.8

Dolomite

A calcium magnesium carbonate, $CaCO_3$ $MgCO_3$, which occurs in a crystalline state and is marketed in micronised form under the trade name 'Microdol' [11]. It is used as a general purpose extender.

Specific gravity 2.87   pH of aqueous suspension 9–10

Gypsum

This is a crystalline calcium sulphate with the formula $CaSO_4.2H_2O$. A purified powdered form was once known as 'terra alba' but this term is no longer used.

Gypsum occurs in considerable quantities in Great Britain and is processed by crushing, grinding, washing and drying. The material, however, is used to a very limited extent in paints. The chief interest in gypsum lies in its relationship to plaster of Paris.

Plaster of Paris

When gypsum is heated to 120°–130°C (248°–266°F), part of the water of crystallization is lost and plaster of Paris is obtained. The latter forms the basis of many commercial plasters used for interior walls. When these plasters are mixed with water, re-hydration sets in with the formation of a dense matrix of gypsum crystals. The reaction is the reverse of that taking place in the formation of plaster of Paris and the two can be represented thus:

$$\underset{\text{Gypsum}}{2CaSO_4.2H_2O} \underset{\text{Water}}{\overset{\text{Heat}}{\rightleftharpoons}} \underset{\text{Plaster of Paris}}{2CaSO_4.H_2O} + 3H_2O$$

Plaster of Paris is known as a *hemihydrate plaster* and the formula is frequently written as $CaSO_4.\frac{1}{2}H_2O$.

Calcium silicate

An extender characterized by very fine particle size. It is produced by precipitation from solutions of a calcium salt and an alkali silicate, followed by filtration, washing and drying.

Properties and uses of calcium silicate

Calcium silicate possesses the following physical characteristics:

Specific gravity 2·1	Oil absorption 140
Refractive index 1·47	Mean particle size range 30–50 mμ (millimicrons)
pH of aqueous extract 10	

It is used as a flatting and thickening agent in solvent-based paints. The high pH of the aqueous extract makes calcium silicate an attractive additive to primers for steelwork.

The material is marketed under the trade name 'Microcal' [12].

Toxicity of calcium silicate

Calcium silicate is classed as non-toxic, but the dust is a 'nuisance particulate'.

COMPOUNDS OF ALUMINIUM

China clay

This is also known as kaolin and is a hydrated aluminium silicate to which the formula $Al_2O_3.2SiO_2.2H_2O$ is usually assigned.

It is produced by the decomposition of felspar, $K_2O.Al_2O_3.6SiO_2$, which is one of the components of granite. The latter consists of a matrix of crystals of quartz, felspar and mica. Breakdown of the granite and subsequent decomposition of the felspar leaves china clay mixed with fairly coarse silica and some mica.

Large deposits of china clay occur in the St Austell area of Cornwall.

Processing of china clay

The china clay is quarried by jets of high-pressure water and forms a slurry from which the large pieces of granite and quartz settle out. The slurry is then subjected to levigation to separate silica and mica, and the various grades obtained are filtered, dried and powdered. The grades of finest particle size are often called colloidal china clays. Specially coated grades are also available.

Properties of china clay

China clay possesses the following physical characteristics:

Specific gravity 2·6
Oil absorption 30–50 according to grade
Refractive index 1·56
Mean particle size range 1 μm downwards in finer grades. Coarser grades 1 μm up to 10 μm

The ultimate particles are flat plates (Fig. 4.1).

It is marketed as a fine white crystalline powder which is practically transparent in organic media. Like many other clay minerals it possesses base-exchange properties. Acid sites are present on the clay particles and neutralization of these sites produces the so-called 'alkaline clays'. These are characterized by ready dispersibility in water to give deflocculated dispersions.

Both faces and edges of the lamellar clay crystals contain electrically active sites leading to attraction between adjacent particles and the formulation of a flocculated structure. This structure is formed when china clay is used in paint and constitutes one of its most valuable properties.

Uses of china clay

The fairly high oil absorption combined with fine particle size makes it a useful flatting and thickening agent for low-sheen paints such as undercoats, eggshell and matt finishes. It is also used in the formulation of some types of 'high build' paints. The ability to form a flocculated structure, as described above, is

Courtesy: Paint Research Association

Fig 4.1 Photomicrograph of china clay (× 18 000)

utilized in many paint formulations where settlement is likely to be troublesome. Any settlement is likely to be bulky, soft, and readily reincorporated.

Calcined china clay

Calcination of china clay at 950°–980°C (1742°–1796°F) results in loss of combined water and a fundamental change in crystal structure [13]. The calcined product is harder but less dense than the original material and shows little tendency to induce structure formation in paint systems. During calcination, air voids are created within the particles and these voids confer a degree of light scattering and consequent opacity on the product.

Calcined china clays possess high oil absorption values and are of especial interest for undercoats, flat alkyd enamels, and emulsion paints.

Identification tests

China clay has a soapy feel when rubbed between the fingers. When heated with concentrated hydrochloric acid and evaporated to dryness (preferably several times) it is decomposed. The residue is taken up with dilute hydrochloric acid and filtered. A gelatinous mass of silica remains on the filter and the filtrate gives the normal reactions of aluminium.

Toxicity of china clay

The material itself presents no known toxic hazard but the dust can be classed as a 'nuisance particulate'.

Slate powder

This is an impure hydrated aluminium silicate obtained as a dark grey or greenish grey powder from slate cutting and trimming operations.

It is very hard and coarse and is useful in dark coloured fillers and stoppers because it facilitates rubbing down and does not clog the paper.

Slate powder is cheap and is used as an inert filler in a number of industries.

Mica

This term covers a group of hydrated aluminium silicates which occur in large masses and which are readily split into very thin sheets. These are converted to powder either by dry processing or under water, the latter giving the finer product for paint purposes. The particles are very thin plates and are spoiled by overgrinding. They are sometimes subjected to micronizing and so mica is offered to the industry as dry-ground (coarsest), wet-ground, and micronized (finest).

The type supplied to the paint industry is a hydrated potassium aluminium silicate the composition of which can be represented by the formula $K_2O.2Al_2O_3.6SiO_2.2H_2O$, but many variants occur.

Properties and uses of mica

Mica possesses the following physical characteristics:

Specific gravity 2·80–2·85 Refractive index 1·58
Oil absorption 75

It is an inert material and its use in paint is based on the property of the particles to leaf or to orientate themselves parallel with a paint surface and to overlap. This property confers improved moisture resistance and anti-penetrating properties and, in some cases, hard settlement of the pigment in the liquid paint is prevented.

Toxicity of mica

Finely ground mica in the form of dust presents a certain hazard, but this is not as serious as with quartz or asbestos.

Bentonite

Bentonite is a clay obtained from the United States and consists basically of a hydrated aluminium silicate. It is assigned the general formula $Al_2O_3.4SiO_2.2H_2O$, but the aluminium is often replaced to varying degrees by magnesium and sodium. The characteristic property of bentonites is their ability to absorb several times their own weight of water and then set to a stiff gel. They are therefore used as thickeners in certain types of water paint.

Bentonites possess the property of undergoing ion exchange reactions with certain organic bases to produce materials which gel in organic liquids, but not in water. These are discussed later.

Aluminium hydroxide ('Alumina hydrate')

Aluminium hydroxide releases water vapour when heated to about 200°C (392°F) giving the monohydrate. This reaction is useful in flame-retarding paints in which the compound is sometimes used as an extender. It is marketed in the micronized form.

Aluminium hydroxide is used also as a base for certain organic lakes. In this case it is generally used as a freshly prepared suspension.

Aluminium silicate

An extender of which the particles are aggregates (0.1–1.0 μm diameter) of extremely small units of average diameter of about 15 nm [14]. It has a number of uses including viscosity control, anti-settlement and as a partial replacement (4–6% by weight) for titanium dioxide in alkyd and emulsion paints.

Specific gravity 2.1

Sillitin [15]

This was formerly known as Neuburg Chalk and consists of hexagonal lamina (0.02 μm thick) of Kaolinite containing fine quartz particles in the lamina. It shows exceptional anti-settling properties and, for this reason, is useful in metal primers. It is used also in electrodeposition where the migration of the particles in the electric field is analagous to that of pigment dyestuffs. This results in uniform deposition.

COMPOUNDS OF MAGNESIUM

Magnesium carbonate

This material occurs naturally as the mineral magnesite, $MgCO_3$, but this form of the carbonate is very seldom used in paint. It also occurs in conjunction with calcium carbonate in the mineral dolomite which is used in the micronized form Microdol [12] as a general purpose extender.

Precipitated magnesium carbonate

The magnesium carbonate used in paint is prepared by precipitation and is a hydrated basic carbonate. It can be manufactured from either magnesite or dolomite and has been represented by the formula $MgCO_3.3Mg(OH)_2.11H_2O$ (amongst others).

Properties and uses of precipitated magnesium carbonate

Precipitated magnesium carbonate possesses the following physical characteristics:

Oil absorption 120	Refractive index 1·7
Specific gravity 2·9	

It is a very fine powder which has found use as a flatting agent, especially in undercoats and flat finishes. In recent years it has been displaced to a considerable extent by the hydrated silicas.

Identification tests

On heating it is decomposed and leaves a residue of magnesium oxide, MgO. It is readily soluble in dilute mineral acids with evolution of carbon dioxide and formation of a solution giving the normal reactions for magnesium ions.

Asbestine

This is a fibrous type of extender which has been used in the paint industry for many years. The fibrous nature has been of value in preventing dense settlement of pigments and for imparting a 'matted' structure in primers for surfaces containing hair-line cracks.

However, the structural similarity between asbestine and asbestos has led to a search for an alternative material. A number of 'asbestine substitutes' are available. Among these 'Lubestine' [16] has been found to be a useful replacement.

Talc

This is also known as 'french chalk' and, on account of the 'greasy' feel, the rock has traditionally been known as 'soapstone'. It is a hydrated magnesium silicate and can be represented by the formula $Mg_3H_2(SiO_3)_4$ but the ratio of magnesium to silica in commercial deposits can vary considerably. This can be due, in part at least, to the presence of magnesium carbonate (magnesite) in many natural deposits, and this results in a 'spread' of properties.

Properties and uses of talc

Talc has the following characteristics:

Specific gravity 2·65–2·85	Refractive index 1·59–1·64
Oil absorption 25–35	pH of aqueous suspension 9–9·5

Ground talcs contain a variety of particle shapes, but lamellar particles predominate. Particle sizes vary from 5 to 10 μm but micronized grades are available and these possess flatting and anti-settling properties.

Talc is a useful general purpose extender for solvent-based paints. It is hydrophobic and consequently disperses without difficulty. In water, however, talc has a great tendency to flocculate.

Identification tests

The 'greasy' and 'soapy' feel coupled with examination by microscope is usually sufficient, but the micronized grades are difficult to identify by the latter technique. Warming with dilute hydrochloric acid will dissolve magnesium carbonate (with evolution of CO_2) but decomposition of the hydrated magnesium silicate requires boiling with the concentrated acid. The solution will give the normal reactions for magnesium ions; the silica will remain as an insoluble residue of silicic acid.

Toxicity of talc

As a magnesium compound talc is regarded as non-toxic. The dust, however, is classed as a 'nuisance particulate'.

Note: Care should be taken in the use of 'fibrous talc' which may contain a proportion of fibrous tremolite, a material with proven carcinogenic effects in the human body [17].

SILICA

The paint industry uses silica obtained from natural sources as well as manufactured hydrated silicas. The former tend to be very hard and coarse and their use today is limited to certain specific types of product.

Natural silicas

Quartz

This is a crystalline material and is the commonest form of natural silica. It occurs in the form of crystalline masses and as spheroidal quartz sands. The main use of the extender is in fillers and stoppers but coarser grades are used in resin-based screeds and flooring compounds. A range of coloured aggregates for flooring systems is made by coating spheroidal quartz sand with pigmented epoxy resin [18].

Properties of quartz

Powdered quartz possesses the following physical characteristics:

Specific gravity 2.8 Refractive index 1.53

Flint

The ground form of this material is sometimes used in fillers and stoppers. It is not quite as hard as quartz.

Kieselguhr

This is a member of the family known as diatomaceous earths. These consist of the siliceous remains of diatoms and other creatures which were deposited in primordial seas over long periods of time. The deposit is amorphous, readily powdered and highly absorbent. It is therefore used as a flatting and consistency-control agent in paints and as a filter aid in oil processing.

Celites [19]

These comprise a range of diatomaceous earths and are characterized by excellent colour. They are marketed in various grades ranging from 2 to 20 μm mean particle size, and with oil absorption from 75 to 110. They are used for flatting and for viscosity control in paints.

Diatomite

This is an indigenous diatomaceous hydrated silica obtainable in a range of particle sizes with oil absorptions from 60 to 100. The specific gravity is 1·95 and refractive index 1·46. Diatomite is used as a flatting agent and it is claimed to improve intercoat adhesion.

Toxicity of natural silicas

The properties are controlled by the crystalline form. Amorphous grades are classified as 'nuisance particulates', but crystalline quartz is more hazardous and can lead to silicosis.

Manufactured silicas

These are generally of very fine particle size and low specific gravity. They are marketed under a number of trade names [20]. A typical member is produced by flame hydrolysis of silicon tetrachloride.

Properties and uses of manufactured silicas [21]

Manufactured silicas possess the following physical characteristics:

Average particle diameter 7–40 nm
pH of aqueous suspension 3.6–4.5
BET surface area 110–380 sq. metres per gram

These silicas are chemically inert and, by virtue of their large surface areas, they are used on a considerable scale for flatting and viscosity control in solvent type paints. They are used also to boost viscosity in emulsion paints.

Toxicity of manufactured silicas

They are non-crystalline and are classified as 'nuisance particulates'.

References to Chapter 4

[1] Bull, D. S., 'Titanium dioxide pigments by the chloride process'. *Paint & Resin*, Jan./Feb. 1982.
[2] Murley, R. D., Some aspects of the physical chemistry of titanium pigment surfaces. *JOCCA*, 1962, **45**, 16.
[3] Kennedy, D. R. *et al., Trans. Far. Soc.*, 1958, **54**, 119.
[4] Simpson, L. A., Influence of titanium dioxide pigment on the surface roughness of organic coatings. *JOCCA*, 1986, **69**, 232.
[5] Tioxide Group PLC, Stockton-on-Tees, England.
[6] National Lead Industries Inc.
[7] Davidson, S. L., A new versatile lead-free pigment. *JOCCA*, 1975, **58**, 435.
[8] See earlier editions of this book.
[9] 'Calofort' (J. & E. Sturge Ltd); 'Winnofil' (ICI Ltd).
[10] Melbourn Chemicals Ltd.
[11] Norwegian Talc Industries.
[12] Crosfield Chemicals Ltd, Warrington.
[13] ECC International Ltd.
[14] Degussa Bulletin No. 34. Aluminium silicate.
[15] Norman Collins Chemicals Ltd.
[16] Bromhead & Dennison Ltd.
[17] Asbestos Regulations, 1969, H. M. Factory Inspectorate Technical Data, Notes 12 and 13.

[18] 'Caratex' pigmented quartz. Boud Marketing Ltd.
[19] Johns-Manville Ltd.
[20] 'Gasil' (Crosfield Chemicals Ltd); 'Aerosil' (Degussa); 'Cab-O-Sil' (Cabot).
[21] Degussa Bulletin No. 68.

5

Coloured inorganic pigments

The pigments described in this chapter comprise the coloured inorganics used in the manufacture of finishes, undercoats and primers. They contain the red, orange, yellow, green and blue but non-spectrum colours such as crimsons and maroons are included.

THE CHROME PIGMENTS

Lead chromes (CI Pigment Yellow 34, Pigment Red 104)

These all contain lead chromate, either alone or co-precipitated with other lead salts. The colours range from pale primrose to orange and scarlet. Table 5.1 shows the relationship between colour and composition.

Table 5.1

Lead chrome	Composition
Primrose	Lead chromate, lead sulphate and alumina
Lemon	Lead chromate and lead sulphate (relative proportions determine the shade)
Middle	Lead chromate only
Orange	Lead chromate and lead hydroxide
Scarlet	Lead chromate, lead molybdate, and lead sulphate

Properties of lead chromes

There is an increase in depth and redness of colour in passing from the greenish-yellow primrose through lemon and mid to the very reddish orange chrome. Scarlet chrome occupies a rather unique position. It is brighter and cleaner than the orange chromes and possesses much greater tinting strength.

The lead chromes possess the following physical characteristics:

Specific gravity 5·9–6·2 Refractive index 2·4
Oil absorption 12–16

They are all bright pigments with good opacity and high tinting strength. Their general durability is good, but in the past they have shown a tendency to darken slightly on long exterior exposure in paint films. Present-day lead chromes are virtually free from this defect. When exposed in urban atmospheres, however, they are darkened by the presence of hydrogen sulphide and tend to bleach when exposed to sulphur dioxide.

Lead chromes are non-reactive toward paint media and compatible with most other pigments. However, it is considered inadvisable to mix them with pigments containing sulphides such as ultramarine in view of the risk of darkening through lead sulphide formation. When mixed with Prussian Blue they form chrome greens.

Toxicity of lead chromes

As compounds of lead, these pigments are classed as toxic and their use is controlled by the Lead Paint Regulations. Additionally they present the chromate hazard.

On account of their toxic properties, lead chromes and other lead pigments are no longer used in paints for children's toys and for the same reason, they have been phased out of decorative paints.

They continue to be used in some types of industrial maintenance paints where their toxicity is not likely to constitute a hazard (except, possibly, when they are eventually removed). If the paints are to be applied by spray, the Regulations lay down a maximum of 5 percent 'soluble lead' calculated as PbO on the pigment. This is lead which is soluble in 0·25 percent hydrochloric acid (the concentration of this acid which is present in the stomach). The method of determination is described in BS 282:1987 [1], but manufacturers of lead chrome usually declare the figures for their products.

It is probable that in the foreseeable future all paints will be based on leadless pigments.

Basic lead silicochromate

An interesting pigment consisting of a complex lead chromate/lead silicate which is produced when lead chromate is calcined on to a silica core. It is marketed in two grades, differing in particle size, under a trade name [2].

Properties and uses of lead silicochromate

Basic lead silicochromate possesses the following physical characteristics:

Specific gravity 4·0–4·1 Oil absorption 13–18

The pigment is of a dull orange colour and can be used in a wide range of media. As a lead compound it is toxic and its use is controlled by legislation (see Toxicity, above). The principal use of the pigment is in corrosion-inhibiting primers for steelwork. It is claimed to be effective on steel from which rust has not been totally removed.

Identification test

On heating with dilute hydrochloric acid it is partly soluble, giving a solution containing lead and chromate ions (on long boiling the chromate may be reduced to a chromic salt with evolution of oxides of chlorine – these are highly toxic and must not be inhaled).

Zinc chromes (CI Pigment Yellow 36 & 36.1)

The zinc chromes in common use comply with the requirements of BS 282:1987 [1] in which three types are described.

Type 1 A zinc potassium chromate with no specified limits on impurities. It is therefore not suitable for use in metal primers, which is the main use of zinc chromes.

Type 2 A zinc potassium chromate, approximating to the formula $3ZnCrO_4.K_2CrO_4.Zn(OH)_2$. It contains an appreciable quantity of soluble chromate but is free from other salts.

Type 3 Zinc tetroxychromate, $ZnCrO_4.4Zn(OH)_2$. Contains very little soluble chromate and is free from other salts.

Type 2: Zinc chrome

Properties and uses of zinc chromate

A greenish-yellow pigment with poor opacity, zinc chromate has the following physical characteristics:

Specific gravity 3·4–3·5 Oil absorption 14–18

It dissolves readily in dilute mineral acids and reacts with many organic acids.

The main use of zinc chrome is in primers for steel and for light alloys. The protection of steel is thought to result from the presence of soluble chromate in the primer film. Moisture permeating the film carries chromate ions to the steel surface when they polarize the anodic areas of potential corrosion cells. Primers used for this purpose usually contain iron oxide in addition to zinc chrome to give the primer the necessary opacity.

A different mechanism, although also dependent on the presence of soluble chromate, is thought to operate in primers applied to aluminium and aluminium alloys. Unpainted aluminium carries a film of adherent oxide which protects the metal (except under severe conditions). The oxidizing action of the chromate maintains the oxide film and repairs any damage.

Identification tests

Zinc chrome is soluble in dilute acids and alkalis. The solution in acid gives the normal reactions for zinc, potassium and chromate ions. When the pigment is warmed with water a yellow 'bleed' (soluble chromate) is apparent.

Type 3: Zinc tetroxychromate

This resembles the Type 2 material in colour and poor opacity. It also is soluble in dilute mineral acids and reacts with a number of organic acids. The soluble chromate content, however, is very low and is irrelevant to the principal use of the pigment in etch primers. Here it is dispersed in an alcoholic solution of polyvinyl butyral resin. Phosphoric acid is added to confer the 'etching' properties, but film formation from etch primers is a complex process.

Identification tests

Zinc tetroxychromate shows less 'bleed' in water than does the Type 2 pigment. Like the latter it is soluble in dilute acids and alkalis. The solution in acid gives the normal reactions for zinc and chromate ions.

Toxicity of zinc chromes

The hazards are associated with the chromate radical, and great care must be taken in handling soluble dichromates and chromic oxide, CrO_3, during the manufacture of zinc chrome. Skin ulcers as well as attack on the nasal membranes can result from contact. A hazard also arises from zinc chrome dust, and adequate protection must be provided when handling the pigment. The Toys (Safety) Regulations, 1974[3] lay down a limit of 250 parts 'soluble' chromium per million parts of dry paint film.

Barium chromate (CI Pigment Yellow 31)

This pigment has the formula $BaCrO_4$ and can be prepared by precipitation from solutions of barium chloride and a chromate. It is a very weak, greenish-yellow pigment and is used in stoppers for use on light alloys.

It is soluble in dilute hydrochloric acid, the solution giving the usual reactions for barium and chromate ions.

Toxicity of barium chromate

Soluble barium compounds are toxic and the ready solubility of barium chromate in dilute hydrochloric acid places the pigment in the 'toxic' class.

Strontium chromate (CI Pigment Yellow 32)

Strontium chromate is a brighter colour than the barium compound. It has been found very suitable for use in water-soluble stoving primers for application by spray, dip or electrodeposition.

Identification tests

Strontium chromate dissolves in dilute hydrochloric acid on warming to give a solution containing strontium and chromate ions.

Toxicity of strontium chromate

The hazards associated with the manufacture and use of strontium chromate are associated with the chromate radical and are similar to those associated with zinc chrome (*q.v.*).

CADMIUM COLOURS (CI Pigment Yellow 37, Pigment Red 108)

The term covers a number of colours ranging from a pale primrose to lemon, orange, scarlet, crimson and maroon. The paler colours consist of cadmium sulphide, but the deeper shades are co-precipitated cadmium sulphide and cadmium selenide.

Cadmium sulphide exists in two crystalline forms, the α form (Pale lemon shade) and the redder β form (Deep lemon shade). Intermediate shades consist of mixtures of the two forms, produced by varying the conditions of precipitation.

The deeper colours consist of co-precipitated cadmium sulphide and selenide, the different shades being obtained by varying the ratio of sulphide to selenide. Generally, in the orange and scarlet shades the ratio of CdS to CdSe is about 100 to 12; in the maroons it is about 100 to 40.

Following precipitation and washing, cadmium pigments are dried and then calcined in an inert atmosphere at about 500°C to develop the colour and heat resistance.

Properties and uses of cadmium colours

The cadmium colours possess the following physical characteristics:

	Yellows	Deep shades
Oil absorption	32–35	22–28
Specific gravity	4·2–4·3	4·5–4·7

They are bright colours, although the red shades are a little dull compared with the majority of red organic colours. Although attacked by concentrated mineral acids they are resistant to organic acids, dilute mineral acids and alkalis. They are non-bleeding in paint media, possess good opacity, and are stable to heat up to 500°C (932°F). The fastness to light is good, but fading on weathering has been observed with tinted whites in alkyd/amino stoved finishes.

Cadmium pigments are expensive and this limits their use as replacements for the toxic lead chromes. Their main use is in heat-resisting paints.

Identification tests

Cadmium colours are resistant to heat and alkali but are decomposed on heating with concentrated hydrochloric acid. Hydrogen sulphide or a mixture of hydrogen sulphide and hydrogen selenide is evolved and recognized by the characteristic smell. The solution gives the normal reactions for cadmium ions.

Toxicity of cadmium colours

The solubility of cadmium pigments is very low but, in handling these materials, care must be taken to avoid inhalation of dust. Masks should be worn.

A danger arises in welding through films or burning-off paint containing cadmium pigments. This can lead to the formation of toxic cadmium oxide, CdO.

TITANIUM NICKEL YELLOW

This is a very interesting member of the rather limited range of heat-resisting pigments.

It is prepared from the titanium hydroxide precipitate produced in the sulphate process (Chapter 4). The slurry is placed in a vat, treated with salts of nickel and antimony, filtered and calcined. The product is ground and graded.

Properties and uses of titanium nickel yellow

The pigment is a compound of titanium, nickel and antimony with specific gravity 4·5 and oil absorption 22 to 25. It has a dullish yellow self-colour but gives very bright and clean tints on reduction with white. The fastness to light is very good, even in the pale tints. It is also fast to solvents, paint media, acids and alkalis.

A very valuable property is its fastness to heat; it is stable up to 500°C (932°F). In this respect it rivals the cadmium yellows.

The principal use of this pigment is as a stainer for heat-resisting paints.

YELLOW AND RED OXIDES OF IRON

Natural oxides

Iron oxide is one of the most abundant mineral oxides in the earth's crust. It is widely distributed and has been used as a pigment for many centuries. If the deposits are to be economically workable as pigments they must be of good colour, uniform, free from contamination, and must contain a reasonable quantity of iron oxide. The colours of the natural oxides range from yellow to black, and the impurities are usually silica together with compounds of aluminium and calcium.

Principal ores of iron

1 **Limonites** or hydrated oxides in which the constitution can vary from $2Fe_2O_3.H_2O$ to $Fe_2O_3.4H_2O$. These comprise the yellow ochres, the siennas and the umbers.

2 **Haematite,** which consists basically of Fe_2O_3 and constitutes the red oxides. It is the principal source.
3 **Magnetite (or lodestone)** A magnetic black oxide with formula Fe_3O_4.
4 **Micaceous oxide of iron** So called because the particles are lamellar, i.e. in the form of thin plates, and in this respect the particles resemble those of mica. It is a greyish-black colour.
5 **Siderite or Spathic ore** Mainly ferrous carbonate and used for smelting.
6 **Iron pyrites,** FeS_2, used as a source of sulphur dioxide in sulphuric acid manufacture.

The hydrated oxides

Ochres

Ochres are actually coloured clays and consist of mixtures of hydrated oxide of iron with mineral silicates. The composition and colour vary considerably from one source to another. Generally the colour is described as yellow but is dirtier and redder than the yellow of the paler lead chromes or of the cadmium sulphides.

Ochres are widely distributed, and good quality materials are produced in France and South Africa, among other places.

Processing of ochres This follows the normal pattern for natural pigments – grinding, washing, drying, disintegrating and grading. Much of the yellow ochre supplied to the paint industry today has been micronized.

Properties and uses of ochres In iron oxide pigments, properties such as tinting strength and opacity vary with the iron oxide content. The latter figure for ochres can range from 16 to 60 percent and there is therefore a very wide variation in pigment properties. The specific gravities average 2·8 to 3·0 but the oil absorptions range from 20 to 40.

Much of the ochre used by the paint industry today is the micronized variety and is used as a stainer for gloss paints, undercoats and emulsion paints when it confers a pleasant reddish cream tint. It is, however, being displaced by the stronger but yellower Ferrite Yellow.

Siennas

These are brownish yellow in colour and contain a small quantity (about 1 percent) of manganese dioxide. They are obtained principally from Italy and Sicily. Like the ochres, siennas vary in composition, the iron oxide content ranging from 40 to 70 percent.

Siennas are usually micronized and in this form are used as stainers. They are semi-transparent in oil and were much used in scumble stains when graining was more fashionable.

The material in the condition as mined is known as Raw Sienna. If this is calcined, dehydration of the hydrated iron oxide takes place to produce the orange-red Burnt Sienna. The physical characteristics of the two forms are as follows:

	Raw Sienna	Burnt Sienna
Specific gravity	3·1–5·0	3·5–5·0
Oil absorption	35–45	35–45

Umbers

These are dark brown in colour and contain up to 15 percent of manganese dioxide. The best grades have been obtained from Cyprus, but the composition varies, the iron oxide content ranging from 30 to 50 percent.

They are semi-transparent in oil and were formerly valued as pigments for scumble stains. The micronized forms are sometimes used as paint stainers.

Like the siennas, the umbers are marketed in two grades, Raw and Burnt, the latter being a very rich reddish-brown. Their specific gravities range from 3·5 to 4·5 and oil absorption from 32 to 50.

Red oxides

Deposits of ferric oxide of varying quality occur in many parts of the world, but few English deposits are suitable as pigments.

Extensive deposits occur in Spain from which considerable quantities are imported into this country. The ground, levigated and dried ore contains 80 to 90 percent of iron oxide and is frequently blended with other grades for the market. Spanish oxide is a bright colour and has been popular in paint manufacture for many years.

A bluer shade of natural red oxide is provided by the Persian Gulf material known, in short, as Gulf Red. This is a very bright rich colour and was formerly known as Indian Red, but this term is now reserved for certain grades of manufactured red oxides.

A fault of the natural oxides is lack of uniformity, and so pigment manufacturers offer blends which are consistent in quality and, by micronizing, the texture and particle size are also closely controlled.

The natural oxides vary in specific gravity from 3·5 to 5·0 and the oil absorption ranges from 12 to 30.

Identification tests The natural yellow and red oxides of iron dissolve on boiling in concentrated hydrochloric acid to give solutions with the normal reactions for ferric ions. With the siennas and umbers the solutions also contain some manganese ions. In all cases there is an insoluble siliceous residue, which is absent from manufactured oxides.

Synthetic yellow and red oxides

There are two general processes used for the manufacture of these pigments – precipitation and calcination. In each case the basic raw material is crystalline ferrous sulphate, $FeSO_4.7H_2O$, known as copperas or green vitriol. This is obtained as a by-product in the tin-plate and galvanizing industries where the steel sheets are pickled in sulphuric acid.

Another source of ferrous sulphate is the sulphate process for the manufacture of titanium dioxide from ilmenite (Chapter 4). Large quantities are obtained as a by-product from this process.

Precipitation process

A solution of ferrous sulphate is treated with a solution of an alkali to precipitate ferrous hydroxide which is then oxidised to the ferric compound. Conditions of precipitation and oxidation can be controlled to yield either ferrite yellow or red oxide.

Ferrite yellow (CI Pigment Yellow 42)

The formula $Fe_2O_3.H_2O$ is often assigned to ferrite yellow but the form FeO.OH is more commonly used. However, the chemical composition can vary with conditions of manufacture.

Ferrite yellow (Fig. 5.1) is a soft pigment with specific gravity 4·0 and oil absorption 50 to 60. It is cleaner and brighter than ochre and possesses much greater tinting strength as a result of the higher iron oxide content (about 90 percent as Fe_2O_3). It has excellent lightfastness and is completely insoluble in oils, resins and solvents. On heating it reddens as a result of loss of combined water and is converted to red oxide, Fe_2O_3.

Courtesy: Paint Research Association

Fig 5.1 Photomicrograph of yellow iron oxide (× 15 000)

It is widely used as a stainer in gloss paints, undercoats and aqueous paints.

Identification test On strong heating it is converted to red oxide, Fe_2O_3. It is soluble in hydrochloric acid to give a clear solution and leaves no residue. The solution gives the usual reactions for ferric ions.

Red oxide (CI Pigment Red 101)

Compared with the natural red oxides, the synthetic type is brighter, cleaner and has a higher tinting strength due to higher Fe_2O_3 content (over 99 percent). The opacity, which is also proportional to the ferric oxide content, is also higher.

The pigment has a specific gravity of 5·0 and an oil absorption of 30. It is fast to light and insoluble in oils, solvents, and resins. It is resistant to all alkalis and dilute acids but attacked by hot concentrated acids. It is also heat resistant.

Synthetic iron oxide is used in air-drying and stoving finishes, undercoats and primers, emulsion paints, plastics, and cements.

Identification tests Boiling with concentrated hydrochloric acid gives a

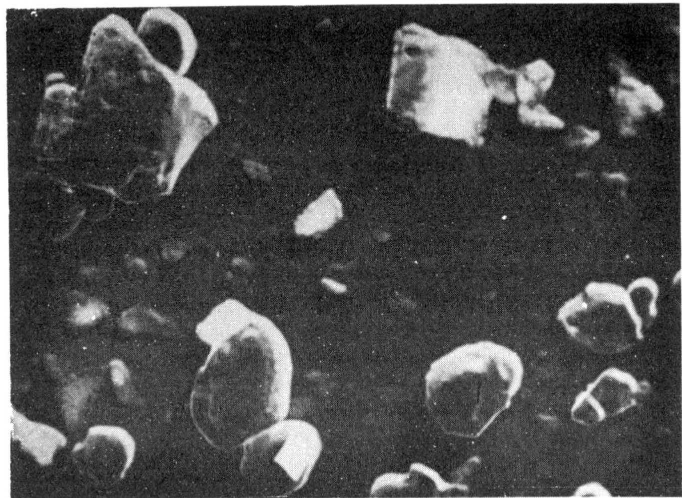

Courtesy: Paint Research Association

Fig 5.2 Photomicrograph of red iron oxide (× 30 000)

clear solution showing the usual reactions of ferric ions. It leaves no residue and so differs from the natural material.

Transparent oxides

These are pigments of extremely fine particle size which, when dispersed in paint media, give transparent films on curing (see p. 26). It is well known that oxides of iron absorb ultraviolet radiation and so protect the binder in which they are dispersed. These transparent oxides are, therefore, ideally suited for use in light-fast varnish stains for exterior woodwork.

Calcination process

This process also utilises ferrous sulphate crystals, $FeSO_4.7H_2O$ as the basic raw material and is used, in the main, for the deeper shades of iron oxide. The ferrous sulphate crystals are dehydrated to $FeSO_4H_2O$ and this is then calcined at a temperature determined by the grade of oxide required.

At a calcination temperature of about 600°C the product is the yellowest and strongest of the calcination oxides known as Turkey Red.

If the temperature is increased to 900°C, the resulting pigment is deeper, bluer and weaker than Turkey Red. It is known as Indian Red.

At higher temperatures (up to 1000°C) the product is deeper still and is known as Purple Oxide or Purple Brown.

Properties and uses

The calcination oxides possess specific gravities of 4·5 to 5·0, and the oil absorptions range from 20 to 25.

Turkey Red is a bright soft pigment with good opacity and tinting strength. Indian Red is deeper, bluer and harder and possesses lower opacity and tinting strength. These properties are developed still further in the Purple Oxides.

These pigments are sometimes used in gloss paints and undercoats. In primers, however, their use is somewhat risky due to the possibility of the presence of sulphate which is extremely difficult to remove by washing. They are fast to light and heat and are insoluble in all oils, resins and solvents. They are also resistant to alkali and acids except boiling concentrated acids, and even here the attack is slow.

Identification tests

Boiling (usually prolonged) with concentrated hydrochloric acid yields a solution showing the usual reactions for ferric ions.

Bauxite residue

The mineral Bauxite, which is one of the chief sources of aluminium, contains appreciable quantities of iron oxide. After extraction of the aluminium, the residue is ground and sold under the name Bauxite Residue. It contains 60 to 70 percent Fe_2O_3, the balance consisting of siliceous matter and possibly traces of alumina.

It is sometimes used in primers for steelwork, as the protective properties appear to be superior to some of the natural red oxides.

RED LEAD (CI Pigment Red 105)

Manufacture of red lead

Red lead is manufactured by calcining litharge (PbO) in an oxidizing atmosphere at about 350°C (662°F). Reaction is slow but cannot be accelerated since decomposition of the red lead takes place below 500°C (932°F). Accurate temperature control is therefore necessary. The reaction can be represented by the equation:

$$6PbO + O_2 \underset{470^\circ C}{\overset{350^\circ C}{\rightleftharpoons}} 2Pb_3O_4$$

The final product always contains some free litharge, the amount of which can be controlled by the degree of oxidation.

Properties of red lead

Red lead is an orange-red pigment with the following physical characteristics:

Specific gravity 8·8	Refractive index 2·42
Oil absorption 7–8	Mean particle size range 3–12 μm

The pigment is marketed in three grades conforming to the requirements of BS 217[4]. These are:

Type 1 Non-Setting Red Lead
Type 2 Ordinary Red Lead
Type 3 Red Lead for Jointing Purposes

The three types differ in respect of the free PbO content, and Table 5.2 shows the composition of some of the present-day red leads:

Table 5.2

	Type 1	Type 2	Type 3
Total oxides of lead	Not less than 99.5%	Not less than 99.5%	Not less than 99.5%
Pb_3O_4	Not less than 95%	Not less than 72% and Not more than 90%	Not less than 43% and Not more than 72%
Free PbO	Balance	Balance	Balance

The free PbO content determines the use of the pigment. A high proportion, as in Type 3, will react rapidly with linseed oil and set to a hard mass. Type 2 reacts more slowly, but paints made with this type can be stored for no longer than 24 hours. Type 1 is used in paint manufacture and gives stable paints with good working properties.

Red lead is a reactive pigment and its chemical reactions support the accepted constitution $2PbO.PbO_2$. When dispered in linseed oil or other oxidizing media lead soaps are formed by reaction between the PbO and acidic by-products produced during the oxidation process. These soaps are thought to play an important part in the protection afforded to steelwork by red lead primers. The traditional medium for red lead primers has been linseed oil, but although the protective qualities have been excellent, the primers were slow-drying and remained soft for a considerable time. Other quicker-drying media have been used, including alkyds, epoxy esters, and chlorinated rubber.

Red lead is a very toxic pigment, its manufacture and use being controlled by the Lead Paints Regulations, 1927, and subsequent amendments. As a result of this high toxicity the use of the pigment is steadily declining, a process which has been accelerated by the introduction of the non-toxic zinc phosphate.

Identification tests

When strongly heated, red lead decomposes and then melts to an orange liquid which sets to a yellow solid (litharge). In the decomposition oxygen is evolved and the reaction is the reverse of the process of manufacture.

On treatment with dilute nitric acid the PbO portion of the molecule is dissolved out, leaving a brown residue of PbO_2. Addition of a few crystals of a reducing agent such as sodium nitrite or oxalic acid reduces the PbO_2 and gives a clear solution of lead nitrate. This gives the normal reactions for lead ions.

Lead cyanamide, $PbCN_2$

Lead cyanamide is a yellow pigment with specific gravity 6·8 and oil absorption 21 to 23. The extract obtained on treating 5 g of the pigment with 100 ml of distilled water is alkaline, the pH being 9·3.

It is a very reactive pigment and forms lead soaps with drying oils. When made into paints it does not settle in the tin and the paints show good brushability.

The pigment is very effective in corrosion-inhibiting primers for steelwork and this is probably due, in part at least, to its highly alkaline nature. It has also been used in wood primers and undercoats and can be used in drying oil, oleoresinous or alkyd media.

As a lead compound it is, of course, subject to the Lead Paint Regulations.

INORGANIC BLUE PIGMENTS

There are only three pigments of importance in this class. These are Ultramarine, Prussian and Cobalt blues.

Ultramarine Blue (CI Pigment Blue 29)

The natural material, a mineral known as 'lapis lazuli', occurs in Iraq, Tibet and China and has been used as a pigment for many centuries. It was prized for its brilliance and purity of tone and although the use of lapis lazuli as a paint pigment was discontinued many years ago, it has continued to be used in artists' colours until very recent years.

Lapis lazuli possesses a very complex constitution containing silicates of lime, alumina and sodium together with sulphides of the latter.

Early in the nineteenth century several observers reported a blue compound in certain blast furnace slags and this was found to give similar reactions to lapis lazuli. This observation stimulated further research which led ultimately to the modern process for the manufacture of Ultramarine.

Manufacture of ultramarine blue

The raw materials employed include sulphur, soda ash, siliceous materials such as china clay, silica, etc., and a carbonaceous material such as rosin. These are mixed and heated in a muffle furnace for some days. The product is ground, washed, dried and graded.

Variations in shade of blue are obtained by altering the proportions of the ingredients. Rose and violet shades are produced by treatment of the blue pigment with hydrochloric acid gas at about 275°C (524°F).

Properties and uses of Ultramarine Blue

Ultramarine Blue is a bright pigment and is one of the reddest blues available; the colour is not easily matched with other pigments. It has the following physical characteristics:

Oil absorption 30–35	Mean particle size diameter 0·1–3·0 μm
Specific gravity 2·35	Refractive index 1·51

As a consequence of the method of manufacture, the heat resistance is excellent. It is insoluble in all paint media and solvents, is fast to alkali (except in concrete), but is readily attacked by mineral acids.

Ultramarine Blue possesses good lightfastness, but loss of colour can occur if it is exposed in a paint film to a moist acidic atmosphere. It is strongly hydrophilic and disperses readily in water, but it can migrate into water from oil or other organic medium. 'Ultra blue' is therefore used mainly in aqueous media, but it is also employed in organic media where a cationic or non-ionic surfactant assists wetting and helps to stabilize the dispersion. It is widely used in screen inks.

The constitution of Ultra blue is related to that of the zeolites and ion-exchange reactions are possible. By the action of silver nitrate, for example, sodium in the blue can be replaced by silver to give a yellow-silver ultramarine.

Toxicity of Ultramarine Blue

Ultramarine Blue possesses no known toxic hazard, but the dust can be classed as a 'nuisance particulate'.

Identification tests

The pigment is unchanged on ignition or on treatment with alkalis. It is decomposed by dilute mineral acids, giving hydrogen sulphide and a precipitate of sulphur and leaving a siliceous residue. This test differentiates between Ultramarine and Prussian Blue, the latter being decomposed by alkali but not by dilute acids.

Prussian Blues (CI Pigment Blue 27)

These are iron compounds of the hexacyanoferrate radical $[Fe(CN)_6]''''$ and contain also an alkali metal or ammonia. They are known in the United States as Iron Blues. In this country there are two types known as Potash or Non-bronze Blues and Ammonia-soda or Bronze Blues.

Potash or non-bronze blues

These are usually considered to be forms of ferric potassium hexacyanoferrate, $KFe[Fe(CN)_6]$ and can be prepared by either the Indirect or Direct process. These blues precipitate in a very finely divided condition which renders them very difficult to filter. They are therefore made in as coarse a condition as possible by precipitation at or near 100°C (212°F).

Indirect process of manufacture

A hot solution of potassium hexacyanoferrate is added to a hot acidified solution of ferrous sulphate. A white precipitate of ferrous potassium hexacyanoferrate is formed:

$$FeSO_4 + K_4Fe(CN)_6 \rightarrow FeK_2Fe(CN)_6 + K_2SO_4$$

This is then oxidised to the blue ferric compound (Potash blue)

$$6FeK_2Fe(CN)_6 + KClO_3 + 3H_2SO_4 \rightarrow \underset{\text{Potash blue}}{FeKFe(CN)_6} + 3K_2SO_4 + KCl + 3H_2O$$

Direct process of manufacture

In the Direct process, a ferric salt is used and the potash blue is precipitated:

$$FeCl_3 + K_4Fe(CN)_6 \rightarrow \underset{\text{Potash blue}}{FeKFe(CN)_6} + 3KCl$$

This is not as widely used as the Indirect process. The latter is cheaper and gives a brighter product.

Ammonia-soda or bronze blues

These differ from the potash blues in that the ammonium radical replaces the potassium and that sodium hexacyanoferrate is used in the manufacture.

They are usually manufactured by the indirect process which follows that outlined above for Potash blue. A hot 10 percent solution of sodium hexacyanoferrate is added to a hot 10 percent solution of ferrous sulphate in the presence of ammonium sulphate:

$$FeSO_4 + Na_4Fe(CN)_6 + (NH_4)_2SO_4 \rightarrow Fe(NH_4)_2.Fe(CN)_6 + 2Na_2SO_4$$

The white precipitate is then oxidized with sodium chlorate in the presence of sulphuric acid:

$$6Fe(NH_4)_2.Fe(CN)_6 + NaClO_3 + 3H_2SO_4$$
$$\rightarrow \underbrace{6FeNH_4.Fe(CN)_6}_{\text{bronze blue}} + 3(NH_4)_2SO_4 + NaCl + 3H_2O$$

The pigment is then filtered, washed and dried at 50°C to 60°C (122° to 140°F).

Properties of Prussian Blues

In view of the difficulties involved in the drying and subsequent grinding and dispersion processes, considerable work has been done on the 'flushing' of Prussian Blue. This has proved a difficult operation in view of the hydrophilic or water-attracting properties of the blue. It shows no tendency to migrate into the oil phase. However, by the use of cationic or certain non-ionic surface active agents it is possible to change the character of the pigment surface. By this means Prussian Blue has been made easier both to filter (as a result of flocculation) and to flush. In the latter case the blue is converted to the hydrophobic (water repelling) condition.

The terms 'non-bronze' and 'bronze' refer to the degree of metallic bronze exhibited by paint or ink films containing the respective pigments. The bronzy appearance seems to be associated with the presence of 'rafts' of pigment particles in the paint or ink surface and is a characteristic of the ammonia-soda blues. These blues are used in printing inks. In paints the non-bronze potash blues are used since bronziness is unacceptable.

The specific gravity of Prussian Blue is 1·97 and the oil absorption, which is related to particle size, varies from 70 to 160. The extreme fineness of the particles results in a high tinting strength, but the opacity in paint media is poor.

It possesses good lightfastness and is resistant to solvents and resins. In the earlier types of oil paint, Prussian Blue sometimes lost its colour on storage, due, presumably, to reduction to the colourless potassium ferrous ferrocyanide. On exposure of the paint film to the air, however, the original colour was restored by oxidation.

Prussian blue is resistant to mineral and organic acids but is decomposed by alkali (reactions which are the reverse of those of ultramarine blue). It is, therefore, unsuitable for use in paints to be applied to alkaline surfaces such as cement render, concrete or lime plasters.

Mixtures of non-bronze blues with lead chromes constitute the Chrome or Brunswick Greens (*below*). Potash blues were the standard blues for paints until the introduction of the phthalocyanine blues (Chapter 7) and the amount now used is comparatively small.

Identification tests

The reactions with acid (stable) and alkali (decomposed) are generally considered sufficient.

Toxicity of Prussian Blues

Prussian Blue is considered to present a very low health hazard and this is supported by a long history of freedom from ill effects in usage.

Fire risk

This can be a serious hazard. When Prussian Blue burns, the highly dangerous hydrocyanic (prussic) acid can be evolved.

Other varieties of Prussian Blue

Brunswick Blue A reduced blue produced either by precipitation of the Prussian Blue on to a base such as blanc fixe or barytes or by addition of the base to the Prussian Blue suspension before filtration. The blue is then more easily filtered, gives less trouble in drying, and is more readily dispersed. However, the presence of the base is often not desired.

Milori Blue This term is sometimes applied to the best and palest potash blues which are completely devoid of bronze.

Chinese Blue A Prussian Blue with a high bronze lustre.

Cobalt Blue (CI Pigment Blue 28)

Cobalt oxide forms solid solutions with some other metallic oxides. These are highly coloured and possess great resistance to heat, light and chemicals.

Cobalt Blue is sometimes known as Thénard's Blue and is a solid solution of cobalt and aluminium oxides. It is produced by calcination at 1000° to 1200°C (1832° to 2192°F) of mixtures of the oxides, hydroxides, or salts of the two metals.

Properties of Cobalt Blue

Chemical analysis figures of Cobalt Blue correspond with the composition 15 percent CoO, 81 percent Al_2O_3, 4 percent H_2O. It is a very bright and clean colour and almost transparent when dispersed in paint media. It is valued for its colour but more particularly for its extreme stability. It is completely resistant to light, heat, acids and alkalis, but is expensive.

Uses of Cobalt Blue

Cobalt Blue is used occasionally in paints but is too expensive and lacking in opacity for general use. It is used as a pigment for artists' paints.

Identification

A good pointer is provided by the extreme stability. It produces a fine blue borax bead, which serves to differentiate it from Phthalocyanine Blue.

Cerulean Blue

A solid solution of cobalt oxide, tin oxide and silica. It is paler and greener than Cobalt Blue and possesses the same stability as the latter. It is too expensive for general use.

Cobalt Violet

This term is applied to a range of purples and violets produced by calcination of cobalt phosphate alone or in admixture with other metal phosphates. They are extremely stable but are weak tinctorially.

GREEN PIGMENTS

For a number of years the green pigments in general use in the paint industry were confined to two groups of colours, both of which contained the metal chromium. These were (1) the chrome (or Brunswick) greens and (2) the oxide and hydrated oxide of chromium. The latter two have always been used for special purposes, but the chrome greens have been widely used in both decorative and industrial paints.

However, on toxicity grounds their use in decorative paints has now been discontinued. They continue to be used in certain industrial locations where a health risk is not considered likely to arise during application, exposure or ultimate removal.

Chrome greens (CI Pigment Green 15)

The pigments are basically mixtures of lead chromes and Prussian Blue and, as mixtures, they exhibit the properties of the two components. They form a range of colours which have been standardized and are set out in BS 381C [5]. They are:

Grass green	British Standard Colour 218
Brilliant green	British Standard Colour 221
Light Brunswick green	British Standard Colour 225
Mid Brunswick green	British Standard Colour 226

By the addition of certain other pigments such as raw sienna, red oxide and/or black oxide, a range of bronze greens is produced. These are generally deeper and redder than the corresponding chrome greens and are set out in BS 381C. Their general properties are similar to those of the chrome greens.

Manufacture of chrome greens

It is generally agreed that the most satisfactory chrome greens are the so-called 'struck' greens which are made by precipitating or 'striking' the lead chrome in the presence of Prussian Blue. In this way the lead chrome is virtually precipitated on to the blue and yields a pigment from which the blue has less tendency to separate in the paint film (see Flooding and floating, *below*).

Properties and uses of chrome greens

The relative proportions of lead chrome and Prussian Blue can vary widely (the blue content can vary from 2 to 35 percent) and consequently figures for oil absorption and specific gravity can also show considerable variation.

Chrome greens are non-bleeding in paint media and exhibit good lightfastness, opacity, and durability when exposed in paint films. In industrial and urban atmospheres containing sulphur dioxide, surface 'blueing' takes place as a result of reduction of the chromate radical. They are sufficiently heat-resistant to withstand normal stoving conditions, i.e. $\frac{1}{2}$ to 1 hour at 120°C (248°F).

The Prussian Blue ingredient is attacked by alkali, and consequently chrome greens cannot be used in paints for alkaline conditions.

They are subject to the restrictions of the Lead Paints Regulations and on account of toxicity they are no longer used in decorative or toy paints but continue to be used in certain industrial maintenance paints where the toxicity does not present a serious hazard. 'Leadless' greens for use in decorative and similar paints are based on mixtures of organic yellows and phthalocyanine blue or green.

Flooding and floating

Depending on conditions of air circulation and humidity, paint films containing chrome greens can dry to a darker shade than the liquid paint due to an increase in Prussian Blue in the surface. This increase in concentration of the blue appears to be associated with the type of solvent evaporation and with convection currents in the fluid paint film.

If the deepening of surface colour is uniform the change is known as 'flooding' and this occurs under conditions of 'diffusive' solvent evaporation when disturbing influences such as draughts are absent.

More often, however, the surface has a mottled appearance and close examination will show the presence of Bénard cells. These are often hexagonal in shape and show differences in colour concentration between the centre and edges. These cells result from convection currents set up in the drying film, together with an 'ebullioscopic' type of solvent evaporation from the surface. This occurs when air movement removes the solvent vapour and equilibrium conditions are not reached.

'Floating' can be converted to the uniform and less objectionable flooding by the addition of certain silicone oils at the rate of 1 part in about 40 000 parts of paint.

The introduction of a degree of thixotropy or structure by a thixotropic medium, hydrogenated castor oil, Bentone [6] or extenders such as Celite [7] or prepared silica [8] will reduce the film mobility and so reduce the intensity of colour change. A complete change of pigment is often the only satisfactory remedy. Flooding and floating can be exhibited by other mixtures of pigments which differ in specific gravity and surface characteristics. Greys based on titanium dioxide and carbon black can be particularly troublesome; blues made from mixtures of titanium dioxide and phthalocyanine blue can exhibit the defects, but this mixture responds more readily to remedial measures.

Floating can occur with some single pigments, for example, with toluidine red in a long-oil alkyd medium, but the effect is not usually noticeable.

Identification tests

As mixtures of lead chromes and Prussian blue, chrome greens show the reactions of both these pigments.

Toxicity of chrome greens

The toxic hazards are largely those associated with the lead chromes (p. 79). The toxicity of Prussian Blue is very low (p. 93).

Chromium oxide (CI Pigment Green 17)

This consists, as the name suggests, of anhydrous chromium oxide, Cr_2O_3.

It is a sage-green colour and is dull and chalky compared with the chrome greens. It possesses the following characteristics:

Specific gravity 5·2   Oil absorption 10

The opacity is very high as is also the resistance to light, heat, acids and alkalis. It is extremely durable and is used under any conditions where resistance is more important than a bright colour. It can safely be used in water paints and finds wide use in colouring concrete and roof tiles, plastics; printing inks, and ceramics.

Identification tests

Chromium oxide is characterized by general inertness and colour. Fusion with a mixture of anhydrous sodium carbonate (5 parts) and sodium peroxide (1 part) converts it to sodium chromate which can be extracted with water. Acidification followed by addition of hydrogen peroxide gives the fugitive blue colour which is characteristic of chromate and hence of chromium.

Toxicity of chromium oxide

Chromium oxide presents a very low toxic risk.

Guignet's Green or Viridian (CI Pigment Green 18)

This is a hydrated oxide of chromium and is generally assigned the formula $Cr_2O_3.2H_2O$.

It is a much brighter pigment than chromium oxide and possesses the following physical characteristics:

Specific gravity 3·2 Oil absorption 63

It has very poor opacity but is fast to light, alkali and all types of paint media. The resistance to acids and heat is not as high as that of chromium oxide. It is unaffected by cold mineral acids but dissolves in hot concentrated hydrochloric acid. On strong heating it is decomposed at about 250°C (482°F). Lack of opacity and rather high cost limits its use in paints, but it is a satisfactory stainer for water paints. However, it is gradually being replaced in these paints by organic pigments.

Identification tests

It is turned black on very strong heating. It is soluble in hot concentrated hydrochloric acid to give a solution of chromium chloride. This gives the normal reactions of chromium ions. Fusion with a mixture of anhydrous sodium carbonate and sodium peroxide gives sodium chromate (see Chromium oxide).

References to Chapter 5

[1] BS 282; 389; 1987. Specification for lead chromes and zinc chromes for paints. British Standards Institution.
[2] 'Oncor' (National Lead Industries Inc.).
[3] H. M. Stationery Office. *See also* BS 3443: 1968. Code of safety requirements for childrens toys and playthings. British Standards Institution.
[4] BS 217: 1986. Specification for red lead for paints and jointing compounds. British Standards Institution.
[5] BS 381C: 1980. Specification for colours for identification, coding and special purposes. British Standards Institution.
[6] Steetley Chemicals (NL Industries Ltd).
[7] Johns-Manville, Denver, Colorado.
[8] 'Aerosil' (Degussa); 'Gasil' (Crosfield Chemicals Ltd).

6

Black, metallic, and miscellaneous pigments

BLACK PIGMENTS

The bulk of the black pigments used in the modern paint industry consists of elementary carbon (in a fairly pure state as carbon black, lamp black and synthetic graphite or mixed with siliceous matter in natural graphite), black oxides of iron (natural and manufactured) and micaceous oxide of iron.

Carbon blacks(CI Pigment Black 7)

These are derived from petroleum gases and oils and are classed as Channel or Furnace blacks according to the method of production.

Channel black The basis of the channel process is the burning of hydrocarbon gases using specially designed burners in a limited supply of air so that combustion is incomplete and smoky flames result. In the original process (from which the name is derived) the smoky flames impinged on lengths of channel iron on which the carbon was deposited and from which it was removed by scrapers.

In the modern 'gas black' process [1] petroleum oil is used in place of gas and the smoky flames impinge on cooled rollers from which the carbon is removed by scrapers.

The quality of the black can be controlled by the shape of the flame, distance from channel surface, amount of air, and quality of the gas.

When first produced, channel black is very fluffy due to entrained air and gaseous by-products resulting from incomplete combustion. It is therefore compressed or converted to beads by either wet or dry pelletizing.

Wet pelletizing The carbon black is passed into a cylinder fitted with a central rotating shaft to which spokes are attached. Water is added at the same time and the carbon black forms pellets which are then dried.

Dry pelletizing The carbon black particles are caused to adhere together by passing the black slowly through rotating drums.

The efficiency of the channel process is very low and prevention of smoke emission into the atmosphere has been found very difficult. For these reasons and the fact that considerable improvements have been effected in the furnace process, furnace blacks are being used to replace channel blacks in a number of applications.

Furnace blacks The older furnace blacks were produced by combustion of gases in a limited supply of air in a refractory-lined furnace. A single large flame was employed and the products, containing the carbon black, were quenched with water before passing to bag filters or electrostatic precipitators. The resulting carbon black was used only in the rubber industry.

The modern process [2] uses a similar type of furnace but employs oil instead of gas (although mixtures are sometimes used) and by varying the conditions a wide range of carbon blacks can be produced, covering the grades required for paints and printing inks as well as rubber. A high proportion of the carbon black consumed at the present time is made by the furnace process.

Oxidized blacks carry carboxyl and other groups on the surface and show improvements in certain characteristics, notably flow properties and ease of dispersion.

Properties of carbon blacks

The physical characteristics of the three main grades of carbon black used in paints and inks are summarized in Table 6.1 [1].

Carbon blacks are classified according to the intensity of blackness which is quoted on the Nigrometer scale (Table 6.1) on which lower figures indicate more intense blackness. The intensity of blackness increases with decreasing particle size.

Carbon blacks are hygroscopic and, if exposed to a moist atmosphere, can absorb up to 15 percent of moisture, in consequence of which they are difficult to disperse in organic media. The oxidized types are an exception.

As a result of the almost total absorption of incident light and the fine particle size, carbon blacks have very high opacity. In a brushing, air-drying paint, 25–35 grams of carbon black per litre is generally sufficient to give a completely opaque film.

Table 6.1

Black	Mean particle size range mμ	Blackness (nigrometer)	Surface area BET M^2 gram	Oil absorption	pH of aqueous suspension
Channel (gas)	13–30	63–84	100–500	80–150	2.5–4.0
Furnace	15–70	76–94	80–300	70–140	2.4–8.0
Lamp	50–120	93–102	20–50	70–140	6.0–9.0

Notes (a) mμ = nm (nanometre) = 10^{-9} metre.

(b) Each type of black is supplied in a number of grades (degrees of blackness), the particle sizes and surface areas of which fall generally within the ranges quoted.

(c) Figures for surface area by other methods, e.g. from average particle diameter, are often appreciably lower than the BET figures owing to the 'porous' nature of carbon black 'particles'. These particles appear to consist of agglomerates of much smaller units, giving the particles a 'porous' structure. The ultimate units appear to have a distorted crystalline structure akin to graphite [3].

They are completely fast to light and chemicals (with the exception of concentrated nitric acid) and to the temperatures which can be tolerated by most paint systems.

When used in paints, carbon blacks tend to adsorb materials from the medium. Thus an air-drying black paint gradually loses its drying properties on storage (unless the paint contains an additive, e.g. Nuact [4] to prevent this) as a result of the adsorption of part of the drier metals by the carbon black. The drying properties are recovered to some extent by re-milling the paint. On prolonged storage, black air-drying paints also show an increase in viscosity due to adsorption of lower molecular weight fractions of the polymer.

The fine particle size of carbon blacks is accompanied by high tinting strength. When mixed with white, the tone of the resulting grey is related to the particle size of the black. The lowest particle sizes give brownish tones but coarse blacks give bluish greys. For this reason lamp black (*below*) has long been preferred for tinting as it gives clean bluish greys but similar results are obtained with some of the coarser furnace blacks.

Uses of carbon blacks

Carbon blacks are used in conventional paint systems and in inks. Care should be taken, however, with some chemically-cured systems to ensure that the pigment does not impair the curing reaction.

They are marketed in several physical forms ('fluffy', dense, beads) and also dispersed in 'chips' (in plasticized cellulose nitrate), in plasticizers, and in stainers for organic and aqueous paints.

Identification tests

They are inert to all solvents and acids (with the exception of concentrated nitric acid when vigorous oxidation can ensue) and alkalis. On ignition they burn, leaving no appreciable ash. They are differentiated from lamp black by the more intense self colour.

Lamp (Vegetable) black

The term 'vegetable black' was originally given to the product obtained by burning vegetable oils in a limited supply of air. It was marketed in two grades, known as vegetable and lamp blacks, but the two terms are now used for the one product.

Manufacture of vegetable black

The present-day method consists in incomplete combustion of liquid petroleum products contained in a cast-iron pan situated under a flue. The lamp black is collected by cyclone and filters. The particle size can be varied by varying the conditions of combustion.

Properties of vegetable black

Lamp black has the following physical characteristics:

Specific gravity 1·8–1·9	BET surface area 20–50 m^2/gram
Oil absorption 70–140	pH of aqueous dispersions 6–9
Nigrometer value 93–102	Mean particle size range 50–120 mμ

It is coarser than carbon black and the colour is duller and greyer. When mixed with whites it gives clean bluish greys and is used as a tinting pigment rather than as a self colour.

Lamp black is fast to light, paint media, chemicals (with the exception of concentrated nitric acid when vigorous oxidation can take place) and to the temperatures which can be tolerated by most paint systems. It can contain up to 2 percent of oily matter and is therefore difficult to disperse in water but fairly readily dispersed in organic media.

Uses of vegetable black

Lamp black is used as a stainer in paints. It is also used in rubber and inks.

Identification

The dull and grey self colour serves to distinguish lamp black from carbon black. It is wetted by water only with great difficulty, and on ignition it burns and leaves no appreciable ash.

Graphite

This material is a crystalline form of carbon and is obtained from natural sources (chiefly from Sri Lanka) or is manufactured from a high-carbon type coal. It is also known as 'black lead' or 'plumbago'. The particles are in the form of hexagonal plates.

Natural graphite is ground, washed with dilute acid, then with water, and dried. The quality is assessed on the carbon content and is very variable.

Manufactured graphite is prepared from anthracite by the Acheson process. The anthracite is finely powdered and packed tightly into crucibles. These are then heated to 1200° to 1500°C (2192° to 2732°F) in absence of air and in an electric furnace. There is a 98 percent conversion to graphite. The success of the process depends on the availability of a cheap source of electricity and it is therefore worked extensively in Canada.

Properties of graphite

Graphite is a dark grey material with a metallic lustre and a characteristic slippery feel. The specific gravity varies, the natural being about 2·5 and the manufactured material 2·25.

It has good opacity and the plate-like structure and characteristic surface give it great spreading power. When made into paint the particles leaf with their planes parallel to the surface and overlapping. The film then possesses high resistance to moisture penetration. Graphite also has a tendency to *grip* metal surfaces, resulting in good adhesion.

Uses of graphite

The uses of graphite are mainly in protective paints for structural steelwork either in finishes or, mixed with red lead, in primers.

Identification

Graphite has a characteristic metallic appearance and slippery feel. On ignition the carbon burns and leaves, with natural material, a white siliceous ash. With manufactured graphite, very little ash is obtained.

Iron oxide blacks

Natural black oxide of iron

This material is also known as Magnetite or Black Magnetic Oxide and has the formula Fe_3O_4. It is very coarse and dense and is used in fillers and stoppers but not in paints.

Precipitated black oxide, Fe_3O_4

The manufacture of this pigment is on similar lines to that of the yellow and red oxides. A solution of ferrous sulphate is treated with alkali to precipitate ferrous hydroxide. This is then oxidized to the black oxide by blowing air through the suspension in open tanks:

$$6Fe(OH)_2 + O_2 \rightarrow 2Fe_3O_4 + 6H_2O$$

The conditions of oxidation are very carefully controlled. The pigment is then filtered, washed well, and dried.

Properties and uses of precipitated black oxide

Precipitated black oxide of iron has a specific gravity of 4·7 and an oil absorption of 28 to 30. It has a fine soft texture and moderate tinting strength. It is used in undercoats where carbon black can lead to increase in viscosity, and sometimes in oil-type finishes.

The lightfastness is good as is also the fastness to solvents and paint media generally. The heat resistance is satisfactory for normal stoving conditions, i.e. 120°C (248°F) for half an hour, but exposure to air for prolonged periods at high temperatures leads to some reddening of the colour. It is fast to alkalis but is soluble in hot concentrated hydrochloric acid.

Identification

Ignition in air results in oxidation to red Fe_2O_3:

$$2Fe_3O_4 + \tfrac{1}{2}O_2 \rightarrow 3Fe_2O_3$$

It is soluble in hot concentrated hydrochloric acid and the solution gives the normal reactions for ferric ions.

Micaceous iron oxide

This is a unique physical form of iron oxide, the particles of which are glistening flakes resembling mica. It occurs in a number of widely distributed locations but deposits vary in their suitability for use as a paint pigment.

The desirable plate-like particles are often mixed with other oxide of a powdery nature and this detracts from the value of the material. Supplies from Austria and France have been found satisfactory [5].

The pigment as supplied is a greyish or greenish black, dense and coarse product, the average particle diameter of which can be as great as 60–70 μm.

When incorporated into paint and the latter is applied to a surface the flakes become orientated with their major planes parallel to the plane of the paint surface. This occurs, not only on flat areas, but also on sharp edges [6]. The result of this orientation of the particles is increased resistance to moisture penetration and shielding of the binder from degradation by UV radiation [7]. The result is a very durable film.

Properties and uses of micaceous iron oxide

Micaceous iron oxide possesses the following physical characteristics:

Specific gravity 4·7
Oil absorption 20–25

It is insoluble in all paint media and resistant to most chemicals.

Micaceous iron oxide occupies a well established place as a major ingredient in protective paints for structural steelwork where protection is the primary requirement. It is used in a variety of media including oleo-resinous (tung oil/phenolic), chlorinated rubber and two-pack epoxies. Some of these are discussed in Chapter 21.

Identification

The characteristic appearance is often sufficient. Chemical identification as an iron oxide involves solution in concentrated hydrochloric acid followed by dilution and test for ferric ions.

METALLIC PIGMENTS

Aluminium (CI Pigment Metal 1)

Finely divided aluminium powder appeared at about the beginning of this century and its use as a paint pigment developed gradually from that time. Considerable impetus was given to its use by the introduction of the paste pigment, so avoiding the dust nuisance which was, and is, a feature of the powder. Today both types are in use, but the consumption of powder in paint is very small compared with that of the paste.

Aluminium pigment is an interesting material in that the particles are in the form of small flakes which possess a high metallic lustre and exhibit the property of leafing in organic media. The particles orientate themselves with their planes parallel to the surface of the paint film.

Manufacture of aluminium

The early method consisted of beating the powdered metal into flakes by multiple hammer mills in the presence of a suitable lubricant, the most widely used being stearic acid.

Modern methods employ ball mills in which the flattening of the metal particles into flakes is effected by the cascading action of the steel balls. The mill is charged with the powdered aluminium, stearic acid and white spirit. Quality of product and output are determined by the size of charge of steel balls, charge of aluminium and white spirit, and speed of rotation of the mill. The pigment slurry from the mill is filter-pressed and the press cakes, containing 66–68 percent of metal, comprise the aluminium paste of commerce. If powder is required, the white spirit is removed under vacuum at as low a temperature as is practicable.

Types of aluminium pigment

The pigment is marketed in both paste and powder forms, and each of these is available in leafing or non-leafing types.

Aluminium paste

The leafing and non-leafing types are graded according to particle size into the following grades.

Standard A general purpose grade, very widely used, and containing approximately 67 percent pigment.

Polished or Brilliant (leafing type only). This is designed to give a more mirror-like finish than the standard. The paste usually contains about 75 percent pigment.

Naphtha base Standard A paste containing 67 percent aluminium in which the solvent is naphtha in place of white spirit. It is useful when the paint medium is incompatible with white spirit.

Ink and Fine Lining These comprise the finest particle-size pigments.

Aluminium powder

This is also marketed in leafing and non-leafing types and in various grades based on particle size.

Properties and uses of aluminium

Paste, Leafing Type The outstanding property of this type is the ability of the particles to leaf in organic media. The flakes lie close to the surface, and overlapping results in an almost continuous bright metallic film. The leafing properties depend on the presence of a thin layer of stearic acid on the surface of the particles and are destroyed by the presence of lead driers in the paint.

In spite of the presence of a layer of stearic acid on the surface of the particles, aluminium pigment is very reactive and will react with organic acids present in the paint medium. This can result in loss of leafing power and brilliance, but hydrogen can also be liberated, resulting in pressure development in the tins. The acid value of the medium used should therefore be as low as possible and should not, in any event, exceed 4.

As is well known, aluminium is amphoteric and so is attacked by both acid and alkali. The presence of the stearic acid causes an 'induction period' which is followed by a vigorous reaction.

Aluminium is non-toxic and the leafing paste is widely used in decorative and metal priming paints and in industrial finishes. In conjunction with silicone resins it is used in paints for high-temperature resistance. It is also used in bituminous aluminium paints where it increases the light - and heat - reflectance of the surface.

Paste, Non-Leafing Type There are certain paints where the continuous metallic lustre is not required but where the presence of the flake particles is desirable. These include primers for woodwork and the industrial finishes where special effects are required, for example, polychromatic and hammer finishes. The non-leafing paste is preferred for these materials.

Powder, Leafing The properties are generally similar to those of the paste, but the finishes produced are somewhat brighter but less smooth. For this reason, and the dust nuisance, the powders are now used to a very limited extent.

Powder, Non-Leafing The uses of this material are similar to those of the paste type, but the latter is used whenever possible to avoid the dust hazard.

Identification

Aluminium pigment has a characteristic appearance and when heated strongly in the air will burn to the oxide. It is soluble in acid and alkali. The acid solution gives the normal reactions for aluminium ions.

Lead

Metallic lead is produced in a very fine state of division, but on account of the great affinity for oxygen, it is supplied as a dispersion in water or in an organic medium. The finely divided metal is strongly hydrophobic and so flushing from the aqueous suspension into an organic medium presents no difficulties.

Properties and uses of lead

It is a very finely divided pigment with a mean particle size of less than one micrometre. Its main use is in corrosion-inhibiting primers for steelwork and here it has been successfully used in a number of media of different types including linseed oil, isomerized rubber and epoxy resins.

The mechanism by which lead protects iron has created a great deal of interest. Electrode potential measurements on the metals in the massive state indicate that metallic lead should be cathodic to steel and when in contact should stimulate corrosion. However, Whitby [8] has shown that in a film of metallic lead primer, the lead is anodic to steel and could therefore protect by sacrificial corrosion. This mechanism would require metal contact and this has been found to be unnecessary for the effective functioning of metallic lead primers. Satisfactory results have been obtained when the metallic lead content of the pigment has been as low as 50 percent.

According to Wild [9], metallic lead pigment absorbs oxygen very readily from its environment and therefore probably protects by inhibiting the cathodic reaction. This 'oxygen barrier' idea would explain the efficiency of metallic lead in media other than drying oils such as isomerized rubber, chlorinated rubber, and epoxy resins.

In drying oils and other oxidizing media it is possible that some passivation of the steel surface takes place as a result of the formation of lead soaps of scission products, as suggested by Mayne and van Rooyen [10] for red lead.

Metallic lead is a highly toxic pigment and its use is subject to the Lead Paints Regulations. Application of primers containing metallic lead is by brush or hand roller only, but on account of its high toxicity the use of the pigment is declining.

Identification

Metallic lead is dissolved by nitric acid and the solution gives the normal reactions for lead ions.

Stainless steel flake

This pigment is available in both dry and paste forms. When used in paints, the films possess exceptional hardness and very good corrosion-resisting properties. It has high opacity and is used in situations where corrosive or chemical attack is likely to be severe.

Bronze powders

This term covers a range of alloys of copper and zinc containing a small quantity of aluminium. The paler and brighter shades are marketed under the name 'gold powder'.

Manufacture consists in exposing the molten alloy to a jet of high pressure air which produces nodules of various sizes. These are converted into flakes by either a stamping process or ball mill, graded into sizes and finally polished. Their specific gravity (8.3–8.7) is considerably higher than that of aluminium (2.7) and, although they possess leafing properties, these are less pronounced.

In general the bronzes are very sensitive to small amounts of acidity in the medium and tarnish rapidly on exterior exposure. They are used therefore for interior decoration and for this purpose are often dispersed in straight polymer solutions such as polyvinyl acetate or polystyrene.

Identification tests

The appearance of these pigments is characteristic. When warmed with 50 percent nitric acid (this should be done very cautiously) bronze powder dissolves. The solution gives the reactions of copper and (usually) zinc ions.

Zinc

This is marketed as a blue-grey powder under the name *Zinc Dust* containing 95 to 97 percent metallic zinc. The balance is zinc oxide formed by oxidation of the finely divided metal.

Manufacture of zinc dust

Metallic zinc (spelter) is vaporized and the vapour is led into large condensers. Under the conditions prevailing, the metal condenses as a dust which is removed by suitable methods. By rigid control of conditions the particle size distribution can be controlled to within fairly narrow limits.

Properties of zinc dust

Zinc dust is supplied in two grades known as Standard and Superfine. These compare as follows:

	Standard	Superfine
Specific gravity	7·1	7·1
Average particle size range	5–9 μm	2·25–3·0 μm

The particles are spherical in shape. The pigment has very good opacity but very low tinting strength. It is very reactive.

Uses of zinc dust

In the paint industry the use of zinc dust is largely confined to the protection of steel surfaces. Primers for steelwork contain between 92 and 95 percent metallic zinc in the dry film and are therefore designated *zinc rich*. The following types of binder have been employed satisfactorily: polystyrene, isomerized and chlorinated rubbers, epoxies, acrylics, and vinyls.

The mechanism by which zinc-rich primers protect steel is electrochemical and is akin to the protection afforded by zinc in galvanized iron. For this to be effective it is essential that there be actual contact between the steel surface and the metallic zinc. The surface of the steel must therefore be perfectly clean. There must also be contact between the particles of metallic zinc in the film, so that an adequate supply of metal is available. If therefore the metal content of the zinc-rich film falls below 92 percent, sufficient binder is present to isolate the particles. If the metal content is above 95 percent insufficient binder is present to provide the necessary adhesion.

When iron and zinc are in contact in a corrosive environment, the zinc becomes anodic to the iron and is preferentially corroded. The action is therefore described as *sacrificial protection*. This topic is discussed more fully in Chapter 16.

In the United States and to some extent in this country, zinc dust (80 parts) is mixed with zinc oxide (20 parts) in drying oil or oleoresinous media to produce corrosion-inhibiting primers for general use on steel. The performance of these paints is of a high order but the mechanism of action has not been completely elucidated.

Inorganic zinc silicate primers

Zinc dust will react with silicates of the alkali metals to produce films which are hard and adherent. They were used for many years to protect structural steelwork but the films were very slow in curing under humid conditions. Present day inorganic zinc silicate primers are based on the reaction between zinc dust and a prehydrolysed silicon ester.

The systems are two-pack and require moisture to cure to a hard, adherent and impervious film of zinc silicate containing metallic zinc. Such primers give excellent protection to steel under severe conditions, provided the steel has been thoroughly cleaned. It is generally blast-cleaned to SA 3 of the Swedish Standard [11] or first quality to BS 4232 [12].

Identification tests

On heating in air, zinc dust burns to form the oxide (yellow when hot, white when cold). It reacts vigorously with dilute hydrochloric acid to give a solution exhibiting the normal reactions of zinc ions.

Toxicity of zinc dust

Metallic zinc presents no significant toxic hazard, but the dust is classified as a 'nuisance particulate'. Zinc oxide is formed by welding or flame-cutting through zinc-coated metal and this can cause 'zinc fume fever'. The effects are unpleasant but transitory.

MISCELLANEOUS PIGMENTS

Calcium plumbate

Calcium plumbate differs from the other pigments containing lead, with the single exception of red lead, in that the lead occurs in the anion. The composition is closely related to that of red lead, as can be seen when the two are compared:

Calcium plumbate	$2CaO.PbO_2$
Red lead	$2PbO.PbO_2$

A calcium/lead plumbate, $CaO.PbO.PbO_2$, also exists.

Manufacture of calcium plumbate

Commercial calcium hydroxide and litharge are heated together in the proportions of two molecular weights of the former and one molecular weight of the latter. The optimum temperature is about 700°C (1292°F) and the time of heating 8 to 10 hours.

Properties and uses of calcium plumbate

The calcium plumbate of commerce is a cream or light buff-coloured soft powder. The colour varies with the purity of the raw materials, and the use of pure lime will give a white product. The colour can also be influenced by slight variations in the ratio of lime to litharge.

The specific gravity is 5·7 and oil absorption 10 to 17. It has high opacity and is readily dispersed in oil media on account of its alkaline reaction. Cold water extraction yields about 0·5 percent soluble matter, essentially calcium hydroxide. It is thought that the alkaline nature contributes to the rust-inhibiting properties of the pigment.

Calcium plumbate is used extensively in corrosion-inhibiting primers for steelwork and can be employed in a number of media such as linseed oils, stand oils, long oil alkyds and chlorinated rubber. The mechanism whereby calcium plumbate protects steel is discussed by Read [13], who considers it probable that an inhibitive film is laid down at both anodic and cathodic areas on the steel. This is supported by Evans [14] who has described rust inhibition by some calcium compounds.

In addition to the use on steelwork, calcium plumbate primer is one of the few primers which show satisfactory adhesion to new galvanised iron. It has also given satisfactory results on timber.

Identification

The aqueous extract shows an alkaline reaction. Treatment with dilute nitric acid yields a solution of calcium ions and a brown residue of PbO_2. This is filtered off, redispersed in dilute nitric acid, and a few crystals of sodium nitrite or oxalic acid added. The lead dioxide is reduced to the monoxide, which dissolves to a colourless solution giving the normal reactions for lead ions.

Toxicity of calcium plumbate

As lead compounds, the manufacture and use of calcium plumbates are controlled by the Lead Paint Regulations. By the same token it is probable that, in due course, calcium plumbate will be phased out.

'Modified' barium metaborate pigments

Two white pigments derived from barium metaborate are marketed under the trade names 'Butrol 23' and 'Busan 11-M1' [15]. The former is a corrosion-inhibiting material, claimed to be the equal of zinc chrome, and suitable for use in both solvent based paints and etch primers. Busan 11-M1 also possesses corrosion inhibiting properties but is used also for fire retardance and control of mould and bacteria.

Aluminium triphosphate

A white non-toxic pigment consisting of aluminium phosphate modified with zinc and silicate and marketed under the trade name 'K-White' [16]. As a corrosion inhibiting pigment it is claimed to be superior to both zinc phosphate and zinc chrome. Two grades are available, designated K82 and K84 and differing mainly in water solubility (1 percent and 0.5 percent respectively). Both grades are used in solvent based media for corrosion protection but K84 is used also with water-soluble resins and in etch primers as a replacement for zinc chrome.

Molybdate pigments

Two white corrosion-inhibiting molybdate pigments are marketed under the trade name 'Molywhite' [17]. Basic zinc molybdate (Molywhite 101) is employed in solvent-based primers. Calcium zinc molybdate (Molywhite 212) is designed for water-thinned (electrocoating) and latex paints. Both pigments are used widely in the USA but less so in the UK.

Nacreous pigments

The particles of these pigments are platelets which exhibit an iridescence or lustre, often described as 'pearlescence' or 'mother of pearl'. This effect occurs in some fish scales and in the shells of some marine animals. Production of pigments with the desired reflectance properties requires a base of laminar structure such as that provided by mica. The pearlescent effect is obtained by deposition of layers of transparent metal oxides (possessing a higher refractive index) on to the surface of mica [18]. Nacreous pigments can be used to enhance the appearance of other transparent pigments but the effect is lost when mixed with opaque pigments [18].

LUMINOUS PIGMENTS

Generally the colour of material objects is the result of selective absorption and reflection of wavelengths in the visible spectrum (Chapter 2). Certain substances, however, possess, in addition, the property of absorbing in the invisible ultraviolet waveband and re-emitting part of the energy at longer wavelengths within the visible spectrum. Such materials may be of two types.

Fluorescent pigments The luminosity exists only as long as the substance is exposed to the exciting radiation and disappears almost immediately on its removal.

Phosphorescent pigments These store energy from the exciting radiation so that light emission persists after this is removed. These pigments possess an after-glow which persists for short or long periods depending on their composition.

Fluorescent pigments

A widely used type contains fluorescent dyestuffs in a hard resin base. They are thermoplastic and possess softening points between 100°C and 120°C (212° and 248°F).

They are daylight-activated and are used in high-visibility paints, but as they are translucent the paints are applied over a flat white ground coat.

The performance of the paints can be improved by a coat of protective lacquer (PVA, acrylic, or cellulose) but the exterior durability is limited to about 12 to 18 months.

An inorganic type of fluorescent pigment is used in the coatings applied to the interior surfaces of fluorescent electric lamps. Among the materials used are calcium and magnesium tungstates, zinc orthosilicate and cadmium borate.

Phosphorescent pigments

This class includes the materials commonly known as luminous pigments. They are, in the main, sulphides of metals, of which the most important are zinc, calcium, strontium, barium, and cadmium.

The phosphorescence is obtained by preparation of the compounds in the form of *activated* crystals, the activity being produced by minute traces of compounds of copper and manganese; some of the rare earth metals are also useful in improving the luminosity. These activating materials are known as phosphorogens and appear to exist as solid solutions in the crystals of the parent sulphide.

In the manufacture of luminous pigments scrupulous care has to be taken to exclude traces of materials which can destroy the luminosity. Chief among these are compounds of lead, iron and chromium.

The process of manufacture depends on the preparation of the sulphide in a pure state. This is dried and converted to the 'phosphor' or active material by calcination with the activating agent, a flux, and a reducing agent. The flux comprises a eutectic mixture of salts of sodium or potassium, e.g. borax, sodium chloride, sodium sulphate or potassium sulphate, all of which must be pure.

The temperature of calcination varies with the nature of the metal and is generally between 800°C and 1100°C (1472° and 2012°F). After calcination the powder is disintegrated only, care being taken to avoid contact with any 'poisonous' metals. Fine grinding reduces the luminosity, and the particle size range is from 50 to 150 μm.

The colour emitted by the luminous pigment depends on the composition and on the nature of the activating phosphorogen. The individual sulphides give the following colours:

Zinc sulphide	Yellowish-green
Cadmium suphide	Red
Calcium sulphide	Violet
Strontium sulphide	Blue/Green
Barium sulphide	Yellow

Variations can be obtained by combinations of the individual pigments.

Moisture destroys the luminosity and they should, therefore, be stored under dry conditions.

Paints produced from these pigments contain the minimum amount of binder and have limited durability.

References to Chapter 6

[1] Degussa Ltd. Pigment Blacks brochure. Columbian (UK) Ltd. Columbian Raven Blacks brochure.
[2] McKeown, J., *JOCCA*, 1987, **70**, 26.
[3] Dollimore, D., The characterisation of carbon black surfaces. *JOCCA*, 1971, **56**, 616.
[4] 'Nuact' (Durham Raw Materials).
[5] Bishop, D. M., Micaceous iron oxide pigments. *JOCCA*, 1981, **64**, 57.
[6] Wiktorek, S. and Bradley, E. G., The orientation of micaceous iron oxide particles in organic coatings applied to edges. *JOCCA*, 1986, **69**, 172.
[7] Carter, E., Micaceous iron oxide pigment in high performance coatings. *Pol. Paint and Col. J.*, April 2, 1986.
[8] Paint Research Association Technical Paper No. 96.
[9] Wild, G. L. E., *JOCCA*, 1965, **48**, 43.
[10] Mayne, J. E. O. and van Rooyen, *J. Appl. Chem.*, 1954, **4**, 384.
[11] SIS 05 59 00 ; Swedish Standards Institution.
[12] BS 4232. Surface finish for blast-cleaned steel for painting. British Standards Institution.
[13] Reid, N. J., *JOCCA*, 1958, **41**, 352.
[14] Evans, U. R., An Introduction to Metallic Corrosion, 3rd Edn, Edward Arnold, 1979.
[15] Buckman Laboratories.
[16] Bromhead and Denison Ltd.
[17] Sherwin Williams Chemicals (Kingsley and Keith Ltd).
[18] Houseman, G., Colouration and colour enhancement of inks by nacreous pigments. *JOCCA*, 1987, **70**, 329.

7

Organic pigments

The organic pigments offer a wider range and a greater variety of shades than do the inorganics. Although some are inferior in opacity, they possess good staining power, softness of texture and brilliance. A number of brilliant dyes and pigments can be obtained from plant or animal sources, but these have now been almost entirely superseded by the synthetic types. However, a few continue to be used for special purposes and are listed, together with others of historic interest, in Appendix C.

Sources

The aromatic hydrocarbons, derived largely from the petroleum industry, constitute the raw materials of a large proportion of the organic pigments. The hydrocarbons are converted to the dyes and pigments in stages involving the formation of the following classes of compounds.

Primaries These are the first-stage products derived from the hydrocarbons. Examples are nitrobenzene, toluene sulphonic acid, and nitronaphthalene.

Intermediates are derived from the primaries by further reaction which can involve more than one stage, e.g. aniline from nitrobenzene, acetoacetanilide from aniline and acetoacetic ester.

Intermediates are converted directly to pigments or dyestuffs and the synthesis of these can be represented thus:

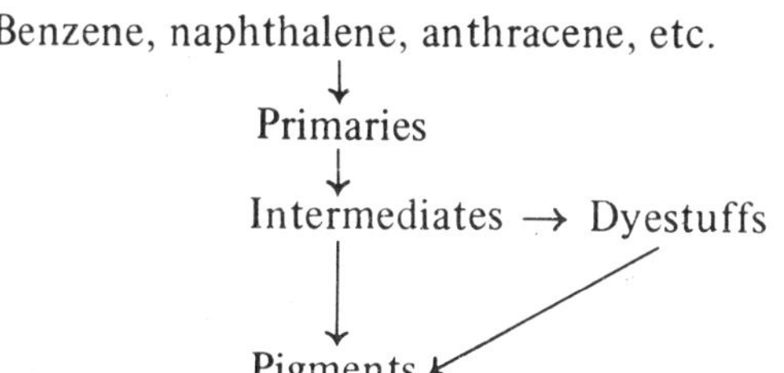

Colour-forming groups

The colour of organic pigments depends on the presence of certain groups in the molecule. These are known as *chromophores* and the commonest are the following:

$$>C{=}C<, \quad >C{=}O, \quad >C{=}S, \quad >C{=}NH, \quad >C{=}N-,$$

$$-N{=}N-, \quad -N{=}O, \quad -NO_2.$$

Some materials contain, in addition, salt-forming groups known as *auxochromes*. These are important in pigment manufacture. The principal salt-forming groups are:

$$\underbrace{-OH, \quad -SO_3H, \quad -COOH}_{\text{acidic}} \qquad \underbrace{-NH_2, \quad -NHR, \quad -NR_2}_{\text{basic}}$$

The molecule without the auxochrome, but containing the chromophore, is known as the *chromogen*.

General classification of organic pigments

It is possible to use a very broad classification and to divide organic pigments into pigment dyestuffs, toners, and lakes.

Pigment dyestuffs contain no salt-forming groups and comprise a wide variety of chemical types. These include azo compounds, metal complexes and polycyclic compounds.

Toners are produced by precipitation of acid dyestuffs by treatment with salts of certain heavy metals (mainly calcium, barium and manganese) or of basic dyestuffs with acidic reagents. Examples are Lithol toners and 'Fanal' type pigments. The term 'toner' is sometimes applied to pure pigment dyestuffs.

Lakes are 'reduced' pigments prepared by precipitation on to white bases such as alumina (aluminium hydroxide), blanc fixe, china clay, etc. In some cases a reactive base such as alumina forms a co-ordination complex with the dyestuff and becomes an integral part of the pigment. Madder lake is of this type.

CHEMICAL CLASSIFICATION OF ORGANIC PIGMENTS

The great number of organic pigments known fall into a comparatively small number of chemical groups. Classification is by chromophore (as in the nitroso and azo groups) or by general structure (as in the anthraquinones). The following groups are the most widely known, but the number of examples quoted is but a small fraction of the whole range available.

NITROSO GROUP

This group contains only one member of interest to the paint industry.

Pigment Green B (CI Pigment Green 8) is an interesting iron co-ordination compound produced from the bisulphite compound of α-nitroso-β-naphthol. The structural formula is indicated below, and it appears likely that the iron atom is complexed with three naphthol radicals, only one of which is shown:

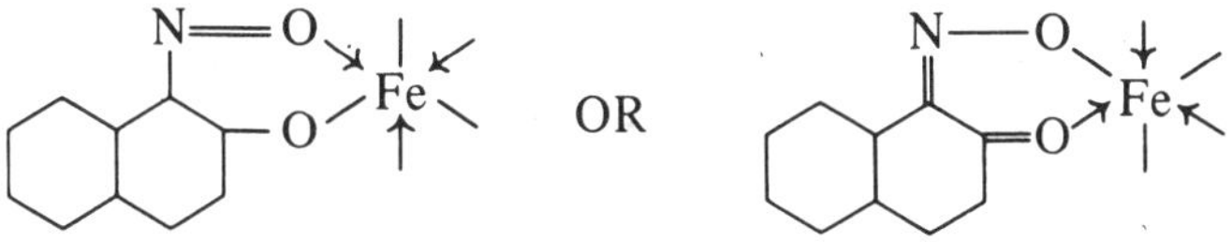

OR

In the pure state, Pigment Green B is an intense dark but dull green. When reduced with white, or blended, it produces pleasant green shades. The alkali resistance is good and it is therefore used in distempers and water paints.

NITRO GROUP

This group is now almost obsolete but is included because the method of preparation of the best-known member, Lithol Fast Yellow GG, is of interest.

Lithol Fast Yellow GG is made by a condensation reaction between one molecule of formaldehyde and two molecules of *p*-chloro-*o*-nitraniline, giving the compound:

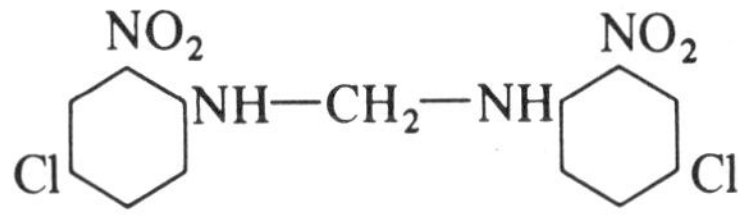

NO_2 NO_2

$NH-CH_2-NH$

Cl Cl

Such a reaction is not common in pigment manufacture, but condensation reactions are used in the synthesis of large molecules of high lightfastness (further details are given later in this chapter).

Lithol Fast Yellow GG was used as a stainer for water paints, but it has now been replaced by the Hansa yellows.

AZO GROUP

This is the largest group, in respect of both numbers and quantity used in the paint industry. The members all contain the chromophore $-N=N-$ and can be pigment dyestuffs, toners, or lakes.

Preparation of azo pigments

The process involves two stages: (i) the conversion of an aromatic primary amine into a diazonium compound by diazotization; and (ii) the coupling of the diazonium compound with a naphthol or second amino compound to give the dyestuff or pigment.

Diazotization

When an aromatic primary amine is treated with nitrous acid in the presence of excess mineral acid at 0°C (32°F), the following reaction takes place:

$$ArNH_2 + O=N-OH + HCl \rightarrow Ar\overset{+}{N}\equiv\overset{-}{N}Cl + 2H_2O$$

Diazonium compound

The nitrous acid is produced *in situ* by addition of sodium nitrite solution to the solution or suspension of amine in mineral acid. Most diazonium compounds are unstable above 5°C (41°F) and so the temperature must be kept as near 0°C as possible. Ice is usually added for this purpose.

According to Hughes, Ingold and Ridd [1] the most probable machanism involves the formation of nitrosyl chloride from the nitrous and hydrochloric acids:

$$O=N-OH + HCl \rightarrow O=N-Cl + H_2O$$

This then reacts with the aromatic amine to give the diazonium compound

$$ArNH_2 + O=N-Cl \rightarrow Ar\overset{+}{N}\equiv\overset{-}{N}Cl + H_2O$$

The rate of diazotization is largely controlled by the strength of the amine and this is reflected in the solubility in the acid solution. Strong bases, such as aniline and toluidine, dissolve readily and are diazotized rapidly. Weaker bases, such as 3-nitro-4-toluidine, are diazotized as suspensions in mineral acids. They dissolve slowly as the diazotization proceeds to give soluble diazonium compounds. Very weak bases, e.g. 2,4-dinitraniline, will not diazotize in dilute mineral acid and a special method has to be employed (*below*). Special methods are also used in cases where both base and diazonium compound are insoluble in dilute acid. Tobias (2-naphthylamine-1-sulphonic) acid is of this nature and the diazotization is described later.

In practice, one or other of the following methods is used.

Direct method The base is either dissolved or suspended in dilute hydrochloric acid at 0°C (32°F), sodium nitrite solution added, and the whole stirred until the reaction is complete. This is indicated by the absence of blue coloration when spotted on starch/potassium iodide paper.

Nitrosyl sulphuric acid method Very weak bases such as 2,4-dinitraniline are dissolved in concentrated sulphuric acid together with sodium nitrite. Diazotization takes place by the action of nitrosyl sulphuric acid:

$$NO_2\text{-}C_6H_3(NO_2)\text{-}NH_2 + O{=}NHSO_4 \longrightarrow NO_2\text{-}C_6H_3(NO_2)\text{-}\overset{+}{N}\equiv\overset{-}{N}HSO_4 + H_2O$$

The mixture is poured into ice and water to give a soluble diazonium compound.

Reverse method This is used when both base and diazonium compound are insoluble in water, as with Tobias acid (mentioned above). The acid is dissolved in sodium hydroxide solution and the sodium nitrite added. The mixture is then added slowly to hydrochloric acid and ice at 0°C, when the insoluble diazonium compound is formed. The insoluble diazonium salts are more stable than the soluble type and will withstand filtration and washing followed by redispersion before coupling.

Coupling

When a solution of a diazonium compound, such as that made from 3-nitro-4-toluidine (*above*) is added to a weakly alkaline solution of 2-naphthol (also called β-naphthol), coupling takes place in the 1-position by electrophilic substitution:

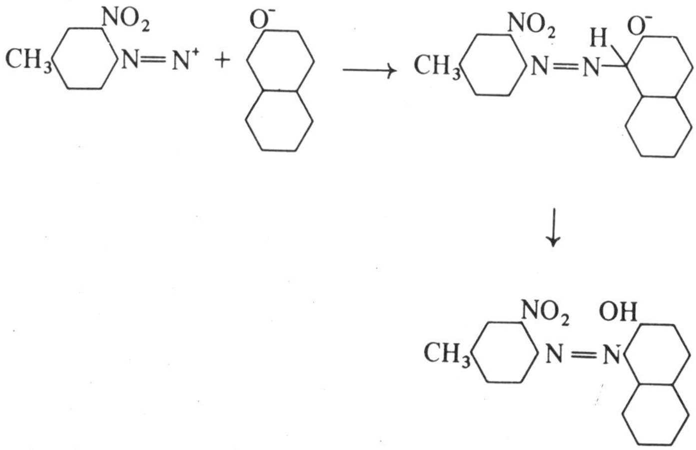

NO_2 O^- NO_2 H O^-

CH_3 $N{=}N^+$ + ⟶ CH_3 $N{=}N$

↓

NO_2 OH

CH_3 $N{=}N$

The product is the brilliant red pigment known as Helio or Toluidine Red (CI Pigment Red 3). 2-naphthol will couple slowly in slightly acid solution if it is precipitated in a fine condition immediately before coupling.

The position in which coupling takes place is determined by the recognized mechanisms. With 2-naphthol, no coupling takes place if the 1-position is occupied. With 1-naphthol, coupling takes place in the 4-position, but in the 2-position if position 4 is occupied.

The coupling of a diazonium compound with an amino compound is carried out in weakly acid solution. Thus, if the diazonium compound used above is added to a suspension of acetoacetanilide in dilute acetic acid, coupling takes place according to the scheme

$$CH_3C_6H_3(NO_2)N{\equiv}N^+ + CH_3{-}C(OH){=}CH{-}CO{-}NH{-}C_6H_5 \longrightarrow CH_3C_6H_3(NO_2)N{=}N{-}C({=}C(OH){-}CH_3){-}CO{-}NH{-}C_6H_5$$

The product is the bright reddish-yellow pigment dyestuff, Arylamide Yellow G, also known as Hansa Yellow G (CI Pigment Yellow 1).

Red azo pigment dyestuffs

The Toluidine Red described above is a typical member of this class and is one of the most widely used.

An extensive range of reds of varying shade and lightfastness is made by selection of base and coupling component. Thus Paranitraniline Red ('Para' Red) (CI Pigment Red 1) and Permanent Red 2G (CI Pigment Orange 5) are made from diazotized 4-nitraniline and 2,4-dinitraniline respectively, coupled with 2-naphthol.

A further group of azo pigment dyestuffs is obtained by coupling diazotized amines with the anilides or toluidides of β-oxynaphthoic (2-naphthol-3-carboxylic) acid (generally referred to as BONA). The latter are known as Naphthol AS or Brenthol bases and have the general formula

$$C_{10}H_6(OH)CO.NH{-}C_6H_5$$

These bases couple in a similar way to β-naphthol and produce pigments of good lightfastness and which are generally deeper in shade than those made from β-naphthol. The following are some of the best-known members of this class.

Permanent Red F4RH (CI Pigment Red 7), made from 5-chloro-2-toluidine and BONA 5-chloro-2-toluidide.

Permanent Red FRL (CI Pigment Red 9), made from 2:5-dichloraniline and BONA *p*-toluidide.

Permanent Bordeaux F2R (CI Pigment Red 12), made from 5-nitro-2-toluidine and BONA-*o*-toluidide.

Properties of red azo pigment dyestuffs

The red azo pigment dyestuffs are bright soft pigments with good light-fastness in mass, but they show considerable variation in lightfastness when reduced with white. They are resistant to dilute acids and alkalis. Generally they are fast to oils and solvents, but a few, among which Para Red is notorious, tend to bleed, and films containing these pigments cannot be over-painted with white or pale colours without discoloration.

The heat resistance is moderate and most members will tolerate stoving at 120°C (248°F) for half an hour. Above this temperature, performance is irregular; for example, Toluidine Red will tend to volatilize or turn brown.

These red azo pigment dyestuffs are hydrophobic and flush readily from aqueous dispersion into organic media. If, therefore, they are to be used in emulsion paints they should be converted to the hydrophilic condition by treatment with an anionic or suitable non-ionic surface-active agent.

The specific gravities of these pigments are in the range 1·4 to 1·5.

Yellow and orange azo pigment dyestuffs

This is a smaller group than the red, but the members are bright colours with good tinting strength. They are non-toxic and in recent years the yellow members have replaced lead chromes in many types of paint. The following are the principal members of the group.

Arylamide (Hansa) yellows

These are prepared by coupling diazotized amines with acetoacetanilide or its derivatives under weakly acid conditions. Thus Arylamide Yellow G (CI Pigment Yellow 1) is prepared from diazotized 3-nitro-4-toluidine by coupling with acetoacetanilide suspended in dilute acetic acid. The reaction takes place according to the scheme set out on p. 118. It is the reddest of the arylamide yellows. Other members of the group of interest in paint manufacture are

Arylamide Yellow 5G (CI Pigment Yellow 5) from 2-nitraniline and acetoacetanilide

Arylamide Yellow 10G (CI Pigment Yellow 3) from 4-chloro-2-nitraniline and acetoacet-2-chloranilide.

Properties of the arylamide yellows

Generally there is a gradation in properties from the reddest and strongest Yellow G to the greenest and weakest Yellow 10G. The lightfastness also increases in passing from the G to the 10G. This change from redder to greener one accompanied by an increase in lightfastness is associated with the introduction of chlorine into the pigment molecule. This effect of halogen introduction is fairly general in organic pigments.

The arylamide yellows are all bright and soft pigments with fairly good fastness to light, moderate tinting strength and poor opacity. They will withstand stoving temperatures up to 120°C but show a tendency to 'bleed' in some media resulting in the formation of 'bloom' on the stoved films.

They are hydrophobic and flush readily from aqueous dispersion into organic media. When used as stainers for emulsion paints they must therefore be converted to the hydrophilic condition by treatment with an anionic or suitable non-ionic surface-active agent. They are also used as stainers in solvent-based paints and for making very bright greens with phthalocyanine blue.

The specific gravities of the arylamide yellows are in the range 1·4 to 1·5.

Benzimidazolone pigments

Research directed toward improving the tinting strength and insolubility of the arylamide yellows resulted in the discovery of an interesting new group of azo pigments [3]. They are produced by using the compound as the coupling component.

$CH_3.CO.CH_2.CO.NH$ (benzene ring) NH NH > CO

The benzimidazolone yellow possesses the general structure (A) which can be written in form (B) which illustrates ring formation by hydrogen bonding and the formation of – CO.NH – groups which, it is thought, are responsible for the improved insolubility and strength.

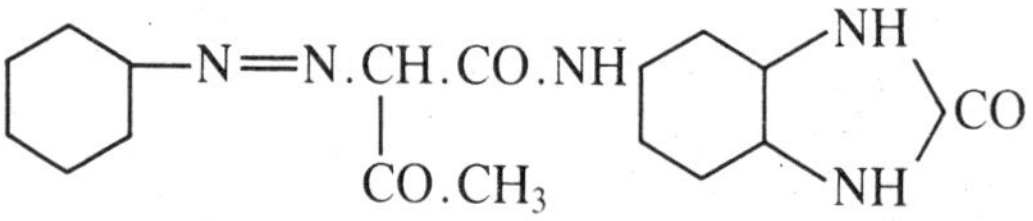

N=N.CH.CO.NH NH CO NH
CO.CH_3 (a)

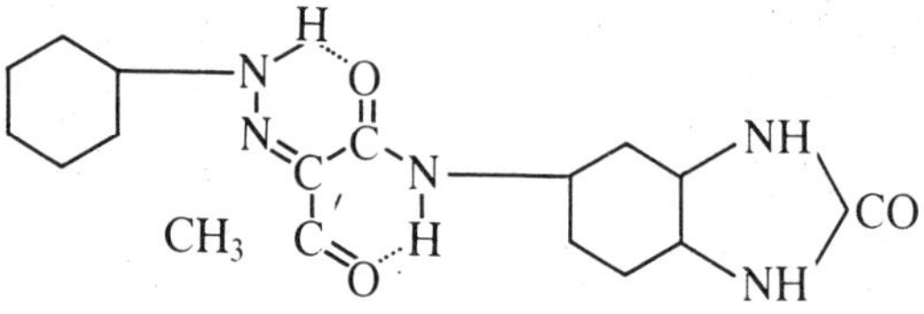

H N O N C C N CH_3 C H O NH CO NH (b)

Benzimidazolone yellow

A range of shades, varying from yellow through red to brown, can be produced by varying the substituents in the diazo component.

Benzidine yellows

These are prepared by coupling tetrazotized benzidines with aceto-acetanilide or its derivatives. The most commonly used base is 3:3′-dichlorobenzidine. This is tetrazotized and coupled with acetoacetanilide according to the scheme

CH_3 CH_3
CO Cl Cl CO
CO— CH_2 + N=N Cl ⟨ ⟩—⟨ ⟩ Cl N=N + CH_2— CO
NH NH

↓

CH_3 CH_3
CO Cl Cl CO
CO—CH—N=N⟨ ⟩—⟨ ⟩N=N—CH—CO + 2HCl
NH NH

Properties of benzidine yellows

These yellows possess fair opacity and appreciably greater tinting strength than the Arylamide yellows. The lightfastness varies with the nature of substituent atoms, the chlorine substituted types being the best.

Pyrazolone type These are made from benzidines and derivatives of Pyrazolone. The best-known member of the class is Permanent Orange G.

Permanent Orange G is made by coupling one molecule of tetrazotized 3:3′-dichlorobenzidine with two molecules of 1-phenyl-3-methyl-pyrazolone:

Cl Cl
CH_3—C—CH_2+ N=N Cl ⟨ ⟩—⟨ ⟩ N=N Cl + CH_2—C—CH_3
N CO CO N
N N

↓

Cl Cl
CH_3—C—CH—N=N⟨ ⟩—⟨ ⟩N=N—CH—C—CH_3
N CO CO N
N N

Properties of permanent orange G

A very bright orange pigment dyestuff with good lightfastness but poor opacity.

Azo condensation pigments

In general the lightfastness of azo pigments, as well as their resistance to heat and solvents, tends to improve with increase in size and complexity of the molecule. The synthesis of these large molecules proved difficult. For example, disazo compounds were not readily produced on attempting to react a coupled product with further diazonium salt [2]. Success was achieved by Geigy who devised a method for combining two mono-azo pigment molecules by condensation. Azo derivatives of 2-oxynaphthoic acid can be converted to acid chlorides, and these can then be condensed with diamines to produce a large molecule. The scheme below illustrates the general principle.

A range of azo pigment dyestuffs of this and other types are now being produced [3]. They show, in general, good lightfastness and resistance to heat and a reduced tendency to bleed in solvents. In spite of their improved properties, azo-condensation pigments have not been used to a great extent in paints, mainly on account of poor colour strength and high cost. They are, however, used widely in plastics.

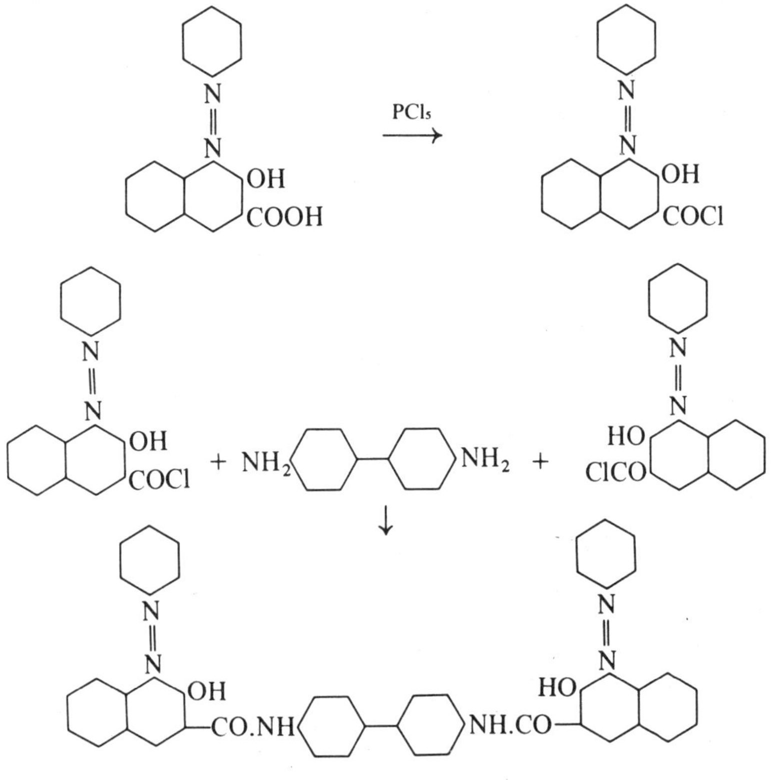

N=N, OH, COOH $\xrightarrow{PCl_5}$ N=N, OH, COCl

N=N, OH, COCl + NH_2–C₆H₄–C₆H₄–NH_2 + N=N, HO, ClCO

↓

N=N, OH, CO.NH–C₆H₄–C₆H₄–NH.CO, HO, N=N

Azo toners and lakes

These all contain one or more salt-forming groups of the acidic type such as sulphonic ($-SO_3H$) and/or carboxyl ($-COOH$). These groups can be present in either or both of the coupling components. The best known member of the Toner class is **Lithol Red R** (CI Pigment Red 49) which is made by coupling diazotized β-naphthylamine-α-sulphonic (Tobias) acid with β-naphthol. The reaction follows the scheme

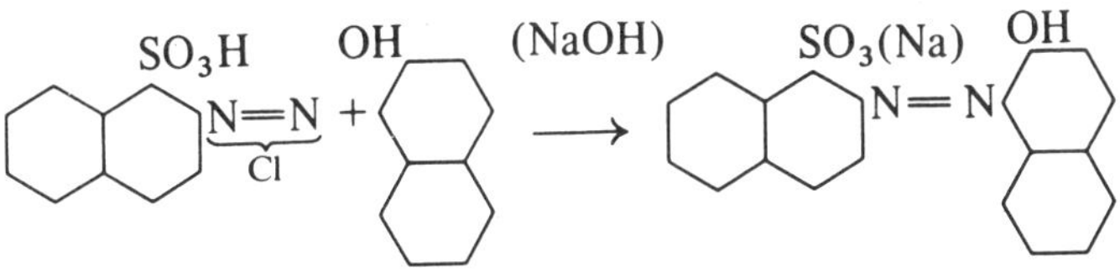

SO_3H N=N Cl + OH (NaOH) ⟶ $SO_3(Na)$ N=N OH

The manufacture of Lithol Red Toners provides an excellent example of the general procedure and is therefore described in more detail.

Manufacture of Lithol Red toners

1 Diazotization of Tobias acid Both the acid and its diazonium compound are insoluble in water and so the method of reverse diazotization is employed.

Tobias acid (1 mol) is dissolved in the necessary sodium hydroxide solution and a solution of sodium nitrite (1 mol) added. The mixed solution is cooled to 10°C (50°F) and added slowly and with stirring to $2\frac{1}{2}$ mol of hydrochloric acid in ice and water. The whole is stirred until no nitrous acid is detectable, indicated by the absence of a blue colour when spotted on starch-potassium iodide paper.

The diazonium compound can be used either as the suspension as prepared or, since it is fairly stable, it can be filtered, washed to remove soluble salts, and re-dispersed in cold water for coupling.

2 Coupling with β-naphthol The suspension of the diazonium compound is added slowly and with stirring to a solution of β-naphthol (1 mol) in sodium hydroxide (1 mol) and sodium carbonate ($\frac{1}{2}$ mol) at 10°C (50°F). Coupling takes place according to the scheme outlined above to produce the sodium salt of the dyestuff. This is only slightly soluble in water and can either be filtered off or used in suspension for further reaction.

3 Conversion to toners Lithol toners are the calcium or barium salts of the Lithol dye prepared as above and are produced by the addition of a solution of the metal salt to a suspension of the sodium derivative. The toner is formed according to the scheme

SO_3Na N=N OH $+ \frac{1}{2}CaCl_2 \rightarrow$ $SO_3\frac{Ca}{2}$ N=N OH $+ NaCl$

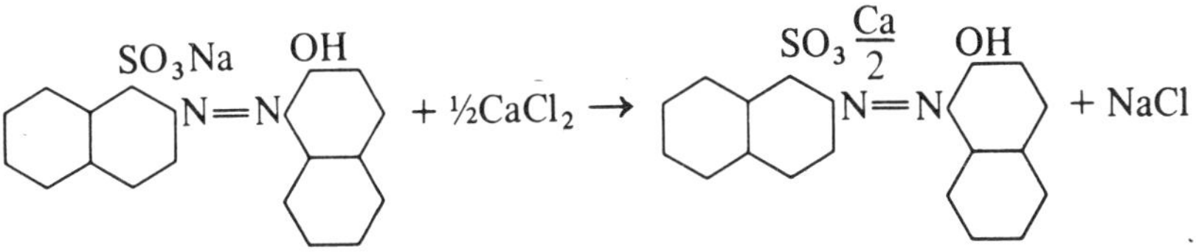

The suspension of the toner is heated to 90° – 95°C (194° – 203°F) to develop the brilliance and strength of the pigment. It is then filtered, washed thoroughly to remove soluble salts, and dried.

This process is used for the blue-shade Calcium Toner (Lithol Toner BS) and for the yellow-shade Barium Toner (Lithol Toner YS).

Other toners are prepared by the same general procedure and the following are some of the best-known.

Permanent Red 2B (CI Pigment Red 48) from 2-chlor-4-aminotoluene-5-sulphonic acid and β-oxynaphthoic acid. It is made as the barium or calcium toner.

Permanent Red 4B (CI Pigment Red 57) from 4-aminotoluene-3-sulphonic acid and β-oxynaphthoic acid. The calcium toner is also known as **Lithol Rubine BK** and as **Crimson Toner**.

Lake Bordeaux B from Tobias acid and β-oxynaphthoic acid. The calcium and manganese toners are used in cellulose lacquers.

General properties of Azo toners

The opacity of these pigments is superior to that of the pigment dyestuffs and they possess high tinctorial strength. With the exception of the Lithol toners, which are poor, the lightfastness of the class is fairly good but is not up to that of the pigment dyestuffs. In pale tints they all tend to fade. Generally they show good heat and solvent resistance and are used in air-drying and stoving paints.

Rosinated toners

If an alkaline solution of rosin (virtually a solution of sodium abietate) is added to the suspension of sodium lithol before addition of the heavy metal salt, the precipitated pigment will consist of an intimate mixture of the Lithol toner and an abietate or rosinate of the heavy metal. Such materials are known as Rosinated toners and can contain from 10 to 30 percent of metal rosinate. They show a greater degree of brilliance than the straight toner and are of comparable strength.

Rosinated Lithol toners are used in printing inks for the following reasons: (i) brilliance of self-colour, (ii) softness of texture, (iii) ease of dispersion in oils and other organic media, and (iv) moderate cost. They are seldom used in paints since Lithols lack the necessary fastness to light.

A number of other toners are also made in rosinated forms, e.g. **Permanent Red 2B** and **Permanent Red 4B**.

Lakes from azo colours

Many of these are reduced toners and can be prepared by forming the toner in presence of the base. The bases used include barytes, blanc fixe, alumina, alumina/blanc fixe, china clay and alumina/china clay. The ratio of base to toner can be varied at will.

Other lakes are formed by combination of a reactive dyestuff with a base containing alumina. An interesting example of this type is afforded by **Pigment Scarlet 3B** where the dyestuff is soluble in water and is produced by coupling diazotized anthranilic acid with β-naphthol-3:6-disulphonic acid. The dye solution is added to a suspension of alumina/blanc fixe, often in the presence of a sulphonated oil and a zinc soap. The process has to be controlled carefully to give the brightest product.

Pigment Scarlet 3B is used in paint and cellulose lacquers.

Nickel azo yellow (CI Pigment Green 10)

An olive-yellow pigment produced by coupling diazotized 4-chloroaniline with 2,4-dihydroxyquinoline followed by conversion to the nickel co-ordination compound. It has the following constitution:

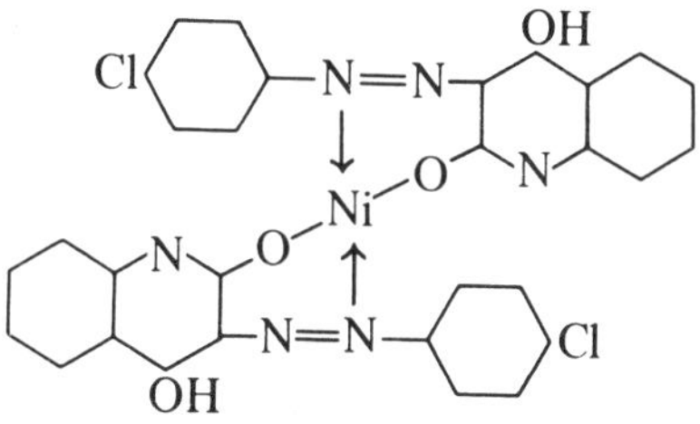

Cl—C₆H₄—N=N— … OH, N, O → Ni ← O, N, OH … —N=N—C₆H₄—Cl

It possesses good fastness to light and heat and to most organic solvents.

Other nickel-complex pigments as well as complexes of copper and cobalt have been described by Kaul [3]. These metallic complexes lack opacity when dispersed in paint media but show very good durability. The opacity can be improved by blending with suitable inorganic pigments and such mixtures are used in car finishes.

PHTHALOCYANINE PIGMENTS

The phthalocyanines comprise an important class of pigments ranging in colour from deep blue to yellowish green. The first member of the class was observed as an impurity in a batch of phthalimide being made at the works of Scottish Dyes Ltd in 1928. The material was examined by Linstead [4] who established its constitution as an iron phthalocyanine. The corresponding compounds of other metals were synthesized, and the most interesting was a deep-blue material containing copper. In the compound the atom of copper is linked to form a 16-membered ring of alternate carbon and nitrogen atoms. The copper atom is combined with the four adjacent nitrogen atoms by two covalent and two co-ordinate bonds:

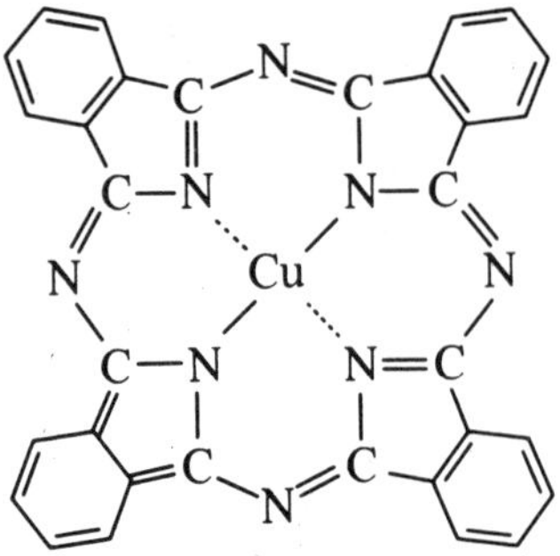

The structure shows a molecule in which the only deviation from symmetry is the presence of one *o*-quinonoid ring. This symmetry, combined with resonance within the relatively complex molecule, contributes to the outstanding chemical and physical stability.

Many other metals have been used in place of copper, but the compounds lack the brilliancy and chemical stability of the copper compound. The latter possesses a clean brilliant tone, high tinting strength and lightfastness, and is very inert chemically. It forms the basis of the phthalocyanine range.

Preparation of copper phthalocyanine

Several methods are available, but there are two main types, (a) the dry (fusion) process and (b) the wet (solvent) process.

In the dry process, phthalonitrile and cuprous chloride are heated together in the presence of metallic copper. Phthalimide is obtained as a by-product. Other dry processes involve heating phthalic anhydride with a copper salt in the presence of urea.

The wet process consists in heating phthalic anhydride, cupric chloride, urea and a catalyst in trichlorobenzene.

The crude copper phthalocyanine obtained by these methods is further processed to produce the pigmentary form. This process consists either in dissolving in cold concentrated sulphuric acid followed by pouring into hot water to precipitate the blue, or in milling with sodium chloride in presence of a solvent. Both processes yield a product of fine particle size and different crystal form from the crude.

Properties of copper phthalocyanine

Copper phthalocyanine exists in several crystal forms, but two only are of practical interest. These are designated as α (the redder) and β (the greener) forms. The crude compound possesses the β-form but is converted to the α-modification on further processing [5]. Early α-forms tended to revert to the larger crystals when in contact with certain aromatic solvents, but this fault has now been overcome and stable α- and β-forms are on the market.

It is a bright pigment of high tinting strength, poor opacity, and outstanding stability and resistance to attack. The pigment is resistant to all acids, with the possible exception of concentrated sulphuric acid in which it is soluble but is reprecipitated on dilution.

Copper phthalocyanine withstands heat up to 200°C (392°F) and sublimes without decomposition at higher temperatures. It is fast to light and is insoluble in all paint media and solvents.

A fault with phthalocyanine blues which has persisted over many years is the tendency to flocculate in paint systems. Flocculation is the formation of clusters of pigment particles after dispersion and can occur in the paint can and in the wet paint film. This fault can be very troublesome if the blue is used to tint a white enamel. Application by brush can break down the flocculates, resulting in blue streaks, while spraying can result in a mottled finish. The introduction of non-flocculating blues has largely corrected this fault and also the tendency to flotation.

The following are the physical characteristics of phthalocyanine blue:

Specific gravity 1·6	pH of aqueous extract 7·0
Oil absorption 46	

Uses of phthalocyanine blues

Their outstanding properties enable them to be used in paints, inks, artists' colours and, in fact, wherever stable blues are required. For this reason they have largely displaced Prussian blue. Very bright greens are produced when they are mixed with arylamide yellows.

Phthalocyanine greens (CI Pigment Green 7 and 36)

The four aromatic rings in the copper phthalocyanine molecule contain 16 replaceable hydrogen atoms. These can be replaced progressively by chlorine or a mixture of chlorine and bromine to give a range of greens. Chlorine alone gives bluish greens, but yellowish greens result when both chlorine and bromine are used to replace hydrogen atoms; for example, 12 hydrogen atoms can be replaced by 8 of chlorine and 4 of bromine or 15 hydrogen atoms by 3 of chlorine and 12 of bromine.

The greens are similar to the phthalocyanine blues in their outstanding resistance properties and are used for similar purposes.

VAT PIGMENTS

Vat dyestuffs are among the fastest to light of textile dyes and some of these have been found to be very useful pigments. The best-known member of the class, **Indigo**, has, however, been little used as a pigment as it lacks brilliance. The interesting members of the class include the following.

Thioindigo derivatives

These comprise a range of colours from pink through red to violet and they all contain chlorine atoms in the molecule. Thus, the constitution of Thioindigo Bordeaux is as follows [5]:

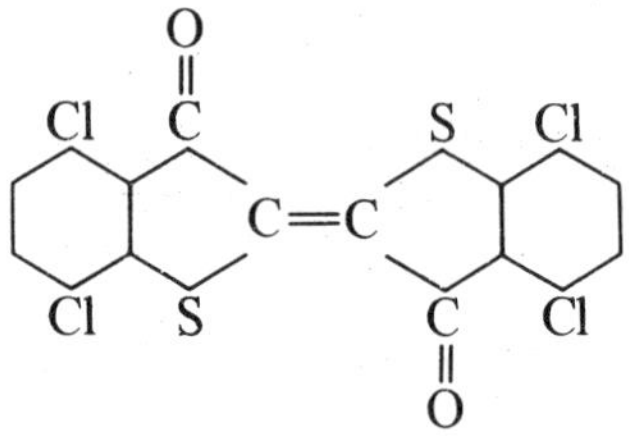

This pigment is very fast to light and solvents, but within the group there is appreciable variation in these properties.

Anthraquinone pigments

Alizarin (1:2-dihydroxyanthraquinone) is a material which has been known for a long time. it was orginally prepared from madder root (*Rubia tinctorum*) in which it occurs as a glucoside, but it is now prepared from the parent anthraquinone. Alizarin yields intensely coloured compounds with a number of metals, but the straight toners are of little interest as they possess poor working properties. More useful pigments are obtained by formation of lakes and these are precipitated on to alumina, usually in the presence of a surface-active agent such as Turkey Red Oil (sulphonated castor oil).

Synthetic Madder Lake (CI Pigment Red 83) is the calcium salt of alizarin precipitated in this way. It is very fast to light and until comparatively recently was used as a standard for lightfastness. Today it is used in artists' colours but finds little use in the paint industry.

Other anthraquinone pigments

A wide range of colours is obtained from the more complex anthraquinone derivatives. Some of these are very fast to light and the constitutions of two of the best-known, **Flavanthrone Yellow** and **Indanthrone Blue**, are depicted below.

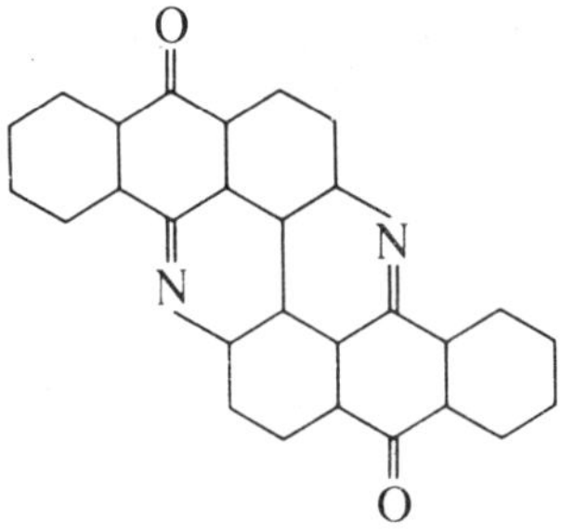

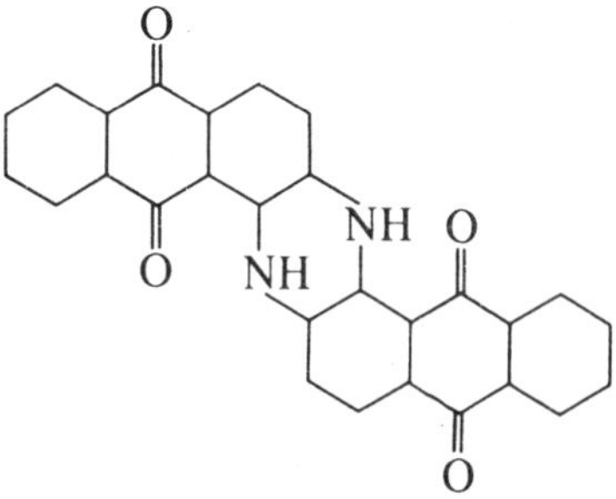

Although the lightfastness of some of these pigments approaches that of phthalocyanine blue, some members exhibit a tendency to bleed slightly in strong solvents.

Still greater complexity of molecular structure occurs in the Perinone and Perylene pigments. Perinone Orange (*below*) possesses very good fastness to light, solvents and heat and is used in car finishes.

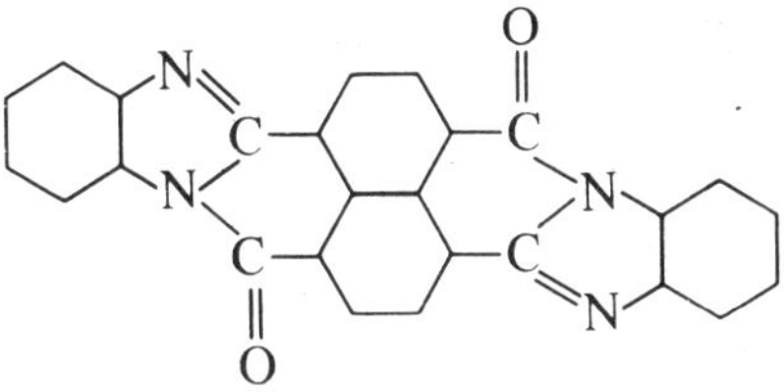

The perylenes are also fast to light and some approach the lightfastness of phthalocyanine blue. They also have excellent heat resistance and for this reason are used in plastics. The constitution of a typical member of the class, Perylene Maroon, is depicted below:

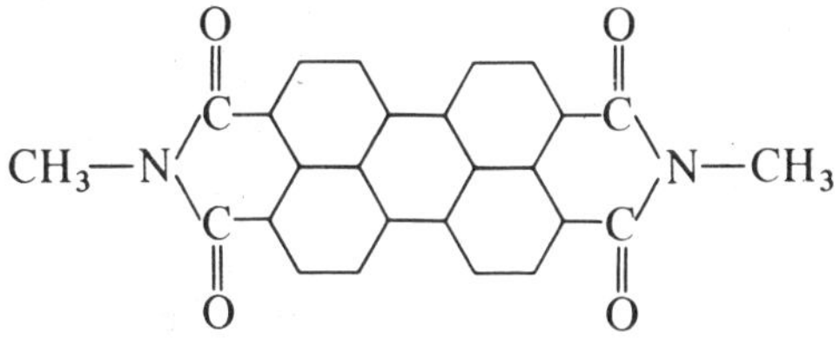

CH_3—N ... N—CH_3

QUINACRIDONE PIGMENTS (CI Pigment Violet 19)

The polycyclic quinacridone molecule (*below*) exists in three stable crystalline forms, two red (α and γ) and one reddish-violet (β). These form the basis of the range of pigments.

The compound was synthesized by du Pont [6] by condensing a dialkyl succinylsuccinate with aniline, followed by ring closure and oxidation:

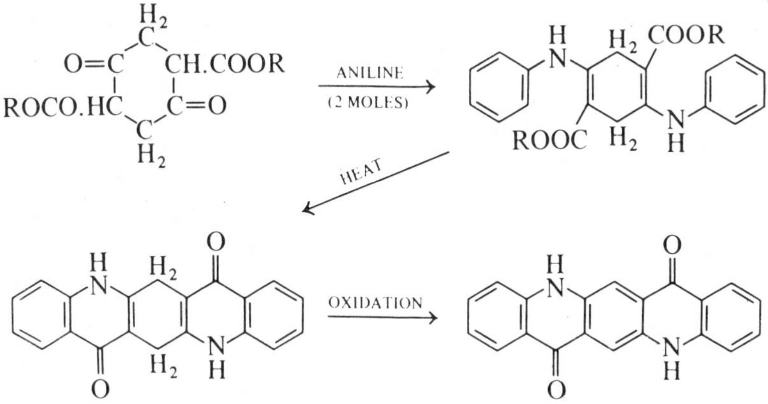

ANILINE (2 MOLES)

HEAT

OXIDATION

Modifications to the basic shades are made by introducing substituents.

The quinacridone pigments possess a lightfastness comparable with that of the phthalocyanine, together with high tinting strength. They are therefore valuable where clean fast tints are required, as on cars and coach work generally.

DIOXAZINE PIGMENTS

The pigments of this class are clean violet shades, and one member, carbazole dioxazine violet (CI Pigment Violet 23), is well known. It has the following constitution:

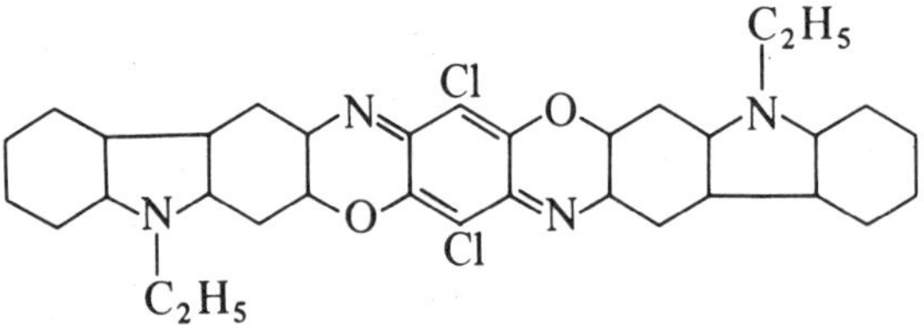

C_2H_5 Cl N O N O N Cl N C_2H_5

This pigment possesses very high tinting strength and good lightfastness both in self colour and in tints. It has a very red tone and is useful for shading the greenish phthalocyanines.

ISOINDOLINONE PIGMENTS

These form a range of compounds produced by condensing two molecules of tetrachloroisoindolinone with an aromatic diamine. They possess the following basic structure:

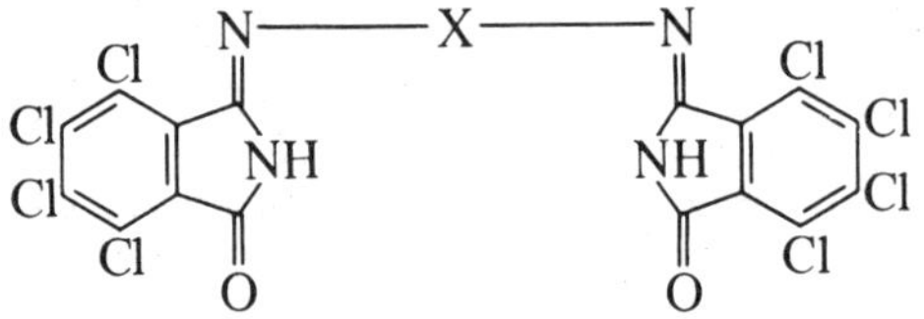

N—X—N Cl Cl Cl NH Cl Cl O NH Cl Cl Cl O

A variety of colours, ranging from yellow to brown, can be obtained by varying the aromatic diamine [7].

Isoindolinone pigments possess outstanding resistance to light, solvents, and weathering.

DPP (1.4-DIKETOPYRROLO-PYRROLE) PIGMENTS

A very interesting recent addition to the range of heterocyclic pigments [8]. It possesses a unique type of structure, as illustrated below, and a wide range of shades ranging from orange-yellow to violet can be produced by varying the substituents at X.

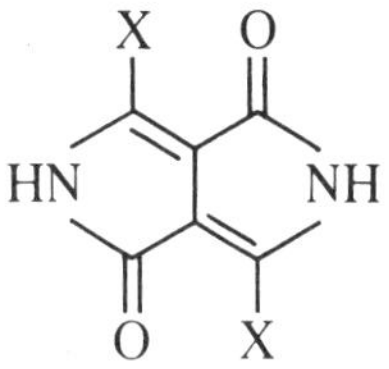

X O

HN NH

O X

DPP pigments are very stable and exhibit high colour strength and resistance to heat with good fastness to light and weathering.

PIGMENTS FROM BASIC DYES

Historical The conversion of basic dyestuffs into pigments by precipitation from acid solutions has been practised for many years. Precipitants used have included sodium phosphate and tannic acid, either alone or with tartar emetic. Lakes were generally made on an alumina base and were very brilliant but possessed poor lightfastness.

With the introduction of the so-called complex acids it was found possible to produce brilliant colours with good tinting strength and much better lightfastness. These were first produced in Germany under the name 'Fanal' colours.

Fanal type pigments

These are toners produced by precipitation of basic dyestuffs with complex acids.

Complex acids comprise phosphotungstic, phosphomolybdic and phosphotungstomolybdic acids and are prepared by addition of strong mineral acid to solutions of the mixed alkali metal salts. Theoretically any proportions could be used, but for best results the ratio of P_2O_5 to MoO_3 is kept within certain limits. The following figures indicate the ratios employed.

Phosphotungstic acid	Commonest ratio $P_2O_5/24WO_3$ Others used are 1/20, 1/16 and 1/12
Phosphomolybdic acid	Commonest ratio $P_2O_5/24MoO_3$ Others used are 1/22, 1/20 and 1/18
Phosphotungstomolybdic acid	Ratios used are $P_2O_5/10WO_3/8MoO_3$ and $P_2O_5 9WO_3/9MoO_3$

For best results there would appear to be a relationship between the number of free acid groups in the complex acid and the number of basic groups in the dyestuff.

The following basic dyestuffs are commonly used:

Methyl Violet	Auramine
Victoria Blue	Rhodamines
Brilliant Green	Setoflavine

In the manufacture of the toner, a salt of the basic dyestuff is dissolved in water and a solution of the complex acid added. The brightness and tinctorial strength of the pigment depend on accurate control of pH, temperature and concentration.

Phosphotungstomolybdic acid is the most commonly used acid and the products are therefore described as *triple salt toners*.

On account of their brilliance and high tinctorial strength the Fanal type colours are used widely in printing inks, but the lightfastness is not quite good enough for use in paints.

PIGMENTS FROM ACID DYESTUFFS

The acid dyestuffs employed are mostly of the azo and xanthene type and are very soluble in water. They are often difficult to precipitate as toners, and generally an absorbing base such as alumina aids complete precipitation. The acid dyestuffs in common use include Pigment Scarlet 3B, eosin, fluorescein, acid greens and erioglaucine. They are precipitated as salts of the heavy metals calcium, barium or lead.

Properties of pigments from acid dyestuffs

The lakes produced from acid dyestuffs are very brilliant and possess high tinting strength. However, the lightfastness is poor and they are very sensitive to alkali. With the exception of Reflex Blue (*below*) and Pigment Scarlet 3B they are now little used in the coating industry.

Reflex Blue

This is prepared from Alkali Blue which is the sodium salt of the monosulphonic acid of phenylated rosaniline. The latter is dissolved in water and the Reflex Blue precipitated under controlled conditions with dilute acid. The product is filtered, washed, and flushed into lithographic varnish.

Reflex Blue possesses high tinting strength and moderate lightfastness. It is used for toning black inks and in metal printing where its good heat-resistance is an asset.

BASES FOR LAKES

Alumina, $Al_2(OH)_6$, is the principal base used for the best-quality lakes. It is prepared by addition of sodium carbonate solution to a solution of aluminium sulphate:

$$Al_2(SO_4)_3 + 3Na_2CO_3 + 3H_2O \rightarrow Al_2(OH)_6 + 3CO_2 + 3Na_2SO_4$$

Alumina is amphoteric and possesses a very soft texture. It is fairly transparent.

Blanc Fixe is a non-reactive base sometimes used for good-quality lakes. The lake is precipitated on to a suspension of the base. Blanc Fixe is sometimes used in conjunction with alumina under the name Alumina/Blanc Fixe.

China Clay forms a permanent suspension in water and is often used as a base for lakes to be supplied in paste form. It is sometimes used with alumina under the name Alumina/China Clay.

Toxicity of organic pigments

The majority are accepted as being reasonably safe. Possible exceptions are barium toners which may contain soluble barium. In this case the necessary measures must be taken to protect the operatives, and a high standard of industrial hygiene must be maintained.

References to Chapter 7

[1] Hughes, E. D., Ingold, C. K. and Ridd, *Nature*, 1960, **166**, 642.
[2] Gaertner, H., Modern chemistry of organic pigments. *JOCCA*, 1963, **46**, 13.
[3] Kaul, B. L., Advances in the science and technology of pigments. *JOCCA*, 1987, **70**, 349.
[4] Linstead, R. P., *J. Chem. Soc.*, 1934, p. 1016.
[5] Smith, F. M. and Easton, F. D., Phthalocyanine pigments: their form and performance. *JOCCA*, 1966, **49**, 614.
[6] Allen, R. L. M., *Colour Chemistry*. Nelson, London, 1971, p. 257.
[7] Geigy, U K Pat. 833 548; 1 093 669.
[8] Ciba-Geigy Pigments, Clayton, Manchester.

8

Solvents and plasticizers

Solvents are used in paints and lacquers to reduce the viscosity or consistency of the material and so facilitate the application of a uniform coating. They must be solvents that are suitable for the oil or resin present. After application the solvent is no longer required and should evaporate completely from the film.

The resins used in lacquers frequently form hard and brittle films and require plasticizers to confer the necessary elasticity and adhesion. These are usually liquids which must have a minimum tendency to volatilize so that they will remain in the film. It is essential that the plasticizer should be completely miscible with the resin, and to this extent it can be regarded as a solvent.

SOLVENTS

Paint and lacquer solvents, with the exception of water, are volatile organic liquids which can be classified into well-defined chemical groups. The solvent power and general physical properties vary over a wide range. A useful but arbitrary classification of solvents is by boiling point (though these are not necessarily related to evaporation rates), thus:

Solvents	*Boiling point*
Low boiling point	Below 100°C (212°F)
Medium boiling point	Between 100° and 150°C (212° and 302°F)
High boiling point	Between 150° and 250°C (302° and 482°F)
Plasticizers	Above 250°C (482°F)

The important properties of paint solvents are the following:

1 Solvent power
2 Rate of evaporation
3 Boiling point and distillation range
4 Flash point and inflammability
5 Toxicity.

The following properties are also of importance as indicators of purity–specific gravity, colour, moisture content, acidity or alkalinity, and non-volatile matter.

General properties of solvents

Solvent power

The film-forming materials used in paints and lacquers are, with a few exceptions, organic in nature and possess no crystalline structure. When treated with suitable solvents they dissolve to give solutions the viscosity of which increases with increasing concentration. There is no solubility limit, as in inorganic solutions, but problems of application limit the concentrations used. These solutions produce continuous films on evaporation of the solvent.

It is well known that certain types of resin or polymer dissolve only in certain types of solvent ('like dissolves like') and it was early realized that the polarity of the molecules of each was an important factor. Polar compounds are unsymmetrical in molecular structure and can contain groups which cause changes in the electron distribution. Typical examples are the nitro ($-NO_2$), carbonyl ($=CO$) and hydroxyl ($-OH$) groups. Molecules containing such groups exhibit a permanent dipole moment, indicating the presence of separated charges.

One result of this polarity is the tendency of such molecules to form associated groups by attraction of opposite poles. A consequence of this association is the fact that the viscosity of polar liquids is generally greater than that of non-polar liquids of similar molecular weight.

The effect of polarity in the solute is illustrated by cellulose nitrate which dissolves readily in polar solvents such as esters and ketones but is insoluble in non-polar hydrocarbons. On the other hand, some of the weakly polar drying oils are insoluble in polar alcohols but soluble in non-polar hydrocarbons.

In both polar and non-polar solvents there are variations in solvent power or strength between individual types. Thus in the non-polar solvents, the aromatics give solutions of resins of lower viscosity (at the same concentration) than those in aliphatic hydrocarbons.

Miscibility of solvents

When two liquids, A and B, are brought into contact, mixing will occur only if the molecules of species A are able to move freely among those of species B. If, however, the attractive forces between A and A or B and B are greater than those of A for B, then the two liquids will be immiscible.

The attractive forces between the molecules in a liquid are responsible for the internal cohesion of the liquid, and the intensity of this is known as the 'cohesive energy density' [1]. It can be defined as the energy required to separate the molecules in unit volume of liquid or the cohesive energy per unit volume. It is related to the latent heat of vaporization in the following way:

$$\text{Cohesive energy density} = \frac{\Delta H - RT}{M/D}$$

where ΔH = latent heat of vaporization
R = gas constant
T = absolute temperature
M = molecular weight
D = density

It can also be defined as the energy required to separate the molecules in 1 cm^3 of liquid, i.e.

$$\text{Cohesive energy density} = \frac{\Delta H_v}{V}$$

where ΔH_v = molar heat of vaporization in calories
V = molar volume in cm^3

Solubility parameter

This term was suggested by Hildebrand and Scott [2] for the square root of the cohesive energy density, i.e. $(\Delta H_v/V)^{\frac{1}{2}}$, and has been assigned the symbol δ.

Solvents which possess similar solubility parameters possess similar internal cohesions and, in the majority of cases, solvents are miscible, provided the solubility parameters do not differ by more than one unit. Similarly, polymers dissolve in solvents of similar solubility parameters, but, as in the case of solvents, exceptions to this general rule are encountered.

This problem was investigated by Burrell [3] who suggested that hydrogen bonding played an important part in cohesive energy density, and also that solubility parameter values should be a range rather than single figures. However, difficulties were encountered initially in assigning quantitative values to the hydrogen bonding factor, as explained by Burrell [4]. The idea that three types of intermolecular forces – dispersion, hydrogen bonding and dipole moment – should be included in solubility figures was suggested by Hansen [5] and by Crowley *et al* [6]. They proposed a 'three dimensional' solubility parameter. The parameters originally suggested have been modified into the present day 'Hansen Solubility Parameter' which comprises a hydrogen bonding component (∂h), a polar component (∂p) and a non-polar component (∂d). This method of defining solvent power is widely accepted.

Hydrogen bonding is low in aliphatic and aromatic hydrocarbons (which are also non-polar), moderate in ethers, esters and ketones (which contain polar and non-polar groups) and high in alcohols, acids and amines (where the polar component predominates).

Solubility of polymers

Figures for amorphous polymers are determined experimentally by ascertaining the solvent or solvent mixture in which the polymer exhibits maximum solubility. The figure assigned to the polymer is that of the solvent or solvent mixture used. Values for solvent mixtures are calculated on a volume basis. Graphical analysis of resin solubilities has been employed by Teas [7].

'Crystalline' polymers, e.g. PTFE (polytetrafluoroethylene) or polyethylene, are not amenable to the treatment described above since they are insoluble in normal solvents at ambient temperature. If they are heated above their melting points in the presence of certain solvents they will dissolve, but separate on cooling.

Measurement of solvent power (traditional method)

Kauri-butanol value

This test is applied to solvents to be used in varnishes and paints. The solvent to be tested is put into a burette and added slowly to 20 g of a solution of Kauri gum in butanol (100 g Kauri to 500 g butanol). The solution is contained in a flat-bottom flask which is placed on a printed sheet of paper. The end-point is reached when the precipitated gums causes the outline of the print to appear blurred. The volume of solvent added (in millilitres) is the Kauri-butanol (KB) value. A good solvent is indicated by a high KB value.

Dilution ratio

This is important with solvents for cellulose nitrate which are divided into three classes.

1 True solvents.
2 Latent solvents which are not solvents for cellulose nitrate but which increase the solvent power of the main solvent.
3 Diluents which are not solvents for cellulose nitrate but which are added to reduce cost or to assist the solution of other ingredients.

The dilution ratio can be defined as the tolerance of a solution of cellulose nitrate in a given solvent for a given diluent and is determined in the following way.

A solution of cellulose nitrate is made up in the given solvent and titrated with the diluent in question until a permanent slight precipitate is formed. The titration can be repeated after addition of a further quantity of solvent, but the concentration of cellulose nitrate at the end should not be less than 8 percent. The dilution ratio is the volume of diluent added divided by the volume of solvent used.

The concept of latent solvent applies also in systems other than cellulose nitrate lacquers. A mixture of xylol and *n*-butanol is a much more effective solvent than xylol alone for stoving finishes, especially for the alkyd-amino type.

Rate of evaporation

This is of fundamental importance in the choice of a solvent for a particular use, and with solvents of the same type, e.g. the petroleum distillates, it is roughly proportional to the distillation range. This does not hold, however, with solvents of different classes. The fundamental factors which control evaporation are the specific heat of the material and the latent heat of evaporation, and these can differ considerably in different classes of solvents. As an example, the evaporation rate of tetrachlorethane with a boiling point of 147°C (296.6°F) is considerably higher than that of water with a boiling point of 100°C (212°F).

A simple and rapid method for evaluation of evaporation rate is to place 0.5 ml of the solvent on a No. 1 Whatman filter paper. This is suspended from the edge, away from direct draughts, and the time noted for complete evaporation. Ethyl ether is frequently taken as a standard for fast evaporating solvents, but other standards, e.g. 2° xylol, can be used.

This test is known as the filter-paper spot method and is useful also for detecting the presence of oil or other non-volatile (or less volatile) contaminant. Oils and greases form a permanent spot on the paper.

It is often desirable to measure the rate of loss of solvent under conditions similar to those in a drying paint film. For this purpose a suitable instrument is the Shell Liquid Film Evaporometer [8] in which the rate of solvent loss from resins or plasticizers can be measured. The instrument also enables an analysis to be made of the solvent mixture present at various stages during the evaporation.

Distillation range

The boiling points of pure liquids at a specific pressure are constant and can be used as a method of identification. If 100 ml of such a liquid is distilled in the standard apparatus [9] the whole of the sample will distil at a constant temperature.

Commercial solvents, however, are seldom pure and when distilled by the previous method, the temperature indicated by the thermometer will rise throughout the distillation. The range of temperature shown is known as the 'distillation range' and is used both as a method of identification and as an indication of the purity of a given solvent. A small piece of copper foil is usually placed in the distillation flask. The degree of staining of the foil is an indication of the amount of corrosive sulphur compounds present.

In the determination of distillation range, the temperatures taken are (i) at the appearance of the first drop of distillate, (ii) at the completion of 95 ml of distillate.

Flash point and flammability

With the exception of the chlorinated hydrocarbons, organic solvents are flammable, and a fire risk is therefore associated with their use. When these liquids are heated, the concentration of vapour in the air above will increase until it will flash on application of a flame. The lowest temperature at which this occurs is known as the flash point.

The determination of flash point is carried out by one of two standard methods. The Abel (closed cup) apparatus is used for liquids which flash below 49°C (120°F) and the method of use is described in BS 2000, Part 33, 1982. The method is now embodied in The Highly Flammable Liquids and Liquefied Petroleum Gases Regulations, 1972, Schedule 1. It is also described in BS 3900, Part 8 (Danger Classification by Flash Point) and Part 9 (Determination).

A flash tester is used for the rapid determination of flash point using small quantities (2 cm^3) of material and is described in BS 3900, Parts A13 and A14.

For liquids with flash points above 49°C (120°F), the Pensky-Martens apparatus is used. This is described in BS 2000, Part 34, 1983.

The flash points of mixtures of two solvents of differing flash point can lie between the two or can be below either of the individual flash points.

This is exhibited by mixtures of xylol and *n*-butanol. Flash points of solvents can be raised by mixing them with small quantities of the non-flammable chlorinated solvents.

If the temperature is high enough, a mixture of flammable vapour and air can ignite without application of a flame or spark. The temperature at which this takes place is known as the *auto-ignition* temperature.

Explosive limits

Mixtures of flammable vapours and air are capable of exploding on ignition if the concentration of vapour lies within certain limits. These are known as the Lower (LEL) and Upper (or Higher) Explosive Limits and are expressed as a percentage, by volume, of the vapour in air. Below the LEL there is insufficient solvent vapour and above the upper figure there is insufficient oxygen for an explosion to occur. In practice the concentration of flammable vapour should be kept below the LEL.

Legislation relating to solvents

A number of Regulations are in force which define and lay down conditions covering the storage, use and transport of flammable liquids. These are, in the main, petroleum products but amendments are in force relating to other flammable liquids.

The Petroleum (Consolidation) Acts, 1928 (amended by SI 1244 1984) applies to any petroleum spirit including benzene, toluene, xylene, etc. which has a flash point (Abel closed cup) below 73°F (22.8°C). It excludes acetone, ether, ethanol and methylethyl ketone.

The Petroleum Mixtures Order SI 993 1929 (amended by SI 1244 1984) extends the Regulation to include mixtures of petroleum and any other material, the flash point of the mixture being below 73°F (22.8°C).

The Highly Flammable Liquids and Liquified Petroleum Gases Regulations SI 917 1972 controls the storage, use and labelling for transport of highly flammable liquids. These are defined as liquids which give off a flammable vapour at a temperature below 32°C (Abel closed test) and, in addition, support combustion when tested by the method described in Schedule Z of the Regulations.

The enforcement of these Regulations and the issue of storage permits is generally in the hands of the Local Authority.

EEC Directives

In Substances Directive 67/548/EEC the degrees of flammability of solvents are designated as follows:

Extremely flammable	Flash point below 0°C (and Boiling point below 35°C)
Highly flammable	Flash point below 21°C
Flammable	Flash point between 21°C and 55°C (both points inclusive)

The Solvents Directive 73/173/EEC, amended by 86/781/EEC covers 'any preparation intended for use as a solvent which consists only of petroleum mixtures listed in the annexe and containing, in addition, any Extremely Flammable, Highly Flammable or Flammable substance'. A similar Directive 77/128/EEC amended by 81/916/EEC applies to paints, putties, sealants, etc. The EEC Directives apply to the labelling and transport of these flammable materials throughout the member countries.

Toxicity of solvents

The majority of solvents can dissolve and so remove the natural oils from the skin leaving the latter prone to dermatitis and other infection. The skin should be protected when solvents are handled; barrier cream should be applied or gloves should be worn. On prolonged contact some solvents can be absorbed into the system causing internal disorders of various types.

It is extremely difficult to prevent inhalation of solvent vapour by paint makers and applicators. Although the common paint solvents such as white spirit and xylol are not considered to be highly toxic, inhaling the vapours even in low concentration over a prolonged period can result in headache and/or nausea. Limits have therefore been laid down for the maximum permissible concentration of vapours to which an operative can safely be exposed over an 8-hour working period. These concentrations are always well below the LEL and are expressed as Occupational Exposure Limits (OEL). These are time-weighted average figures and vary from one class of solvent to another and among individual members within a group. Figures are laid down in Guidance Note No. 40 published by the Health and Safety Executive; these replace the previously used TLV figures published by the American Conference of Governmental Industrial Hygienists (ACGIH). Guidance Note No. 40 is updated annually and since figures are liable to amendment in the light of experience they are not quoted in this book. Reference should be made to the current Guidance Note.

The disposal of solvent waste is controlled by the 1974 Control of Pollution Act and the 1980 SI 1709 Control of Pollution (Special Waste) Regulation. These apply to any waste containing specified substances including 'Hydrocarbons and their oxygen, nitrogen and sulphur compounds'. Control of waste disposal is vested in the Local Authority.

For a more detailed treatment of toxicity as well as the fire hazard associated with paint materials the reader is recommended to the section entitled 'Hazards to the Paint User' by Dr H. Brunner in Hess' *Paint Film Defects*, 3rd edn, Chapman & Hall, 1979.

Other solvent properties

Specific gravity

This can be determined by several methods. For rough determinations, a hydrometer can be used. In this case it is essential to note the temperature and to use a wide vessel to avoid *wall effects*. More accurate measurements can be made with the specific gravity bottle or the Westphal balance.

Colour

With the exception of the heavy naphthas, solvents have very little colour. Any intensification of colour (or appearance of colour) in solvent which is normally water-white is an indication of contamination.

Moisture

An indication of the presence of moisture can be obtained by shaking the solvent with anhydrous copper sulphate, which turns blue in the presence of moisture, or by the addition of a miscible solvent which has no tolerance for water. Examples of the latter are benzene and petroleum ether. In this test the presence of moisture is shown by the separation of droplets.

Moisture can be determined quantitatively using the Karl Fischer method [10].

Acidity or alkalinity

If the solvent is miscible with water or industrial methylated spirit, a solution in one of these is titrated with standard acid or alkali using phenolphthalein as indicator. Other solvents are shaken with neutral distilled water and the aqueous extract titrated.

Non-volatile matter

The presence of non-volatile matter is often indicated in the filter-paper spot test for evaporation rate. It is determined quantitatively by placing a known quantity of solvent in an open dish and evaporating in a steam bath or in a steam-heated oven. Any non-volatile residue can then be weighed after cooling.

Classes of solvents

Petroleum hydrocarbons

A considerable number of very useful and widely used solvents are obtained either by distillation of crude petroleums or by thermal cracking of petroleum gases. The products obtained by distillation are mainly paraffin hydrocarbons containing small amounts of aromatics. The thermal cracking processes are designed to give solvents with high aromatic contents.

Low-boiling petroleum fractions

Petroleum ethers (ligroins)

These are marketed in several grades according to boiling range. The three most common ranges are 40° to 60°C (104° to 140°F), 60° to 80°C (140° to 176°F), and 80° to 120°C (176° to 248°F). The specific gravities range from 0·645 to 0·676 and the aromatic contents from 1 to 5 percent.

Special boiling-point solvents

These are produced in six or more boiling ranges between about 30° and 160°C (86° and 320°F). They are generally referred to as SBP 1,2,3, etc. and can vary considerably in aromatic content. The solvent power generally increases with aromatic content, and members of this range are used as rubber solvents as well as for lacquer formulation.

Flash points below 23°C (73°F):Highly Flammable Liquids.

White spirit

This solvent was introduced as a paint thinner in about 1885 by Samuel Banner and for many years it was regarded as a substitute for turpentine. The term 'turpentine substitute' persists in some quarters, but the term 'white spirit' is universally accepted in the British paint industry.

White spirit is produced from the petroleum fraction distilling between the low boiling fractions (the SBPs and petrols) and the burning oils. The boiling range is mainly between 150° and 190°C (302° and 374°F), although a small percentage may boil up to 210°C (410°F). This high-boiling fraction evaporates slowly at room temperature and should therefore be as small as possible. The actual composition and especially the aromatic content varies with the process and source of petroleum, but supplies generally conform to the following requirements of BS 245:1976 [11].

Boiling range	Up to 155°C (311°F)	10% max.
	Below 195°C (383°F)	90% min.
	Max. boiling point	210°C (410°F)
Flash point 34°C (93°F) minimum		

In addition the material must be neutral, free from objectionable sulphur compounds, and the residue on evaporation must not exceed 0.005 percent.

The characteristic odour of white spirit is due mainly to impurities and aromatics. A special grade of odourless white spirit with a higher flash point (46°C, 115°F) is available, but due to the removal of aromatics the solvent power is not as good as the normal grade.

White spirit possesses moderate solvent power and is miscible with most organic solvents. It dissolves raw and refined vegetable oils as well as light and medium stand oils. It is not, however, a good solvent for highly polymerized stand oils and blown oils. In this case the addition of a proportion of a stronger solvent such as an aromatic or dipentene is necessary to prevent separation on standing.

Generally white spirit is a poor solvent for resins, and its use as the sole solvent in oil/resin combinations such as oleoresinous varnishes or alkyd resins is limited to the long-oil types of media. As the proportion of oil decreases, stronger solvents have to be introduced, and the short-oil types of alkyd will tolerate little, if any, white spirit. Comparatively few resins are soluble in white spirit; examples that are soluble are rosin, ester gum, soft and modified phenolic resins, and Damar resin. White spirit also dissolves asphaltic bitumen and petroleum pitch.

In addition to the extensive use of white spirit as a paint thinner, large quantities are consumed in the manufacture of polishes, and for cleaning purposes.

Flash point of commercial white spirit: 41°C (105°F).

Distillate

This can be regarded as a type of white spirit containing a higher proportion of high-boiling hydrocarbons. The boiling range is about 150° to 275°C (302° to 527°F) and consequently the material contains an appreciable *tail*. When used in paints this tail holds the film open much longer than normal white spirit does, and distillate is sometimes used to lengthen the wet-edge time of brushing finishes and undercoats.

VM & P naphtha (Varnish Makers' and Painters' naphtha)

This is a petroleum fraction marketed in the United States. It is somewhat similar to white spirit.

Aromatic hydrocarbons

Toluol (commercial toluene), $C_6H_5CH_3$

This is the most extensively used diluent for cellulose nitrate lacquers. It is not a solvent for cellulose nitrate, but solutions in true solvent will tolerate greater additions of toluol than of other hydrocarbons. This has an important bearing on the cost of the lacquer. Toluol is a solvent for a large number of resins and is miscible with drying oils and most other solvents. It does not dissolve PVC, copals, or shellac.

The range of commercially available toluols is described in BS 805 (Toluenes): 1977.

Flash point 4°C (39·2°F). Class 1. Highly flammable liquid.

Xylol (commercial xylene)

This is a mixture of the three isomers in which the meta-isomer predominates:

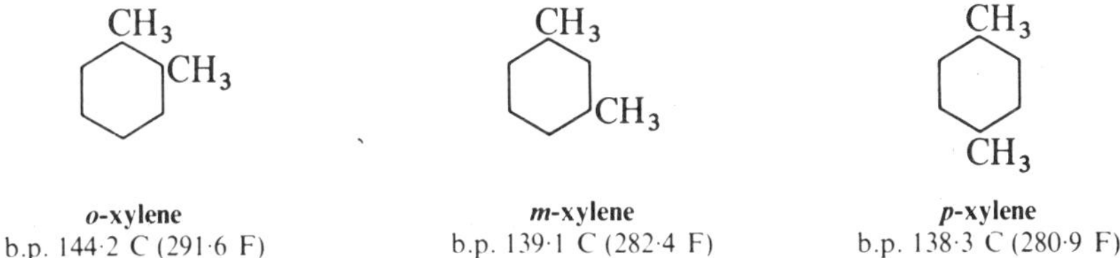

o-xylene	*m*-xylene	*p*-xylene
b.p. 144·2 C (291·6 F)	b.p. 139·1 C (282·4 F)	b.p. 138·3 C (280·9 F)

In some grades small quantities of other aromatic solvents are also present.

Xylol is now one of the most important solvents used in the paint industry. This is the result of its high solvent power for a very wide range of resins used in both stoving and rapid air-drying coatings, together with a convenient rate of evaporation. The solvent power is increased appreciably on addition of 10 to 20 percent *n*-butanol, and this mixture is very widely used in stoving finishes. The rate of evaporation is too great for use in brushing finishes.

Xylol is used as a diluent in cellulose nitrate lacquers and as the main solvent in lacquers based on polystyrene, polymethylmethacrylate and chlorinated rubber. It is also an excellent solvent for asphaltic bitumen and petroleum pitch.

In BS 458:1977 (Xylenes) there are three principal grades based on boiling range, i.e. the difference in temperature between 5 percent and 95 percent distillation. Thus for 3° xylene the range is 3°C lying between 137·5° and 144·5°C. For 5° and 10° xylenes the boiling ranges lie between 137° and 145·5°C, and between 135° and 148°C, respectively. The specific gravity range for 3° xylene is 0·860 to 0·875, but in this and the other grades the specification includes low-gravity material.

The flash point of 3° xylene is 24.4°C (76°F), but mixtures with *n*-butanol (flash point 36·7°C, 98°F) can flash below 22·8°C (73°F) and are then subject to the restrictions on highly flammable solvents laid down in the Petroleum Act. Xylol is classed as a highly flammable liquid.

High aromatic solvents from petroleum

These range in specific gravity from 0·833 to 0·895 and contain high percentages (80 to 93) of aromatics, principally the isomers of trimethyl benzene. They are supplied in graded boiling ranges between 100° and 210°C (212° and 410°F) and have flash points above 32°C (89°F).

They possess high solvent power and the odour is faint and not unpleasant. These solvents are marketed under proprietary names [12] and have largely replaced the coal-tar solvents.

Solvent naphtha

A solvent which is sometimes referred to as 90/160 naphtha from the fact that 90 percent distils below 160°C (320°F). It is used as a solvent in bituminous paints and in anti-fouling compositions.

Terpene solvents

Turpentines

These solvents are obtained from the resinous materials present in pine and other coniferous trees. If a cut is made in the trunk of one of these trees a gummy material is exuded and can be collected by cups attached to the trunk. Distillation of this material yields the volatile 'gum' turpentine. Steam distillation of the wood or of the stumps of pine trees yields 'wood' or 'stump' turpentine.

Another variety known as sulphate turpentine is obtained as a by-product in the wood-pulp industry and is obtained chiefly from Sweden.

For many years turpentine was the standard thinner for paints and varnishes, but since the introduction of alkyds and other synthetic resins together with supply difficulties during World War II, it has been replaced by other solvents.

The composition of turpentine is related to that of dipentene and pine oil and for this reason a survey of its properties is included.

Composition and properties The turpentines consist of mixtures of members of the terpene group of unsaturated cyclic hydrocarbons. The mixtures consist chiefly of α-pinene, β-pinene, carene, sylvestrene, and dipentene and the properties vary with the source of the turpentine. α-pinene is the main constituent of gum turpentine, whilst the wood turpentines consist mainly of dipentene.

The turpentines are clear colourless liquids with a characteristic pungent odour. Long exposure to the vapour can cause headache and nausea.

They evaporate uniformly but, due to the unsaturation, a degree of oxidation and polymerization takes place at the same time. The product (linalool) so formed is non-volatile but miscible with drying oils and resins and therefore becomes part of the film.

The physical characteristics of the turpentines are as follows:

	Gum turpentine	Wood turpentine
Specific gravity	0·862–0·872	0·859–0:875
Distillation range	150°–180°C (302°–356°F)	150°–170°C (302°–338°F)
Refractive index ($D_{20°C}$)	1·469–1·478	1·463–1·483
Flash point	30°–37°C (86°–99°F)	34°–36°C (93°–96°F)
Non-polymerizable material	Not more than 11%	Not more than 16%

The turpentines are good solvents, and the Kauri-butanol value ranges from 50 to 90. In this respect wood turpentine, which consists largely of dipentene with small amounts of pinene, is a better solvent than the gum variety.

Dipentene

This solvent is prepared by a modification of the distillation process for wood turpentine. It has a higher distillation range (175° to 195°C, 347° to 383°F) and an appreciably lower rate of evaporation. It is an excellent solvent for many synthetic resins and is often used to improve the wet-edge time of decorative undercoats and finishes. It also possesses some anti-skinning properties. The physical characteristics of dipentene are:

Specific gravity	0·844–0·850
Boiling range	175°-195°C (347°–383°F)
Refractive index ($D_{20°C}$)	1·472
Flash point	54°C (130°F)
Kauri-butanol value	105

Dipentene forms coloured complexes with some metals, and the green coloration resulting from its reaction with cobalt is sometimes apparent in white enamels.

Pine oil

Pine oil is usually obtained with turpentine and is separated by fractional distillation. The chief constituent is terpineol. It has a higher boiling range (200° to 230°C, 392° to 446°F) than turpentine and a very much lower rate of evaporation. The specific gravity ranges from 0·932 to 0·945 and the refractive index ($D_{20°C}$) from 1·475 to 1·483.

Pine oil has an agreeable odour and is used as a re-odorant in paints and disinfectants. It also possesses some germicidal properties.

Alcohols

Methyl alcohol (methanol), CH_3OH

Large quantities of methyl alcohol are produced by reacting carbon monoxide and hydrogen at high pressures and in presence of a catalyst. The reaction temperature is about 300°C (572°F).

$$CO + 2H_2 \rightarrow CH_3OH$$

The mixture of carbon monoxide and hydrogen is obtained from methane present in petroleum gases. The methane is reacted with oxygen at high temperature (about 1400°C, 2552°F):

$$CH_4 + \tfrac{1}{2}O_2 \rightarrow CO + 2H_2$$

An alternative method utilizes a high-temperature reaction between carbon dioxide and hydrogen:

$$CO_2 + 3H_2 \rightarrow CH_3OH + H_2O$$

In both these processes the methyl alcohol is condensed out of the system.

Methanol is readily oxidized to formaldehyde and this constitutes one of its principal uses. It has a limited use in varnishes and stains. The physical properties are:

Specific gravity (20°C)	0·792
Boiling range	95% between 64° and 67°C (147° and 153°F)
Flash point	6°C (42°F)

Methyl alcohol has pronounced toxic properties and is used as a denaturant for the ethyl alcohol in methylated spirit. It is a cumulative poison and can lead to giddiness, unconsciousness and in extreme cases to impaired sight or permanent blindness.

Ethyl alcohol (ethanol), C_2H_5OH

Two general methods are used for the production of ethyl alcohol, the choice being determined by economic considerations.

Where there is an ample supply of fermentable material such as potato starch or molasses the fermentation process is employed. Enzymes contained in malt convert the starch into sugars, and these are converted to ethyl alcohol and carbon dioxide by the yeast enzymes. The alcohol is removed from the final liquor by distillation.

The second general method uses ethylene as the raw material and is widely used in the USA. There are two methods by which ethylene can be converted to ethyl alcohol.

In the older method ethylene is dissolved in 95 percent sulphuric acid under pressure and forms a mixture of mono- and diethyl sulphates. On hydrolysis, ethyl alcohol is produced and is separated by distillation.

The more modern method is to pass a mixture of ethylene and steam over a catalyst at 300°C(572°F) and at high pressure. The reaction is a reversible one:

$$CH_2 = CH_2 + H_2O \leftrightharpoons C_2H_5OH$$

and the alcohol is condensed out of the system.

Properties and uses Pure anhydrous ethyl alcohol has the following characteristics:

Specific gravity (20°C)	0·7937
Refractive index	1·3619
Boiling point	78·3°C(172·9°F)
Flash point	14°C (57°F)

Anhydrous alcohol is not usually encountered outside the research laboratory. The material used as a solvent is known as Industrial Methylated Spirit (IMS) and consists of ethyl alcohol (95 percent by volume) denatured with 5 percent methyl alcohol. Grades of IMS are available ranging from 61° overproof (91·84 percent by volume) to 74° overproof (99·24 percent by volume).

The sale and use of Industrial Methylated Spirit is very tightly controlled by the Customs and Excise Authorities.

Ethyl alcohol forms azeotropic (constant boiling point) mixtures with a number of other solvents and with water. The mixture with water boils at 78·15°C(172·67°F) and contains 95·57 percent ethyl alcohol and 4·43 percent water.

Industrial Methylated Spirit dissolves a number of natural resins including shellac, with which it forms several spirit varnishes. It also dissolves some synthetic resins such as polyvinyl acetate and cyclohexanone resin. It is completely miscible with water, hydrocarbons and castor oil.

n-propyl alcohol (n-propanol), $CH_3CH_2CH_2OH$

This is obtained as a by-product in the American Hydrocol process for production of hydrocarbons by the interaction of carbon monoxide and hydrogen in presence of a catalyst. It can also be obtained by the hydration of propylene.

The physical constants of the pure alcohol are:

Specific gravity (20°C)	0·8052
Refractive index	1·3860
Boiling point	97·2°C (207°F)
Flash point	22°C (71·6°F)

It forms azeotropic (constant boiling point) mixtures with a number of other solvents and with water. The latter mixture boils at 87·7°C(189·9°F) and contains 71·7 percent *n*-propyl alcohol and 28·3 percent water.

n-propanol dissolves rosin, shellac, ester gum and castor oil. It is sometimes used as a substitute for ethyl alcohol.

Iso-propyl alcohol (isopropanol), $(CH_3)_2CHOH$

Isopropanol is prepared by the hydration of propylene, $CH_3CH = CH_2$. The latter is dissolved in strong sulphuric acid under pressure to give a mixture of isopropyl sulphates. The mixture is then hydrolysed to give isopropyl alcohol. The overall reaction can be represented by the equation

$$\begin{matrix} CH_3 \\ \ \ \searrow \\ \ \ \ \ CH \\ /\!\!/ \\ CH_2 \end{matrix} + \begin{matrix} OH \\ | \\ H \end{matrix} \longrightarrow \begin{matrix} CH_3 \\ \ \ \searrow \\ \ \ \ \ CHOH \\ \ \ \nearrow \\ CH_3 \end{matrix}$$

The crude alcohol obtained in this way is purified by azeotropic distillation.

Pure isopropyl alcohol has the following physical constants:

Specific gravity (20°C)	0·7925
	(commercial isopropanol 0·787)
Refractive index	1·3776
Boiling point	82·4°C (180·3°F)
Flash point	12°C (53°F)

Isopropanol has a somewhat stronger odour than ethyl alcohol and, like the latter, is miscible with water in all proportions. It is very similar to ethyl alcohol in solvent properties and has been used extensively as a substitute for ethanol. It is not a solvent for cellulose esters but possesses latent solvent properties, in that small additions increase the solvent power of the true solvents. In admixture with small quantities of esters it dissolves cellulose nitrate.

n-butyl alcohol (n-butanol), $CH_3(CH_2)_2)CH_2OH$

The older process of manufacture is dependent upon the fermentation of maize starch and molasses by a special enzyme. Some acetone and other alcohols are produced at the same time. Distillation then yields a product containing about 98 percent *n*-butyl alcohol, the balance being small amounts of acetone and other (chiefly ethyl) alcohol.

There are two modern processes for the production of *n*-butanol. One of these is based on the 'Oxo' process [13] in which carbon monoxide and hydrogen are added to olefines in the presence of a cobalt catalyst to give aldehydes. Under these conditions propylene yields *n*-butyraldehyde together with some isobutyraldehyde:

$$CH_3CH = CH_2 + CO + H_2 \rightarrow CH_3CH_2CH_2CHO$$
$$(+ \text{ some } CH_3CH(CH_3)CHO)$$

High-pressure hydrogenation of the aldehydes, using a cobalt catalyst, then yields the alcohol:

$$CH_3CH_2CH_2CHO + H_2 \rightarrow CH_3CH_2CH_2CH_2OH$$
$$(+ \text{ some } CH_3CH(CH_3)CH_2OH)$$

The *n*-butanol is then purified by distillation.

The alcohol can also be obtained by hydrogenation of crotonaldehyde, which is itself prepared from acetaldehyde. The reaction is carried out at 180°C (356°F) using a nickel-chromium catalyst:

$$CH_3CH = CH.CHO + 2H_2 \rightarrow CH_3CH_2CH_2CH_2OH$$

n-butanol possesses the following physical constants:

Specific gravity (20°C)	0·811
Boiling point	118°C (244·4°F)
Refractive index	1·3992
Flash point	37°C (98°F)

It dissolves a number of natural resins and the following synthetics: ureaformaldehyde, melamine-formaldehyde, polyvinyl acetate. It is miscible with hydrocarbon solvents and linseed oil.

n-butanol is used extensively in conjunction with xylol in stoving finishes, where it appears to act as a very efficient latent solvent. It is also used as a damping agent for cellulose nitrate and is used in high-flash and brushing lacquers where it also increases the blush resistance.

Secondary butyl alcohol (methylethylcarbinol), $CH_3CHOH.C_2H_5$

This is prepared from butylene, $C_2H_5CH = CH_2$ by a method similar to that used for isopropyl alcohol (*q.v.*). It has the following physical constants:

Specific gravity (20°C)	0·806
Refractive index	1·397
Boiling point	99·0°C (210°F)
Flash point	21°C (70°F)

It is extensively used as a medium boiling-point solvent in lacquers and is a latent solvent for cellulose nitrate.

Diacetone alcohol, $CH_3COCH_2C(OH)(CH_3)_2$

A colourless, odourless liquid with the following physical properties:

Specific gravity (20°C)	0·942
Refractive index	1·424
Boiling point	167°C (333°F)
Flash point	67°C (153°F)

Diacetone alcohol tends to decompose slowly on heating or long storage, especially in the presence of alkali. The main product is acetone. It is a good solvent for cellulose nitrate and is a useful high-boiling solvent for cellulose acetate lacquers.

Esters

Esters are produced by the general reaction between an alcohol and an acid according to the equation

$$R.COOH + R'.OH \rightleftharpoons R.COOR' + H_2O$$

The reaction is a reversible one and ultimately a state of equilibrium is reached. If however the water is removed by the use of a small quantity of concentrated sulphuric acid or other powerful dehydrating agent, the esterification will proceed to completion.

Some esters can be synthesized from unsaturated hydrocarbons in petroleum gases; for example, ethyl acetate can be synthesized according to the following scheme:

$$CH_2{=}CH_2 + H_2O \rightarrow CH_3CH_2OH \rightarrow CH_3CHO + H_2$$

ethylene — ethyl alcohol — acetaldehyde

acetaldehyde (2 mols, Tischenko reaction) $\downarrow$ $CH_3COOC_2H_5$

ethylene —acetic acid + catalyst, 150°–170°C (302°–338°F)$\rightarrow$ $CH_3COOC_2H_5$ ethyl acetate

The esters prepared from the lower alcohols are volatile liquids, some of which are used in cellulose ester lacquers. They have high solvent power and the different esters are generally similar in properties, the main difference being in the distillation range and rate of evaporation. For a given alcohol, the boiling point of the ester increases with increasing size and complexity of the molecule of acid used. The same general rule holds for a given acid with increase in the alcohol molecule.

The lower esters possess characteristic fruity smells which in some cases can be very pungent. Their main use is in cellulose ester lacquer, and the following esters are among the most widely used.

Ethyl acetate, $CH_3COOC_2H_5$

A low boiling solvent of high solvent power, which is now being replaced gradually by methylethyl ketone which is appreciably cheaper. The British Standard Specification [14] for ethyl acetate specifies specific gravity (20°C, 68°F) 0·901 to 0·904. Boiling range—a minimum of 95 percent between 76° and 78°C (169° and 172°F).

n-butyl acetate, $CH_3COOC_4H_9$

This is the most widely used medium-boiling solvent for cellulose nitrate lacquers. It is sufficiently volatile to leave the film readily but does not cause chilling of the film with consequent blushing. It is an excellent solvent for many other resins and oils but is not used to any degree outside cellulose nitrate lacquers. In these it is gradually being replaced by methyl isobutyl ketone which is cheaper and possesses greater solvent power. The specific gravity at 20°C (68°F) is 0·879 to 0·882 and the boiling range from 124° to 128°C (255° to 262°F).

Oxo-acetates

These are hexyl and heptyl acetates marketed under the trade names Exxate 600 and Exxate 700 respectively [15]. They are powerful high boiling solvents suited particularly to use in high solids paints [16].

Ethyl lactate, $CH_3.CH(OH)COOC_2H_5$

This is a very efficient high-boiling solvent for cellulose ester lacquers. It has high solvent power for other resins and will tolerate the addition of considerable amounts of diluent. It imparts good flow to the lacquer and high gloss to the film, but the slow evaporation rate causes the film to remain soft for some time.

Ethyl lactate is also used in small quantities in stoving lacquers where it assists *secondary flow* in the stove and eliminates any film imperfections. The specific gravity at 20°C (68°F) is 1·032 to 1·038; boiling range, 95 percent between 135° and 160°C (275° and 320°F).

Ethers

Ethers are produced by the elimination of a molecule of water from two molecules of alcohol by the use of a powerful dehydrating agent such as concentrated sulphuric acid. The two alcohol molecules may be identical or different.

$$R.OH + HO.R' \rightarrow R\text{—}O\text{—}R' + H_2O$$

The lowest member of the group is dimethyl ether, $CH_3.O.CH_3$ which is a gas. *Diethyl ether*, $C_2H_5.O.C_2H_5$ (ordinary ether) is a very volatile liquid with a boiling point of 34·5°C (94·1°F), and is highly flammable. It is used in laboratories as an extraction solvent.

As the size of the alkyl group or groups increases, the ethers become less volatile. They are generally inert (apart from flammability) toward the majority of chemicals and are useful solvents for oils and a number of natural resins.

Di-isopropyl ether, $(CH_3)_2CH.O.CH(CH_3)_2$

Used industrially for low-temperature and other extractions. It is less volatile than diethyl ether and is a solvent for rosin and ester gum, oils, hydrocarbons and rubber. It is not a solvent for synthetic resins.

Di-isopropyl ether is a dangerous material as it forms very unstable peroxides on exposure to the air and these explode on heating. Their formation can be minimized by storage in the dark and addition of reducing agents such as hydroquinone or catechol or by storage over 20 percent sodium hydroxide solution.

The formation of unstable peroxides on exposure to light and air is a property of most of the lower ethers.

Glycol ethers and ether-esters

These derivatives of ethylene glycol have been found to be more toxic than was previously thought and their use in coatings has been discouraged. Propylene glycol derivatives, which show considerably lower toxicity, are being used as replacements. These are excellent solvents for a wide range of resins, are compatible with aliphatic and aromatic hydrocarbons and the majority are miscible with water. In addition to their use in solvent type paints they are used as coalescing solvents in emulsion paints and as co-solvents in water-soluble paints.

The physical characteristics of the most widely used members are set out below [17].

Table 8.1

Solvent	Specific gravity	Boiling point °C at 760 mm	Flash point °C*	Evaporation rate (Butyl acetate = 100)
Methoxy propanol	0·920	120	35	70
Ethoxy propanol	0·896	132	42	46
Methoxy propyl acetate	0·970	146	50	31
Ethoxy propyl acetate	0·946	158	54	19
isoButoxy propanol	0·873	161	60	14
isoButoxypropoxy propanol	0·906	220	110	3
Methoxypropoxy propanol	0·951	187	79	3
Dipropylene glycol monomethyl ether	0·945	176	73	—

*Pensky Martens closed cup

Ketones

Ketones are a class of compound characterized by the presence of the carbonyl group $\diagdown C=O$ / . The carbon can be attached to two alkyl groups which are the same, as in acetone, $CH_3CO.CH_3$, or different as in methyl ethyl ketone, $CH_3COC_2H_5$. Alternatively the carbonyl group can form part of a ring, such as cyclohexanone (*see below*). Other reactive groups, such as hydroxyl may be present in the molecule as in diacetone alcohol,

$$CH_3COCH_2\text{—}C(OH)(CH_3)_2$$

The lower members of the class are very volatile and highly flammable liquids and are soluble in water. With increasing size of substituent groups the members lose the water-solubility and become less volatile and less flammable. The highest members are solids.

The following ketones are among the most important paint and lacquer solvents.

Acetone (dimethyl ketone), CH_3COCH_3

This is prepared from isopropyl alcohol by passage over a copper catalyst at 500°C(932°F) and at three atmospheres pressure:

$$(CH_3)_2.CHOH \rightarrow CH_3COCH_3 + H_2$$

Acetone is an extremely volatile liquid with a flash point below 0°C. It therefore flashes at normal temperatures and very strict precautions are necessary in its use.

It is one of the most powerful organic solvents and is used as a low-boiling solvent in some cellulose acetate and nitrate lacquers. The great solvent power renders it an attractive ingredient for paint strippers.

Acetone possesses the following characteristics:

Specific gravity (20°C)	0·790
Refractive index ($D_{20°C}$)	1·3599
Boiling point	56·2°C (133°F)
Flash point	– 16·5°C (2°F)

Methyl ethyl ketone, $CH_3COC_2H_5$

Known commercially as MEK, this is prepared by the high-temperature dehydrogenation of secondary butyl alcohol:

$$CH_3.CHOH.C_2H_5 \rightarrow CH_3CO.C_2H_5 + H_2$$

This solvent is similar to acetone in basic reactions but differs in the higher flash point and lower rate of evaporation. The physical characteristics are:

Specific gravity (20°C)	0·805
Refractive index ($D_{20°C}$)	1·379
Boiling point	79·6°C
Flash point	– 7°C(19°F)

MEK is used widely as a resin solvent.

Methyl iso-butyl ketone, $CH_3.CO.CH_2.CH.(CH_3)_2$

Also known as MIBK. It is prepared by the hydrogenation of mesityloxide, an unsaturated ketone prepared by dehydrating diacetone alcohol. The two stages can be represented by the equations:

$$\underset{\text{diacetone alcohol}}{CH_3.CO.CH_2\underset{\mid}{C}(CH_3)_2 \atop \quad\quad\quad OH} \rightarrow \underset{\text{mesityl oxide}}{CH_3CO.CH{=}C(CH_3)_2} + H_2O$$

$$\xrightarrow{+H_2} CH_3CO.CH_2CH(CH_3)_2$$

MIBK possesses the following characteristics:

Specific gravity (20°C)	0·802
Refractive index ($D_{20°C}$)	1·396
Boiling range	95% between 114° and 117°C (237° and 243°F)
Flash point (Abel)	10°C (50°F)

It is a solvent for a very wide range of resins and is extensively used both in stoving enamels and lacquers where it functions as a medium boiling point solvent.

Cyclohexanone

This is a ring compound with the structure

$$CH_2\left\langle\begin{matrix} CH_2-CH_2 \\ CH_2-CH_2 \end{matrix}\right\rangle CO$$

and can be produced by the dehydrogenation of cyclohexanol.

It has the following physical characteristics:

Specific gravity (20°C)	0·950
Refractive index	1·443 – 1·451
Boiling range	95% between 150° and 158°C (302° and 316°F)
Flash point	47°C (117°F)

It is an excellent solvent for a number of resins and is used in cellulose nitrate lacquers, in which it tolerates considerable additions of diluents such as toluene.

Chlorinated hydrocarbons

The hydrogen atoms in both the saturated and unsaturated hydrocarbons can be replaced by chlorine, and many of these substitution products are useful solvents.

As a class they possess the following characteristics.

1 High specific gravity.
2 A low degree of flammability. The highly chlorinated members are completely non-flammable and whilst the lower chlorinated members can be ignited, the heat of combustion is so low that the flame is extinguished.
3 Characteristic odours.
4 Anaesthetic and often pronounced toxic properties.
5 A low heat of vaporization, indicating high volatility in relation to the boiling point.
6 High solvent power for a wide range of organic resins and coatings.

It is unfortunate that the desirable properties of non- or low-flammability and high solvent power are accompanied by toxic properties. The range of application is therefore severely limited, and whenever they are used special precautions must be taken to protect operators from the toxic effects.

The following members of the class are encountered in the surface coatings industry.

Methylene chloride, CH_2Cl_2

A highly volatile liquid with boiling point 41°C (106°F) and specific gravity 1·336. It is a very powerful solvent and is the major component of many non-flammable paint strippers.

Ethylene dichloride, $ClCH_2.CH_2Cl$(1:2-*dichloroethane*)

Boils at 84°C (183°F) and has a specific gravity of 1·44. Like methylene chloride it is a powerful solvent and is used in paint strippers.

Trichloroethylene, $Cl_2C = CHCl$

Boils at 87°C (189°F) and has a specific gravity of 1·48. It is a powerful solvent and evaporates very rapidly. Among its many uses are extraction of oils from crushed seeds, dry-cleaning of fabrics, and recently it has been used successfully as the sole solvent in a dipping paint.

Perchlorethylene, $Cl_2C = C.Cl_2$

This is very similar to trichlorethylene in properties except for the evaporation rate which is lower.

PLASTICIZERS

Plasticizers are used in lacquers or non-convertible coatings which dry by solvent evaporation only, and in some types of stoving finish. Their function is to improve flexibility and extensibility of the film without impairing other properties. They should, ideally, be non-volatile, completely compatible with other film components, and chemically stable. The important classes of plasticizers are the following.

Esters

These comprise the largest class. The boiling points of esters increase with increasing molecular weight until the volatility of the material decreases to the point where it can be used as a plasticizer. Esters with boiling points above 250°C (482°F) have, generally, a very low vapour pressure at normal temperatures.

Table 8.2 shows a few of the many ester plasticizers in common use, with their boiling points and the materials for which they are satisfactory plasticizers.

Table 8.2

Plasticizer	Boiling point (at 760 mm) (101·325 kN/m^2)	Plasticizer for:
Triacetin	280°C(536°F)	cellulose acetate cellulose nitrate
Dimethyl phthalate	284°C(543°F)	cellulose acetate
Dibutyl phthalate	340°C(644°F)	cellulose nitrate ethyl cellulose vinyls
Dioctyl phthalate	230°C(445°F) at 5 mmHg (666·61 Nm^2)	cellulose nitrate, vinyls
Tributyl phosphate	290°C(554°F)	cellulose nitrate
Tritolyl(tricresyl) phosphate	420°C(788°F)	vinyls, cellulose nitrate

Table 8.2 *cont'd*

Plasticizer	Boiling point (at 760 mm) (101·325 kN/m²)	Plasticizer for:
Butyl stearate	365°C(689°F)	cellulose nitrate
Dibutyl sebacate	345°C(653°F)	cellulose nitrate
Tributyl citrate	225°C(437°F) at 5 mmHg (666.61 Nm²)	cellulose nitrate
Butyl oleate	203°C(397°F) at 10 mmHg (1333·22 Nm²)	chlorinated rubber ethyl cellulose

The properties of a range of interesting trialkyl phosphate plasticizers have been described by Quinn [18].

Camphor

Camphor was the first plasticizer to be incorporated into cellulose nitrate, the product being known as 'celluloid'. Chemically it is a ring ketone of the terpene family and is available in both natural and synthesised forms. It is a solid with specific gravity 0·992, melting point 178°C and an appreciable vapour pressure. As a result of the latter there is a steady loss from the plasticized film resulting in progressive embrittlement. For this reason camphor is now used to a limited extent.

Castor oil

First pressings castor oil.
This is used in cellulose nitrate lacquers where permanent flexibility is required, as in leather finishes. It is a non-drying oil, completely non-volatile, and remains permanently in the film. More detail is given in Chapter 9.

Blown castor oil
This is a higher-viscosity material produced by blowing air through the raw oil, and is used as a plasticizer in cellulose undercoats and primers.

Chlorinated waxes

A range of liquid plasticizers derived from hydrocarbon waxes is marketed under the trade name 'Cereclor' (ICI Ltd). They are completely non-volatile and are used as plasticizers for chlorinated rubber, vinyls, and cellulose nitrate.

References to Chapter 8

[1] Scatchard, G., *Chem. Rev.*, 1931, **8**, 321.
[2] Hildebrand, J. H. and Scott, R. L., *The Solubility of Non-electrolytes*, 3rd edn Reinhold, 1950.
[3] Burrell, H., *Off. Dig.*, 1955, **27**, 726.
[4] *Idem*, The challenge of the solubility concept. *J. Paint Tech.*, 1968, **40**, 197.

[5] Hansen, C. J., The three-dimensional solubility parameter. Key to paint component affinities. *J. Paint Tech.*, 1967, **39**, 505.
[6] Crowley, J. D., Teague, G. S and Low, J. W., A three dimensional approach to solubility. Part 1, *J. Paint Tech.*, 1966, **38**, 269; Part 2, *ibid*, 1967, **39**, 19.
[7] Teas, J. P., Graphic analysis of resin solubilities, *J. Paint Tech.*, 1968, **40**, 19.
[8] Elven, Rudd and Tysall, *JOCCA*, 1950, **33**, 520. Loible and Stolton, FATIPEC Congress, 1962, p. 228.
[9] BS 4591 : 1971. Method for determination of distillation characteristics (distillation range and distillation yield). British Standards Institution.
[10] BS 2511 : 1970 (1986). Method for the determination of water (Karl Fischer method). British Standards Institution.
[11] BS 245 : 1976(1987). Specification for mineral solvents (white spirit and related hydrocarbon solvents) for paints and other purposes. British Standards Institution.
[12] 'Aromasol' (ICI); 'Shellsol' (Shell Chemicals); 'Solvesso' (Esso Petroleum).
[13] Waddams, A. L., *Chemicals from Petroleum*, John Murray, London.
[14] BS 553 ; 1965. Ethyl acetate. British Standards Institution.
[15] Exxate solvents. Esso Chemicals Ltd, Southampton.
[16] Taylor, M. A. and de Greef, J. P., Oxo-acetates - new solvents for high solid paint formulations. *JOCCA*, 1985, **68**, 193.
[17] BP Chemicals Ltd. Samuel Banner Ltd.
[18] Quinn, J. A. W., Some newly available trialkyl phosphates. *POCJ*, 1970, p. 949.

9

Drying oils, driers and drying

The oils used in the paint industry are derived mainly from vegetable and, to a much lesser extent, from animal sources. They are esters of glycerol and fatty acids, non-volatile, and unstable at high temperatures.

These oils vary in properties according to the nature of the fatty acids combined with the glycerol, i.e. they may be saturated or unsaturated. In the latter case, the oil possesses the valuable property of setting slowly to a solid and adherent film when spread on a surface and exposed to the air. This film formation is an irreversible process and the film is insoluble in white spirit. This process is known as 'drying' and the oils can be classified into the following groups based on their drying properties:

Drying oils, e.g. linseed
Semi-drying oils, e.g. soya bean, tobacco seed
Non-drying, e.g. castor.

CONSTITUTION OF DRYING OILS

The esters produced by combination of glycerol with fatty acids are known as glycerides. A number of fatty acids are involved, so that the number of possible glycerides is large. Most oils consist of glycerides of several acids in varying proportions.

Glycerol (the pure material is known as glycerine) is a trihydric alcohol derived from propane. It is a colourless, viscous liquid, very hygroscopic and possesses the following constants: boiling point, 290°C (554°F); specific gravity (15°C) 1·265; refractive index 1·473.

As a trihydric alcohol, glycerol is capable of combining with one, two or three acid radicals to form mono-, di- and triglycerides. The constitution of these can be represented as follows, where R is a fatty acid chain:

$$\begin{array}{cccc} CH_2OH & CH_2OCOR & CH_2OCOR & CH_2OCOR \\ | & | & | & | \\ CHOH & CHOH & CHOH & CHOCOR'' \\ | & | & | & | \\ CH_2OH & CH_2OH & CH_2OCOR' & CH_2OCOR' \\ \text{glycerol} & \text{monoglyceride} & \text{diglyceride} & \text{triglyceride} \end{array}$$

Vegetable and animal oils consist mainly of the triglycerides with only small proportions of the mono- and diglycerides.

Fatty acids are a large group of compounds consisting of long hydrocarbon chains of about 16 carbon atoms and upwards, attached to a carboxyl (– COOH) group. They are usually straight chain compounds and are of two types (i) *saturated*, in which the four valencies of each of the carbon atoms are fully satisfied, and (ii) *unsaturated*, which contain one, two or three ethylenic linkages or double bonds. The two types can be represented thus:

Saturated acid – no double bonds
$CH_3CH_2CH_2(CH_2)_{14}CH_2.COOH$
Unsaturated acid
$CH_3CH_2CH = CH.CH_2CH = CH.CH_2CH = CH(CH_2)_7.COOH$

In unsaturated acids containing more than one double bond, the latter may be isolated from each other as in the example preceding, or they may be grouped as closely as possible together and separated by only one C—C linkage. In this case they are said to be 'conjugated'. The following is an acid of this type:

$$CH_3(CH_2)_3.CH = CH.CH = CH.CH = CH.(CH_2)_7.COOH$$

Conjugated unsaturated acid (elaeostearic)

Natural glycerides contain both saturated and unsaturated fatty acids, but it is to the latter that the drying properties are due. The types of fatty acids present in the common drying oils are shown in Table 9.1.

Furthermore, the reactivity and drying properties of conjugated unsaturated acid glycerides is much greater than that of compounds containing isolated double bonds. This accounts for the reactivity of tung oil which contains the conjugated unsaturated elaeostearic acid depicted above.

Drying oils of the linseed type consist of glycerides of fatty acids containing two or three isolated double bonds, whilst semi-drying oils contain acids with only one or two double bonds. This class includes soya bean and tobacco seed oils. Non-drying oils, e.g. castor, consist of glycerides from saturated fatty acids which have no drying properties, or may contain small amounts of acids with one double bond.

The natural vegetable oils rarely consist of pure glycerides, but each glycerol molecule is combined with two or three different acids which may be saturated or unsaturated. The possible combinations are large, and as the esters are not particularly stable it is possible that the changes in properties which occur during storage or treatment are due to a rearrangement of acid radicals between the various glycerides.

Again, partial decomposition of glycerides during ripening or processing can lead to the liberation of free fatty acid. The presence and amount of this acid can influence the behaviour of the oil.

Table 9.1 Fatty acids

Acid	Molecular formula	No. of double bonds	Occurrence
Myristic	$C_{14}H_{28}O_2$	None	
Palmitic	$C_{16}H_{32}O_2$	None	Non-drying oils. Small quantities occur in a number of drying and semi-drying oils.
Stearic	$C_{18}H_{36}O_2$	None	
Arachidic	$C_{20}H_{40}O_2$	None	
Oleic	$C_{18}H_{34}O_2$	1	Majority of vegetable oils
Erucic	$C_{18}H_{34}O_2$	1	Rape oil
Ricinoleic	$C_{18}H_{34}O_3$	1	Castor oil
Linoleic	$C_{18}H_{32}O_2$	2	Most vegetable oils. High percentage in poppy seed, safflower, soya bean, sunflower and tobacco seed.
Linoleic	$C_{18}H_{30}O_2$	3	Linseed, perilla and stillingia oil
Elaeostearic	$C_{18}H_{30}O_2$	3	Tung oil
Licanic	$C_{18}H_{28}O_3$	3	Oiticica oil
Clupanodonic	$C_{22}H_{34}O_2$	4	Small quantity in fish oils

The constitutions of the most commonly encountered unsaturated fatty acids are as follows:

Oleic $CH_3(CH_2)_7CH = CH.(CH_2)_7.COOH$
Linoleic $CH_3(CH_2)_4CH = CH.CH_2.CH = CH.(CH_2)_7.COOH$
Linolenic $CH_3CH_2CH = CH.CH_2.CH = CH.CH_2.CH = CH.(CH_2)_7.COOH$
Elaeostearic $CH_3(CH_2)_3.(CH = CH)_3.(CH_2)_7.COOH$
Licanic $CH_3(CH_2)_3.(CH = CH)_3(CH_2)_4.CO.(CH_2)_2.COOH$

CHEMICAL EXAMINATION OF DRYING OILS

Acid value

The amount of free organic acid in an oil is important in many applications and is expressed as the number of miligrams of potassium hydroxide required to neutralize the acids in one gram of the oil. It is determined by dissolving a known weight of oil in a neutral alcohol/benzene solution and titrating with 0·1 M sodium or potassium hydroxide, using phenolphthalein as indicator. The method is described in detail in BS 6900[1].

Saponification value

When a glyceride is treated with sodium or potassium hydroxide it is saponified or decomposed into the alkali salts of the fatty acids and glycerol. The saponification value is defined as the number of milligrams of potassium hydroxide required to saponify one gram of oil. It is determined by boiling a known weight of the oil with an excess of 0·5 M alcoholic potassium hydroxide. The excess alkali is then titrated with 0·5 M hydrochloric acid, a blank on the alcoholic potash solution being run at the same time. The method is described in detail in BS 6900. The saponification value is a useful check when contamination by a mineral oil is suspected. Mineral oil is not saponified.

Iodine value

The double bonds in an unsaturated glyceride react with iodine to form saturated compounds. The weight of iodine (in grams) required per 100 grams of oil is known as the 'iodine value'. It is a direct measure of the degree of unsaturation and therefore of the drying properties of the oil. The method introduced by Wijs is normally employed in this country and is described in detail in BS 6900. Briefly, the method involves dissolving a known weight of the oil in carbon tetrachloride and then adding an excess of iodine solution (in the form of iodine monochloride). The mixture is kept in the dark for one hour to allow complete reaction and the excess of iodine is estimated by titration with standard sodium thiosulphate solution. A blank experiment on the iodine solution is carried out at the same time.

A low iodine value indicates a poor quality oil or adulteration with a non-drying oil. With oils of the linseed type, containing isolated double bonds, the iodine value gives a correct indication of the degree of unsaturation. With conjugated double bonds, however, addition takes place at points ×:

$$\text{—}\underset{\times}{\text{CH}}=\text{CH.CH}=\underset{\times}{\text{CH}}\text{—} + I_2 \rightarrow \text{—CHI.CH}=\text{CH.CHI—}$$

and the iodine value obtained is therefore lower than the true value.

In the United States the Hanus method is employed which uses iodine monobromide.

Insoluble bromide test

When bromine is added to unsaturated fatty acids containing three or more isolated double bonds the resulting bromides are insoluble in ether. The bromides from fatty acids containing one or two double bonds are soluble in ether. Conjugated double bonds also give soluble bromides. The insoluble bromides can be separated, dried and weighed. The result is expressed as a percentage increase in the weight of the oil, and the following figures are indicative of the values obtained: linseed 34 to 36, soya bean 4, wood oil nil. The test can be used to differentiate between vegetable and fish oils.

Maleic or Diene value

This value indicates the content of acids containing a conjugated double bond. It is based on the Diels–Alder reaction in which conjugated double bonds react with maleic anhydride in the following manner:

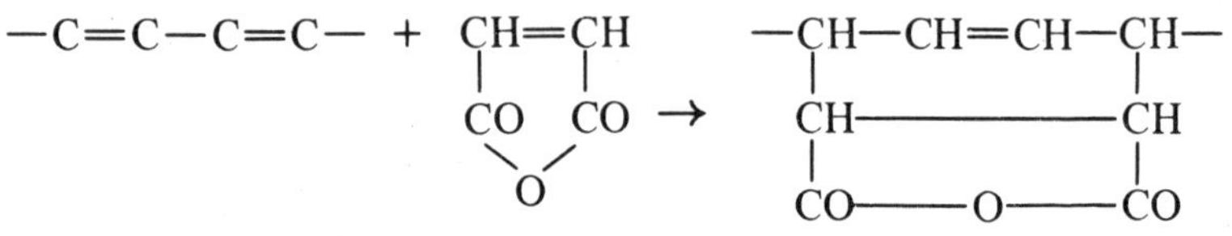

$$
-\mathrm{C{=}C{-}C{=}C}- \;+\; \begin{array}{ccc} \mathrm{CH} & = & \mathrm{CH} \\ | & & | \\ \mathrm{CO} & & \mathrm{CO} \\ & \diagdown\;\diagup & \\ & \mathrm{O} & \end{array} \;\rightarrow\; \begin{array}{ccccc} -\mathrm{CH} & - & \mathrm{CH{=}CH} & - & \mathrm{CH}- \\ | & & & & | \\ \mathrm{CH} & \text{———} & \text{———} & \text{———} & \mathrm{CH} \\ | & & & & | \\ \mathrm{CO} & \text{———} & \mathrm{O} & \text{———} & \mathrm{CO} \end{array}
$$

In the determination, the oil is refluxed with excess maleic anhydride in toluene and the residual maleic anhydride estimated. This is done by conversion to maleic acid by addition of water followed by titration with standard sodium hydroxide solution. A blank determination is done on the maleic anhydride solution, and hence the weight combined with the oil can be calculated.

Acetyl value

If uncombined hydroxyl groups are present, as in castor oil or oils containing mono- or diglycerides, they can be reacted with acetic anhydride or acetic acid to give acetyl derivatives. These acetyl derivatives can be saponified to yield the original compound and acetic acid or its salt. The acetyl value is defined as the number of miligrams of potassium hydroxide required to neutralize the acetic acid produced by saponification of one gram of the acetyl compound.

Hydroxyl value

This is closely related to the acetyl value and is defined as the number of milligrams of potassium hydroxide equivalent to the acetic acid which combined with 1 gram of the hydroxyl-containing oil.

Flash point

The flash points of vegetable oils fall generally within the range 210° to 290°C (400° to 550°F), but the figures for individual oils can vary to some extent with variations in composition, quality and source of the seed, and method of extraction. It can also be affected by the acid value.

The determination of the flash point is carried out in the Pensky–Martens apparatus; figures appreciably lower than those quoted previously are usually an indication of contamination.

PHYSICAL PROPERTIES OF DRYING OILS

Colour

The colour of an oil is of importance as it can influence the colour of a varnish in which it may be incorporated. It can be assessed by comparison with a series of solutions of iodine in potassium iodide as described in BS 6900 [1].

Specific gravity

The specific gravities of the drying oils all fall within a narrow range, generally between 0·925 and 0·933. The bodied oils are a little higher. The specific gravity is not therefore of great value in identification, although it is very useful in formulations involving the oils. It is determined by the usual methods using hydrometer or specific gravity bottle.

Refractive index

The refractive power of an oil is of importance in respect of its relationship to the refractive index of pigments which are dispersed in it. The opacity of a pigment, and therefore of the paint, is a function of the difference between the refractive indices of pigment and oil or other medium.

The refractive index of an oil is best determined on a standard instrument such as the Abbe refractometer in which the angle of total light reflectance from the surface is measured directly. This instrument is described in most physics textbooks.

The range of refractive indices of drying oils is comparatively narrow, from 1·46 to 1·54, but the difference between individual oils is sufficient for use as a test for purity and as an aid to identification.

Viscosity

All liquids flow when a force is applied, some more readily than others. Thin limpid liquids like water or white spirit are said to possess a low viscosity, whilst high viscosity is a characteristic of stand oils. Viscosity can be regarded as resistance to flow and can be defined more precisely as follows.

Suppose two planes, each 1 sq. cm. in area, are immersed in a liquid parallel to each other and 1 cm apart. If the force required to move one plane in a direction parallel to the other at a velocity of 1 cm per second is 1 dyne, then the liquid has a viscosity of 1 poise (after Poiseuille).

On this basis, distilled water has a viscosity of 0·01005 poise or 1·005 centipoise at 20°C. This is used as a standard.

Measurement of viscosity

There are four general methods in use for the measurement of the viscosity of oils by which they can be compared either directly or indirectly with distilled water. These comprise the U-tube (Ostwald), falling-sphere and efflux viscometers, and the bubble tubes. The general principles on which these viscometers operate are as follows.

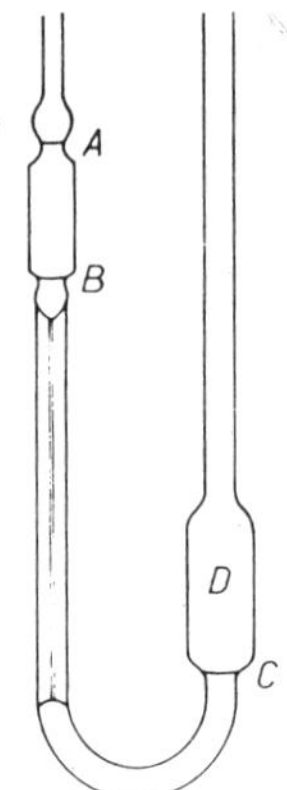

Fig 9.1 U-tube viscometer

U-tube measurement of rate of flow of a definite volume of liquid through a capillary of definite bore. The instrument is illustrated diagramatically in Fig. 9.1. The liquid is introduced by a pipette into the right-hand limb of the instrument and is sucked up into the left-hand limb so that it occupies the section from just above A to just above mark C. The viscometer is suspended in a thermostatically controlled bath and when the correct temperature has been attained, the liquid is allowed to flow back into bulb D, and the time taken for it to fall from A to B is measured.

This viscometer is used for the comparison of the viscosities of two liquids and the results are treated in the following way:

Let η_1 and η_2 be the viscosities of the two liquids
Let ρ_1 and ρ_2 be the densities of the two liquids
Let t_1 and t_2 be the times taken from A to B by the two liquids

$$\text{Then } \frac{\eta_1}{\eta_2} = \frac{(\rho_1 t_1)}{(\rho_2 t_2)}$$

Hence if one of the viscosities is known, e.g. if distilled water is first used in the apparatus, the viscosity of the second liquid can be calculated from this equation.

This method is described in detail in BS 188: 1977 [2].

Falling sphere measurement of the time taken by a metal sphere to fall through a column of liquid of definite length.

The apparatus used consists of a glass tube, 2 to 3 inches in diameter and as long as possible consistent with immersion in a thermostatically controlled bath. Theoretically the tube should be as wide as possible to eliminate wall effects, but in practice volume considerations are important as the liquid must acquire the thermostat temperature in a limited time.

The tube (Fig. 9.2) has two widely separated marks A and B is suspended in a thermostat at 25°C (77°F). The metal sphere C of known radius (about 1 mm) is allowed to fall between the two marks and the time measured.

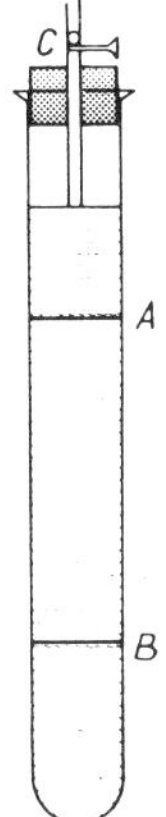

Fig 9.2 Falling sphere viscometer

The principle of the method is illustrated by the following form of the Stokes equation:

$$\eta = \frac{2r^2(D-d)g}{9v}$$

where η = viscosity of the liquid
D = density of sphere
d = density of liquid
g = gravitational constant
v = terminal velocity of sphere.

If S be the distance between the marks and t the time of fall of the sphere, the above equation can be written in the form:

$$\eta = \frac{2}{9}r^2 g\left(\frac{D-d}{S}\right)t$$

If a number of spheres of known radius are used and r^2 is plotted against $1/t$ the result should be a straight line from which η can be obtained. Alternatively, the method can be used for direct comparison of the viscosity of a liquid with that of distilled water. The method is described in detail in BS 188: 1977.

Efflux viscometer [3] This usually consists of a metal cup with parallel sides and with an accurately machined orifice in the centre of the base. The volume is generally 100 cc. The time for the cup to empty through the orifice is noted and the results expressed in seconds. These cups are unsuited to paints which possess any degree of thixotropy, and since a high proportion of paints possess some structure the use of cups has been largely superseded by that of viscometers in which the structure is broken down before the viscosity is measured. Efflux cups continue to be used as control instruments in industrial application of paints which possess near-Newtonian flow properties.

Bubble tubes If a small glass tube, e.g. a test-tube, is almost filled with liquid and a cork inserted leaving an air bubble, on inversion of the tube the bubble will rise at a rate depending on the viscosity of the liquid. The tube can be compared with a graded series of standard viscosity oils sealed in tubes of similar length and diameter. The viscosity of the liquid can then be recorded in terms of the nearest standard tube(s). This method is not highly accurate but is useful for works control.

Sets of standards are produced by the Paint Research Association, Teddington, and numbered 1 to 20. The American Gardner–Holt tubes are based on the same principle and designated by letters A to Z.

Viscosity of oils

The majority of drying and semi-drying oils possess viscosities similar to linseed oil (about 40 centipoises), but tung oil is somewhat higher. The viscosities of stand and blown oils are very much higher, and these are generally designated by viscosity, e.g. 30-poise stand oil.

Viscosity of oil mixtures

When two oils of differing viscosity are mixed in equal proportions, the viscosity of the mixture is given approximately by the relationship:

$$\log\eta = \tfrac{1}{2}(\log\eta_1 + \log\eta_2)$$

where η is the viscosity of the mixture, η_1 and η_2 the viscosities of the individual oils.

When x parts of oil of viscosity η_1 are mixed with y parts of oil of viscosity η_2, then the viscosity of the mixture is given by:

$$(x + y)\log\eta = x\log\eta_1 + y\log\eta_2$$

The accuracy of this expression depends on the difference between η_1 and η_2. If these are widely different the actual viscosity of the mixture is less than that calculated.

Odour

Drying oils possess characteristic odours which are more pronounced in the raw than in the refined oils. Odour is a guide to identification but is of little value in the assessment of purity, unless the adulterant is present in appreciable amount and also possesses a definite odour.

THE DRYING PROCESS

When a film of drying oil is exposed to the air it sets gradually, and ultimately 'dries' to a tough elastic film. Whereas the original oil is soluble in solvents such as white spirit or dipentene, the dried film is not soluble in these solvents and is therefore quite different in nature from the original oil. It is this drying process which enables one coat of paint to be applied over another.

The chemical reactions involved in the drying process are very complex, but a major part is played by atmospheric oxygen and reactions at the double bonds of the unsaturated glycerides. It is thought that after an initial induction period, oxygen is absorbed on the carbon atoms adjacent to the ethylenic groups to form hydroperoxides of the general type $-CH(OOH).CH=CH-$. These hydroperoxides then undergo a series of reactions resulting in the formation of a number of compounds including short-chain carboxylic acids. In addition, free radicals are formed and in their brief existence they induce polymerization by cross-linking between double bonds.

The processes involved in drying are markedly affected by temperature. It is probable that polymerization plays a greater part at high than at low temperatures. In the latter the main process is one of oxidation. The whole drying process takes place much more slowly in the absence of light, so the reactions are, in part at least, photo-catalysed. Drying is also retarded by traces of sulphur dioxide and other impurities in the air [4].

Straight drying oils, such as linseed, dry very slowly, some two to three days being required for raw linseed oil. The process can, however, be accelerated very considerably by the introduction of certain metals in the form of oil-soluble compounds or soaps. The metals used are cobalt, zirconium, manganese and, to a lesser extent, cerium and vanadium. The catalytic activity of these metals is associated with their ability to exert more than one valency. Iron is used as a polymerisation catalyst in some types of high temperature stoving paints.

On account of toxicity lead is no longer used as a drier in decorative paints or in any paints the dry films of which may be contacted by humans, particularly children. It continues to be used in some industrial maintenance paints.

Among the metals investigated as replacements for lead, the most satisfactory was zirconium and mixtures of this metal with cobalt (in the form of salts) are now commonly used in air-drying decorative and industrial coatings and in paints for children's toys.

The action of drier metals does not cease when the film is dry. It continues throughout the life of the film and so contributes to the ultimate embrittlement and breakdown.

The drying effect of the metal compounds appears to be independent of the type of acid radical, provided the solubility in paint media is complete.

Types of drier compounds

Naphthenates 'Naphthenic acid' is the term given to a mixture of acids of high molecular weight obtained from petroleum crudes. The acid value of the material is generally between 170 and 230 and the drier metal salts are supplied to the industry as solutions in petroleum solvents. The solutions contain a specified metal content.

Naphthenate driers can be incorporated into paint or varnish by simple stirring or can be added to varnish during the cooling-down stage. They possess a characteristic and strong smell and this can persist for some time after the paint is applied. For this reason the octoate driers (*see below*) are preferred for decorative work.

Cobalt naphthenate, commonly supplied as a solution containing 6% cobalt, is sometimes used as a sole drier but more often in admixture with zirconium naphthenate. The mixture is marketed under the trade name 'Cozirc 69' (6% cobalt, 9% zirconium) [5].

Cerium naphthenate has been used to some extent as a replacement for lead naphthenate. Like the latter it is used in conjunction with manganese or cobalt, but the action differs somewhat from that of lead in that the film remains soft for a period and then hardens rapidly.

Vanadium naphthenate behaves in a similar manner to cobalt but it has a darkening action on the oil or varnish.

Calcium naphthenate is often added to alkyd gloss finishes in addition to the other driers (cobalt/zirconium). It does not function as a drier but is added for the following reasons. (a) At the pigment dispersion stage it assists in the wetting of many pigments. (b) It reduces the tendency for 'crystalline bloom' to form on the surface of the dry paint film [5].

Octoates (ethylhexoates) are made from the colourless and fairly pure octoic (ethylhexoic) acid. Generally they are superior to the naphthenates in respect of colour, possess little or no smell, and permit the production of low-viscosity solutions of higher metal content than the naphthenates. There is virtually no residual smell after application of paints containing this drier.

EXTRACTION OF OILS

The oil-bearing seed is cleaned and freed as completely as possible from foreign matter and other seeds by screening and winnowing. It is then subjected to the following processes.

Crushing

The seeds are ground to a meal by passage over a vertical three- or five-roll mill. The pressure applied, and therefore the fineness of the meal, is adjusted to the type of seed being processed. If the meal is too fine, subsequent operations are more difficult.

Heating

The meal is then heated to 80° to 90°C (176° to 194°F) with live steam. This serves to disrupt the cells and facilitates removal of the oil, which is separated from the meal by pressing, expelling, or extracting.

Pressing

A measured quantity of the meal is wrapped in filter cloth and placed in an open hydraulic press. The oil is expelled on application of pressure. This type of press has been found satisfactory for linseed, but for meals with a higher oil content, e.g. castor, the cage press is generally used. This consists of a cylinder with perforated walls which is filled with the meal. Application of pressure to the end of the cylinder expels the oil through the perforated walls. This is essentially a batch process.

Expelling

This is carried out in a screw press comprising a worm shaft rotating in a perforated cylinder. The diameter of the cylinder decreases toward the outlet so that the meal is subjected to steadily increasing pressure. The oil is expelled through the perforations and the extracted meal is discharged at the end. Expelling is a continuous process and is the most widely used extraction method. It is well suited to extraction of meal from nuts, for example, in the preparation of tung oil.

Solvent extraction

For this purpose the meal is dried and rolled into flakes. The flakes are then treated with the solvent (petroleum ether or a chlorinated hydrocarbon) in a closed vessel and allowed to stand for some time at a temperature below the boiling point of the solvent. The time allowed is generally about twenty minutes. The solution of oil in solvent is then drained off and the process repeated two or three times.

The counter-current method of extraction is often used. In this, fresh meal meets solvent which has been used in previous extractions, whilst the almost exhausted meal is treated with clean solvent. In this way maximum use is made of solvent and the meal is completely extracted.

The solvent is then distilled off and the oil stored in tanks. The process gives a higher yield than pressing or expelling, but the oil has a higher acid value and is darker in colour. In view of the cost of solvents and solvent losses in the process, an efficient solvent recovery process is essential if solvent extraction is to be economical.

LINSEED OIL [1]

Source

Linseed is the seed of the flax plant which is grown in many parts of the world, but the relative values of the flax and seed depend on climate and local conditions.

Processing

After the seed has been reduced to meal it is heated and the oil is extracted by the impellor method. The yield of oil obtained is about 30 percent of the weight of the seed, but the total oil content of the latter is about 35 percent. The oil so obtained is a thin, brownish-yellow liquid known as raw linseed oil.

Raw linseed oil

The expressed oil has a very characteristic odour, but this, in common with certain other properties, can vary with the source of the seed, the weather conditions during growth, and the degree of maturity when harvested.

The fatty acids present are chiefly linolenic and linoleic with a small quantity of saturated fatty acids. The following figures give the approximate composition of oil from Argentine seed: linolenic 37·5%, linoleic 31·8%, oleic 16·5%, saturated acids 9%, glycerol 4·4%, unsaponifiables 0·8%.

The physical constants of raw linseed oil are set out in Table 9.2. Raw linseed oil is used in the manufacture of glazing putty.

Refined linseed oils

Heat refining

If raw linseed oil is heated slowly, a light gelatinous material separates at between 200° and 250°C (392° and 482°F). This insoluble matter is known as 'break' and consists mainly of phosphatides (see Lecithin, p. 175). Removal of this break is essential if the oil is to be used in any process involving heat treatment. Removal is affected by a filter press designed to collect the filtrate. Oil which has been purified by removal of break in this way is known as heat-refined linseed oil, but the term 'raw' is occasionally – but incorrectly – applied.

Alkali refining

Raw linseed oil generally contains a small amount of free acid (Table 11.1). If the oil is agitated and heated with slightly more sodium hydroxide than is required to neutralize the acids and then allowed to stand, an aqueous layer separates. This contains sodium salts of the fatty acids as well as the break and any water from the oil. It is run off and acidified to give *linseed acid oil.*

The treated linseed oil is washed thoroughly with water and then dried by vacuum or heat. The product is known as alkali-refined or varnish linseed oil (VLO) and the physical constants are given in Table 9.2.

Table 9.2 Characteristics of linseed oils

	Raw	Acid-refined	Alkali-refined	Boiled	Stand oils	Blown oils
Specific gravity (15·5°C)	0·931–0·936	0·930–0·934	0·930–0·934	0·94–0·95	0·95–0·98	0·96–1·0
Viscosity	40 cp.	approx. 40 cp.	approx. 40 cp.	0·8–1·2 poise	3 poise upwards	3 poise upwards
Refractive index (20°C)	1·4800–1·4835	1·4810–1·4825	1·4810–1·4825		1·481–1·490	
Acid value	1–3	7–8	0–0·5	2–4	4–12	4–6
Saponification value	190–195	190–195	190–195			
Iodine value	175–185	175–185	175–185	160–170	130–150	100–140
Unsaponifiable matter	0·7–1·0%			1–2%		
Drying* time	4 days	4 days	4 days	16–24 hours	24–36 hours	24–36 hours

*Without driers

Acid refining

The raw oil is placed in a lead-lined vessel and fairly concentrated sulphuric acid added in a fine stream or spray. About 1 percent on the weight of oil is generally used, and this destroys the break and much of the colouring matter. The acid is then washed out with water and the oil dried by vacuum or heat. The oil is known as acid-refined oil and contains appreciable amounts of free fatty acids (acid value up to 8).

Acid-refined linseed oil is used widely as a grinding medium for pigments, the wetting process being facilitated by the fatty acid groups present. The physical constants are set out in Table 9.2.

Treated linseed oils

The double bonds in unsaturated glycerides will slowly absorb oxygen in the presence of air. The process takes place faster in the presence of drier metals and faster still if the oil is heated in the presence of air or driers. The process is mainly oxidation, but some polymerization takes place at the same time. These processes lead to an increase in viscosity and a decrease in the drying time of the oil.

Boiled oil

Raw (or heat-refined) linseed oil is heated in a vessel through which air is bubbled during the process. Heating is carried out by means of steam coils. Zirconuim and manganese driers are added to the hot oil and the process becomes exothermic. The temperature is then controlled by passage of cold water through the coils. The heating is continued until the specific gravity reaches 0·942 at 15·6°C (60°F). A process of this type in which the temperature reaches a maximum of 137·5°C (279·5°F) produces *dark boiled oil.* If the temperature is kept lower and the manganese drier replaced by cobalt, a *pale boiled oil* results.

The formation of boiled oil is accompanied by an increase in viscosity and specific gravity but a decrease in drying time from 4 days to 12–20 hours. The physical properties are summarized in Table 9.2

Pale boiled oil is frequently added to oleoresinous paints to improve flow and ease of brushing. It can also be added to alkyd-type undercoats for the same purpose.

Blown oil

This is a thickened linseed oil made by blowing air through the heated raw oil in the absence of driers. The oil is heated by steam coils to about 135°C (275°F), air being bubbled through. The reaction is exothermic and is controlled by passing water through the coils. The temperature is then maintained at 125° to 135°C (257° to 275°F) until the desired viscosity is obtained.

Blown oil is a brownish-yellow colour and yellows still further on exposure to light. The physical constants are set out in Table 9.2. Films produced from blown oil possess good initial gloss, but the gloss retention is poor. The higher viscosity grades give good flow and levelling properties to highly pigmented systems and have therefore been used in undercoats and flat wall paints.

The process used for blown oil preparation is the cheapest method of thickening linseed oil and consists mainly of oxidation.

Stand oil [6]

When linseed oil is kept at a high temperature in the absence of driers it slowly thickens and the process can be used to prepare thickened oils of any desired viscosity by cooling at the appropriate moment. The thickening is essentially a polymerization process and the product is known as *stand oil.*

Alkali-refined oil (VLO) is used as the starting material and is heated to 280°C to 300°C (536° to 572°F) until the desired viscosity is reached. This can take from six to eight hours. During the process a number of volatile decomposition products are formed and these can be removed by passing a current of air over the surface of the hot oil (giving *open pot* stand oil) or by passing an inert gas such as nitrogen over the oil. This gives *closed pot* stand oil which is paler than the open pot type and therefore in greater demand.

Stand oils are slow drying. They can be obtained in a wide range of viscosities, from 3 to 200 poise or more. Their basic physical constants are set out in Table 9.2. Stand oils are used as components of varnishes, both 'cooked' and cold-cut, and to improve the levelling properties of undercoats based on alkyd or oleoresinous media.

Litho varnishes are very high-viscosity stand oils of the open pot type. A certain amount of oxidation has taken place in the manufacture, and such oils are admirably suited to lithographic printing.

Enamel oils are mixed stand oils, the best-known consisting of 3 parts linseed and 1 part tung oil. Other ratios, e.g. 9/1 and 5/1, are also used. They are prepared from mixtures of the raw oils in a manner somewhat similar to linseed stand oil, but the temperatures required are a little lower. The enamel oils give films that are harder and more water-resistant and alkali-resistant than the stand oils. They are often used in conjunction with phenolic resins to produce clear and pigmented marine coatings.

TUNG OIL

This oil is also known as 'Chinese wood oil' or simply 'wood oil' and is obtained from two species of the Aleurites family – *A. fordii* found originally in Central China and *A. montana* in southern China and Cambodia. The demand for the oil stimulated work on the propagation of the trees in other parts of the world and today *A. fordii* is being grown successfully in the United States and the Argentine and *A. montana* in Malawi. The former is in the more plentiful supply.

The nuts of these trees yield about 30 percent of a brownish-yellow oil with a characteristic odour. This raw tung oil is more viscous than other raw oils of the linseed type and differs appreciably from these in both composition and properties. The glycerides present in tung oil contain a high percentage of elaeostearic acid, a material containing three conjugated double bonds, and these confer a high degree of reactivity on the oil.

Raw tung oil contains no break, but the trielaeostearin can exist in two isomeric forms, designated α and β. The liquid α-form is the normal state, but under the influence of light it changes to the solid β-isomer. This product is not as easy to handle and is commercially less valuable. The conversion of the raw oil to a stand oil eliminates the tendency to form the β-isomer.

When a film of raw tung oil is exposed to the air it dries to a 'frosted' finish which, under magnification, is seen to be a fine wrinkle. Use is made of this property in the formation of 'wrinkle finishes', but for all other uses the raw tung oil is subjected to heat treatment.

Heat treatment

When tung oil is heated, the increase in viscosity is more rapid than with linseed oil, and if kept at about 280°C (536°F) it forms a stiff gel in a short time. This gelation is irreversible and the gel is insoluble in oils and solvents. If the gelation or polymerization is carried out by immersion of a test-tube (150 mm × 25 mm) containing a sample of the oil in a mineral oil bath at 280°C, the time taken for gelation is an indication of the purity of the tung oil. The end-point is best determined by placing a glass rod in the test-tube and raising it at intervals. When the oil gels, the test-tube and contents can be lifted by the rod. Under these conditions, *Aleurites fordii* oil gels in about 12 minutes but *A. montana* oil takes approximately double this time. Control of the temperature is important as gel times decrease with increasing temperature.

If tung oil is mixed with other oils the gelation time is increased and the gel is soft.

The polymerization reaction is exothermic and in the production of tung oil stand oil rigid control of conditions is therefore essential. If the process gets out of hand, gelation takes place and there is a risk of fire as a result of increase in volume and overflowing of the pot. The oil must be bodied to the correct degree to eliminate the tendency to web or *gas check* when exposed to foul atmospheres from gas or coke flames. A correctly bodied oil is *gas-proof* and will dry to a hard brilliant film.

The tendency to gel on bodying can be checked by the addition of a highly acid resin such as rosin, and there is no risk of gelling when tung oil is bodied in presence of other oils such as linseed. The latter mixture is employed in enamel oils, which are made by heating linseed oil until the desired viscosity increase is obtained, adding the raw tung oil, and then heating at 260°C (500°F) to obtain the required viscosity.

Properties and uses

The physical constants of tung oil are set out in Table 9.3. It is widely used because of its rapid drying and high resistance to water penetration and saponification.

In combination with ester gum or modified phenolic resins it is used in a wide range of air-drying finishes. With 100 percent phenolic resins it provides marine spar varnishes, whilst with coumarone resin it produces a widely used medium for aluminium paints and for primers for alkaline surfaces such as plasters and cement render. In conjunction with phenolic resin it has been used to upgrade the chemical resistance of some alkyds.

DEHYDRATED CASTOR OIL

Raw castor oil is obtained from the seeds of the plant *Ricinus communis* which is widely distributed in tropical and semi-tropical countries. The oil is virtually colourless, non-drying, and has an abnormally high viscosity for a natural vegetable oil. It is soluble in alcohol and is used as a plasticizer and as a lubricant.

The principal acid present – ricinoleic – possesses one double bond and also a hydroxyl group in the molecule. If the oil is heated to about 260° to 280°C (500° to 536°F) in the presence of a catalyst, water is eliminated between the hydroxyl group and the hydrogen of a neigbouring carbon atom. This yields a mixture of two acids each containing two double bonds, but in one case they are conjugated.

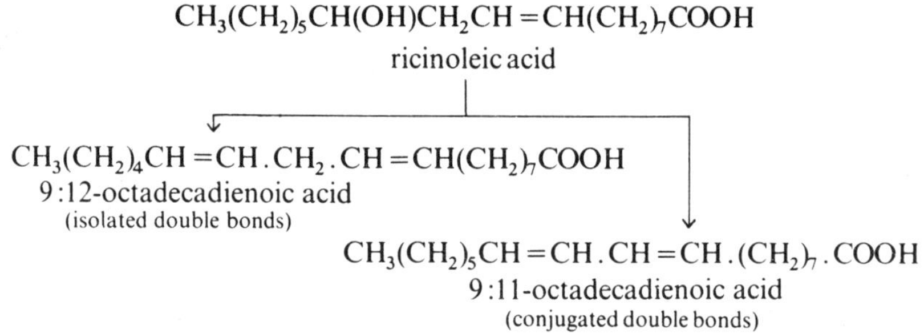

The oil produced in this way is known as dehydrated castor oil (DCO), and by continuing the heat treatment, bodied oils of almost any desired viscosity can be obtained. The reaction is unique in one respect in that the viscosity of the oil actually falls during the first three hours from the original 7 to about $3\frac{1}{2}$ poise and then rises rapidly.

Properties and uses

The physical properties of DCO are set out in Table 9.3. The presence of an acid containing conjugated double bonds results in the oil resembling tung oil in some of its properties. Under certain conditions it will yield a frosted film on drying, and it will gel if heated for a sufficient time. The rates of bodying and gelling are not as great as those of tung oil.

When mixed with driers, DCO dries faster than linseed stand oils, to a tough film which, however, retains a characteristic *after-tack* for some time. It is therefore seldom used as an oil or varnish in air-drying paints, but in the form of varnishes it finds wide use in stoving finishes. It is also incorporated into air-drying and stoving alkyds in which it shows very good non-yellowing properties. In this respect it is far superior to linseed oil.

SEMI-DRYING OILS

Soya bean oil

The soya plant, *Glycine hispida*, is a native of China but is now cultivated on a large scale in the United States, Europe and elsewhere. The ground beans are extracted with solvent and yield 15 to 18 percent of oil, a high proportion of which is used for edible purposes. When intended for paint, the oil is treated to remove the lecithin.

Soya lecithin is a mixture of phosphatides and is marketed in the form of a semi-paste containing 30 to 40 percent of the parent oil. The lecithin is precipitated by addition of about 3 percent water to the warm soya oil, followed by thorough agitation. It is then centrifuged. Lecithin finds wide use in the paint industry as a surface-active agent. It is useful in pigment wetting and the control of flocculation.

Properties and uses of soya bean oil

The constitution and physical properties are set out in Table 9.3. As a semi-drying oil, soya is intrinsically slow drying and the product, after separation of the lecithin, takes about three times as long to dry as does linseed under the same conditions. However, the very small percentage of linolenic acid present results in good colour retention (little yellowing) and the films are far superior to linseed in this respect. The oil shows to greatest advantage when incorporated into alkyds. For this, the fatty acids are first separated and then incorporated into the resin. These resins are frequently described as *linoleic rich.*

Soya modified alkyds are now widely used for non-yellowing white paints and in these the slow-drying characteristics of the fatty acids appear to have been lost. These alkyds are described in more detail in Chapter 10.

Tobacco seed oil

The tobacco plant, *Nicotiana tabacum*, grows in many parts of the world and supplies of oil to this country originate mainly in India. The oil is extracted by pressure or solvent and has a yellowish-brown colour and a characteristic odour. It is bleached before processing for use in the paint industry.

Properties and uses

The constitution and physical properties are set out in Table 9.3. It is a typical semi-drying oil and takes much longer to dry than does linseed under similar conditions. The film remains soft and somewhat tacky but is non-yellowing as a result of the absence of linolenic acid from the triglycerides.

Tobacco seed oil is marketed as bleached alkali-refined oil and as stand oil, both of which are non-yellowing and slow drying. The bulk of the oil, however, is treated to separate the fatty acids and these are then incorporated into alkyds. These alkyds dry satisfactorily and are valued for their non-yellowing properties and durability. They are described in more detail in Chapter 10.

The following semi-drying oils are generally similar to tobacco seed. Their characteristics are set out in Table 9.3.

Table 9.3 Composition and properties of drying oils

	Special acids %	Linolenic acid %	Linoleic acid %	Oleic acid %	Saturated acids %	Iodine value %	Specific gravity	Saponification value	Acid value	Refractive index
Arachis (ground nut)	—	—	13–26	51–71	14–22	85–89	0.916–0.920	188–196	4–10	1.470–1.472
Cashew nut	—	—	7.7	74–77	16.5–18	79–85	0.911–0.918	187–195	1.4–1.6	1.462–1.463
Castor	Ricinoleic 80–85	—	3–4	7–9	10	82–90	0.958–0.969	176–187	1–4	1.477–1.479
Chia	—	41	47	0.7	8.4	200	0.934	192	0.6	1.485
Corn (maize)	—	—	39	43.4	8–10	103–125	0.921–0.928	188–193	4–8	1.467–1.474
Cotton seed	—	—	30–39	33–50	20–25	103–113	0.921–0.932	190–193	0.5–4	1.460–1.475
Hemp	—	24	53	12.6	10	148–155	0.927–0.932	190–193	4	1.477–1.482
Linseed (raw)	—	37–42	31.7–37	9–16	6–9	175–185	0.931–0.936	190–195	1–3	1.480–1.484
Lumbang (candle nut)	—	20.8	39.6	26.2	8.4	155–160	0.920–0.927	190–193	0.5–1	1.477–1.478
Niger seed	—	—	54–57	31–33	10–17	126–134	0.924–0.928	189–198	8	1.467–1.469
Oiticica	Licanic 73–78 Elaeostearic 4–9	—	—	4–6	11–13	179–218	0.966–0.969	188–192	3–15	1.514–1.516
Perilla	—	42.5	32–42	4–10	6–7	193–201	0.932–0.935	190–205	1–6	1.475–1.485
Poppy	—	—	62–65	25–30	7–8	133–169	0.924–0.927	190–198	0.7–3	1.475–1.478
Po-yok	Coupeic 41	—	—	9–10	12	140–157	0.955–0.965	188–192	1–17	1.502–1.516
Rubber seed	—	19.6	31.5	27.3	16	133–143	0.924–0.930	186–195	—	1.466–1.468
Safflower	—	0.7–1	51–67	26–38	6–10	140–150	0.925–0.928	188–194	5–6	1.468–1.469
Soya	—	2–6	50–60	26–30	11–14	125–140	0.923–0.929	188–194	0.3–3	1.468–1.478
Stillingia	—	8–11	52–58	25–26	6–9	160–180	0.936–0.946	204–210	3–10	1.482–1.483
Sunflower	—	—	52–58.5	34–42	6–9	120–135	0.924–0.926	189–194	4–7	1.474–1.479
Tobacco seed	—	—	60–70	16–20	9–10	130–145	0.923–0.925	186–197	1–8	1.474–1.482
Tung	Elaeostearic 72–86	—	—	4–18	4–6	155–170	0.939–0.943	189–197	5–10	1.519–1.522
Walnut	—	10–16	73	17–29	5–9	132–160	0.925–0.927	190–196	—	1.469–1.471

Safflower seed oil The flower is grown in India and Russia and the seeds yield an oil with a high linoleic acid content. The fatty acids are separated and used to produce alkyds with good non-yellowing properties.

Sunflower seed oil is obtained from the seeds of the variety *Helianthus annuus* and is pale yellow. It is mainly used in non-yellowing alkyds.

Poppy seed oil is obtained from the seeds of the opium poppy, *Papaver somniferum*, and is used mainly as a medium for artists' oil colours. The properties of pale colour, slow drying, and good colour retention make it a very suitable medium for artists.

Fish oils

The oils derived from various species of fish differ considerably in their drying properties. Those obtained from fish such as the pilchard and herring dry extremely slowly, but the North American menhaden and sardine give oils with fair drying properties. They all possess very powerful fishy odours.

The poor drying properties result from the presence of appreciable amounts of saturated acids and the fact that the unsaturated acids present dry only slowly. Thus menhaden oil, which is typical of the class, possesses the following composition and properties:

Glycerides of unsaturated acids	72–78%
Glycerides of saturated acids	24–28%
Unsaponifiable matter	1%
Specific gravity	0·929–0·933
Iodine value	150–180
Saponification value	189–193
Acid value	4–7
Refractive index	1·474–1·482

The saturated acids present are those normally found in vegetable oils—myristic, palmitic and stearic, but the unsaturated acids are very different. About 20 percent of oleic acid is present, together with a number of other unsaturated acids, some of which contain 4 to 6 isolated double bonds.

The proportion of unsaturated acids can be increased by a process known as 'segregation'; these acids can be used to replace the vegetable oil fatty acids, in whole or in part, in some types of alkyd and urethane oil.

Tall oil

This is obtained as a by-product in the production of wood pulp by the Kraft process which involves treatment of the coniferous woods with sodium sulphite and caustic soda. In the purification of the wood pulp a residue or 'black liquor' is obtained which contains sodium soaps. This is purified and then acidified to give a dark-coloured crude tall oil. The following figures have been quoted for the composition and properties:

Fatty acids	41–59%
Rosin acids	32–48%
Volatile and unsaponifiable matter	10%

Saponification value	169
Acid value	160

The colour is improved by a decolorizing process or by distillation. The rosin acids can be crystallized out at lower temperatures, and as the product is virtually a mixture of acids it can be converted to salts of drier metals (tallates) or incorporated into alkyds.

MODIFIED DRYING OILS

A number of methods have been suggested for improving the properties of drying oils, and the most interesting of these involve the fatty acids. The methods in most general use are the following.

Segregation The oil is treated with a solvent with which it is not completely miscible, and allowed to stand. Two layers are formed which differ in the ratio of oil to solvent. In addition, a partial separation of saturated from unsaturated glycerides takes place. This separation is reflected in the drying times, one fraction drying more rapidly and the other more slowly than the parent oil. By repeating the process on the fractions a further segregation of glycerides can be effected.

This process has been used to improve the drying properties of soya bean oil by treatment with furfural, whilst liquid propane has been used in the treatment of pilchard oil.

Reaction with unsaturated compounds The unsaturated compounds used are those which react with the double bonds in the unsaturated fatty acid chains. The most widely used compounds are maleic anhydride, cyclopentadiene and styrene.

Compounds with maleic anhydride This material reacts with both conjugated and isolated double bonds in the following manner:

$$\begin{array}{c} | \\ CH \\ \| \\ CH \\ | \\ CH \\ \| \\ CH \\ | \end{array} \quad + \quad \begin{array}{l} CH—CO \\ \| \qquad\quad \rangle O \\ CH—CO \end{array} \quad \rightarrow \quad \begin{array}{l} | \\ CH—CH—CO \\ | \qquad\; | \\ CH \quad\; | \qquad\quad \rangle O \\ \| \qquad\; | \\ CH \quad\; | \\ | \qquad\; | \\ CH—CH—CO \\ | \end{array}$$

Reaction with conjugated double bonds

$$\begin{array}{c} | \\ CH_2 \\ | \\ CH \\ \| \\ CH \\ | \end{array} \quad + \quad \begin{array}{l} CH—CO \\ \| \qquad\quad \rangle O \\ CH—CO \end{array} \quad \rightarrow \quad \begin{array}{l} | \\ CH—CH—CO \\ | \qquad | \qquad\quad \rangle O \\ CH \quad CH_2—CO \\ \| \\ CH \\ | \end{array}$$

Reaction with isolated double bond

The products can be further esterified with glycerol or pentaerythritol to give the so-called maleinized oils. These oils body faster than the parent oils and give films with improved hardness, water resistance and colour.

Compounds with cyclopentadiene Cyclopentadiene reacts with unsaturated drying oils when they are heated together under pressure to produce oils which dry rapidly to a high gloss. A range of cyclopentadienized oils is possible with varying viscosities and drying rates depending on the proportion of cyclopentadiene employed. Cyclopentadiene is assigned the following structural formula.

$$\begin{array}{l} CH{=}CH\diagdown \\ | \qquad\qquad\quad \rangle CH_2 \\ CH{=}CH\diagup \end{array} \quad \text{Cyclopentadiene}$$

Compounds with styrene Styrene or vinyl benzene, $C_6H_5CH = CH_2$, can be heated with unsaturated oils to give interesting products. The properties of these styrenated oils depend on the proportion of styrene used. They dry quickly to hard films but are not widely used in the paint industry. The reaction is used, however, on a considerable scale for the production of styrenated alkyds (Chapter 10).

Mono- and diglycerides of fatty acids These are not normally found in natural drying oils but can be prepared fairly readily from these materials. The oil is heated to 270° to 280°C (518° to 536°F) with glycerol and a catalyst (litharge or alkali) in an inert atmosphere. The product is a mixture consisting mainly of mono- and diglycerides together with a little unchanged triglyceride. Separation can be effected by extraction with methyl or ethyl alcohol.

The monoglycerides are the more important. They are occasionally used as plasticizers for spirit varnishes but their main use is in the manufacture of alkyds. The two free hydroxyl groups are very readily esterified.

Referencсs to Chapter 9

[1] BS 6900. Specification for raw, refined and boiled linseed oil for paint and varnishes. British Standards Institution.
[2] BS 188. Methods for the determination of the viscosity of liquids. British Standards Institution.
[3] BS 3900. Methods of test for paints. Part A6: 1971. Determination of flow time. British Standards Institution.
[4] Holbrow, G. L., Atmospheric pollution; its measurement and some effects on paint. *JOCCA*, 1962, **45**, 701.
[5] Manchem Ltd, Manchester.
[6] BS 4725. Specification for linseed stand oil for paints and varnishes. British Standards Institution.

10

Rubbers, bitumens, pitches, gums and glues

RUBBER DERIVATIVES

Rubber is obtained in the form of an aqueous emulsion, known as 'latex', by tapping trees of the species *Hevea braziliensis* which grows in Malaya, Indonesia and Brazil. The milky liquid is treated with acetic or formic acid to coagulate the rubber, and the resulting spongy mass is rolled and washed to produce white crepe rubber. Other methods of treatment are employed but give products of inferior colour. The white crepe rubber is preferred for the rubber derivatives used in surface coatings.

Chlorinated rubber

The raw rubber molecule contains a number of unsaturated linkages and will react with chlorine to yield a solid product known as chlorinated rubber. Neoprene reacts in a similar way and is used in the process. Both addition and substitution reactions take place, the latter resulting in the formation of hydrochloric acid. In manufacture, the rubber is dissolved in carbon tetrachloride in an acid-resisting vessel fitted with means of heating and cooling. Chlorine is passed into the solution, and on completion of the reaction the solution is washed free from hydrochloric acid with aqueous alkali. Finally the chlorinated rubber is precipitated by addition of methanol, washed and dried. It is marketed as a white powder containing 65–68 percent chlorine [1]. A range of viscosity grades is available, the viscosity being designated by numbers [2]. These represent the viscosity, in centipoises, of a 20 percent w/w solution in toluene.

Properties and uses

It is difficult to remove carbon tetrachloride entirely from chlorinated rubber and the powders usually contain about 3 percent of the solvent [1].

Solvents

It is soluble in aromatic hydrocarbons, chlorinated hydrocarbons, esters, and some ketones. Blends of solvents are usually employed, the composition depending on the method of application of the paint. The use of a proportion of a polar solvent is necessary to prevent 'cobwebbing' in application by conventional spray.

Chlorinated rubber is a non-convertible binder and, as is common with this class of material, difficulties can be encountered in applying a second coat by brush. The solvent in the latter softens the first coat and 'pick-up' or 'lifting' occurs. The inclusion of a proportion of non-solvent such as white spirit reduces this pick-up very considerably.

Heat stability

This is limited and chlorinated rubber coatings should not be used on surfaces likely to exceed 80°–85°C (176°–185°F) for any length of time. Liberation of hydrochloric acid is possible, and it is usual to include a 'stabilizing' agent to absorb the acid. Certain epoxy compounds such as expoxidized soya bean oil are used.

General resistance properties

Chlorinated rubber possesses an extremely low moisture permeability and is therefore used on surfaces which may be permanently in contact with water, such as the submerged areas of swimming pools.

It is resistant to alkali, dilute acids, mineral oils and greases, inorganic chemicals, aliphatic hydrocarbons, and alcohols. It is not resistant to aromatic hydrocarbons, certain esters and ketones, chlorinated solvents, animal oils and fats.

The durability is of a high order and among the uses are marine paints and bottom compositions, road marking paints and paint systems for the protection of exposed steelwork generally.

Compatibility

Chlorinated rubber is compatible with a wide range of other resins, including alkyds, acrylics, hydrocarbon resins and rosin. Blends with polyurethanes are used as adhesives.

It is often blended with certain medium- and long-oil alkyds to upgrade chemical resistance, hardness, and resistance to lubricating oils. The drying times also are reduced. Blends of this nature are used on ships' topsides and superstructures and in primers for chlorinated rubber systems on steelwork.

Blends with certain acrylic resins give tough films with high resistance to chalking under intense solar radiation [1]. They also show good adhesion to galvanized iron and etched zinc surfaces.

Plasticizers

A chlorinated rubber film deposited on evaporation of solvent from a solution is brittle and lacks adhesion. It is therefore plasticized, the amount of plasticizer varying from 20 to 50 percent according to the degree of pigmentation and the end-use of the paint. Plasticizers employed include chlorinated paraffin waxes, many esters including phthalates, phosphates, adipates, sebacates and some drying oils. The most widely used are the chemically inert chlorinated paraffin waxes [3].

Hazards

These are usually associated with the solvents. As mentioned earlier, the chlorinated rubber powder generally contains about 3 percent carbon tetrachloride.

Cyclized or isomerized rubber

When raw rubber is heated with phenol and a catalyst, a change in structure takes place, resulting in a less unsaturated product with an iodine value of 50 to 100.

It is a tough solid material, soluble in aliphatic hydrocarbons and available in a number of viscosity grades. It is compatible with drying oils, varnishes, long-oil alkyds and some medium-oil alkyds. It is not compatible with short-oil alkyds.

Uses

When used alone as a film-former, cyclized rubber requires a plasticizer and is readily compatible with the chlorinated paraffin waxes. Such films require a cobalt drier and show a high degree of chemical resistance, but the resistance to solvents is poor.

The wide range of compatibility of cyclized rubber enables it to be used to upgrade the chemical and water resistance of long-oil alkyds and varnishes, the composition of the mixture being determined by the severity of the conditions. Mixtures with alkyds or varnishes dry to films which show a greater solvent resistance than the cyclized rubber, and these can be over-painted without difficulty.

The general chemical resistance and exterior durability of cyclized rubber is not as high as with chlorinated rubber.

BITUMINOUS MATERIALS

These are generally black or brownish-black materials consisting mainly of hydrocarbons, and soluble in carbon disulphide and certain other solvents of a neutral character. They contain a small amount of free carbon and materials known as asphaltenes to which the colour is largely due.

Bitumens are obtained from natural sources (in which they often occur in association with mineral matter) or as residues from the distillation of crude petroleum oils.

Natural bitumens

Bitumen occurs in admixture with mineral matter, the proportion of which varies from one location to another. Where the mineral content is high, as in the

Trinidad Lake, the product is known as asphalt and is used for road-making. The term asphalt, however, is used rather loosely and is often applied to other types. Materials in which the bitumen predominates and in which the mineral matter content is often very low are of interest for coatings. The following are the most important types.

Gilsonite This is one of the most widely used materials of the class and is obtained from North America. It is a lustrous jet-black brittle solid which gives a brown streak when rubbed on paper and breaks down to a brown powder. When cold it breaks with a conchoidal fracture but softens progressively on heating, finally becoming liquid at about 170°C (338°F). The specific gravity varies between 1·05 and 1·10. It contains very small amounts of mineral matter and is soluble in petroleum solvents, coal-tar solvents and carbon disulphide. Gilsonite is compatible with most of the oils and resins used in varnish manufacture and is used in stoving blacks, black japans and some dipping blacks.

Manjak or glance pitch This is similar in appearance to gilsonite but gives a black streak on paper. It is obtained from both Barbados and Trinidad, the latter type being preferred in view of greater staining power. Trinidad Manjak softens at about 170°C (338°C) but differs from gilsonite in showing limited compatibility with oils. Films produced from Manjak possess good weathering properties.

Grahamite is mined in the southern states of the USA and in Mexico. It is dense black, giving a black streak on paper and is harder and less fusible than gilsonite (softening point 200° to 300°C, 392° to 572°F).

Rafaelite is obtained from the Argentine and is harder than gilsonite. It also possesses a higher specific gravity. The appearance is lustrous black and it breaks with a conchoidal fracture. The melting point is above 200°C (392°F) and the compatibility with drying oils very limited.

Petroleum bitumens

Residual bitumen is obtained from the residues obtained in the distillation of crude asphaltic petroleums. The properties vary with the type of crude material and the conditions of distillation. However, conditions of processing can be closely controlled to give a uniform product from a particular type of crude. They exhibit a wide range of softening points and are very susceptible to changes in temperature, becoming sticky in warm weather.

Blown or oxidized bitumens are prepared by blowing air through semi-liquid petroleum residues at about 280° to 300°C (536° to 572°F). The products are semi-liquids or solids of melting point up to 120°C (248°F). They are less susceptible to temperature changes than the unblown material but have very limited compatibility with the common varnish-making materials.

Petroleum resins or Albino bitumens These are obtained from petroleum oils free from asphaltenes and vary considerably in colour. Many specimens are similar in colour to dark rosin. They are soluble in hydrocarbons and are compatible with drying oils.

General properties of bitumens

The intensely dark colour of bitumens is useful in black finishes but limits their use in coloured paints.

They are very resistant to water penetration and to attack by acids and alkalis. They are, however, very sensitive to changes in temperature and some types soften considerably on slight warming.

When exposed to the air, bitumen films undergo no major chemical change, but oxidation of the surface takes place slowly. The product is harder than the underlying bitumen and tends to craze when changes of temperature occur.

Bitumens are soluble in a wide range of solvents and oils and application of a conventional paint film over one of bitumen will result in the bitumen bleeding into the second coat.

BITUMINOUS PAINTS

This term is used for the products obtained from both petroleum and coal-tar sources, but the petroleum products are the more widely used.

Bituminous paints are designed to take full advantage of the water and chemical resistance of the material and are used as protective coatings on steel and other surfaces where protection is of greater importance than appearance. It is usually necessary to use a blend of bitumens to achieve the desired film properties and these are melted together and then thinned out, usually with aromatic solvents. Drying oils are sometimes incorporated and, in this event, the appropriate quantity of drier is added.

The black colour can be modified by pigmentation, but the range of colours obtainable is limited to greys, dark reds, browns and greens. If petroleum resins are used in place of bitumen, paler shades can be produced but 'clean' colours are not possible.

Dipping Black is usually a 50 percent solution in white spirit of a mixture of gilsonite and residual bitumen. The ratio of these can be varied at will to produce any desired degree of film hardness.

Black varnish consists of a 65 percent solution of a medium-soft coal-tar pitch in heavy naphtha.

Brunswick Black is a quick-drying material consisting of a mixture of gilsonite and rosin dissolved in naphtha. Drying oils are sometimes incorporated.

Bituminous aluminium paint Bituminous paints are used widely to protect metal – chiefly galvanized iron roofs, but with black paints a considerable amount of the radiation is absorbed. If aluminium paste is mixed with the bituminous paint at the rate of about 400 grams of paste per litre (4 lb of paste per gallon) of product, the aluminium *leafs* in the film, producing a bright metallic lustre with a brownish tint. This serves to reflect a high percentage of the radiation and the building is cooler in consequence.

Bituminous aluminium paints are used also on a considerable scale in marine work either as an intermediate coat between primer and anti-fouling composition or as a barrier coat over old, but sound anti-fouling composition before application of fresh coats of composition.

PITCHES

These are fusible solid or semi-solid materials obtained in the distillation of tars from various sources. The source is usually denoted by the prefix.

Coal-tar pitch is a relatively cheap product obtained from the distillation of coal-tar, and the properties vary with the composition of the tar employed. This in turn varies with the type of coal, temperature of distillation, and type of distillation plant.

Coal-tar pitches are compatible with aromatic solvents but are insoluble in aliphatic hydrocarbons, drying oils, and many natural resins. In general they show pronounced acidity.

Certain grades are compatible with epoxy and polyurethane resins and these combinations are widely used as protective coatings for steelwork under highly corrosive or submerged conditions.

Water-gas pitch or oil-tar pitch is derived from tars produced in petroleum distillation. The properties of the product vary with the temperature of distillation, the higher temperatures yielding products similar to some coal-tar pitches. These materials can give coatings of good durability.

Stearine or fatty acid pitches comprise the residues obtained in the steam distillation of fatty acids. The solubility and properties vary according to the nature of the fatty acids. Generally they have low melting points (below 100°C, 212°F) and poor opacity. They are compatible with some of the harder bituminous materials and are sometimes used as softeners. Stearine pitches are sometimes used in black stoving enamels as they harden on heating.

GUMS

The gums are organic materials similar in appearance and origin to the natural resins but differing from them in solubility characteristics and composition. Gums are carbohydrates, soluble in water and insoluble in organic solvents.

Gum arabic is imported mainly from North Africa where it is obtained as an exudation from species of Acacia, the most important being *Acacia Senegal.* It usually contains about 4 percent of impurities and has a specific gravity of 1·35 to 1·50. Gum arabic is soluble in water but is precipitated on addition of alcohol or borax. Addition of aldehydes or phenols renders the gum insoluble in water.

The aqueous solution is neutral but liable to attack by bacteria, and therefore requires the addition of a preservative such as 1 per cent of *p*-chloro-*m*-cresol. Gum arabic is used as an adhesive and a binder in some types of artists' colours. The aqueous solution gives no coloration with iodine solution, a test which serves to distinguish it from dextrin (*below*).

Senegal gum is obtained from *Acacia Senegal* and is less friable and tougher than Gum arabic. It is used widely in artists' water-colours.

Gum tragacanth is obtained as an exudation of the species Astragalus, a shrub which grows wild in Asia Minor. The exuded gum dries on the surface of the stem and forms characteristic horny nodules. It swells to a gel in cold water but goes into solution on addition of alkali. It is used as a binder for artists' water-colours.

Dextrin or British gum is derived from starch by treatment with acid or heating. Flour is moistened with a mixture of dilute hydrochloric and nitric acids and heated to 125°C (257°F), or heated alone to 180° to 200°C (356° to 392°F). The colour of commercial dextrin varies from yellow to brown according to the treatment. It possesses good adhesive properties and is often used as a substitute for Gum arabic. Aqueous solutions of dextrin give a deep blue colour on addition of iodine solution.

Farina is similar in nature to dextrin and is prepared from potato starch. It is used in the preparation of adhesives.

GLUES

The materials in this class have the adhesive properties of gums but differ from them in origin and composition. They belong to the class of complex nitrogenous substances known as proteins and are obtained from animal sources. Before the oil-bound distempers and glue-bound ceiling whites were displaced by emulsion paints, considerable quantities of glues and casein were consumed by the paint industry. Now the quantity used is extremely small, but they merit brief mention.

Gelatin is the term applied to the finest quality glue and is contained in the skins, tissues and bones of animals. The commercial materials are sometimes known as 'Technical Gelatins'. The bulk of the materials are known as glues and are classified according to origin.

Skin glues are manufactured by first heating the skins with lime in baths of increasing strength to remove hair and blood. The skins are then washed thoroughly and extracted with water at progressively higher temperatures and finally in an autoclave at 100° to 110°C (212° to 230°F). The first extracts are the purest gelatin, and with successive extracts the materials become less pure.

Rabbit skin glue is the best known of this type and gives firm gels at low concentrations.

Bone glues are prepared from degreased bones which are treated with cold dilute hydrochloric acid to remove calcium phosphate and other mineral constituents. The residue is washed and then extracted with hot water in a similar way to skin glue.

Size is a weak glue which contains a large proportion of a similar substance, chondrin, derived from cartilage, and which does not form a firm gel. Size forms a very soft jelly and is occasionally used for sealing porous surfaces.

Properties of glues

The best grades of glue are colourless, tasteless and odourless. The specific gravity is about 1·3. Glue swells in cold water and dissolves in hot water, the solution on cooling setting to a gel. The firmness or strength of the gel is taken as a measure of the quality of the glue and is expressed in Bloom values. These are high for stiff gels and low for the softer types.

Glue solutions are very prone to decomposition by bacteria unless treated with an efficient preservative. Materials used for this purpose include *p*-chloro-*m*-cresol and sodium pentachlorphenate.

Casein is a protein present in skimmed milk from which it can be precipitated by addition of dilute mineral acid. The product is known as 'acid casein'. This material swells but does not dissolve in water; it dissolves in presence of an alkali such as ammonia or borax.

Casein is occasionally used as a protective colloid in the styrene - butadiene type of emulsion paints, where it induces good mechanical stability and excellent levelling properties.

Like glue, casein is readily attacked by bacteria unless it is adequately protected by an efficient bactericide.

References to Chapter 10

[1] 'Alloprene', Booklet by ICI (Mond Division).
[2] 'Alloprene' 5, 10, 20 (ICI).
[3] 'Cereclors' (ICI).

11

Synthetic resins

The very wide range of synthetic resins available to both the paint and plastic industries is the result of systematic investigation into the formation of resinous materials, which at one time were regarded as embarrassments in chemical laboratories. Such materials showed no tendency to crystallize and had no sharp melting point. It was realized gradually that these resinous bodies were materials of very high molecular weight and that their formation was associated with the presence of certain groups in the reacting molecules. Furthermore, resin formation depended on the relative quantities of reactants.

POLYMERIZATION

Modern synthetic resins are essentially polymers, i.e. they contain large molecules which have been built up from the *normal* molecules or monomers. The number of monomer molecules which go to a given polymer molecule is not constant, so that in a resin there is a range of polymer sizes or molecular weights. These are found to be distributed round a mean value, in much the same way as the particles in a pigment.

There are two general methods for the production of polymers. These are known as *addition* and *condensation* polymerization. The method employed depends on the type of monomer.

Addition polymerization

Certain types of monomer can be reacted – usually with the aid of a catalyst – to give polymers by simple addition. The reaction usually involves addition at double bonds, and an example is afforded by the polymerization of vinyl acetate. Long-chain polymers are built up by rupture of the double bonds:

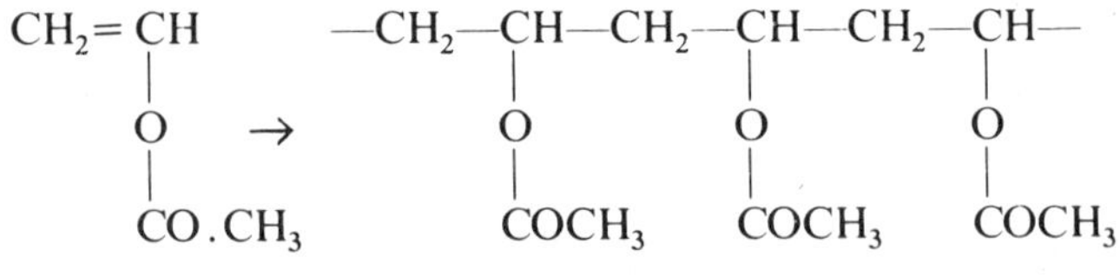

$$\begin{array}{c} CH_2{=}CH \\ | \\ O \\ | \\ CO.CH_3 \end{array} \quad \rightarrow \quad \begin{array}{cccccc} —CH_2— & CH & —CH_2— & CH & —CH_2— & CH— \\ & | & & | & & | \\ & O & & O & & O \\ & | & & | & & | \\ & COCH_3 & & COCH_3 & & COCH_3 \end{array}$$

Long-chain polymers of this nature are thermoplastic.

More than one monomer can take part in addition polymerization, and vinyl acetate can be co-polymerized with, for example, butyl maleate to give a co-polymer:

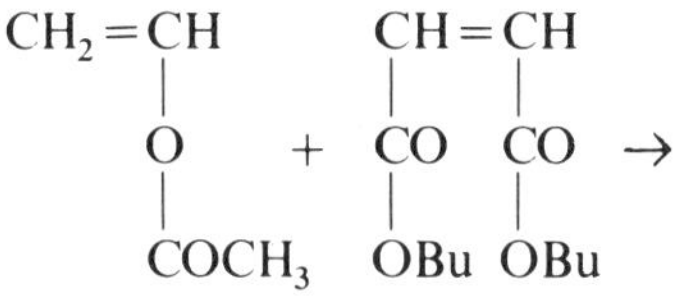

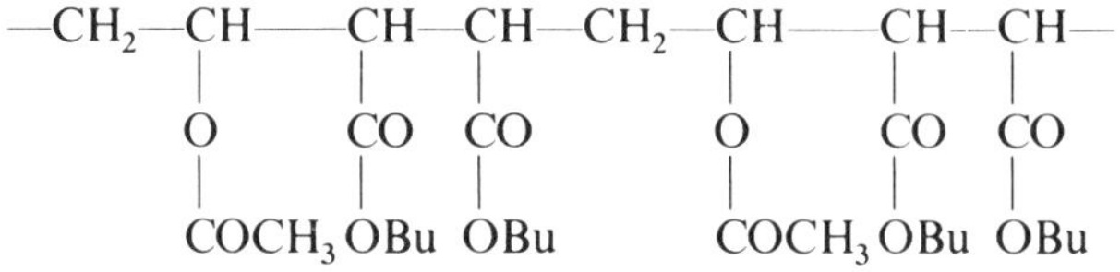

$$\begin{array}{ccccc} CH_2{=}CH & & CH{=}CH & & \\ \vert & & \vert \quad \vert & & \\ O & + & CO \quad CO & \rightarrow & \\ \vert & & \vert \quad \vert & & \\ COCH_3 & & OBu \quad OBu & & \end{array}$$

$$\begin{array}{l} -CH_2-CH-CH-CH-CH_2-CH-CH-CH- \\ \qquad\quad \vert \qquad\quad \vert \qquad \vert \qquad\qquad\quad \vert \qquad\quad \vert \qquad \vert \\ \qquad\quad O \qquad\; CO \;\; CO \qquad\qquad\;\; O \qquad\; CO \;\; CO \\ \qquad\quad \vert \qquad\quad \vert \qquad \vert \qquad\qquad\quad \vert \qquad\quad \vert \qquad \vert \\ \qquad COCH_3\,OBu\;OBu \qquad\quad COCH_3\,OBu\;OBu \end{array}$$

The addition polymerization reaction involving the ethylene group $\rangle C{=}C \langle$ is the basic reaction in the production of vinyl, acrylic and polystyrene resins. However, the ease with which the ethylene linkage is broken to produce polymers varies with the nature of the attached groups. Ethylene and propylene are much more difficult to polymerize than vinyl acetate.

Condensation polymerization

In a very simple reaction such as that between a monobasic acid and a monohydric alcohol, water is eliminated and a simple ester is produced:

$$CH_3COOH + C_2H_5OH \rightarrow CH_3COOC_2H_5 + H_2O$$

No further reaction takes place. If, however, a dibasic acid is used with a dihydric alcohol a more complicated series of reactions can take place:

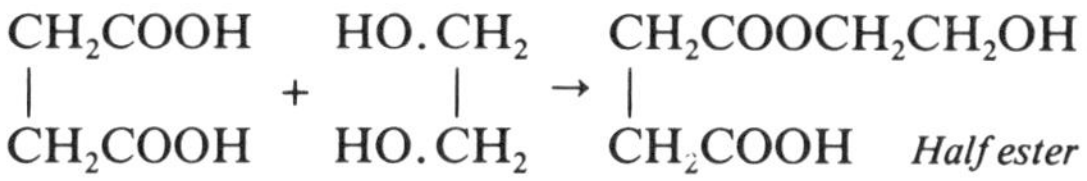

$$\begin{array}{lcccl} CH_2COOH & & HO.CH_2 & & CH_2COOCH_2CH_2OH \\ \vert & + & \vert & \rightarrow & \vert \\ CH_2COOH & & HO.CH_2 & & CH_2COOH \quad \textit{Half ester} \end{array}$$

This half ester can then lose water to give the cyclic ester:

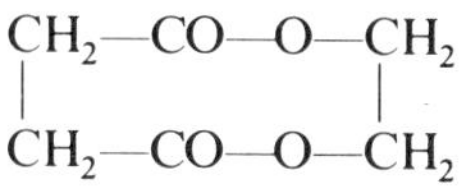

$$\begin{array}{l} CH_2-CO-O-CH_2 \\ \vert \qquad\qquad\qquad\quad \vert \\ CH_2-CO-O-CH_2 \end{array}$$

or it can react with further molecules of either acid or alcohol. In this way quite large molecules can be built up by a series of esterification reactions involving loss of water. This type of reaction is the basis of alkyd formation.

Esterification is but one type of condensation polymerization reaction involving the elimination of water. Another important type is that by which phenolformaldehyde resins are formed. Here the formaldehyde reacts to form a methylene bridge between adjacent phenol molecules. In its simplest form the reaction can be represented as follows:

$$2n\,C_6H_5OH + n.HCHO \rightarrow [-C_6H_3(OH)-CH_2-C_6H_3(OH)-]_n + n.H_2O.$$

Functionality

In a simple esterification such as that between acetic acid and ethyl alcohols, each reactant molecule has one reactive or functional group. The product, ethyl acetate, has no reactive or functional groups and is incapable of reacting further. The functionality is zero.

A dibasic acid such as succinic has two reactive or functional groups and so a functionality of two. Similarly, ethylene glycol has a functionality of two. The half ester produced as the first reaction stage (*above*) also has two reactive or functional groups and therefore a functionality of two.

Maleic acid contains two functional carboxyl groups and also an ethylene linkage. This linkage is capable of addition at each carbon atom, i.e., it has two reactive points, and is bifunctional. The total functionality of maleic acid is therefore four. Table 11.1 shows the functionality of a number of common reactants.

The idea of functionality was first advanced by Kienle [1] as a result of his work in alkyds, and was later developed further by Carothers [2]. The general principles relating functionality and polymer formation can be summarized as follows.

Table 11.1 Functional groups

Material	Formula	Functional groups	Functionality
Glycerol	CH_2OH CH.OH CH_2OH	—OH (3)	3
Pentaerythritol	$C(CH_2OH)_4$	—OH (4)	4
Phthalic acids	$C_6H_4(COOH)_2$	—COOH (2)	2
Adipic acid	$COOH(CH_2)_4COOH$	—COOH (2)	2
Sebacic acid	$COOH(CH_2)_8COOH$	—COOH (2)	2
Maleic acid	CH—COOH ‖ CH—COOH	—COOH (2) >C=C< (2 points)	4*
Fumaric acid	CH—COOH ‖ COOH—CH	—COOH (2) >C=C< (2 pcints)	4*
Formaldehyde	CH_2O or $CH_2(OH)OH$ (hydrate)	—OH (2)	2

Material	Formula	Functional groups	Function-ality
Phenol	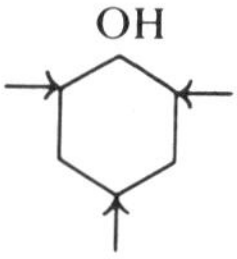 Phenol ring (OH) with arrows marking 2 ortho and 1 para positions	2 ortho reactive 1 para positions	3
Styrene	 Benzene ring with $CH=CH_2$	$>C=C<$ (2 points)	2

*This means a functionality of 2 for esterification and 2 for addition reactions.

A molecule of a monomer containing two functional groups can react with a suitable second bifunctional molecule to give a polymer. Such a polymer will be linear and thermoplastic. The type of polymer, i.e. the molecular weight, will depend on the molecular ratio of the reactants. At a 1:1 ratio a high molecular weight is produced, but an excess of one will stop the reaction and lead to low molecular-weight polymers.

When a bifunctional molecule reacts with one which is trifunctional it is possible for cross-linking to occur, leading to a three-dimensional structure. Such polymers are further cross-linked on heating and yield insoluble and infusible films when stoved.

The basic reaction in alkyd preparation is the reaction between bifunctional phthalic acid and trifunctional glycerine or tetrafunctional pentaerythritol. A condensation polymerization of this type can lead rapidly to insoluble polymers and is discussed under Alkyd resins (*below*).

ALKYD RESINS

The alkyd resins form the largest group of synthetic resins available to the paint industry and the consumption is greater than that of any other resin. They form part of the larger group of polyesters which includes all products of the esterification of polybasic acids with polyhydric alcohols.

Alkyds differ from other polyesters in containing a monobasic fatty acid in the molecule and are often referred to as *oil-modified alkyds*, though the adjective 'oil-modified' is redundant. The nature and often the amount of the monobasic fatty acid determines the drying characteristics of the resin. Saturated fatty acids yield non-drying or plasticizing resins, while drying properties are conferred by unsaturated acids.

Raw materials for alkyds

The following are among the most important in common use.

Dibasic acids

Structure	Name
C_6H_4 with COOH, COOH (ortho positions)	**ortho-phthalic** (used as anhydride m.p. 130 C, 266 F)
C_6H_4 with COOH, COOH (meta positions)	**iso-phthalic** (m.p. >300 C, 572 F)
C_6H_4 with COOH, COOH (para positions)	**terephthalic** (sublimes on heating)

Structure	Name
CH—COOH ‖ CH—COOH	**Maleic** (m.p. 130 C, 266 F) (anhydride m.p. 56 C, 132·8 F)
CH—COOH ‖ COOH—CH	**Fumaric** (sublimes at 200 C, 392 F)
COOH—$(CH_2)_4$—COOH	**Adipic** (m.p. 151 C, 303·8 F)
COOH—$(CH_2)_8$—COOH	**Sebacic** (m.p. 133 C, 271·4 F)

Orthophthalic acid, used in the form of the anhydride, is the most extensively used of these acids.

Polyhydric alcohols (polyols) The two most widely used are glycerol and pentaerythritol, but small amounts of sorbitol and trimethylolpropane are sometimes employed.

Glycerol The general properties of glycerol have been discussed in Chapter 10. It contains two primary and one secondary alcohol group and these react at the same rate with fatty acids to form glycerides. With dibasic acids the primary groups react more readily than the secondary; and it is likely that the former react with phthalic anhydride to give the first stage syrup, and the secondary alcohol groups are available for reaction with fatty acids.

Glycerol can be used in alkyds of all fatty acid contents or oil lengths.

Pentaerythritol, $C(CH_2OH)_4$, is a solid with melting point 253°C (487°F), and is manufactured by reaction between acetaldehyde and formaldehyde in the presence of water and alkali. It contains four primary alcohol groups which are all equally reactive and forms more complex resins than does glycerol. It can be used for medium/long and long oil alkyds only, as the short and medium/short oil types gel rapidly. Compared with the glycerol types, long oil alkyds produced from pentaerythritol dry harder to give more durable films with better moisture resistance. They have a greater tendency to form crystalline bloom in urban atmospheres.

Fatty acids The general properties of the fatty acids have been discussed in Chapter 9.

It has been mentioned earlier that the drying characteristics of an alkyd are determined by the quantity and nature of the fatty acids, and they can be divided into two general groups, drying and non-drying.

The acids employed consist of mixtures derived from the parent oil by hydrolysis of the glycerides. In this way the acids can be separated and, if necessary, purified. An alternative method in which the acids are not isolated consists in heating together the oil and glycerol to produce monoglycerides as in the process of this name (*below*).

Alkyd formation

When two moles of glycerol and three moles of phthalic anhydride are heated together, a syrup is first formed which, on heating further, is converted to a gel. Continued heating causes the formation of a resin which ultimately becomes an insoluble and infusible mass. It is thought that a linear polymer is first formed, and this gradually cross-links to give a very complex structure. This is the type of resin usually formed from a bifunctional acid and a trifunctional polyol and is of no value as a surface-coating resin.

The polymerization to the very large polymers can be checked by the introduction of a monofunctional component into the system. This can be as an alcohol such as cyclohexanol or *n*-butanol, or as a monobasic fatty acid. In practice the latter is used and the fatty acid can be reacted with the first stage or syrup formed from phthalic anhydride and glycerol. On further heating the resin is much less reactive and does not reach the infusible state. This is the basis of the fatty acid method of alkyd manufacture.

Fatty acid process

There are two variants of this process.

1 The phthalic anhydride and glycerol are heated to 180°C (356°F) to the 'first stage' syrup, and molten fatty acids are added to esterify the free hydroxyl groups. Heating is continued at 180° to 220°C (356° to 428°F) until the desired acid value and solubility characteristics are reached. Foaming can be troublesome in this method.

2 The three raw materials, phthalic anhydride, glycerol and fatty acid, are placed in the reaction vessel together with a small quantity of xylol. The vessel is fitted with a condenser to which is attached a water separator of the Dean and Stark type. On heating, the water produced is carried off with the xylol and is separated. The condensing xylol serves to flush the sublimed phthalic anhydride back into the reaction vessel. The amount of water collected is an indication of the progress of esterification, but samples of the resin are removed from time to time for acid value and viscosity checks.

The reaction vessels are generally of stainless steel fitted with stirrer, charge hole, condensing system, and pipes for passing inert gas over the charge. The latter serves to reduce discoloration. Heating can be by immersion heaters or by passing hot oil or other liquid through a jacket surrounding the vessel.

The fatty acid process is being superseded by the alcoholysis process (*below*).

Alcoholysis or monoglyceride process

The formation of an alkyd resin involves esterification of glycerol or pentaerythritol with phthalic anhydride and a fatty acid. In glycerol-type alkyds (but not those from pentaerythritol–see *below*) the necessity for isolating the fatty acids can be avoided by the use of monoglycerides which are then further esterified with phthalic anhydride. The monoglycerides are formed by heating the drying oil with the necessary amount of glycerol to 250° to 280°C (482° to 536°F), when alcoholysis takes place according to the scheme:

$$
\begin{array}{l} CH_2OCOR \\ | \\ CHOCOR \\ | \\ CH_2OCOR \end{array}
\quad + \quad 2 \quad
\begin{array}{l} CH_2OH \\ | \\ CHOH \\ | \\ CH_2OH \end{array}
\quad \rightarrow \quad 3 \quad
\begin{array}{l} CH_2OH \\ | \\ CHOH \\ | \\ CH_2OCOR \end{array}
$$

The phthalic anhydride is added and the reaction completed at 180° to 250°C (356° to 482°F). The plant employed is similar to that used for the fatty acid process and the reaction is assisted by the use of xylol to remove the water.

This process is cheaper than that using the fatty acids, but it is not as versatile. A greater variety in the nature and purity of acids is possible with the fatty acid method.

Alkyds required to contain pentaerythritol as the sole polyol cannot be produced by the alcoholysis or monoglyceride process since the product will contain the glycerol introduced with the monoglyceride.

Dehydrated castor oil alkyds

These differ from other alkyds in their rather unique method of manufacture. The raw materials – raw castor oil, phthalic anhydride and glycerol – are heated together, slowly at first and then rapidly to 260° to 270°C (500° to 518°F). Under these conditions the raw castor oil is dehydrated by the phthalic anhydride, leading to the formation of dehydrated castor oil resins. It is of interest that during the dehydration reaction the viscosity of the mix falls, but rises again when polymerization of the resin commences.

Drying alkyds

These contain unsaturated fatty acids which control, to a large degree, the drying and performance characteristics of the resin. The amount of fatty acid present in the glycerol type can vary from 25 to 80 percent and is generally expressed as a percentage, although the varnish convention of long-, medium- and short-oil is sometimes used. The figures covered by these terms are not precise, but they refer roughly to fatty acid contents of 60 percent upwards, 45 to 60 percent, and 25 to 45 percent respectively.

Pentaerythritol is used only in alkyds with a fatty acid content of about 60 percent and upwards. It tends to gel rapidly when used with lower proportions of fatty acids.

Drying alkyds with fatty acid contents of 60 percent and above are soluble in aliphatic hydrocarbons such as white spirit, and are compatible with drying oils and varnishes. In the range of 45 to 60 percent fatty acid a proportion of stronger aromatic solvent is often necessary, but below 45 percent fatty acids the resins are soluble only in strong solvents such as xylol and naphtha. They are not compatible with drying oils or varnishes.

Driers

The drying times decrease progressively with decrease in fatty acid content, the actual figures depending on the type of acid used. The long and medium types are catalysed satisfactorily by mixtures of lead and cobalt in the form of naphthenates and octoates. The use of lead, however, is being discontinued and the metal has been eliminated from decorative paints; these dry satisfactorily with zirconium/cobalt mixtures. In decorative finishes a proportion of calcium (usually as octoate) is included to check the tendency to 'crystalline bloom' formation on exterior films. The short types of alkyd can be catalysed by cobalt alone, and this type is frequently cured by stoving without driers.

Types of fatty acid employed

Linseed fatty acids produce alkyds with good drying properties and durability but which yellow to a marked degree on exposure. This yellowing appears to be associated with the linolenic acid present. Linseed alkyds based on pentaerythritol dry harder and are more durable than those based on glycerol.

Soya bean fatty acids are widely used and, by reason of their very low linolenic acid content, the tendency to yellow on exposure is much less than with linseed. Alkyds made with these fatty acids and pentaerythritol as the polyol are in general use in the pale shades of decorative enamels.

Safflower, Sunflower and **Tobacco** seeds yield fatty acids free from linolenic acid and containing a high proportion of linoleic acid (like soya bean oil, they are described as 'linoleic rich'), and are used as alternatives to soya acids.

Dehydrated castor oil fatty acids (a mixture of 9:11-octadecadienoic and 9:12-octadecadienoic acids) yield non-yellowing alkyds. These are made in medium- and short-oil lengths and are used in air-drying and stoving (with amino resins) finishes respectively. These fatty acids are also used in some epoxy esters.

Tall-oil fatty acids (essentially linoleic) are derived from the wood-pulp industry. They can now be produced virtually free from rosin acids and yield non-yellowing alkyds exhibiting excellent gloss.

Tung-oil fatty acids are not used alone but often mixed with linseed in alkyds which can be further modified with phenolic resin. Such alkyds dry faster than the linseed type and the film has much greater water resistance. The durability is very good but the film yellows on exposure.

Applications of drying alkyds

Alkyd finishes, in general, show good gloss and gloss retention but are sensitive to moisture both during drying and afterwards. Exterior painting with alkyds during conditions of condensation, which often occur in early summer and autumn, can lead to loss of gloss on account of microrivelling.

When immersed in water for prolonged periods, alkyd films absorb appreciable quantities of water, and the resulting swelling frequently results in blistering. They are therefore not used under conditions of water immersion.

The long and medium types are used in air-drying finishes and undercoats for decorative and maintenance work and in marine topside and weatherwork paints. The medium types are used also in paints for machinery and implements.

Long-oil alkyds are versatile materials, and some are compatible with other types of resin to give films with special properties. Blends with chlorinated rubber, for example, show greater adhesion to clean metal than does chlorinated rubber alone, and are sometimes used in primers for steelwork under chlorinated rubber systems.

Mixtures of some long-oil alkyds and certain vinyl co-polymers show rapid through-drying and are used in undercoats when short overcoating times are required. Care must be taken in the choice of resins and solvents since these mixtures can present stability problems.

Short-oil types find use in quick-drying primers and sometimes in 'half-hour' enamels. They are used also in stoving coatings often with amino resins.

Blends of emulsions of long-oil alkyds and acrylics form the basis of some decorative/preservative treatments for wooden window frames and doors.

Non-drying alkyds

These are produced either by the use of saturated fatty acids in the normal process, or by the polymerization of a bifunctional acid such as adipic with a bifunctional polyol such as ethylene glycol. The former types are the more common, and resins based on castor oil are very widely used. These are made by heating raw castor oil, glycerol and phthalic anhydride at a relatively low temperature (190° to 200°C, 374° to 392°F).

Non-drying alkyds are made from other oils by the fatty acid or alcoholysis process and are usually of short oil length.

They are used as plasticizers for cellulose nitrate lacquers and as film-formers in conjunction with amino resins. In the latter case the films are stoved to effect the necessary cross-linking between the resins.

The non-drying alkyds used in high quality white stoving enamels are frequently based on lauric acid or coconut oil fatty acids.

Vinyl toluenated (VT) alkyds

Vinyl toluene, $CH_3C_6H_4CH = CH_2$ (as well as styrene, $C_6H_5CH = CH_2$) will react with alkyds containing unsaturated linkages when heated in xylol solution in presence of a catalyst (e.g. a peroxide). Addition of vinyl toluene takes place at the double bonds in the fatty acid chain and has the effect of reducing the oil length. The products dry more rapidly and give harder and more chemically resistant films. They can take some time to develop their solvent resistance, and recoating can sometimes be a problem.

Vinyl toluenated alkyds are used for rapid air-dry and stoving finishes in industrial work. The longer oil types (with the lower vinyl toluene content) are less affected by adverse drying conditons than are the straight alkyds, but their durability is not as good.

Silicone alkyds

These are produced by co-polymerizing alkyds with certain silicone intermediates, e.g. low molecular weight siloxanes, so that the product contains up to 30 percent silicone. They are characterized by good drying properties, high gloss, and outstanding weather resistance. Some 'long life' coating systems are based on these resins.

Urethane alkyds

These are discussed on p. 208.

Thixotropic alkyds

Certain modified polyamide resins will combine with conventional alkyds to produce resins with pronounced thixotropic properties.

It is possible, by the use of suitable types and quantities of reactants, to produce resins which can be pigmented to give finishes ranging from full gloss to flat. These are the so-called *jelly* or *dripless* paints which form firm gels at rest but lose their structure and exhibit flow when stirred or applied by brush.

Water-soluble alkyds

One of the most interesting developments of recent years has been the production of alkyd (and other) resins capable of being solubilized in water, but which can be stoved to a water-resistant film. Water-solubility is achieved by the introduction of a number of free carboxyl and hydroxyl groups which are well distributed in the polymer. These groups are neutralized by organic bases, the type of which depends on the type of polymer, but, in general, the tertiary bases are preferred. The products are soluble in water and can be pigmented and applied by spraying or dipping. The film is cured by stoving, when the organic base is volatilized and cross-linking occurs.

POLYESTER RESINS

Whilst alkyd resins are considered to be members of the polyester group, the resins considered here differ from alkyds in several respects but, in common, they can be unsaturated or saturated.

Unsaturated polyester resins

The product obtained from the condensation polymerisation of a glycol such as 1:2-propylene glycol or 1:3-butylene glycol with an unsaturated acid such as maleic (or a mixture of maleic and saturated acid e.g. phthalic) in the form of anhydrides is a linear polymer containing unsaturation in the chain. This is dissolved in a suitable monomer, generally styrene, to produce the final resin. The film is produced by co-polymerization of the linear polymer and monomer by means of a free radical mechanism. The free radicals can be generated by heat, or more usually by addition of a peroxide. In order that curing can take place at room temperature the decomposition of peroxides into free radicals is catalysed by certain metal ions, usually cobalt. **Caution**: the solutions of peroxide and cobalt compound *must not be mixed* under any circumstances but should be added separately to the resin mix and well stirred in. A simple mixture of the two additives is highly explosive.

Films produced by curing in the air can show a residual surface tack for some hours. This is the result of inhibition of curing by the oxygen of the air and is difficult to overcome. One method consists in dissolving wax in the paint, so that the wax forms a skin on the surface of the film. When the latter is dry the wax is removed by polishing. The amount of wax used is of the order 0.01 to 1.0 percent. Wax-free types which are not inhibited by oxygen to the same degree incorporate an allyl group, such as the allyl ether of trimethylolpropane or glycerol, as part of the polyol ingredient.

Properties and uses of unsaturated polyester resins

Unsaturated polyesters give hard, mar-resistant films with good chemical and solvent resistance. The films are characterized by depth of gloss. These resins are used in considerable quantities for wood finishing.

The fact that these polysters cure by a free radical mechanism renders them suited to irradiation curing. This form of cure, in which no heat is produced, is particularly suited to films on wood or board.

Electron-beam curing is used on flat surfaces and it effects cure in less than one second. The more widespread use of electron-beam curing has been hindered by difficulties in the design of plant.

Ultraviolet irradiation curing is used on an increasing scale, particularly for clear coatings on wood or board. Difficulties have been experienced with pigmented coatings. These are discussed under 'Acrylic resins'.

Saturated polyester resins

The main use of these resins is production of polyurethanes by reaction with isocyanates. The reaction involves hydroxyl groups and the components are selected to produce resins with high hydroxyl contents. Saturated dibasic acids such as adipic or phthalic are reacted with polyols such as glycerol, trimethylol propane or 1:3-butylene glycol.

Saturated polyester resins react also with some amino resins on stoving (see *below*).

AMINO RESINS

These are derived by reaction between formaldehyde and either urea or melamine to give urea-formaldehyde or melamine-formaldehyde resins respectively.

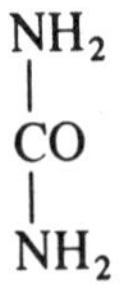

$$\begin{array}{c} NH_2 \\ | \\ CO \\ | \\ NH_2 \end{array}$$

Urea
(m.p. 133 C, 271·4 F)

$$\begin{array}{ccccc} & & N & & \\ & \diagup\!\!\diagup & & \diagdown & \\ NH_2C & & & & C{-}NH_2 \\ | & & & & \| \\ N & & & & N \\ & \diagdown\!\!\diagdown & & \diagup & \\ & & C & & \\ & & NH_2 & & \end{array}$$

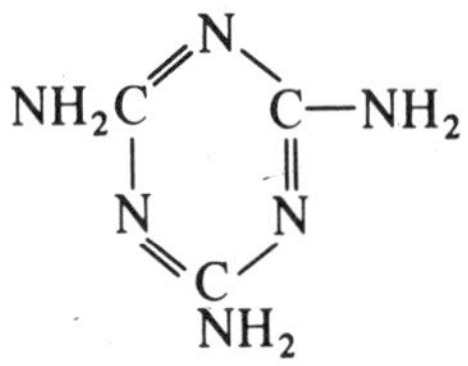

Melamine
(m.p. 350 C, 662 F)

Urea-formaldehyde resins

When urea is heated with formaldehyde solution (formalin) under slightly alkaline conditions, addition takes place to form methylol urea. Either mono- or dimethylol urea is formed according to the ratio of urea to formaldehyde:

$$\begin{array}{l} NH_2 \\ | \\ CO \\ | \\ NH_2 \end{array} + H.CHO \rightarrow \begin{array}{l} NH.CH_2OH \\ | \\ CO \\ | \\ NH_2 \end{array} \quad \text{Monomethylol urea}$$

$$\begin{array}{l} NH_2 \\ | \\ CO \\ | \\ NH_2 \end{array} + 2H.CHO \rightarrow \begin{array}{l} NH.CH_2OH \\ | \\ CO \\ | \\ NH.CH_2OH \end{array} \quad \text{Dimethylol urea}$$

A water-soluble syrup is first formed, and on continued heating to 200° to 240°C (392° to 464°F) hardens rapidly to a colourless hard infusible resin. This resin is valueless as a coating material, but if the condensation is carried out in the presence of a high-boiling primary alcohol such as *n*-butanol, the hydroxyl groups in the methylol ureas are etherified:

$$\begin{array}{l} NH.CH_2OH \\ | \\ CO \\ | \\ NH_2 \end{array} + HO.C_4H_9 \rightarrow \begin{array}{l} NH.CH_2OC_4H_9 \\ | \\ CO \\ | \\ NH_2 \end{array} + H_2O$$

The butylated urea resin produced by polymerization of this ether possesses entirely different solubility properties from the original hard resin. It is soluble in aromatic and some aliphatic hydrocarbons but insoluble in water. It is miscible with some alkyds and some varnishes. The solubility and compatibility with other resins increases with increasing chain length of the alcohol. When the films are cured by stoving, much of the alcohol breaks off and is lost, but some is retained as an integral part of the cured film.

Manufacture of urea resins

Urea and formalin are charged into a stainless steel vessel fitted with stirrer, condensing system, and water collector. The pH of the mixture is adjusted to 8. It is warmed for a short time and then an excess of butanol is introduced. The butanol and water distil into the collector and separate into two layers. The upper butanol layer runs back into the reaction vessel and the water can be tapped off. Such a process is known as azeotropic distillation. The pH of the resin is adjusted to 5 by addition of phosphoric acid, and etherification takes place. The final resin is obtained as a colourless solution in *n*-butanol.

Properties and uses of urea resins

Alkylated urea resins are water white and are compatible with a number of alkyds, both drying and non-drying. The combination of urea resin with non-drying alkyd is used very widely for stoving finishes of almost all types. It appears that combination between the two resins takes place during stoving, probably by etherification of the hydroxyl groups of the alkyd with the butoxy or methylol groups of the urea.

For general purposes a urea:alkyd (solids) ratio of about 1:1 is used and stoved at 120°C (248°F) for 20 to 30 minutes.

Curing of the films at lower or ambient temperatures can be effected by addition of small quantities of mineral or phosphoric acid.

Melamine-formaldehyde resins

A molecule of melamine can react with up to six molecules of formaldehyde to give a series of methylol melamines:

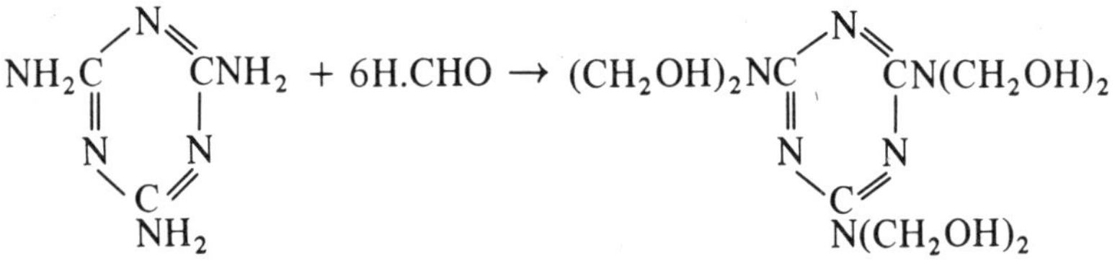

$$C_3N_3(NH_2)_3 + 6H.CHO \rightarrow C_3N_3[N(CH_2OH)_2]_3$$

These methylol melamines are very reactive and the mono-, di-, and tri-compounds tend to react, forming methylene bridges between melamine units. They form ethers with *n*-butanol, and this reaction is generally employed to limit the reactivity and to confer solubility in solvents and other media.

Manufacture of melamine resins

The process employed is similar to that used for urea resins, but conditions are modified in view of the greater reactivity of the melamine.

Properties and uses of melamine resins

Melamine resins are supplied in either *n*-butanol or *n*-butanol/xylol solutions and have limited compatibility with white spirit. They are miscible with drying and non-drying alkyds as well as with some epoxy resins and thermosetting acrylics.

Combinations of melamine resins and non-drying alkyds are widely used in white stoving finishes of domestic equipment and in car finishes. Melamines will tolerate larger quantities of alkyd than will ureas and the films show greater gloss retention, and heat and chemical resistance. The weather resistance and durability are excellent and they are used with thermosetting acrylics in car finishes.

Melamine/alkyd finishes can be cured at slightly lower temperatures than ureas. Like the latter they can be cured at ambient temperatures by addition of small amounts of mineral or phosphoric acid.

The mechanism of cure of alkyd/melamine systems by acid has attracted much attention. A summary and critical examination of the suggested mechanisms has been published by Holmberg [3]. The same author points out [4] that under acidic curing conditions, drying-oil alkyds give harder films than the non-drying types. He explains this by a Diels–Alder cross-linking mechanism.

Cross-linking reactions occur when amino resins are stoved with one of a number of hydroxyl-containing polymers e.g. alkyds, saturated polyesters, epoxies and acrylics. The resulting films show many outstanding properties.

The methylated melamines were a great improvement on the earlier butylated types in respect of compatibility and cost. One of the most widely used resins of this type is the highly versatile hexamethoxymethyl-melamine (HMMM). With epoxy resin this forms the well known Beetle BE 647 [5] which, on stoving, yields films of outstanding gloss retention and resistance to heat, detergents, salt solutions and humidity. Typical applications are can and tube linings, industrial primers and finishes.

Films with outstanding durability, colour and gloss retention are formed on stoving mixtures of HMMM with hydroxyl-modified acrylics e.g. 2-hydroxyethyl methacrylate.

Saturated polyesters react, on stoving, with some amino resins to give high performance industrial coatings. The films show exceptional durability, colour retention and mechanical properties. They are used in coil and strip coatings, appliance finishes and metal decoration.

EPOXY RESINS

Epoxy resins are the products of condensation of epichlorhydrin and diphenylolpropane derivatives. The best known of the latter is bisphenol 2.2-bis (4-hydroxyphenyl) propane, also known as Bisphenol A. The product possesses the following structure:

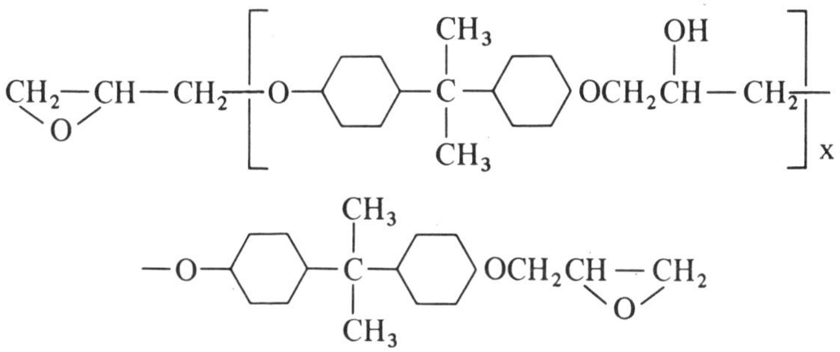

$$\underbrace{CH_2-CH}_{\diagdown O \diagup}-CH_2-\left[O-C_6H_4-\overset{CH_3}{\underset{CH_3}{C}}-C_6H_4-OCH_2\overset{OH}{CH}-CH_2\right]_x-O-C_6H_4-\overset{CH_3}{\underset{CH_3}{C}}-C_6H_4-OCH_2\underbrace{CH-CH_2}_{\diagdown O \diagup}$$

where x indicates the repeating unit. They are formed by condensing the reactants in the presence of alkali at about 100°C (212°C). A range of resins is produced, differing in molecular weight, and they are marketed under the trade names 'Epikote' (Shell Chemical Company) and 'Araldite' (CIBA/Geigy Ltd).

Epoxy resins contain hydroxyl and epoxy groups and the epoxy content is important in the evaluation of these resins. This may be quoted either as *epoxy equivalent*, i.e. the number of grams of resin containing one gram-equivalent of epoxy group, or as *epoxy value*, i.e. the number of epoxy groups in 100 grams of resin. The hydroxyl groups can be esterified with carboxylic acids to give epoxy esters (*below*).

Uncured epoxy resins are either viscous liquids (at lower molecular masses) or thermoplastic solids soluble in esters and ketones and in mixtures of these with aromatic hydrocarbons and some alcohols and propylene glycol derivatives. The following mixtures will illustrate some of the solvent blends used for epoxy resins but it will be realised that many different mixtures are possible. The composition of these will depend on the other ingredients present and the method of application of the paint. Figures represent percentages by weight.

(a)	Diacetone alcohol	25	(b)	Xylol	60
	Toluene	25		Methoxy propanol	10
	Ethanol 95%	25		Aromatic hydrocarbon	20
	Methylethyl ketone	25		Isobutanol	10
(c)	Methoxy propanol	25	(d)	Xylol	20
	Xylol	50		Methylethyl ketone	35
	Aromatic hydrocarbon	25		Toluol	35
				Isobutanol	10

Epoxy resins can be cross-linked by cold-curing or stoving with a range of materials which includes amino and phenolic resins, amines, polyamide resins and isocyanates. The resulting films are very abrasion resistant, resistant to chemical attack and show a high degree of adhesion to metal and other surfaces.

Curing by amino or phenolic resins

These materials cross-link with epoxy resins on stoving to give films of excellent adhesion, chemical resistance and flexibility. Films obtained from the amino resins possess slightly inferior resistance to the phenolic types but are more widely used on account of their better colour. Combinations with melamine resins, particularly hexamethoxymethyl melamine (HMMM) give very pale and resistant films and are used in white finishes for articles such as washing machines. The phenolic types are used where colour is of lesser importance, as in drum linings.

Curing by amines

Primary amines will react with epoxy resins at ambient temperatures to give long-chain polymers according to the general scheme

$$R'NH_2 + \underset{\diagdown O \diagup}{CH_2-CH}.CH_2R'' \rightarrow R'NH.CH_2.\underset{\underset{OH}{|}}{CH}-CH_2R''$$

$$R''CH_2\underset{\diagdown O \diagup}{CH-CH_2} + R'NH.CH_2.\underset{\underset{OH}{|}}{CH}.CH_2R''$$

$$\downarrow$$

$$R''CH_2\underset{\underset{OH}{|}}{CH}.CH_2.\underset{\underset{R'}{|}}{N}-CH_2.\underset{\underset{OH}{|}}{CH}.CH_2R''$$

The use of primary, di-, or triamines such as ethylene diamine or diethylene triamine leads to cross-linking and the formation of extremely resistant films. When the two components are mixed, set and cure of the film takes place at normal temperatures but can be accelerated further by the use of heat. Generally the amount of amine used is about 6 percent on the weight of the resin, and the two are mixed shortly before use. These systems are therefore supplied as two components and the mixture has a limited 'pot-life'. This depends on the type of solvent used, but it is unwise to mix more than is required for one day's working. The pot-life is extended slightly when the resins are pigmented.

The amines used in these reactions are generally toxic and require careful handling. For this reason amine 'adducts' were introduced. In these materials the amino groups are rendered inactive by reaction with a proportion of the epoxy resin to give an adduct which can be handled safely. The amino groups are, however, available for reaction when mixed with additional epoxy resin. Adducts are available based on aliphatic and aromatic amines.

Solvent-free systems By the use of liquid (low molecular weight) epoxy resin and a liquid (amine) hardener it is possible to apply very thick coats and to avoid the faults which can arise from solvent retention in thick films. Drawbacks are high application viscosity and relatively short pot life. This type of coating is used where the absence of solvent is desirable as in painting tank interiors and underwater surfaces.

High solids systems incorporate a small quantity of solvent. This renders application a little easier and tends to lengthen the pot life. The use of this type of coating is increasing since it is environmentally more acceptable and application of fewer coats reduces labour costs.

Curing by polyamides

These resins (p. 205) contain the reactive group —CO.NH— and react with epoxy resins in a similar way to amines, but the reaction is slower and more amenable to control. The films produced are identical in many properties, but whereas polyamides yield the more flexible and water-resistant films, amines and amine adducts give films with higher chemical resistance. These differences are, on the whole, relatively small, and the lower toxicity of polyamides together with the more controllable reaction has resulted in these materials being more widely used.

Curing by isocyanates

The higher molecular-weight epoxy resins are used in order to furnish the maximum number of hydroxyl groups for reaction with the isocyanate. The resulting film shows very high chemical resistance, adhesion, and general mechanical properties.

Epoxy/pitch systems

Epoxy resins can be blended with certain grades of pitch and then cured by amines, amine adducts, polyamides or isocyanates to give products which are considerably cheaper than the straight cured resins and which can be applied in thick coatings to steel and other metals as protection against corrosion. Building materials such as concrete, brick and stone also can be protected.

They are two-pack systems and are used on permanently or intermittently immersed structures such as piles, pier supports and ships' bottoms. They are unsuited to decorative work as the pitch 'bleeds' into the majority of superimposed coatings.

General properties and applications of epoxy resins

The mechanical properties of the films – high adhesion to the majority of clean surfaces combined with high impact and abrasion resistance – render epoxy coatings very suitable for use as floor paints which have to withstand heavy traffic. The solvent-free type is used in building mastics and in resin-bonded screeds for seamless floors.

The high resistance to chemicals and many solvents makes the coatings suitable for use on alkaline surfaces such as concrete, cement render or cement screeds. They are used also in tank lining paints to resist lubricating oils and other petroleum products.

The heat resistance of epoxy films is moderate. They will withstand dry heat to about 150°C but under immersed conditions the maximum temperature is of the order of 80°C.

On exterior exposure epoxy films tend to chalk and so are generally used in primers for the more weather resistant polyurethanes (p. 403).

Large quantities of epoxy resins are used in powder coatings.

Blast primers based on epoxy resins are used in considerable quantities in the constructional and ship building industries. Pigmentation is by zinc powder ('zinc rich' primers), zinc phosphate, iron oxide or iron oxide/zinc chrome.

Epoxy esters

The hydroxyl groups in epoxy resins can be esterified with fatty acids. Acids derived from linseed and dehydrated castor oils are commonly used. The linseed type containing 50–55 percent fatty acid is soluble in white spirit. With driers this type yields durable paint films and the paints are used on storage tanks and other exposed steelwork. They are used also on implements and machinery.

A problem sometimes arises in overcoating undercoats or primers based on epoxy esters. During the drying/curing period the film exhibits, over a period of several hours, a sensitivity to solvent with the result that brush application of a further coat during this period tends to lift the film. Outside this 'sensitive' period, which lasts from about 4 to 16 hours after application, overcoating can proceed normally. In some cases the difficulty can be overcome by a change in the proportions of drier metals.

The dehydrated castor oil type usually contains about 40 percent fatty acids, is soluble in xylol but not in white spirit. It will cure either by air drying (with driers) or by stoving and is used in chemical and abrasion resisting primers, undercoats and finishes for plant and machinery.

POLYAMIDE RESINS

These resins are formed by condensation of polyamines with dimer (or other polybasic) acids, the nature of the product being determined by the functionality rules enunciated by Carothers.

Dimer acids are long-chain aliphatic dicarboxylic acids containing two or more alkyl side chains. They are obtained by polymerization of refined unsaturated C_{18} fatty acids, and the long chains present confer a high degree of flexibility.

Apart from their use in polyamide resins they are used as part-replacements for dibasic or fatty acids in alkyds and epoxy esters.

Polyamide resins have two important applications.

1 They are used as curing agents for epoxy resins (p. 203). For this purpose they are more convenient to handle than the amines, but there are slight differences in the properties of the products.

2 When reacted with alkyd resins they produce the 'thixotropic' alkyds which are used in 'non-drip' decorative paints. The polyamide is added during the final stage of alkyd production, and the structure is thought to result from hydrogen bonding between carboxyl and amine groups.

POLYURETHANE RESINS

The basis of formation of these polymers is reaction of an isocyanate group (-NCO) with compounds containing an active hydrogen atom. The latter attaches itself to the nitrogen atom of the -NCO group. A number of materials contain active hydrogen atoms, and the reactions of an isocyanate group with some of these can be represented as follows..

$$-N=C=O + -OH \rightarrow \underset{\text{urethane 'unit'}}{-\overset{\large H}{\overset{|}{N}}-\overset{\large O}{\overset{\|}{C}}-}$$

1 *Reaction with hydroxyl group*
This reaction is used in the formation of surface films. A di-isocyanate is reacted with hydroxyl-containing polymers such as polyesters, alkyds, epoxies or acrylics.

2 *Reaction with carboxyl group*

$$-N=C=O + HOOC- \rightarrow -NH-CO- + CO_2$$

3 *Reaction with amine*

$$-N=C=O + H_2N- \rightarrow -NH.CO.NH-$$

(substituted urea)

4 *Reaction with water*

$$-N=C=O + H_2O \rightarrow -NH_2 + CO_2$$

This reaction is used in foam production. The amine produced reacts immediately with further isocyanate to give a substituted urea, as in **3**. The substituted urea can react further with free -NCO groups to give a highly cross-linked structure.

Film formation by moisture-cured polyurethanes results from a train of reactions of this nature.

In reaction **1**, the hydroxyl groups may be phenolic or aliphatic, and the latter may be primary, secondary or tertiary. Saturated or unsaturated compounds may be involved.

The isocyanates may be long- or short-chain, aliphatic, or aromatic. Similar comments apply to reactions **2** and **3**. It is therefore apparent that a very great number of technical combinations are theoretically possible.

Materials for polyurethanes

Polyurethanes are two-pack materials, i.e. the material containing the active hydrogen (polyester, alkyd or epoxy) and the isocyanate are packed in separate containers and mixed together immediately before use. The chemical reaction commences at once and consequently such mixed materials have a limited 'pot life'.

Isocyanates

Toluene di-isocyanate (TDI) is widely used in the form of polyisocyanate produced by treatment of the monomer. This, and other polyisocyanates, has a low vapour pressure and usually contains less than 0·7 percent of the highly toxic monomer. Inhalation of the latter can result in serious respiratory effects and can lead to impairment of lung function [5]. The dangers associated with the use of polyisocyanates prompted the Paintmakers' Association to issue a document entitled 'The Hazards arising from the Manufacture of Polyurethane Paints and Varnishes'. The toxic effects are confined to the isocyanate raw materials. The resulting polyurethanes are not hazardous (apart from solvent), provided no unreacted isocyanate is present. Films produced from aromatic polyisocyanates yellow on exposure; the aliphatic compounds do not show this defect.

Resins

The resins most widely used for reaction with di-isocyanates are saturated polyesters, the general requirements being a high hydroxyl value and an acid value as low as possible to avoid liberation of carbon dioxide.

The nature of the film produced will depend on the chemical structures of both polyester and isocyanate. Soft films are formed from polyesters based on aliphatic acids, but harder and sometimes brittle films result from the use of alkyds based on phthalic and terephthalic acids. The type of polyol used in the resin also affects the film properties. Glycerol and pentaerythritol give hard films by cross-linking, but diols such as 1,4-butanediol or ethylene glycol give much softer films.

In view of the very wide range of properties, especially flexibility and hardness, required for different types of application and surface, manufacturers find it convenient to produce a limited range of polyesters of widely differing characteristics. The desired properties are then obtained by mixing the polyesters in the appropriate ratios.

Solvents

The polyesters used are soluble in esters and ketones, and a balanced mixture of low-, medium-, and high-boiling types is generally used. In some cases toluene or xylene is used as a diluent to reduce costs. Alcohols must be avoided. The di-isocyanates are generally dissolved in esters. All solvents should be dry to avoid reaction between water and the isocyanate group, leading to formation of carbon dioxide.

Film formation

Two-pack type

This comprises the largest class and is used for both clear and pigmented coatings. The ratio of isocyanate to polyester is important in that low ratios of isocyanates give softer films.

Calculations of reacting ratios can be difficult with polyesters which may contain more than one polyol. The di-isocyanates are fairly pure and their equivalent weight is known. The polyester is calculated on the basis of an experimental value known as *isocyanate equivalent*. This is the weight of polyester (in grams) which reacts with one gram-equivalent of isocyanate. The blend of polyesters is chosen to give the desired properties, and the isocyanate required is calculated from their isocyanate equivalents.

Polyester and isocyanate are made up in separate containers and mixed in the appropriate ratio shortly before use.

In pigmented finishes the pigment is dispersed in the polyester, and most pigments can be used safely. However, basic pigments such as red lead, zinc oxide and zinc chrome as well as Prussian Blue cause premature gelling of the system.

Additions of cellulose acetobutyrate or ethyl cellulose are often made to improve the flow properties and to avoid *cissing*.

Properties and uses

The curing of polyurethane films will take place at temperatures down to 0°C, a property which gives them an advantage over the epoxies. Depending on the type of polyester or polyol and hardener, films can be produced in a wide range of hardness and flexibility. It is usual for manufacturers to stock a limited range of polyesters and polyols and to blend these according to requirements. The films are characterised by good adhesion to most clean surfaces but for exterior exposure, where they exhibit excellent weather resistance, they are generally applied over an epoxy primer. The films tend to yellow if aromatic isocyanates are used but retain their colour with aliphatic isocyanates. Other film properties are good resistance to moisture, chemical attack and many solvents.

Moisture-cured type [b]

The polyether triols (prepared by reaction of glycerol with polypropylene glycol) contain secondary hydroxyl groups and react more slowly with isocyanates than do the normal polyesters. If these are mixed with an excess of di-isocyanate a reaction occurs giving virtually a solution of a prepolymer. Films of this material when exposed to the atmosphere react with the moisture present and cross-linking takes place.

These materials are marketed as solvent-free liquids which can be applied directly, with thinning if necessary, to a variety of surfaces. The films are very hard, adherent, glossy and show high abrasion resistance. They are used on wooden floors and also on concrete and similar floors where their alkali resistance is an advantage.

Pigmentation inevitably introduces traces of moisture which have to be removed. This can be affected by addition of an isocyanate 'moisture scavenger' [7] which is added to the paint immediately after manufacture. Reaction with the moisture (with evolution of CO_2) takes place before the cans are closed.

Blocked urethanes

These are prepolymers in which an isocyanate has been reacted with a 'blocking agent' to produce a product from which the blocking agent is driven off on stoving. The regenerated isocyanate can then react with hydroxyl-containing materials [8]. Among the blocking agents used are phenol, methylethyl ketoxime, caprolactam and diethyl malonate.

One-pack urethane alkyds

These are prepared by forming a mixture of mono- and di-glycerides from a drying oil and reacting the free hydroxyl groups with a di-isocyanate. The product is soluble in white spirit and cures by the normal drying process, usually with the addition of cobalt drier. One-pack urethane alkyds dry to hard and tough films, but the overall resistance is not as high as that of the two-pack type of polyurethane. The film hardness and resistance to solvents and chemicals is superior to normal alkyds, and they can be used to upgrade these materials.

The tough, hard wearing properties of the film has resulted in their wide use as clear varnishes for marine work. In pigmented form they are used where impact and abrasion resistance is necessary as, for example, in paints for kitchen or office furniture.

VINYL RESINS

Vinyl resins are straight-chain thermoplastic polymers produced by addition polymerization of compounds containing the vinyl group $CH_2=CH-$. The polymers can be made from straight monomers to give homopolymers or from two or more monomers to give co-polymers. In this way a very large number of co-polymers is possible, for example by polymerizing vinyl acetate with other unsaturated monomers. The polystyrene and acrylic resins are also members of this class.

Methods of polymerization

Two methods are employed: (a) *solution polymerization* in which the monomer is dissolved in a suitable solvent, and (b) *emulsion polymerization* in which the monomer is emulsified in a non-miscible solvent, usually water. In each case, a catalyst, e.g. benzoyl peroxide, is present. Polymerization is catalysed also by heat, irradiation or light.

Emulsion polymerization is cheaper than the solution method and is also amenable to a greater degree of control. The process can be stopped at any stage of the polymerization and, in addition, emulsions can be produced in any desired particle-size range.

Polymers which are to be used in solution form can be separated by filtration after acidification of the emulsion. The product is dried and dissolved in a suitable solvent. Considerable quantities of the polymer emulsions are used directly in the manufacture of latex emulsion paints.

The following are the principal types of vinyl resin.

Polyvinyl acetate (PVA) is a colourless resin with a low softening point (about 38°C, 100°F), non-toxic and thermoplastic. It is supplied in a number of grades differing in degree of polymerization and, to some extent, solubility characteristics. The lower grades are soluble in ethyl alcohol, esters, ketones, and aromatic solvents. They are insoluble in white spirit. Films produced from solutions of the resin tend to be brittle and lack adhesion on smooth surfaces. They are usually plasticized with up to 20 percent ester-type plasticizer such as dibutylphthalate, but gradual loss of plasticizer by volatilization or absorption into a porous substrate will lead ultimately to embrittlement. The plasticized forms possess improved adhesion, but all films of polyvinyl acetate tend to absorb appreciable quantities of moisture.

The polyvinyl acetate chain has the following structure:

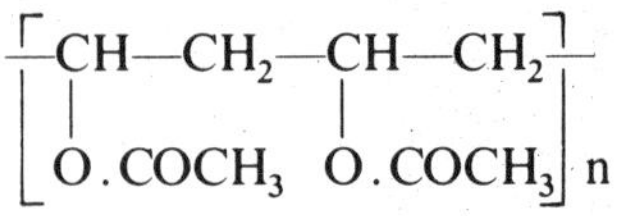

$$\left[\begin{array}{cccc} -CH- & CH_2- & CH- & CH_2- \\ | & & | & \\ O.COCH_3 & & O.COCH_3 & \end{array} \right]_n$$

Polyvinyl alcohol This is produced by the incomplete hydrolysis (88–89 percent) of polyvinyl acetate. It is used in paper coatings, strippable coatings and in release agents for cast resin moulds.

Polyvinyl formal Reaction of polyvinyl alcohol with formaldehyde yields polyvinyl formal:

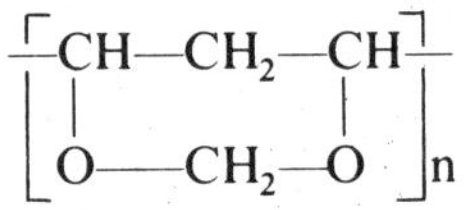

$$\left[\begin{array}{ccc} -CH- & CH_2- & CH- \\ | & & | \\ O- & CH_2- & O \end{array} \right]_n$$

In combination with phenol-formaldehyde resins, polyvinyl formal is used in wire enamels.

Polyvinyl butyral This is produced in a similar way to polyvinyl formal, using butyraldehyde in place of formaldehyde:

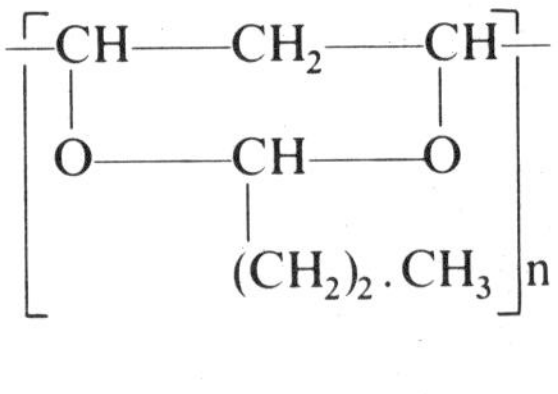

$$\left[\begin{array}{ccc} -CH- & CH_2- & CH- \\ | & & | \\ O- & CH- & O \\ & | & \\ & (CH_2)_2.CH_3 & \end{array} \right]_n$$

Polyvinyl butyral is used in etch and some types of blast primer. Both butyral and formal resins show very good adhesion to most surfaces.

Polyvinyl chloride (PVC) is produced from vinyl chloride ($CH_2 = CH.Cl$) by one of the recognized methods of polymerization, i.e. bulk, emulsion, or suspension. The operation must be carried out under very close control in view of the recognized carcinogenic properties of vinyl chloride.

It is a colourless polymer with a high Glass Transition Temperature (Tg) of approximately 80°C (176°F). (Tg is the temperature at which a polymer changes from a hard rigid state to a soft plastic condition.)

PVC has very limited solubility in common solvents, and in surface coatings it is used either as a suspension (plastisol or organosol) or as a co-polymer (*below*).

Plastisols consist of PVC dispersed in a suitable plasticizer. They are usually pigmented and yield thick coats which are stoved at 170°–193°C (338°–379·4°F). At this temperature the PVC absorbs the plasticizer and becomes sufficiently soft to form a continuous film. Such films are extremely hard and resistant to most reagents. On bare metal the adhesion is sometimes suspect, and they are often applied over an epoxy primer.

Organosols are dispersions of PVC in plasticizer and thinner. They are pigmented and can give thinner and smoother coats than the plastisols. They require similar stoving treatment.

The solids content of organosols is higher than that of vinyl solutions, and they are replacing these solutions in the internal coating of metal containers.

Both organosols and plastisols are used in the coil coating of metal strip.

Vinyl co-polymers

Vinyl acetate can be co-polymerized with a wide range of other unsaturated monomers to give a variety of products with any desired degree of hardness or flexibility. The need for added plasticizer, so essential with the homopolymer, polyvinyl acetate, is eliminated. Among monomers commonly used are acrylates, maleates, and caprates. Emulsion polymerization is the common method of production.

In the decorative field very considerable quantities of vinyl co-polymer emulsions are used. These have, for a number of years, been based on the co-polymers mentioned above, but a definite improvement was made when ethylene was co-polymerized with vinyl acetate. These co-polymers, containing 8–25 percent of ethylene, have proved to be superior to other co-polymers in durability, resistance to reagents, and non-yellowing properties. An additional advantage, particularly in cold countries, is in film formation, i.e. coalescence of the dispersed polymer. This will take place down to 0°C (32°F).

A range of co-polymers of vinyl chloride and vinyl acetate containing 86–90 percent vinyl chloride is available [9]. These polymers are thermoplastic, non-reactive, very resistant to moisture penetration and to chemical attack. Some members have been partially hydrolysed in order to introduce polar groups to enhance the adhesion to smooth metal and other surfaces. The non-hydrolysed material is generally suspect in this respect.

Vinyl chloride/vinyl acetate co-polymers are soluble, to varying degrees, in aromatic hydrocarbons, esters, ketones and chlorinated hydrocarbons. The solutions generally contain 15–30 percent solids so that multicoat systems are required to give a reasonable film thickness.

They are used in paints for the protection of chemical and process plants, particularly if these are in exposed situations. Some members are compatible with certain long oil alkyds to which they are added to upgrade chemical resistance and to shorten the drying time and overcoating interval.

POLYSTYRENE RESINS

Styrene or vinyl benzene, $C_6H_5CH = CH_2$, is the basic monomer for this type of resin. It is manufactured by reacting benzene and ethylene at high temperature and pressure in presence of a catalyst. Ethyl-benzene is produced which is then dehydrogenated to yield styrene:

$$C_6H_6 + CH_2 = CH_2 \rightarrow C_6H_5CH_2CH_3$$
$$C_6H_5CH_2CH_3 \rightarrow C_6H_5CH = CH_2 + H_2$$

The product is then fractionally distilled in vacuum, sulphur being present to prevent polymerization of the styrene. This and other methods of synthesis, as well as the properties and polymerization of styrene, have been described in the literature [10].

Vinyl toluene, $CH_3C_6H_4CH = CH_2$, which consists of a mixture of the meta and para isomers, is manufactured in a similar manner. During the preparation a proportion of the ortho isomer is produced. This is separated by fractional distillation under vacuum.

Properties of styrene

It is a mobile liquid with a characteristic odour and polymerizes readily. Premature polymerization is inhibited by addition of a very small quantity of *p*-tertiary butyl catechol or hydroquinone.

Polymerization of styrene

Polymers of styrene itself are of little interest to the paint industry, but styrene forms a number of interesting and useful co-polymers. The co-polymerization reaction generally requires free-radical initiation and can be carried out either in solution (in aromatic hydrocarbon or carbon tetrachloride) or in emulsion. The exclusion of air is advisable, since oxygen retards the reaction.

Emulsion polymerization is the most widely used method, the free radical initiation being effected by either a redox system or a persulphate.

Co-polymers of styrene

Co-polymers of styrene and acrylic acid derivatives form the basis of many thermosetting acrylic systems (see p. 213).

The co-polymers of vinyl toluene and acrylic esters, in solution form, are used as binders in exterior masonry finishes. They show good durability. Styrene becomes an integral part of polyester finishes which are described on p. 197.

One of the most interesting co-polymerization reactions of styrene and, more particularly, vinyl toluene, is with alkyd resins. These are described on p. 196.

ACRYLIC RESINS

These are closely related to the vinyl resins. The acids acrylic, $CH_2 = CH.COOH$ and methacrylic $CH_2 = C(CH_3)COOH$ as well as their esters can be polymerized to form long-chain thermoplastic resins:

$$CH_2 = \underset{\underset{\displaystyle CH_3}{|}}{\overset{\overset{\displaystyle COOCH_3}{|}}{C}} \rightarrow -CH_2-\underset{\underset{\displaystyle CH_3}{|}}{\overset{\overset{\displaystyle COOCH_3}{|}}{C}}-CH_2-\underset{\underset{\displaystyle CH_3}{|}}{\overset{\overset{\displaystyle COOCH_3}{|}}{C}}-CH_2-$$

Methyl methacrylate **Polymethyl methacrylate**

Polymers of methyl methacrylate and other esters containing alkyl side chains are hard brittle solids but those from the 'straight' esters are soft and flexible. By co-polymerising mixtures of the two types it is possible to produce polymers with any desired degree of hardness and flexibility. These are the so-called '100% acrylics'.

Acrylic acid, as well as the straight and branched esters, can be co-polymerised with other unsaturated compounds such as styrene, vinyl toluene, vinyl acetate or vinyl chloride to give a variety of interesting and useful products.

Polymerisation of the acrylic monomers takes place readily by a free radical mechanism and is catalysed by heat and oxidising agents such peroxides. The reaction can be carried out by either solution or emulsion methods.

Properties of acrylic polymers and co-polymers

Generally these are colourless, thermoplastic solids soluble in esters, ketones and aromatic solvents. They are insoluble in aliphatic hydrocarbons and are resistant to water, alkali, mineral oils and greases. They also possess good lightfastness, adhesion and electrical properties. The exterior durability is very good, especially the co-polymers with styrene and vinyl toluene which are used as binders for masonry finishes. The co-polymers with vinyl acetate are used in emulsion paints.

Co-polymers of methyl methacrylate with butyl acrylate or ethyl hexyl acrylate have proved successful in acrylic gloss paints [11].

Some of the softer members can be blended with chlorinated rubber and a small proportion of chlorinated paraffin to give tougher films than are obtained with chlorinated paraffin as the sole plasticiser. The films show better colour retention, durability and overcoating properties [12]. Good adhesion to galvanized and etched zinc surfaces is also claimed.

Thermosetting acrylic resins

The acrylic resins described above are long-chain thermoplastic materials and so not suited to stoving procedures. It is, however, possible to introduce reactive groups, spaced along the polymer chain, which can be cross-linked with other groups to give complex three-dimensional structures. Groups introduced include carboxyl, hydroxyl, amino, and epoxy.

Thermosetting acrylic resins are, in general, based on the monomers methacrylic ester, acrylic ester, and styrene. Other monomers are incorporated to provide the reactive groups. The nature and relative amounts of monomers control the flexibility and other properties. For example, methacrylates and acrylates are used for maximum durability. The incorporation of styrene or vinyl toluene yields finishes resistant to detergents and household chemicals, particularly when cross-linked with epoxy resin (see *below*).

Carboxyl groups can be introduced by the use of a proportion of methacrylic acid. Cross-linking of the product with epoxy resins gives a film of outstanding chemical resistance, hardness, and adhesion.

Hydroxyl groups are introduced by using hydroxyalkyl methacrylates or acrylates. The resulting resin can be cross-linked with hexamethoxymethyl melamine to give films of outstanding weather-resistance and gloss retention.

They are used as stoving finishes on cars or can be used as acid catalysed lacquers.

Acrylic resins incorporating acrylamide can be reacted with formaldehyde and butanol giving a butoxymethyl acrylamide monomer, containing the group.

$$—CH_2—\overset{|}{C}H—CO—NH.CH_2OC_4H_9$$

This group is similar to that in a butylated amino resin and will cross-link with other materials containing free —OH groups such as alkyds and epoxies. When cross-linked with the latter, the resulting films have outstanding resistance to fats, oils and household chemicals. They are, therefore, used on domestic appliances.

Acrylic systems used for the finishing of furniture are suited to curing by ultra violet radiation. Whilst this form of cure has always proved successful with clear coatings, in the earlier years difficulties were encountered with pigmented coatings, particularly the high build types. Developments in photo-initiators have now largely overcome this problem [13].

SILICONE RESINS

These are semi-inorganic high polymers containing both silicon and carbon, and from these elements it is possible to prepare a large number of silicone polymers according to the number and types of groupings. They can be 'tailored' for a great many specific purposes and are marketed as resins (solutions), oils, and rubbers.

Formation of silicones

The chemical reactions involved in the formation of silicone resins are many and complex. The following scheme indicates the general reactions involved:

$$\text{Silica } (SiO_2) \rightarrow \text{silicon} \xrightarrow[\text{halide + Cu}]{\text{alkyl or aryl}} \text{organic chlor-silanes e.g. } R_1SiCl_3,\ R_2SiCl_2$$

$$\xrightarrow{\text{hydrolysis}} \text{Silanols, e.g. } R_1Si(OH)_3,\ R_2Si(OH)_2 \xrightarrow{\text{heat}} \text{Silicones}$$

Properties of silicones

The nature of the alkyl and/or aryl groups determines, to a large degree, the properties of the resin. Methyl silicones are useful as waterproofing agents but find no application as paint materials. The introduction of a phenyl group confers heat resistance, but pure phenyl silicones are too brittle and fuse readily. All-round properties of a high standard are exhibited by co-polymers of methyl and phenylsilicone. These are far superior to the methyl or phenyl resins in respect of heat resistance, flexibility, electric properties, and durability.

Silicone resins are supplied in solutions in aromatic solvents, and can be thinned with similar solvents as well as esters and ketones. They are not soluble in white spirit.

Silicone resins will not cure at ambient temperatures and require heating to 200° to 250°C (392° to 482°F) for about two hours for complete cure. The process can be accelerated by the use of the naphthenates or octoates of zinc, cobalt or iron.

The films produced can vary from rubbery to glass-like hardness. The softer types are the more resistant to heat and weathering, but to obtain the necessary mechanical properties the two types can be blended.

Uses of silicones

Silicone resins are used as the sole binder in paints to withstand high temperatures, particularly when pigmented with aluminium. Such paints require a curing period of 2 hours at 200°–220°C (392–428°F) after which they will withstand prolonged exposure to temperatures up to 430°C (806°F). With other heat-resisting pigments the service temperature is lower, about 280°–300°C (536°–572°F).

Binders consisting of co-polymers of alkyds and silicones, containing up to 30 percent silicone, are now being used for gloss paints with outstanding weather-resistance – see p. 196.

References to Chapter 11

[1] Kienle, R. H., *JSCI*, 1936, **55**, 229T.
[2] Mark, H. and Whitby, G. S. (Eds), *The Collected Papers of W. H. Carothers on High Polymeric Substances*. Interscience, New York, 1950.
[3] Holmberg, K., The role of the Diels-Alder reaction in the curing of drying oil alkyd-melamine systems. *JOCCA*, 1978, **61**, 356.
[4] *Idem*. The mechanism of the acid catalysed curing of alkyd-melamine systems. *Ibid*, 359.
[5] BIP Chemicals Ltd.
[6] *see also* Hughes, A. H., Polyurethane resins in moisture curable surface coatings. *JOCCA*, 1986, **69**, 10.
[7] 'Additive T1' (Bayer).
[8] Stieveter, P. C., Advances in environmentally acceptable polyurethanes. *JOCCA*, 1987, **70**, 262.
[9] 'Vinylite' resins (Union Carbide Ltd.).
[10] Warson, H., *PPCJ*, 1971, Sept. 17, 259.
[11] Sykes, D., *PPCJ*, 1987, Nov. 25, Suppl. p. 12.
[12] Alloprene booklet. ICI (Mond Division).
[13] 'Irgacure' (Ciba – Geigy).

12

Cellulose ester and ether products

Cellulose comprises the bulk of the fibrous tissue of all plants, and the purest form is obtained from cotton in which it forms a protective fibre round the seeds. It is a member of the carbohydrate class and the composition of the *repeating unit* can be represented by the formula

$$C_6H_7O_2(OH)_3 \quad \text{or} \quad \left[\begin{array}{c} \quad\quad CH_2OH \\ \quad\quad | \\ \quad\quad CH-O \\ \quad / \quad\quad \backslash \\ -O-CH \quad\quad\quad CH- \\ \quad \backslash \quad\quad / \\ \quad\quad CH-CH \\ \quad\quad | \quad\quad | \\ \quad\quad OH \quad OH \end{array}\right]$$

but the actual molecular weight is very high. The unit depicted above contains three alcohol groups, one of which is primary and the remaining two are secondary. The range of esters and ethers derived from cellulose is produced by reactions of these hydroxyl groups, and it is theoretically possible to produce mono-, di- and tri-derivatives. In practice, mixtures are usually obtained.

Esters of cellulose have been known for a considerable time and in 1855 a patent was taken out by Parkes in which he suggested the use of nitrocellulose solutions as protective coatings. He later suggested the use of camphor as a plasticizer and this led to Hyatt's introduction of celluloid in 1869.

At first the use of cellulose esters as paint media was severely restricted by lack of solvents of suitable evaporation rate. The introduction of amyl acetate in 1882 provided a solvent of reasonably slow evaporation, but the high viscosity of the solutions necessitated considerable dilution before application, resulting in very thin films. In addition the plasticizer, camphor, slowly volatilized from the film, leading to embrittlement and breakdown. The production of low-viscosity cellulose nitrate and the introduction of a wider range of solvents and plasticizers led to the mass-production of cellulose nitrate lacquers, and over the last fifty years very large quantities have been produced.

CELLULOSE ESTERS

Cellulose nitrate

This is very often - but erroneously - called 'nitrocellulose'. It was first produced by Schoenbein in 1845, who treated cellulose with a mixture of nitric and sulphuric acids.

The modern method of manufacture employs the same reaction, the sulphuric acid acting both as a dehydrating agent to absorb the water produced and as a catalyst. The cellulose used is either cotton linters, which are short fibres attached to the seed and remaining after the longer fibres have been removed for spinning, or purified wood-pulp. The linters are subjected to preliminary treatments to remove fragments of seed and other impurities and are often given a mechanical treatment to open out the fibres before nitrating.

The cellulose is then treated in stainless steel vessels with a mixture of nitric and sulphuric acids, and the temperature controlled carefully until the desired degree of nitration is attained. The mixture is then run into a stainless steel centrifuge where the bulk of the acid liquor is separated. The crude cellulose nitrate is drowned in water and then subjected to several washings with fresh water. Removal of acid impurities is essential for stability of the material in use, and after washing with water it is heated in slightly acidified water to about 140°C (284°F). This serves to release any sulphuric acid entrapped in the fibres and also reduces the viscosity of the product. The cellulose nitrate is then washed with hot water and centrifuged.

The material still contains about 30 percent water but cannot be dried owing to the dangerous nature of the dry compound. It is therefore 'dehydrated' by replacing the water by methylated spirit, butanol or isopropanol. The replacement of water by alcohol is carried out either by forcing the alcohol through the wet cellulose nitrate cake in a press or by spraying the material with the alcohol while it is being spun in a centrifuge.

The use of the higher alcohols has the advantage of more complete dehydration as they are less hygroscopic than the lower members. Nevertheless the material is never completely anhydrous. Cellulose nitrate then reaches the user 'damped' with one of the above alcohols; it should not be allowed to dry out after the drums are opened.

Properties of cellulose nitrate

The degree of nitration depends on conditions, and cellulose nitrates are divided into three classes according to the nitrogen content:

10·5–11·2% Low nitrogen (L)
11·2–11·8% Medium nitrogen (M)
11·8–12·2% High Nitrogen (H)

The nitrogen content has an important bearing on the choice of a cellulose nitrate for a given purpose. The most widely used are those with a high nitrogen content, whilst the low-nitrogen grades tolerate large amounts of alcohol and are used in paper lacquers.

The viscosity of a cellulose nitrate solution is of great importance and is influenced by the type of material and the nature of the solvent blend. It is generally measured in the form of a standard solution in aqueous acetone, the solvent consisting of 95 percent acetone and 5 percent water by volume. The concentration of the solution varies with the grade of cellulose nitrate being tested, and viscosity ranges are designated by letters H (high), M (medium), L (low) and X (extra low). The accepted method of designating a cellulose nitrate gives information on type, nitrogen content and viscosity. For example, a material designated DHX 30/50 is a dense (D) type with high nitrogen content (H) and extra low viscosity (X) lying between 30 and 50 poise.

Testing

The quality of cellulose nitrate must be carefully controlled to ensure successful results. The following tests are among the most important.

Nitrogen content The older method consisted of measurement of the volume of nitrogen evolved when a known weight of the material is decomposed with concentrated sulphuric acid in a Lunge nitrometer. In the modern method the cellulose nitrate is hydrolysed by sulphuric acid in glacial acetic acid solution. The liberated nitric acid is estimated by titration electrometrically with standard ferrous ammonium sulphate solution in sulphuric acid.

Viscosity The solution of standard concentration in aqueous acetone is placed in a wide vertical tube in a thermostatically controlled bath at 20°C (68°F). The viscosity is then determined by the falling sphere method (p. 164).

Stability At normal temperatures a properly prepared cellulose nitrate is stable, or at least decomposes at a rate that can be ignored. However, any slight tendency to instability, i.e. decomposition, is accelerated by heat, and resistance to heat is therefore used to assess stability. It is usually carried out by the Bergmann and Junk test in which the cellulose nitrate is heated for 2 hours at 132°C (270°F) and the acid decomposition products are absorbed in water. These are estimated by titration.

Volatile content When lacquers are prepared from 'damped' cellulose nitrate the alcohol content must be taken into consideration. It is usually about 30 percent but is likely to vary. The figure is determined by heating a known weight of the damp material in an aluminium dish in an oven at 60° to 65°C (140° to 149°F). Heating is continued until the weight is constant, and high temperatures must be avoided. The dry material should be re-damped as soon as the estimation is completed.

CELLULOSE LACQUERS

Cellulose lacquers are not simply solutions of cellulose derivatives, but also contain plasticizers to give the necessary film flexibility, and resins to increase the gloss and adhesion. They may also contain non-solvents or diluents to control the cost. A large proportion of the lacquers used is pigmented.

The performance is controlled by the type of cellulose derivative used and by the cellulose - plasticizer - resin ratio. The large number of materials available renders it possible to design cellulose lacquers for a very wide range of uses. Some of the general principles involved in lacquer formulation are discussed below.

Cellulose nitrate lacquers

Cellulose nitrate is the principal component of the film in most cases and there is a wide variety of types available. A low-viscosity type permits the formulation of high-solids lacquers leading to high-build films. Higher viscosity cellulose nitrate gives lacquers of lower solids at spraying viscosity, but the solids can be increased by the use of suitable resins. These lead to greater gloss and hardness of the film, and the types used are discussed below.

Ethyl cellulose is compatible with cellulose nitrate in all proportions and can be incorporated to reduce flammability and increase flexibility.

Resins These are used to improve the gloss and adhesion of the film. The film produced from cellulose nitrate alone is a semi-gloss and has poor adhesion. The resin must have good compatibility with cellulose nitrate and plasticizer in all proportions, and the amount used varies considerably.

Damars These are extensively used on account of good solubility, the dewaxed type being the most suitable. They have a pale colour and confer good gloss and adhesion. Damars are insoluble in alcohol and so cannot be used in the low-nitrogen type spirit lacquers.

Sandarac and **mastic** can be regarded as resins of the same type as damar but have the advantage of ready solubility in alcohol.

Dewaxed shellac is used in the spirit type of lacquer where it confers good adhesion, toughness and flexibility. Such lacquers are not suitable for exterior exposure.

Ester gums The glycerol types are completely compatible with cellulose nitrate and soluble in the solvents used. The pentaerythritol types are much less compatible. Low acid value in the ester gum is essential, and the plasticizer content is very important. Generally the proportion of ester gum is high for interior finishes and low in finishes for exterior exposure.

Maleic resins show good compatibility and produce very hard films. These must be well plasticized to avoid brittleness. They are not recommended in lacquers for exterior exposure.

Cyclohexanone resins possess no acidity and are useful in clear lacquers for metals. They have good compatibility and confer good gloss and adhesion on the film.

Alkyds have replaced a number of the natural resins such as kauri, Congo copal, and rosin. Both drying and non-drying types are used, but the latter find widest use. The addition of a non-drying alkyd plasticizes the film and confers good gloss and durability. The films are also resistant to alcohol and to heat.

The castor-oil modified types are used in greatest quantity, but other non-drying types are used for special purposes. The general requirements of such an alkyd are stated [1] to be: fatty acid content 30 to 40 percent; phthalic anhydride 40 to 50 percent; acid value 5 to 25.

Plasticizers

A characteristic of cellulose esters is considerable swelling on addition of solvent before actual dispersion or solution. When a film of the ester solution dries, the process is reversed, resulting in a contraction of the film and a tendency to pull away from the surface. The function of a plasticizer is to correct this tendency and to confer elasticity on the film without destroying the hardness. The ratio of plasticizer to cellulose ester is important and generally is higher for the low-viscosity grades. If a proportion of ethyl cellulose is used in the lacquer, the amount of plasticizer can be reduced.

The ideal plasticizer for high-quality lacquers would be freely soluble in the solvents used, completely compatible with the ester and resin, non-volatile, and chemically inert. With such a material the film would retain its elasticity indefinitely. In practice it is difficult to obtain the desired properties with one plasticizer, and blends are generally used. Even so, the ideal requirements are never completely met.

A great many plasticizers are available and the following are in common use.

Dibutyl phthalate	Tricresyl phosphate
Dioctyl phthalate	Triphenyl phosphate
Dimethylglycol phthalate	Trichlorethyl phosphate
Tributyl citrate	Castor oil

Solvents (cf. Solubility Parameter, p. 136)

The volatile portion of a lacquer usually consists of a blend of liquids, and these are conveniently divided into three groups, based on solvent power.

True solvents form the major portion of the volatile constituents.

Latent solvents are not solvents for the principal film-formers, but increase the solvent power of the other volatile ingredients.

Diluents are not solvents for the principal film-formers but are added to reduce cost. They may also improve the solvent power of the mixture for the resin and plasticizer.

The range of true solvents is not extensive, and typical solvents for cellulose nitrate are esters, ketones and glycol ethers. Examples of the first two classes are:

Esters	*Ketones*
ethyl acetate	acetone
isopropyl acetate	methyl ethyl ketone
n-butyl acetate	methyl isobutyl ketone
ethyl lactate	methyl cyclohexanone

Evaporation rate

The rate of evaporation of a solvent is one of its most important properties and is usually evaluated against ether. A solvent which evaporates too rapidly can cause severe chilling of the film to a temperature below the dew point; and moisture deposition on the soft film can lead to blushing. Also, if the true solvent leaves the film before the latent solvent or diluent, precipitation of the film-former can take place.

Viscosity

The viscosity of a solution of cellulose nitrate depends on the concentration and grade of ester used, the composition of the solvent mixture, and the temperature. The viscosity falls with increasing temperature as is usual. The effect of solvents is not as simple. With simple solvents, viscosity of the solution is lower with low molecular weight solvents than with those of high molecular weight. A mixture of solvents generally gives a lower viscosity than the same proportion of a single solvent. Thus a mixture of alcohol and butyl acetate gives a solution of much lower viscosity than butyl acetate alone. The alcohol here is acting as a *latent solvent*. Alcohols alone are not solvents for cellulose nitrate but they possess a latent solvent power when mixed with other solvents. In practice an alcohol is usually present as the damping agent for cellulose nitrate and its presence is therefore an advantage.

Diluents

In order to keep the cost of a lacquer as low as possible, only sufficient solvent is used to dissolve the cellulose nitrate, the final consistency being attained by dilution with an aromatic or petroleum hydrocarbon. These are not actual solvents but are compatible with the mixture. The proportion of diluent used will depend on its compatibility with the active solvent and this is determined by the dilution ratio. This is given by the expression

$$\text{Dilution ratio} = \frac{\text{Volume of diluent which will just precipitate the dissolved cellulose nitrate}}{\text{Volume of solvent used}}$$

The determination is carried out by adding the diluent to a solution of cellulose nitrate in the specified solvent until the cellulose nitrate commences to precipitate and will not redissolve in the mixture. The value depends on the type of cellulose nitrate and the concentration. This is generally 10 percent. The dilution ratios in Table 12.1 are quoted by Durrans [1].

Table 12.1

Solvent	Toluene	Xylene	Petroleum	Butanol
Acetone	4·5	3·9	0·6	7·0
Amyl acetate (*n*)	2·2	2·2	1·4	7·3
Butyl acetate (*n*)	2·7	2·4	1·4	7·3
Butyl phthalate	2·8	2·7	1·7	8·0
Butyl tartrate	10·6	7·7	1·4	15·0
Di-acetone alcohol	3·1	2·9	0·5	7·8
Ethyl lactate	5·6	4·8	0·7	10·2
Glycol mono-ethyl ether	5·0	4·7	1·0	6·9
Methyl ethyl ketone	4·5	3·3		
Tricresyl phosphate	3·3	4·2	0·7	4·9

Pigmented lacquers

The dispersion of pigments in cellulose nitrate lacquers follows broadly the same lines as for gloss paints. The pigment content is generally lower, and the lower viscosities of the media can result in troublesome settlement of heavy pigments. The desired properties of the pigment are maximum opacity, freedom from any tendency to bleed in lacquer solvents, resistance to ultraviolet light and freedom from toxicity (this applies to spray paints and any finishes for children's playthings).

The majority of pigments used in gloss paints can also be used in cellulose nitrate lacquers, but care should be exercised with the following, which are either unsuitable or show the defects mentioned.

Zinc oxide	Reactive. Poor opacity but gives good gloss and colour. Can be used for interior work.
Ultramarine Blue	Poor strength and opacity.
Phthalocyanine Blue	Used mainly for tinting. When used alone shows tendency to give bronzy films of poor gloss.
Hansa Yellows	Tendency to bleed.
Ferrite Yellows	Poor gloss. Used for tinting.
Helio Red	Tendency to bleed.
Maroons	Many show poor exterior durability.
Lead Chromes Chrome Greens	Toxic.

Dispersion of the pigment can be effected in a ball or pebble mill, or the pigment can be dispersed in plasticizer alone on a triple roll mill after premixing. A slow evaporating solvent can be used to ease the material over the rolls, but the process is slow.

Chip method This method is now widely used and consists in dispersing the pigment in a plastic cellulose dough in a heavy mixer. Dispersion of the pigment is effected by high shear forces resulting from high power input and high consistency. The dispersion process is carried out in a Bridge–Banbury mixer which incorporates a pair of bladed rotors turning in a closed chamber. The latter is jacketed for heating or cooling. The material to be processed is forced between the blades by hydraulic ram, and dispersion is effected in a short time.

In the chip process, the damp cellulose nitrate, pigment, and solvent (often a solvent plasticizer) are charged into the mill. When dispersion is complete the mass is passed through sheeting rolls and reduced to a thickness of about 2 mm ($\frac{1}{16}$ inch). The sheets are then broken up into chips.

Several companies now specialize in the supply of chips to the industry, and their use makes the manufacture of pigmented lacquers a relatively simple operation.

Cellulose acetate

Cellulose acetate was first introduced in 1865 by Schützenberger who prepared it by treating cellulose with acetic acid and acetic anhydride. The formation of cellulose triacetate can be represented by the equation

$$C_6H_7O_2(OH)_3 + 3CH_3COOH \rightarrow C_6H_7O_2(OCOCH_3)_3 + 3H_2O$$

The triacetate is of little interest on account of limited solubility, but by controlled hydrolysis part of the acetate groups can be split off, yielding a product soluble in a number of solvents and useful for lacquer manufacture. It is usual to assess cellulose acetate by the acetyl value, which is the weight (grams) of acetic acid liberated on hydrolysis of 100 grams of ester.

Manufacture of cellulose acetate

Two methods are in use at the present time.

Dreyfus method This process involves a pretreatment of purified cotton linters or wood pulp with acetic acid containing a small amount of sulphuric acid. This serves to break down the cellulose structure and to increase the reactivity. The product is then added to a mixture of glacial acetic acid and acetic anhydride, the temperature being kept at 4° to 5°C (39·2° to 41·0°F). When the reaction is complete, a clear viscous solution of triacetate is obtained which is then subjected to hydrolysis or 'ripening'. Water is added gradually and the whole then heated to 60° to 70°C (140° to 158°F) to effect the hydrolysis. For lacquer manufacture an acetyl value of 54 to 55 is considered suitable.

Methylene chloride process Dry cellulose is treated first with acetic anhydride and then with a mixture of acetic anhydride, methylene chloride and sulphuric acid. Rigid control of conditions, especially temperature, is essential. When the reaction is complete, some water is added and the mixture hydrolysed to the desired degree. The sulphuric acid is then neutralized by sodium acetate, and the methylene chloride recovered by distillation. The cellulose acetate is precipitated from the acetic acid solution by addition of water, washed well, and dried.

Properties of cellulose acetate

The cellulose acetates are white granular materials with specific gravities of about 1·36. They are thermoplastic and possess high but indefinite melting points. In the molten condition they decompose rapidly.

The general properties are determined largely by the degree of acetylation and the viscosity. They are soluble in a limited range of solvents, the most important of which are methyl acetate, ethyl acetate, acetone, methylethyl ketone, methylcellosolve acetate, carbitol acetate and diacetone alcohol. In addition they dissolve in mixtures of ethyl alcohol (10 parts) with methylene chloride or ethylene chloride (90 parts).

Films of cellulose acetate are very much less flammable than those of cellulose nitrate but have inferior water resistance. They are very resistant to degradation by sunlight and are very resistant to petrol, oils and greases. They are not resistant to alkali. For best results, plasticizers such as triacetin or phthalates are incorporated in quantities up to about 30 percent.

Cellulose acetates are used in lacquers for wood, metal, paper and leather; in strippable coatings, and in paint strippers.

Cellulose acetobutyrate

This material is in some ways very similar to cellulose acetate but possesses several advantages. It is more resistant to water, is soluble in a wider range of solvents, and is compatible with a greater number of plasticizers and resins.

Cellulose acetobutyrate can be prepared in methylene chloride solution using either a mixture of acetic and butyric anhydrides, or a mixture of acetobutyric anhydride, acetic anhydride and a small quantity of sulphuric acid.

The ratio of butyric to acetic anhydride can be varied to give a range of products in which flexibility, solubility and compatibility increase with increasing butyric content. They are used in lacquers for wood, wire, paper and plastics.

Safety Regulations

The highly flammable nature of cellulose nitrate and of the solvents used in cellulose lacquers has resulted in the issue of statutory regulations governing the manufacture, storage and use of these lacquers.

Cellulose nitrate is classed as explosive if the nitrogen content is greater than 12·3 percent. The type used in lacquers contains between 10·5 and 12·3 percent nitrogen and is supplied damped with at least 25 percent of butanol, isopropanol or ethanol.

Cellulose lacquers and thinners containing hydrocarbon and other solvents are controlled by the Highly Flammable Liquids and Liquefied Petroleum Gases Regulations, 1972. The regulations control the storage, use and labelling of highly flammable liquids which are defined as substances which give off a flammable vapour at a temperature less than 32°C (89·6°F) (Abel) and, in addition, support combustion. The method of test is described in Schedule 2 of the Regulations. These Regulations extend the classification of highly flammable liquids from a flash-point limit of 22·77°C to 32·22°C.

Regulations also exist controlling the labelling of containers and the conveyance by road of flammable liquids.

CELLULOSE ETHERS

Ethyl cellulose

If purified cellulose is treated with concentrated sodium hydroxide solution an *alkali cellulose* is produced by reaction with the hydroxyl groups:

$$C_6H_7O_2(OH)_3 + 3NaOH \rightarrow C_6H_7O_2(ONa)_3 + 3H_2O$$

The sodium compound can be reacted with an alkyl chloride such as ethyl chloride under appropriate conditions to give a tri-ether:

$$C_6H_7O_2(ONa)_3 + 3C_2H_5Cl \rightarrow C_6H_7O_2(OC_2H_5)_3 + 3NaCl$$

The method of manufacture is based on these reactions, the conditions being rigidly controlled to ensure the correct degree of ethylation. The tri-ether is of no interest as it has poor solubility and compatibility and is not thermoplastic. The type found most suitable contains between 2 and 2·5 ethoxyl groups per cellulose 'unit' which is equivalent to 42 to 48 percent ethoxyl content.

When the desired degree of ethylation has been reached, the excess of ethyl chloride is removed and the ethyl cellulose precipitated by addition of water. The material is washed free from chloride and dried.

Properties of ethyl cellulose

Ethyl cellulose is a white granular powder with specific gravity 1·14. It softens at about 140°C (284°F) and melts at between 200°C and 220°C (392°

and 428°F). The degree of flammability is very low. It is resistant to dilute acids and to alkalis.

Films of ethyl cellulose are tough and flexible, retaining the latter property over a wide range of temperature. The films are transparent and transmit ultraviolet light without discoloration. The water permeability is superior to that of cellulose acetate but inferior to that of cellulose nitrate.

Ethyl cellulose has the widest range of solubility of all the cellulose derivatives. It is soluble in alcohols, ketones, esters, aromatic and chlorinated hydrocarbons. The range of solubility is controlled by the degree of ethylation, reaching a maximum when about 2·5 of the three available hydroxyl groups are combined. It is also compatible with many drying oils, varnishes, and synthetic resins.

Uses of ethyl cellulose

Ethyl cellulose is incorporated into oleoresinous varnishes to improve drying characteristics, hardness and water resistance. It is also used in flowback lacquers, cable lacquers and coatings for foil and paper.

Water-soluble cellulose derivatives

These materials absorb water and slowly form gels or 'solutions' of high viscosity at low concentrations. With the exception of sodium carboxymethyl cellulose (SCMC) they are ethers which are tasteless and non-toxic. In addition to their extensive use as thickeners and protective colloids in latex emulsion paints they are used in foodstuffs. The most widely used are the following.

Methyl cellulose A methyl ether of cellulose which is obtained by methylating alkali cellulose with dimethyl sulphate or methyl chloride. The degree of etherification can vary and this controls the properties. When the number of methoxy groups per anhydroglucose unit lies between 1·3 and 2·4 the product is soluble in water, but with increasing methoxy groups it becomes soluble in organic solvents [2].

Aqueous solutions of methyl cellulose are non-ionic and have surface tensions lower than that of water; consequently they form foams on vigorous agitation. The solutions form firm gels which coagulate reversibly on heating and require protection against bacterial attack.

When used as a thickener and protective colloid in emulsion paints the films show good moisture resistance and sizing properties, but in brushability and storage stability they are inferior to hydroxyethyl cellulose and SCMC.

Hydroxyethyl cellulose An ether in which the degree of etherification is approximately 8·5 hydroxyethyl groups per 10 anhydroglucose units. It produces firm gels and is soluble in both hot and cold water. The solutions are non-foaming and show excellent tolerance to electrolytes.

In latex emulsion paints it gives good scrub resistance, brushability, flow out and sizing properties on porous surfaces. In common with other cellulose ethers it requires protection against bacteria.

Sodium carboxymethyl cellulose (SCMC) This is prepared by reacting alkali cellulose with sodium monochloracetate. The degree of substitution is of the order of 7 carboxymethyl groups per 10 anhydroglucose units. The solutions in water are soft gels and exhibit pseudoplastic flow. They are anionic with a pH of 7·0 and are stable between pH 4 and pH 10.

SCMC gives precipitates with trivalent and some divalent metals, and when used as a protective colloid in emulsion paints it is usual to add a sequestering agent such as Calgon [3]. The emulsion paints show excellent brushing and storage properties but are inferior in other respects to those based on methyl and hydroxyethyl cellulose. Like the latter, SCMC solutions require protection against bacterial attack.

The choice of cellulose thickener for a specific purpose will be influenced by the relative importance of the desired properties. Each product has strong features as well as weaknesses, and a relative assessment is necessary before deciding on the type to be used.

References to Chapter 12

[1] *Durrans' Solvents*, 8th Edn, revised by E. H. Davies. Chapman & Hall, London, 1971.
[2] 'Celacol' water-soluble cellulose ethers (British Celanese Ltd).
[3] Albright & Wilson Ltd.

13

Hard resins and varnishes

The hard resins described below are derived from natural and synthetic sources and are used mainly in varnishes and/or lacquers.

NATURAL RESINS

Natural resins are obtained from many varieties of trees which exude a viscous liquid when the bark is damaged. This exuded *oleoresin* then hardens and acts as a protective for the damaged area. With the exception of rosin which is derived from pines, natural resins are obtained from tropical forest trees.

These resins are insoluble in water but dissolve to varying degrees in organic solvents. This property serves to distinguish them from the natural *gums* which, although obtained from certain species of trees, are soluble in water.

Classification

Natural resins are classified according to origin into *recent* and *fossil* types.

Recent resins are obtained from living trees by making incisions and then collecting the exuding liquid in containers attached to the trunk. The depth of cut varies with the type of tree. Damar occurs in the bark and requires shallow cuts, whereas Manila occurs in the sapwood and requires deeper incisions. Resins obtained in this way are soft and are used in spirit varnishes.

Fossil resins occurs as lumps in the ground and were exuded by trees which have long since rotted away. The cost of collection of fossil resins has now rendered them uneconomic for most purposes, and their use has been discontinued except for a few very specialized applications. The traditional 'copal varnish' has been replaced by other types such as linseed/tung oil/phenolic. In view of their historic interest the fossil resins are listed in Appendix E.

Composition

Resin-producing trees contain a viscous liquid or oleoresin which is exuded when the tree is cut. This liquid contains a proportion of essential oil which volatilizes on exposure to the air, and the residue hardens slowly by oxidation and polymerization.

Natural resins, in general, are very stable and not readily attacked by acids or alkalis. They are insoluble in drying oils. On heating they soften and then liquefy, but further heating causes decomposition.

Properties and identification

Appearance

They generally break with a conchoidal fracture, but a number possess a characteristic *shape* and appearance. In many cases an examination of the structure as appearing under the microscope can give useful information. For this test the resin face is polished and then etched with dilute alcoholic potash solution.

Colour

This can vary from white to very dark brown and the resins can be transparent or can exhibit varying degrees of opacity. The commercial assessment of a resin is based very largely on colour, but size of lumps and freedom from impurities are also taken into account. In one or two instances, for example, Accroides and Dragon's Blood, the colour associated with the resin was, for many years, considered a valuable asset, but these resins are now little used.

Odour

Most resins possess a characteristic odour which is intensified on slight warming.

Hardness

Natural resins differ considerably in this property. Since the majority of recent resins (with the possible exception of rosin) are used in lacquers or spirit-type varnishes, hardness is of great importance.

Melting point

With one or two exceptions, natural resins do not possess sharp melting points but soften and liquefy gradually over a range of temperature. The determination can be carried out by placing the powdered resin in a capillary tube sealed at one end and attached to the bulb of a thermometer in a heating bath. As the temperature is raised slowly, two points are noted: (i) the softening point at which the particles soften and coalesce, and (ii) the fusion point at which the material becomes liquid. The difference between these points depends to a large degree on the complexity of the resin.

The softening point of resins can be determined by the Ring and Ball softening point method set out in BS 6782: Part 6 [1].

The sample is contained in a horizontal ring suspended in water which is heated slowly. The softening point is the temperature at which the sample sags through a specified distance. The apparatus is illustrated in Fig. 13.1. An 800 ml beaker A supports a brass plate B from which are suspended two brass strips C and D. D is solid but C contains a small hole for thermometer E and two holes of larger diameter to accommodate the shouldered brass rings F. These rings have projecting brass pins on the inside to prevent steel balls G from dropping through. Water is used as the heating liquid.

In the determination, the brass rings are placed on a non-adherent surface such as an amalgamated brass plate, warmed, and molten resin is run in to fill them. When completely cold they are placed in the apparatus and the standard steel balls placed in position on the surface of the resin. The beaker is heated slowly and at the softening point the resin will flow or sag through the ring. The softening point is taken as the temperature at which the resin touches plate D.

Solubility

Natural resins are all insoluble in water and show a wide diversity and range of solubilities in organic solvents. The term *solubility* in this context indicates mutual dispersibility rather than the formation of a true solution of resin in solvent. Many resins absorb considerable amounts of solvent and swell before going into solution.

Table 13.1 shows the relative solubilities of a number of natural resins.

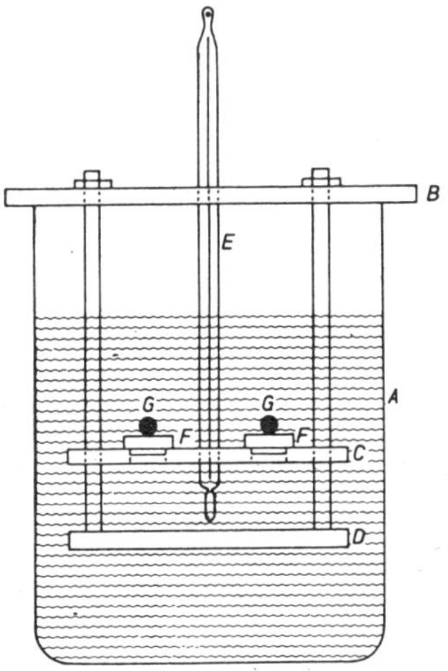

Fig 13.1 Ring and ball apparatus

Table 13.1 Solubilities of natural resins
S = soluble P = partly soluble I = insoluble

Resin	Alcohols	Esters	Ketones	Hydro-carbons	Mixed solvents in which resin is soluble
Rosin (colophony)	S	S	S	S	—
Mastic	S	S	S	S	—
Damar	P	S	P	S	Alcohols + esters Alcohols + hydrocarbons
Shellac	S	I	S	I	Alcohols + esters
Sandarac	S	I	S	I	Alcohols + esters
Soft Manilas	S	I	P	I	Alcohols + esters Alcohols + hydrocarbons

Insoluble matter

This is important where resins are to be used in clear lacquers. It can be determined either by allowing a resin solution of known concentration to settle in a calibrated tube and then reading off the volume of insoluble matter, or by filtering off the insoluble matter, drying and weighing.

Acid value

In natural resins appreciable quantities of the resin acids are present in the free state and the acid values are therefore generally high. It is common practice to carry out the determination using standard alcoholic potash or to add an excess of alcoholic potash and then to determine the excess by titration with standard acid [2].

Iodine and saponification values

Owing to the complex nature of natural resins, iodine and saponification values are of little use in identification.

Refractive index

This varies from 1·515 to 1·540 and is of little value in identification.

Spot-tests for resins

Natural resins give characteristic colours in one or more of the following tests.

Storch–Lieberman test (for rosin)

The sample is warmed with an equal volume of acetic anhydride, allowed to cool, and the acetic anhydride layer separated by means of a pipette. Addition of a drop of 50 percent sulphuric acid gives a fugitive bluish-violet coloration if rosin is present.

Storch–Morawski test

This is a modification of the Storch–Lieberman test and is useful for identifying different resins. In the test, about 2 g of resin are dissolved in 10 ml of acetic anhydride by heating. The solution is cooled, placed in a white dish and treated with a drop of sulphuric acid. The results obtained with different resins are indicated in Table 13.2.

Table 13.2

Resin	Storch–Morawski	Halphen–Hicks
Rosin	Bluish-violet (fugitive)	Green, becoming blue and finally purple
Manila	Pink	Weak green, becoming blue and finally purple
Damar	Deep red	Brown, becoming maroon
Sandarac	—	Lilac, becoming violet
Shellac	Pale green	No coloration

Halphen–Hicks test

Two solutions are prepared:
(i) Phenol (1 vol.) in carbon tetrachloride (2 vol.)
(ii) Bromine (1 vol.) in carbon tetrachloride (4 vol.)
A small sample of the resin is dissolved in solution (i) and a drop placed in a depression on a porcelain tile. A drop of solution (ii) is placed in an adjacent depression and the two drops are covered with a watch glass. Colorations are produced on the surface of the resin solution by the bromine evolved from solution (ii). The colours vary to some degree with different samples of the same resin, but Table 13.2 indicates the general type of result obtained.

Natural resins in current use

The natural resins in current use comprise the following.

Manila copal resins

Although the term 'Manila' covers a range of resins, the material in common usage is the soft spirit-soluble type. The quality varies, some consignments containing appreciable quantities of dust and other insoluble matter. The solutions in alcohol, therefore, have to be clarified before use. These spirit varnishes are used for insulating purposes and for the manufacture of road-marking paints. The following figures are typical for the important properties:

Specific gravity 1·06–1·08	Softening point 45°C (113°F)
Acid value 145	Melting point 120°C (248°F)

Damar

A type of resin obtained from several trees of the Dipterocarpus family which is common in Malaysia, Sumatra and Borneo. The resins range from pure white to black and from transparent to opaque. They are designated by the locality of origin or port of shipment.

Damars are soft resins which melt at about 110°C (230°F) and have an acid value of 20 to 35. They are soluble in petroleum and aromatic hydrocarbons but only slightly soluble in alcohols and ketones. The solutions are generally turbid due to the presence of insoluble polymerized fractions.

They give flexible films and are therefore used in paper varnishes. They are also used extensively in cellulose nitrate lacquers to improve gloss and adhesion. For this purpose the resin must be free from any insoluble matter and a purified variety known as *dewaxed* damar is used. This is manufactured by dissolving the natural resin in toluol and adding alcohol carefully to precipitate and coagulate the polymerized material. This is filtered off and the purified damar obtained by evaporation of the solvent.

Rosin or colophony

Rosin, sometimes known as colophony, is the residue remaining in the stills when the oleo-resin from pine trees is distilled for the preparation of terpene solvents. The molten rosin is discharged while still hot (about 155°C, 311°F), strained, and filled into metal drums where it solidifies. The ratio of rosin to volatile solvent obtained in this way is about four to one. The product is often referred to as 'gum rosin'.

Wood rosin is obtained by extraction of the old stumps of pine trees which are reduced to chips and then treated with steam to separate volatile constituents. The chips are next treated with petroleum solvents to extract the rosin and the latter is obtained on evaporation of the solvent. Wood rosin is generally of inferior colour to gum rosin and contains more unsaponifiable matter.

Sources

The major part of the world's supply of rosin is obtained from the United States where the Naval Stores industry handles rosin and its derivatives. Smaller quantities are obtained from France, Portugal and Spain. A certain amount is also produced in India.

Classification

Rosin varies considerably in quality and colour, ranging from virtually colourless and transparent through shades of yellow to brown and finally almost black. The American rosin is graded by reference to a series of coloured cubes and designated by letters, e.g. WW (water-white), WG (window-glass), X (extra pale). In France the paleness of the rosin is denoted by one or more As, up to six, the latter being the palest.

Constitution

Rosin contains about 90 percent of acids, largely in the free condition with a small amount of resenes and esters. The acids are abietic, dextropimaric, and laevopimaric, the structures of which are as shown.

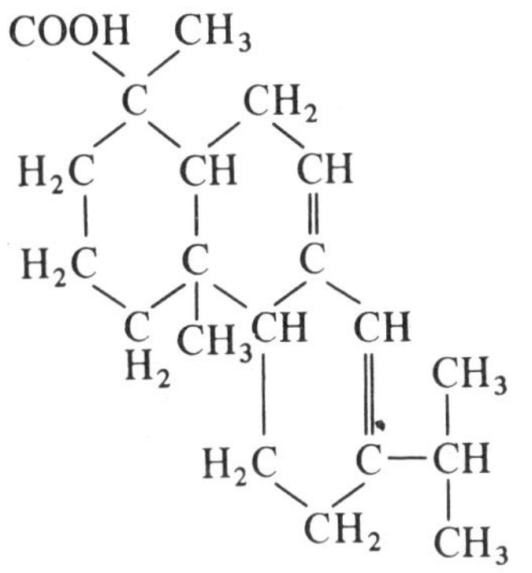

Abietic acid

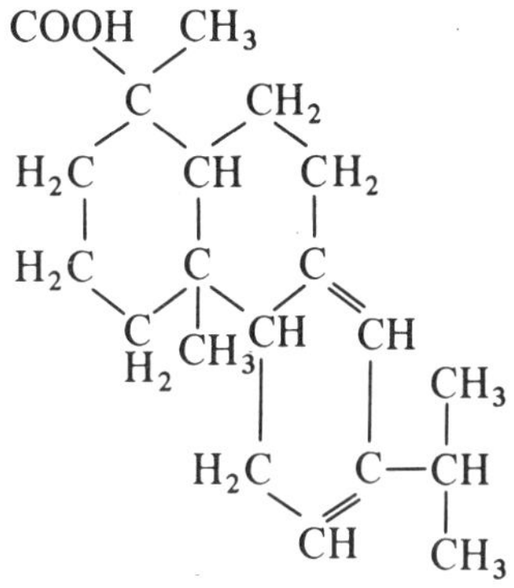

Laevopimaric acid

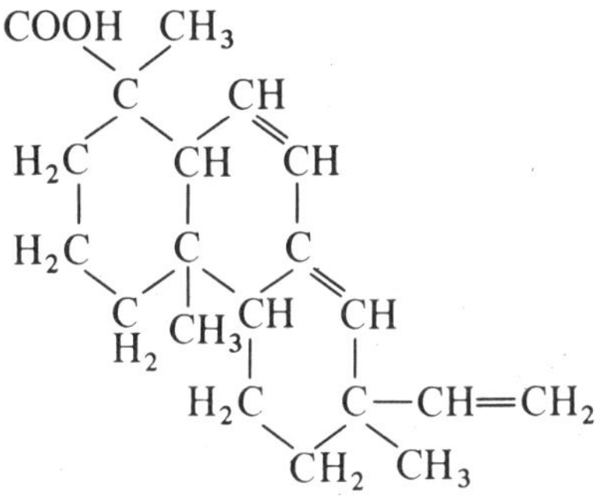

Dextropimaric acid

The proportion of abietic acid to its isomers varies according to the source of the rosin. Their unsaturated nature accounts for the fact that on exposure to the air rosin oxidizes slowly and becomes partly insoluble.

Properties

When cold, rosin is a brittle solid and small pieces can be powdered by the fingers. On warming it softens, emits a characteristic smell and melts at 110°C (230°F). The general characteristics are set out below. As a result of the high acid content, rosin forms esters (ester gums) with glycerol or pentaerythritol and also forms metal rosinates with many metals. Sodium rosinate is soluble in water, but the salts of heavier metals such as calcium and zinc are insoluble in water, though they dissolve in many organic media.

Rosin also reacts with maleic anhydride in a manner similar to unsaturated oils to produce adducts known as 'maleic resins'. These are used extensively in varnishes and lacquers.

The majority of organic solvents dissolve rosin when it is fresh, i.e. from the drum, but long exposure to the air results in the formation of insoluble oxidation products.

Specific gravity 1·04–1·08 Acid value 150–180

Softening point 80°C (176°F)	Saponification value 145–195
Melting point 110°C (230°F)	Iodine value 160–250 (usually about 230)

Uses

When made into a varnish with tung oil, rosin checks the gelling tendency of the oil; the resulting varnish is quick drying and is sometimes used for machinery paints. Considerable quantities of rosin are consumed in the manufacture of modified phenolic resins.

The intense reactivity of rosin is due to the presence of a carboxyl group and two double bonds. If one or other of these is modified, considerably improved products are obtained.

Metallic salts (rosinates)

If rosin is heated with an aqueous solution of sodium hydroxide or sodium carbonate a cloudy solution of sodium rosinate (largely sodium abietate) is formed. Sodium rosinate can be used as a source of other metal rosinates. Lead, cobalt and manganese rosinates are precipitated when a solution of a salt of the metal is mixed with sodium rosinate solution. The rosinates of these metals were formerly used as driers but have now been replaced by naphthenates and octoates for two reasons: (i) the latter are more readily soluble in paint media, and (ii) the presence of rosin detracts from the durability of the paint film.

Calcium rosinate (limed or lime-hardened rosin). This is the most widely used metal rosinate and can be prepared by either fusion or solution processes. In the former the rosin is melted and heated to about 230°C (446°F), preferably in an aluminium pot to prevent discoloration. Pure lime is then added, the quantity being about 6 to 7 percent of the weight of the rosin. This is insufficient to neutralize the rosin completely, as this is undesirable, the solubility of the rosinate decreasing with high metal contents. Theoretically, about 9 percent of lime would be required for complete reaction.

The solution method consists in dissolving the rosin in white spirit and heating the solution with calcium hydroxide. Calcium rosinate is soluble in white spirit.

When calcium rosinate is used in the manufacture of wood oil varnishes, it is usual to lime the rosin in the varnish pot before addition of oil to the hot melt.

With a 6 to 7 percent calcium content the melting point of the calcium rosinate is about 125°C (275°F) and the acid value about 40.

Zinc rosinate can be prepared by heating rosin with zinc oxide, using methods similar to those for calcium rosinate. The reaction can be very vigorous and is modified by using a mixture of zinc oxide and lime.

Ester gums

These are produced by esterification of rosin with either glycerol or pentaerythritol. The glycerol ester has been known the longer and is produced by melting rosin in an aluminium or stainless steel vessel, adding 10 to 12 percent of glycerol and keeping the temperature at about 280°C (536°F) until the desired acid value is reached. This usually lies between 6 and 12. The palest types of ester gum are produced in closed vessels through which a stream of carbon dioxide is passed.

The pentaerythritol esters are produced in a similar way but at about 300°C (572°F). These esters give harder and more durable films than the glycerol types.

One of the principal uses of ester gum is in the preparation of varnishes with tung oil. These have been used extensively in gloss paints in the past, but their use today is limited to certain types of paint such as machinery finishes.

Ester gums are added to cellulose nitrate lacquers to improve gloss and adhesion. They are components of some modified phenolic resins and can be reacted with maleic anhydride.

Other rosin esters

Rosin reacts with lower alcohols such as methyl and ethyl alcohols to produce viscous liquids which are sometimes used as plasticizers. A semi-solid is produced by esterification with ethylene glycol.

Hydrogenated rosin

The degree of unsaturation of rosin is reduced by treatment with hydrogen in the presence of a catalyst. This hydrogenated rosin is less reactive and does not tend to oxidize on storage. It can be esterified with glycerol to yield exceptionally pale ester gums.

Polymerized rosin

Rosin will react with concentrated sulphuric acid to produce a polymer of higher viscosity. This will also produce spirit varnishes of higher viscosity than those produced with normal rosin. Polymerized rosin can be converted to ester gums and these yield varnishes of increased hardness and durability.

Lac (shellac)

This resin occurs on the twigs and branches of certain trees and is a product of the life cycle of a particular type of insect. India produces about 95 percent of the total world consumption and the most important trees involved are the Kusum (*Schleichera trijuga*), the Palas (*Butea frondosa*) and the Ber (*Zizyphus jujuba*).

The larvae of the insect *Lacifer lacca*, which is related to the cochineal, swarm on the thin twigs of the trees and puncture the bark in order to feed on the sap. They excrete a resin which forms a protective coating over their bodies. Breathing takes place through ducts in the resinous layer, the ducts being lined with wax which is also excreted by the insect. The twigs become thickly encrusted with a dark red resin, the composition of which varies to some extent with different host trees. At the appropriate season the twigs are severed and collected. They constitute *sticklac*, the crude raw material for subsequent processing.

Seed lac Removal of the crude resin from the twigs is carried out by a process of rolling and sifting. The crude resin is then placed in shallow tubs and well kneaded with warm water to remove the water-soluble colouring-matter and other impurities. The process is repeated and the purified resin is dried quickly in the sun to give the granular seed lac. The product varies in colour from pale yellow to brown, due to a water-insoluble colouring-matter, and is graded according to size and colour. Mechanical methods of extraction are now superseding the above traditional method.

Shellac For many years it has been customary to add a small quantity of orpiment (arsenic sulphide) to seed lac to improve the brilliance before further processing. Rosin is also added to lower the melting point.

In the manufacture of shellac, the treated seed lac is placed in a tubular bag, about 50 mm in diameter and 6 to 9 metres (20 to 30 ft) long. This is held before a fire, when the molten lac exudes through the walls of the bag and drops on to a warm surface. The bag is slowly twisted in opposite directions from each end to assist the exudation of the resin. The plastic mass so obtained is worked out into a thin sheet about 1200 mm (4 ft) square, allowed to cool and broken up into flakes. These constitute commercial shellac, the colour of which ranges from pale yellow to brown. It is marketed in four main grades, known as Lemon (pale yellow), Fine Orange, T. N. Orange (free from orpiment and rosin), and London T. N. (brown and containing about 3 percent rosin).

Button lac differs from shellac in physical form only. The hot lac is allowed to fall on to a cold surface where it solidifies to flat discs ('buttons').

Garnet lac The residue left in the bag after extrusion of the lac is known as 'kiri' and contains the less fusible constituents and traces of dye. The kiri is removed, dried, and broken into pieces. This material is not used in the paint industry, the main outlets being a binding agent for abrasive wheels, and insulation. Mechanical methods for working up the seed lac produce good quality lacs and these are often known as 'garnet lac'.

Dewaxed lac For many purposes, such as French polish for example, the presence of wax in the shellac is an advantage, but for many spirit varnishes it is undesirable. It can be removed by dissolving shellac in alcohol to give an approximately 30 percent solution, adding half the volume of white spirit, and warming to 80°C (176°F). Two layers separate, with the wax dissolved in the white spirit.

White (bleached) lac is prepared in the following manner. The lac is dissolved in sodium carbonate solution, and sodium hypochlorite solution is added slowly to decompose the colouring-matter. The solution is filtered to remove the wax and the white lac precipitated by addition of dilute sulphuric acid. It is then coagulated by heating, washed thoroughly with hot water, and moulded into hanks while still plastic; or cooled, dried and coarsely powdered. The hanks usually contain about 25 percent moisture and the powder form about 6 percent.

Bleached lac is used in the manufacture of colourless spirit varnishes, but the solubility in alcohol decreases appreciably on long exposure to the air.

Lac wax is recovered in the production of white (bleached) lac. It is a hard material and is used in some kinds of polish.

Composition of lac

The constitution of lac is very complex. It contains several substances of differing degrees of hardness and solubility and, in addition to two or more resins, contains the alcohol-insoluble wax. Two of the constituent acids have been isolated. These are aleuritic acid, present to the extent of 40 to 50 percent, and shellolic acid which is present in small quantity.

Properties and uses of lac

Lac is a hard brittle resin which softens at about 65°C (149°F) and melts at about 80°C (176°F). The physical and chemical characteristics are set out below. It is freely soluble in alcohols and some ketones but insoluble in aliphatic and aromatic hydrocarbons, esters, and drying oils. The alcohol solution of normal refined lac is turbid, due to the presence of insoluble wax, but solutions of the dewaxed resin are clear.

The principal use of shellac in the surface coatings industry is in the manufacture of spirit lacquers and polishes. These give films which are hard, adherent, and abrasion resistant. The insolubility of lac in other media is utilized in knotting, which is an alcohol solution of lac. It is used as a sealer on new wood to prevent rosin exuding from knots into the paint film and causing discoloration. For this purpose the lac should be free from rosin.

Lac films possess good insulating properties and the resin is widely used for this purpose. It is, however, very brittle and requires a plasticizer. For this purpose methylcyclohexanol phthalate is often used. The lac films possess poor water resistance and perish rapidly on exposure to the weather. Water absorption results in a whitening of the surface which is followed by breakdown.

If alcohol solutions of lac are stored in metal containers, darkening takes place as a result of attack on the metal. This can be prevented by the use of a small quantity of ethyl aniline phosphate.

Characteristics of shellac

Softening point 65°C (149°F)	Specific gravity 1·15–1·20
Melting point 80°C (176°F)	Iodine value 18–20
Acid value 65–75	Saponification value 225–230

Rosin adulteration is indicated by an appreciable increase in iodine value.

Other resins

Mastic is obtained as an exudation when cuts are made in the stem and branches of a Mediterranean shrub *Pistachia lentiscus* and is marketed in the form of pale yellow 'tears'. It is brittle and breaks with a conchoidal fracture. When placed in the mouth it softens readily and in this respect differs from Sandarac which is similar in appearance. It is completely soluble in alcohol, acetone and aromatic hydrocarbons but insoluble in white spirit. The latter property distinguishes it from Damar which is soluble.

Mastic is used in paper varnishes and for varnishing pictures. An alcohol solution is used as a fixative for pencil and crayon drawings.

Sandarac An exudation from the tree *Callitris quadrivalvis*, a native of North Africa. Trees of similar type occur in Australia and yield a resin of practically identical character. The resin is marketed as yellowish-white tears, similar in appearance to mastic, but it powders and does not soften when placed between the teeth. It is soluble in alcohol and yields harder films than mastic. The main use of sandarac is in varnishes for charts and labels.

SYNTHETIC HARD RESINS

Phenolic resins

The formation of resinous materials by the reaction between phenol and formaldehyde has been known for many years and was put to commercial use by Baekeland in the early years of this century. He produced the first plastic, 'Bakelite', and so stimulated a study of the reaction between various phenols and formaldehyde which has produced many valuable resins.

The reaction between phenol and formaldehyde alone is slow, but takes place rapidly in the presence of acid or alkali catalysts. The two types of catalyst produce different products.

Acid catalyst

When phenol and formaldehyde (formalin) solution are boiled in the presence of 0·5 percent of a strong mineral acid, reaction takes place fairly rapidly and results in the formation of two distinct layers. The upper aqueous layer is run off and the residual layer heated to 150°C (302°F) to remove all volatile matter. On cooling, a hard, brittle, amber-coloured resin is produced. The reaction can be represented thus:

$$n\,C_6H_5OH + (n-1)HCHO \rightarrow \left[-C_6H_3(OH)-CH_2-C_6H_3(OH)-CH_2-C_6H_3(OH)-CH_2- \right]_n$$

The molar ratio of formaldehyde to phenol is less than one, and any excess phenol is removed in the final heating. Resins produced in this way are known as *novolacs* and are thermoplastic. They are widely used in the plastics industry but find little application in paints as they are insoluble in oils and hydrocarbon solvents.

Alkali catalyst

If phenol and formaldehyde are heated together with the addition of about 1 percent sodium hydroxide or lime, reaction takes place with the formation of a resin of somewhat different constitution from that produced under acid conditions. The reaction can be presented thus:

$$n\,C_6H_5OH + (n+1)HCHO \rightarrow HOCH_2-C_6H_3(OH)-CH_2-C_6H_3(OH)-CH_2\cdots C_6H_3(OH)-CH_2OH$$

The molar ratio of formaldehyde to phenol is greater than one, and reaction can also take place in the para position. The product contains a range of molecules of differing molecular complexity and this is reflected in considerable variation in viscosity and reactivity.

These resins, known as *resoles*, are brown syrups, soluble in alcohol and are thermosetting, i.e., on further heating they react to produce insoluble and infusible products. They are used in the plastics industry, but their insolubility in oils and hydrocarbons results in little application in the paint industry.

If, however, ammonia is used in the preparation, resins of the following type are produced:

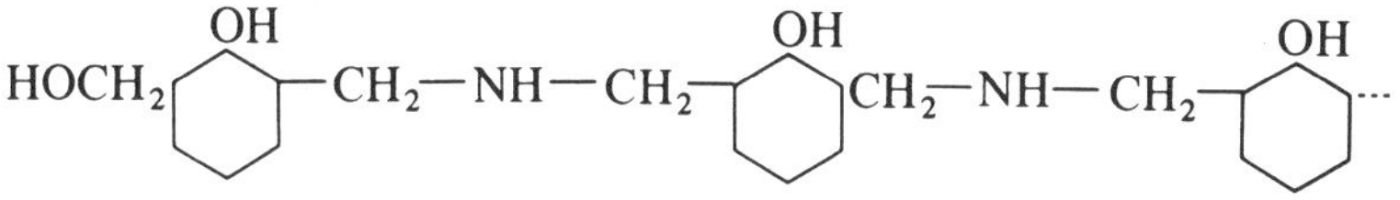

$$HOCH_2-C_6H_3(OH)-CH_2-NH-CH_2-C_6H_2(OH)-CH_2-NH-CH_2-C_6H_4(OH)\cdots$$

Such resins are sometimes used in stoving lacquers for cans, but their greatest use is in varnish-making as they are soluble in oils.

Modified phenolic resins

The difficulties encountered in the early days in producing oil-soluble phenolic resins led to the examination of their compatibility with other resins. It was found that the resoles reacted with rosin to give products which could be esterified with glycerol, and the resulting resins were soluble in oils. These modified resins usually contain 10 to 25 percent of the resole, and their general properties depend on the type of phenol used. Phenols substituted in the para position by tertiary butyl and octyl groups give very useful varnish resins. The greatest resistance to yellowing is provided by the use of diphenylol propane. All these modified resins possess melting points appreciably higher than that of rosin.

Properties

The rosin-modified phenolic resins are brittle solids with melting points between 100° and 160°C (212° and 320°F). The unesterified types are highly acid (AV 100 to 120) and are soluble in alcohols, hydrocarbons and drying oils. They react rapidly with basic pigments. Esterification yields resins of acid value 15 to 20 which are insoluble in alcohols and which vary in solubility in hydrocarbons and oils. Yellowing on exposure is a general property of these resins, but the least discoloration is shown by resins based on diphenylol propane.

They are used widely in varnish-making in conjunction with linseed, enamel oils, and wood oil. With the latter they retard the gelling tendency and give varnishes with high water-resistance.

Generally, varnishes made on these rosin-modified phenolic resins are used in a wide range of paints including decorative undercoats, primers, and marine paints. They are also used in certain types of printing ink.

Pure oil-soluble phenolic resins

It has been shown that the rate at which phenolic resins, based on substituted phenols, cross-link to form insoluble and infusible products is controlled by the nature and position of the substituent group. Para-substituted phenols react most slowly and, furthermore, the rate decreases with increase in the size of the group. As the groups become large, not only does the rate of reaction fall but the resin develops definite solubility in oils. Such para-substituted phenols can be reacted to give either novolacs or resoles, and these in turn are known as *non oil-reactive* and *oil-reactive* respectively.

Non oil-reactive resins

These are commonly based on *p*-phenyl-phenol and are prepared by the normal acid-catalysed method. They are soluble in aromatic hydrocarbons and possess softening points of about 100°C (212°F). Varnishes are made very easily by heating the resin with the drying oil, and the films produced are durable and possess a high degree of alkali and water resistance. They are used for spar varnishes and in machinery paints.

Oil-reactive resins

These are made from *p*-tertiarybutyl and *p*-octyl phenols by reaction with formaldehyde in the presence of an alkali catalyst. Sufficient sodium hydroxide is present to dissolve the phenol, and the resin is subsequently precipitated with acid. It is then washed thoroughly and heated in vacuum to remove all volatile materials.

Properties and uses

The oil-reactive phenolics are soluble in hydrocarbons and soften at 75° to 80°C (167° to 176°F). When heated with oils, reaction occurs through the reactive methylol ($-CH_2OH$) groups. This reaction is vigorous with conjugated oils such as dehydrated castor and wood oils and yields varnishes with a higher degree of resistance than the oil-soluble type. These show very high resistance to water and alkalis and are used as spar varnishes and in chemical-resistant finishes.

These resins also react with rosin and ester gum to produce the rosin-modified phenolic resins.

Cashew nutshell liquid

This is a reddish-brown viscous liquid which consists of approximately 90 percent anacardic acid (a derivative of salicylic acid) and 10 percent cardol (a monohydroxy phenol). On strong heating anacardic acid is decarboxylated and converted into cardol.

The liquid can be polymerized to give resins soluble in both aliphatic and aromatic hydrocarbons and possessing very good resistance to acid and alkali.

Maleic resins

These are produced by reaction between maleic anhydride and rosin, utilizing the Diels–Alder addition reaction. Addition of maleic anhydride to a conjugated unsaturated system takes place according to the general scheme:

```
R'—CH                              R'—CH—CH—CO
   ||         CH—CO                   |   |    \
   CH         ||    \                 CH  |     \
   |     +    ||     O   →            ||  |      O
   CH         ||    /                 CH  |     /
   ||         CH—CO                   |   |    /
R''—CH                             R''—CH—CH—CO
```

A number of conjugated unsaturated systems can be used, and if maleic anhydride is reacted with rosin, addition to the abietic acid takes place according to the following scheme:

```
COOH  CH3
    \ /
     C      CH2
   /   \  /    \
H2C     CH      CH                         CH—CO
 |      |       ||                  +      ||     \
H2C     C       C                          ||      O
   \   / | \  /    \                       CH—CO  /
    C    |  CH      CH
    H2  CH3 |       ||
           H2C      C—CH(CH3)2
              \   /
               CH2
```

```
                 COOH  CH3                          CO
                     \ /                      CH—      \
                      C      CH2            /           O
                    /   \  /    \          /    \       |
          →      H2C     CH      CH———————       \      |
                  |      |       |                \     |
                 H2C     C       C                 \   CO
                    \   / | \  //  \                CH—
                     C    |  CH     CH            /
                     H2  CH3 |      |            /
                            H2C     C————————————
                               \  /   \
                                CH2    CH(CH3)2
```

The product can then be esterified with glycerol or pentaerythritol, and in practice the three ingredients, rosin, maleic anhydride and polyhydric alcohol are sometimes heated together until reaction is complete.

Properties and uses

The properties of these resins depend on the amount of maleic anhydride used, the type of rosin, and type of polyhydric alcohol. Not more than 12 percent of maleic anhydride is used, as resins approaching this composition have high melting points and low solubility. Pentaerythritol gives harder films than glycerol.

Maleic resins are soluble in hydrocarbon solvents, esters, ketones and drying oils. Their ready solubility renders them suitable for *cold-cut* varnish-making, but they are often dissolved in oils by heating. They give pale varnishes, and with linoleic-rich oils produce non-yellowing whites and tints. They are also used in stoving lacquers for metal decorating.

The chemical and alkali resistance of the films is poor, but for normal exposure, when they are frequently blended with alkyds, the performance is satisfactory. They are often added to cellulose nitrate lacquers to improve gloss and adhesion.

Maleic resins bear some resemblances to three other types of resin.

1 **Ester gum** The maleics behave as ester gums which have been hardened by the maleic-rosin adduct.

2 **Rosin-modified phenolics** Maleics blend with oils with equal ease and give varnishes with many similar properties. However, the maleics do not possess the same resistance to alkalis and chemicals, but they are superior in respect of colour retention.

3 **Alkyds** The resemblance here is through the use of a polybasic acid, a polyol and an unsaturated monobasic acid.

Coumarone-indene resins

Resins made by the polymerization of coumarone and indene mixtures were the first synthetic resins to be used in varnish manufacture. They are produced by catalytic polymerization of coumarone and indene which occur in the heavy naphtha fractions of coal-tar distillation.

CH ‖ CH O — **Coumarone**

CH ‖ CH C H_2 — **Indene**

Polymerization takes place through the double bond and is usually catalysed by concentrated sulphuric acid, but other catalysts are sometimes employed. The formation of polymers from coumarone can be represented in the following scheme, and that from indene is similar:

CH=CH O → —CH–CH—CH–CH—CH–CH— O O O

The indene is sometimes separated and converted to polyindene resin, but commercial coumarone resins are usually a mixture of the two.

These resins are marketed in a range of colours from pale amber to dark brown and from soft gums to brittle solids. The melting points range from 90°C to about 150°C (194° to 302°F) according to the degree of polymerization. They yellow on exposure to sunlight, are very inert and resemble natural resins in a number of properties. Thus, they are insoluble in alcohols and soluble in esters, acetone, and aromatic hydrocarbons. The solubility in aliphatic hydrocarbons varies from partial to complete according to grade.

Coumarone-indene resins are soluble in drying oils and can be used in cold-cut varnishes or can be dissolved in oils by heating. Such varnishes have very low acid values (<1) and so are used in the manufacture of metallic paints such as aluminium and bronze, where the brilliance and leafing properties of the metals are destroyed by small amounts of free acid.

Varnishes incorporating coumarone-indene resin and wood oil show high resistance to attack by alkali and are therefore used in primers for alkaline plasters, asbestos, and cement render.

Petroleum resins

These are chemically inert resins produced from resinifiable aromatic streams using Friedel–Craft catalysis, followed by neutralisation and washing with water [3]. The solvent is then removed by evaporation. The main constituents of these resins are derivatives of styrene and indene. They are soluble in aliphatic and aromatic hydrocarbons, are compatible with a wide range of other resins. Other useful properties are resistance to ultraviolet radiation and an acid value of less than 0.1. These resins are used to improve hardness and gloss in finishes whilst the low acid value renders them suitable for use in oleoresinous media for metallic paints.

Polyimide resins

These resins are of particular interest by reason of their outstanding heat-resisting properties. They are formed by a condensation polymerization of the dianhydride of a tetracarboxylic acid with an aromatic diamine:

$$n.O\begin{matrix}\diagup CO-\\ \diagdown CO-\end{matrix}C_6H_2\begin{matrix}-CO\diagdown\\ -CO\diagup\end{matrix}O + n.H_2N.Ar.NH_2 \rightarrow$$

Pyromellitic anhydride

$$O\begin{matrix}\diagup CO-\\ \diagdown CO-\end{matrix}C_6H_2\begin{matrix}-COOH\\ -CO-\end{matrix}\left[NH.Ar-NH-CO-C_6H_2\begin{matrix}(HOOC-)\ -COOH\\ -CO-\end{matrix}\right]_{n-1}NHArNH_2$$

Polyamic acid

A solution of polyamic acid is applied to the article, dried and then stoved at a temperature in excess of 150°C (302°F). Dehydration takes place with formation of polyimide resin:

$$\text{Polyamic acid} \longrightarrow \text{O}\!<\!\!\begin{matrix}\text{CO}\\\text{CO}\end{matrix}\!\!>\!C_6H_2\!<\!\!\begin{matrix}\text{CO}\\\text{CO}\end{matrix}\!\!>\!\left[\text{N.Ar.N}\!<\!\!\begin{matrix}\text{CO}\\\text{CO}\end{matrix}\!\!>\!C_6H_2\!<\!\!\begin{matrix}\text{CO}\\\text{CO}\end{matrix}\!\!>\right]_{n-1}\text{N.Ar.NH}_2 + 2nH_2O$$

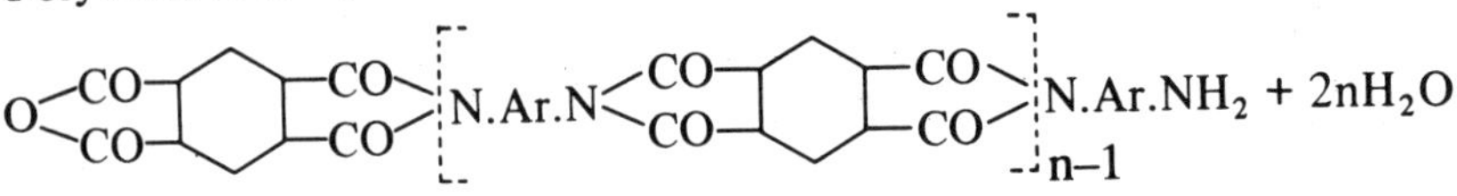

Polyimide resin

By the use of different dianhydrides and diamines a range of resins can be produced. Many of these are stable for long periods at 300°C (572°F) and at higher temperatures for shorter periods.

VARNISHES

The term varnish is applied to a class of clear liquids of 2 to 3 poise viscosity which, when exposed to the air in thin coatings, dry to tough, glossy films. Whilst traditional varnishes are referred to as 'oil' and 'spirit' types, modern varnishes are frequently solutions of alkyds, polyurethanes, polyesters, and so on. In this section the traditional types are considered.

Oil varnishes

Oil varnishes are combinations of drying oils and resins which are thinned to application viscosity with volatile solvent. Driers are incorporated to catalyse the drying process, and this will be controlled largely by the amount of oil present or the ratio of oil to resin. This ratio, by weight, is known as the *oil length* and can be varied considerably. Varnishes are described as *long, medium*, or *short* oil length, and these have roughly the following oil/resin ratios:

Long oil	3/1 to $4\frac{1}{2}$/1
Medium oil	2/1 to 3/1
Short oil	1/3 to 2/1

In the USA oil lengths are expressed in gallons (US) of oil per 100 lb resin, so that a '10 gallon' oil length varnish would consist of 10 US gallons of oil to 100 lb resin. (Note: 1 US gallon = 0·833 Imperial gallon = 3·78 litres.)

Some of the short-oil varnishes retain a surface tack for some time and have been used as adhesives for gold leaf. Such varnishes are, in consequence, known as *goldsizes*.

The main use of oil varnishes has been the finishing of wood. The wood was either stained or grained before varnishing, although in some cases, where the wood presented a naturally attractive appearance, the varnish was applied without the preliminary staining. As a result of changing fashions in the use of wood and the growth of mass-produced furniture and wooden articles generally, oil varnishes for interior work have been largely replaced by lacquer type finishes, notably cellulose nitrate, polyurethanes and polyesters.

The term 'mixing varnish' has long been applied to oil varnishes with are used as paint media. Before the widespread use of alkyds, these mixing varnishes were the standard media for most gloss paints and undercoats, but today their uses are limited to certain metallic and maintenance paints.

Although oil varnishes are now manufactured on a reduced scale and the use of fossil resins is practically obsolete, a short account of these materials will, it is hoped, be of interest.

Oil varnishes contain the following four major ingredients.

Resins, which provide the hardness and gloss of the film. Both these properties increase with increasing resin content, but the durability decreases. Natural resins, especially the fossil copals, were formerly used extensively (*see below*), but these resins are no longer economically viable. Tung oil varnishes for special purposes were often made with ester gum or limed rosin. Resins used in modern varnishes include pure and modified phenolics, maleics, coumarone, and hydrocarbon types.

Drying oils contribute the elasticity, toughness, and general durability. These properties improve with increasing oil content, but the effect is controlled by the type of oil. The commonly used oils are linseed (as VLO, stand oil, or enamel oil) and tung oil. Varnishes often contain both oils.

Driers catalyse the conversion (by oxidation and polymerisation) of the oil to the solid state. They are added as solutions of the naphthenates or octoates of zirconium, cobalt and, sometimes, manganese. Lead drier is no longer used in varnishes.

Thinners are added for the sole purpose of reducing the oil/resin complex to a suitable consistency for application. They should evaporate completely from the applied film. White spirit is the usual thinner for long-oil varnishes, but the shorter types, and especially some tung oil varnishes, require a proportion of stronger solvent such as xylene, naphtha, or aromatic petroleum solvent ('Shellsol', 'Solvesso' or 'Aromasol') [4].

Varnishes from fossil resins

The industrial production of durable oil varnishes was originally based on the combination of linseed oil with African hard fossils such as Animi and Sierra Leone copals. In their natural condition these resins are not soluble in drying oils but can be made soluble by heating for some time at a temperature just above the fusion point. This results in a partial depolymerization during which about 20 to 30 percent of the weight of the resin is lost in volatile decomposition products. This process is known as *gum running*.

Congo copal has been used on a large scale by varnish manufacturers but has also been run in bulk, allowed to cool, and marketed as *fused copal*, or esterified with glycerol and marketed as *copal ester*.

The varnish-making operation consists of the following successive steps: (a) running the resin, (b) incorporating the run resin with the linseed oil, (c) cooling, dilution with thinner and addition of driers, (d) clarification and maturing. Clarification is necessary since the natural resins always contain a certain amount of extraneous matter.

Properties and uses of copal varnishes

The drying times of these varnishes vary with the oil length, the long-oil types taking from 16 to 20 hours and the short-oils from 1 to 2 hours.

The term 'copal varnish' has been used in the decorating trade for the best-quality varnish. Long-oil types are used for exterior work and medium-oil types for interior decoration. The long-oil types were formerly used as carriage varnishes, and for many years were the standard type of medium for hard-gloss paints. Other types of varnish made on copal resins are the short-oil types known as 'hard church oak', 'flatting varnish', and the goldsizes.

Varnishes from modified phenolic resins

Modified phenolic resins are soluble in drying oils at comparatively low temperatures, so the varnish-making process is relatively simple. Again, the wide range of rosin-modified phenolic resins available enables varnishes to be designed to meet a wide spectrum of requirements.

Some of the softer types of modified phenolic resin are soluble in white spirit in the cold and can be made into varnishes by solution in this solvent followed by addition of bodied oil and driers. This process is known as *cold-cutting*.

With the harder types, the resin and part of the linseed oil are heated to 250° to 260°C (482° to 500°F) until a drop forms a clear bead on cooling on glass. A second portion of oil is added and the test repeated. The process is continued until all the oil has been added and the varnish gives a clear solution in white spirit. It is then cooled and thinned, and driers are added in the normal way.

The softer types of modified phenolic resin can be treated in a similar way, but the oil and resin can be blended in fewer steps.

These varnishes are very similar in their general properties to those made from Congo copal but they are more easily made.

Tung oil is often incorporated into these varnishes along with the linseed oil and is best used in the form of enamel oil. These oils contain 20 to 25 percent tung oil co-polymerized with linseed oil. Processing is similar to that for linseed oil, but the varnishes have appreciably greater water resistance.

Tung oil varnishes

Tung (or wood) oil can be used in varnish-making, either in the form of a stand oil (these are available in a range of viscosities) or as the cheaper raw tung oil. The stand oil can be blended with the softer modified phenolic resins or with coumarone resins at 200° to 220°C (392° to 428°F), but harder resins require a higher temperature, with its attendant risk of gelation. To reduce this risk the resin is best cooked with linseed oil and oiled out with the tung-oil stand oil. It is then cooled rapidly and thinned.

It is more economical to use raw tung oil and to effect the bodying during the processing, but this method requires close control and experience. The raw tung oil is mixed with half the resin and heated rapidly in an open pot to 280°C (536°F). This temperature is necessary to ensure that the tung oil is *gas proof*, i.e. that the film on drying will be unaffected by gases from combustion. It must be borne in mind that tung oil will gel in 12 to 20 minutes at 280° to 285°C (536° to 545°F), so all operations must be carried out smoothly and without delay.

When the temperature has reached 280°C the pot is moved away from the fire, and after about 2 minutes the remainder of the resin, in small pieces, is added rapidly, with stirring. The temperature of the mixture falls to 250° to 260°C (482° to 500°F) at which temperature bodying is completed. It is then cooled rapidly and thinned.

Many tung-oil varnishes contain a proportion of linseed oil which is often useful as a *chill-back* if the tung-oil bodying tends to get out of control.

Tung oil–pure (100 percent) phenolic varnishes

The two types of pure phenolic resin behave differently towards tung oil, the resole type being reactive and the novolac type being non-reactive but soluble in the oil.

Reactive resole resins The manufacture of varnishes from these resins is effected by heating the tung oil to 170° to 180°C (338° to 356°F), at which point the resin is added and the whole stirred until the resin is dissolved. The temperature is then raised to 230°C (446°F) when reaction occurs accompanied by foaming. The intensity of foaming varies from one resin to another, and care must be taken to regulate the temperature so that the reaction proceeds slowly, otherwise there is a risk of the pot boiling over. The melt is held for about 25 to 30 minutes, the end-point being indicated by lessening or cessation of foaming. It is then cooled and thinned.

It is of interest that with reactive resins a gas-proof varnish is obtained at the relatively low temperature of 230°C (446°F).

Non-reactive novolac resins These resins dissolve in tung oil but no reaction occurs between them, so bodying involves polymerization of the oil only. The varnishes are made in the normal way for tung-oil media, but the thinners must contain a high percentage of aromatic hydrocarbons as the resins have limited solubility in white spirit.

Properties and uses of tung oil–phenolic varnishes

These varnishes are all characterized by higher water resistance than linseed oil types, the highest water resistance and the quickest dry being shown by the 100 percent phenolic types. Of these, varnishes from the reactive types are the most popular. They are used as spar varnishes and as media for certain types of maintenance paints, for example, those pigmented with micaceous iron oxide and used on structural steelwork.

Tung oil–coumarone varnishes

These are usually made at about 2/1 oil length and are conveniently manufactured from bodied tung oil. The oil and resin are heated to 220°–230°C

(430°–450°F) until a uniform melt is obtained, cooled rapidly, and thinned with white spirit. The alkali resistance of these varnishes is of a high order and they are used in primers for alkaline surfaces such as plasters, cement render, and asbestos board. The low acid value of these media renders them very suitable for use in metallic paints.

Physical properties of oil varnishes and films

Colour This is influenced by the colour of the oils and resins used and also by the care taken in manufacture. Excessive heating is a frequent cause of dark colour. The measurement of the colour of varnishes is carried out by methods similar to those used for oils (Chapter 10).

Viscosity Varnishes are generally applied by brush and the brushing properties are determined largely by the viscosity. If this is too high, brush-drag will be experienced and the varnish will be difficult to spread; if too low, the varnish will form runs on vertical surfaces and the film will be excessively thin. It is found that a viscosity of 2 to 3 poises at 25°C (77°F) is generally satisfactory for a brushing varnish. Viscosity measurements are carried out by the same methods as those used for oils (Chapter 10).

Drying time The drying time of a varnish is dependent on the composition but, in addition, it is influenced by film thickness, temperature and humidity of the air, nature of the substrate, and the amount of light reaching the surface. The methods for the determination of the drying times of varnishes are the same as those used for paints and are described in BS 3900, C2 and C3 [5].

Flexibility One of the functions of the oil in a varnish is to contribute flexibility, a property which enables a varnish film to accommodate the dimensional changes in the substrate without cracking. This is particularly important on exterior woodwork when expansion and contraction imposes great stresses on the varnish. The *cold-check* test is often applied to varnishes for woodwork. A wooden panel coated with the varnish in question is subjected to a number of cycles consisting of exposure to warm (60°C, 140°F) and cold (−10°C, 23·3°F) conditions. Varnishes which lack flexibility will be found to crack after a few cycles.

The general methods for the determination of the flexibility of varnishes are identical with those used for paints [6].

Hardness This property, like many others, cannot be assessed in isolation. The property which is important from a practical viewpoint is the resistance of the film to mechanical injury, and this depends on a combination of hardness and toughness. The methods for measuring these are identical with those used for paint [7].

Moisture absorption of films All oleoresinous varnishes absorb water to some extent and the amount present in the film will be in equilibrium with the moisture content of the air (or relative humidity). Moisture absorption is related, to a large degree, to the presence of water-attracting (hydrogen bonding) groups, such as hydroxyl and/or carboxyl, in the film as well as water-soluble oxidation products. It is also affected by the nature and amount of resin present. Tung oil/phenolic varnishes absorb considerably less moisture than, for example, a clear alkyd film whilst chlorinated rubber, which contains no water-attracting groups absorbs very little.

These facts are illustrated by the following figures:

Varnish	*% water absorbed at 20°C*
Tung oil/phenolic	2.5
Alkyd	6.2
Chlorinated rubber	0.8

Measurement of water absorption can be carried out by casting films on weighed glass panels. When the varnish is dry the panels are re-weighed and immersed for the required time in distilled, tap or salt water.

Figures quoted for moisture absorption do not reflect exterior durability. The absorption of moisture is not a rapid process and is of importance only when films are allowed to remain wet for prolonged periods. This can occur under marine conditions but tung oil/phenolic varnish has, nevertheless, been the standard marine varnish for many years. Alkyd resin films are little affected by showers of rain and the paints show good exterior durability.

Ultimate breakdown of alkyd paint films is due to the combined effect of moisture and radiation (see p. 443). Chlorinated rubber is usually employed in the form of pigmented coatings and these can show very good durability. Moisture permeability indicates the ease with which moisture will pass through a film and may or may not be accompanied by absorption. The least permeable films are those containing no water-attracting groups such as chlorinated rubber or vinyls. Measurement of film permeability can be carried out using a Payne cup. A detached varnish (or paint) film is secured firmly over the open end of a metal cup containing a desiccating agent. The cup is weighed and then exposed in a humid atmosphere. It is re-weighed at invervals over a specified time. The increase in weight gives the amount of moisture which has passed through the film. The film permeability is expressed as grams (moisture)/100 cm^2/24 hours at 20°C.

Both moisture absorption and film permeability are affected to a marked degree by the amount and nature of any pigment present.

Non-volatile content This can be determined by weighing 1 to 2 grams of varnish into a circular flat-bottomed metal dish 75 mm in diameter and heating for 3 hours at 105° to 110°C (221° to 230°F). It is then cooled and weighed. Difficulties are sometimes encountered with this method due either to skinning of the sample or to retention of solvents by certain types of resin. The method is useful for checking and correcting solvent losses which take place during the thinning of varnishes.

Flash points and **specific gravities** of varnishes are determined by the methods used for solvents and oils respectively.

Spirit varnishes

These consist of solutions of resins in volatile solvents, and dry entirely by evaporation. The resin film remains soluble in the parent solvent and is sometimes termed a *non-convertible* coating.

Shellac is the most widely used resin in this class and is dissolved in Industrial

Methylated Spirit (IMS) by placing the resin and solvent in a wooden barrel churn. This is rotated until the resin is in solution. A number of spirit varnishes and 'polishes' are made from shellac alone or in admixture with other resins. The two members in current use are French polish and knotting.

French polish consists of a 25–30 percent solution of orange shellac in IMS. It is used for polishing good-quality furniture, especially antiques.

Knotting, a 35–40 percent solution of orange shellac in IMS, is used, as the name suggests, for sealing knots in new timber. The purpose is the prevention of exudation of resinous materials which would soften and discolour paint films.

Shellac solutions in ethanol are highly acidic and can cause rapid corrosion leading to discoloration if packed in tinplate containers. The corrosion can be prevented by addition of a small quantity of ethylaniline phosphate.

Polyvinyl acetate is soluble in IMS to give water-white solutions of zero acid value. Such solutions require a plasticizer but yield films of limited weather resistance. They are used as coatings on polished metals and also as media for bronze and aluminium paints for interior use.

References to Chapter 13

[1] BS 6782. Binders for paints. Part 6: 1987. Method for determination of softening point (ring & ball). British Standards Institution.

[2] *ibid.* Part 3. Method for determination of acid value. (q.v.)

[3] 'Unilene' Petroleum resins (Marlow Chemicals Ltd, Sevenoaks).

[4] 'Shellsol' (Shell Chemicals); 'Solvesso' (Esso Petroleum); 'Aromasol' (ICI).

[5] BS 3900. Methods of test for paints. Part C2, Surface drying time. Part C3, Hard drying time. British Standards Institution.

[6] *ibid.* Part E1, Bend test.

[7] *ibid.* Part E2, Scratch test.

14

The process of pigmentation

The basic operation in the manufacture of a paint is the dispersion of the particles of pigment (often mixed with extender) in a paint medium to produce a uniform, stable system.

The techniques and associated theory applied to non-aqueous and water-soluble paints differ from those for aqueous emulsion paints, and the two will therefore be considered separately.

NON-AQUEOUS PAINTS

The process of pigment dispersion comprises three stages which are inter-related and can take place simultaneously.

1 Breakdown of aggregates to smaller units and individual particles.
2 Wetting of these units and particles by the paint medium.
3 Stabilization of the resulting dispersion.

Stage 1: Breakdown of aggregates

This is essentially a mechanical process involving pigment and medium. It is often referred to as 'grinding', although little breakdown of the individual particles takes place. The mills employed are described in Chapter 15.

It is not always necessary to effect a complete breakdown of the aggregates to individual particles; for each type of paint there is an 'acceptable degree of dispersion' which, in production control, is assessed by the use of the 'fineness-of-grind' gauge (Fig. 14.1).

The gauge consists of a steel plate on which are machined one or two wedge-shaped channels either 50 micrometres (0·002 in.) or 100 micrometres (0·004 in.) deep at one end and tapering to zero at the other. Calibration can be in micrometres, inches^{-3}, or in Hegman units of 0–8 or 0–10, increasing from the deep end. On the 0–8 Hegman scale, acceptable figures for undercoats are about 6 and for finishes at least 7·5 and preferably 'off scale'. Readings are

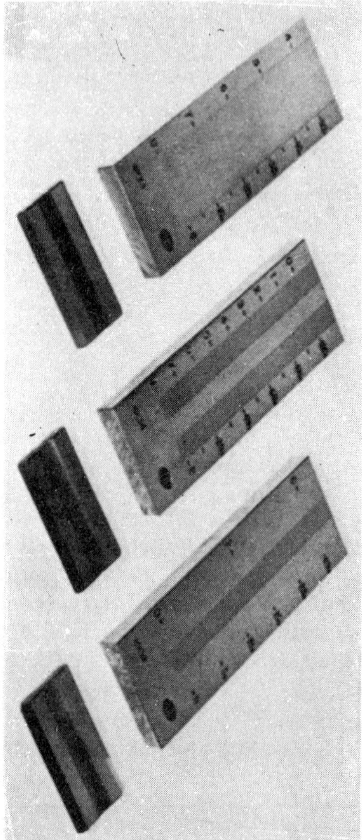

Fig 14.1 Fineness-of-grind gauge

taken by placing a drop of the pigment dispersion in the deep end and then drawing it along the channel with a doctor blade held vertically. The film is viewed at grazing angle and the point noted where the surface shows the general appearance of aggregates or 'bittiness'.

Whilst the fineness-of-grind gauge is a useful instrument for use in production control of dispersion, the information provided is insufficient for investigational work. Some knowledge is often required of the state of dispersion throughout the system, for example, particle size distribution, flocculation, etc. Among methods employed to obtain this information are determination of (a) tinting strength and (b) particle-size distribution by sedimentation rates, optical and electron microscope and centrifugal methods. A critical summary of these methods has been published by Sheppard and Cope [1] and by Carr [2].

The economic aspects of pigment dispersion have been examined by Varley and Bower [3] who point out that whilst certain factors such as raw material costs, labour, services, etc. are not amenable to control, the cost of producing the degree of dispersion required for a given product should be watched carefully. There is little to be gained by expending energy in dispersing beyond the practical requirements. The same authors list those paint properties which are improved and those which suffer by an increase in the degree of dispersion. Improvements are obtained in colour strength, brightness, gloss and transparency; on the other hand, opacity, lightfastness, weathering and flow properties are adversely affected.

Stage 2: Wetting of pigment particles

The part played by wetting in pigment dispersion has attracted considerable attention as in evidenced by the number of publications, a selection of which is given at the end of the chapter [4]. The following treatment is intended as an introduction to the subject.

The term 'wetting' is generally accepted as referring to the replacement of the pigment/air or pigment/moisture interface by a pigment/medium interface.

All solids adsorb a certain amount of moisture and, in addition, air is held within the aggregates of pigment particles. The adsorbed film is often held very tenaciously and, with finely divided powders, can be of considerable area.

The process of wetting involves a number of factors, predominant among which are (a) the surface tension or, more correctly, the adhesion tension at the solid/liquid interface, and (b) the nature of the pigment surface.

Interfacial tension

The adhesion tension between pure liquids and solids can be studied by use of the classical contact angle.

Contact angle A drop of liquid in contact with a solid surface will assume a shape that is governed by the surface tension at the liquid-solid interface. Suppose a drop takes the shape illustrated in Fig. 14.2, which depicts a cross-section through the drop and the surface.

The contact angle, θ, lies between the solid surface and the tangent to the liquid surface at C and is invariably measured in the liquid.

If γ_1, γ_2 and γ_3 are the surface tensions at the liquid/air, liquid/solid and solid/air interfaces respectively, then the horizontal forces acting at C will be in equilibrium, or

$$\gamma_3 = \gamma_2 + \gamma_1 \cos \theta$$

The work of adhesion, *W*, is given by

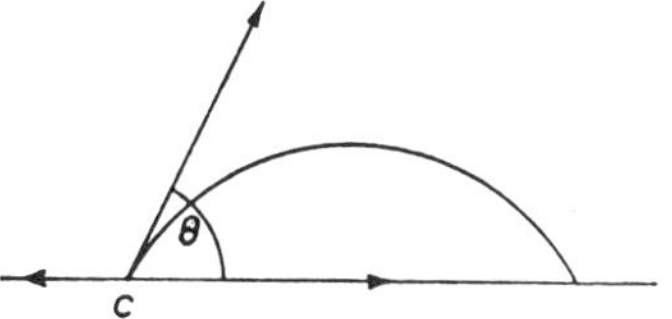

Fig 14.2 Section through a drop of liquid

$$W = \gamma_3 + \gamma_1 - \gamma_2$$
$$\therefore W = \gamma_2 + \gamma_1 \cos\theta + \gamma_1 - \gamma_2$$
$$= \gamma_1 (1 + \cos\theta)$$

In other words, the work of adhesion, which is a measure of the 'wettability' of a powder by a given liquid, can be obtained from a knowledge of the surface tension of the liquid and the contact angle between the liquid and the solid surface.

Measurement of the contact angle between a pigment surface and a liquid presents difficulty, but suitable methods have been described by Crowl [5]. This technique is really applicable only to pure liquids, but, according to Crowl, it can be applied to dilute paint media and shows some correlation with rates of dispersion in the early stages of ball milling.

Generally, however, more practical methods for assessing wettability are desirable and these inevitably must involve a consideration of the characteristics of the pigment surface.

Nature of the pigment surface

It is generally accepted that pigments can be designated hydrophilic or hydrophobic according to their tendency to attract or repel moisture, respectively. These terms are, however, relative and there is no sharp line of demarcation between the two classes. For example, rutile titanium dioxide is more hydrophobic than anatase, but they are both hydrophilic in comparison with the arylamide yellows. The majority of inorganic pigments are hydrophilic and disperse more readily in water than in organic media. Most organic pigment dyestuffs are hydrophobic and are not readily dispersed in water. In fact, if a dispersion in water is mixed with oil, flushing takes place, i.e. the pigment migrates to the oil phase.

This hydrophilic or hydrophobic nature of pigments can be changed by surface treatment. Some inorganic pigments are coated to render them hydrophobic and so more readily wetted by organic media, whilst the hydrophobic organic pigment dyestuffs are made hydrophilic for use in aqueous emulsion paints.

Pigment powders possess large surface areas per unit of weight. This is related to the particle size distribution and varies from one type of pigment to another. Surface areas are usually measured by the well-known nitrogen adsorption method (p. 33), but for many practical applications involving comparisons of pigments, oil absorption determinations are a sufficient guide.

Also related to particle size is the surface energy which increases with decreasing particle size. Much of the energy is concentrated at the edges of the pigment crystals.

Most pigments, by virtue of their surface energies, tend to adsorb certain molecular species from paint media [6]. The degree of adsorption can vary widely and depends on both pigment and medium. Alkyd resins contain a proportion of polar molecules, the polarity arising from free carboxyl and hydroxyl groups, and they comprise a wide range of molecular weights. The lower molecular weights are relatively more polar and tend to be preferentially adsorbed on certain pigment surfaces. The effect of this adsorption of medium by the pigment can be threefold.

1 At the dispersion stage, it assists in displacing the air and/or moisture at the pigment surface and so promotes the wetting process.
2 It coats the particles with an envelope of resin. This reduces the inter-particle attraction which can lead to flocculation and it also prevents dense settlement of the pigment on standing.
3 If this dispersion is stored for any considerable time or, often, when pigments of very fine particle size are involved, the adsorption can result in a progressive increase in viscosity and, in some cases, gelling can be the final result. For example, if carbon black or phthalocyanine blue are milled into alkyd at high concentration, rapid gelling can take place. The viscosity increase and ultimate gelling result from the withdrawal of the lower molecular weight species from the alkyd solution. These normally act as solvents for the higher molecular weight fractions, the solubility of which becomes progressively lower until gelation occurs.

In addition to adsorption of molecules of the medium, some pigments adsorb drier metals, resulting in a progressive increase in the drying time of the paint. Carbon black is notorious in this respect, as well as in causing progressive increase in viscosity of paints in which it is the major pigment. Titanium dioxide also has been known to adsorb driers, but to a lesser degree than carbon black.

In cases of drier adsorption, re-milling of the paint will restore much, but not all, of the original drying property.

Instances occur when the wetting and dispersion of a pigment in a medium is extremely slow, as indicated by an extended milling time. This is often attributed to 'poor wetting by the medium' or a 'difficultly wettable pigment'. In such cases the use of an appropriate surface-active agent can be of assistance. Cationic or non-ionic types are employed in non-aqueous media. Both types are polar and tend to be adsorbed on pigment surfaces, so inducing wetting.

According to Stephen [7] the pigment/solvent relationship is important in the dispersion process and to ensure subsequent dispersion stability. The correct solvent balance also is required to prevent unnecessary flocculation. He considers that hydrogen bonding, HLB values of surfactants as well as solvent and resin characteristics are involved.

Surface-active agents (surfactants)

A very large number of surfactants is available for assisting pigment dispersion and for stabilising the resulting dispersions. They can be classified into a relatively small number of general types but within these types there are innumerable modifications and many surfactants being marketed are mixtures of two or more types. In the following the basic characteristics of the main types are considered and no attempt is made to include the very many modifications.

Anionic surfactants

These are water-soluble and used only in aqueous media. They are described on p. 340.

Cationic surfactants

These are generally salts of primary, secondary or tertiary amines and the activity resides in the cation. They can be subdivided into (a) ammonia types and (b) quaternary ammonium salts.

Ammonia type Fatty acid amines or diamines form salts with free fatty acids to give products of the general type

$$\left[H_3\overset{+}{N}.R\overset{+}{N}H_3\right]\ \left[R'COO^-\right]_2$$

The cation is reactive and is adsorbed strongly on negatively charged surfaces such as oxidized metals, glass and paper. It will displace previously adsorbed moisture, and cationics of this type are used to assist the dispersion of pigments in organic media. The polar groups of the cation are attracted to the pigment surface, displacing the moisture, and the hydrocarbon chains are directed away from the surface. The latter becomes lyophilic and is readily wetted by the organic medium.

Strongly hydrophilic pigments are difficult to flush but can be made to do so by addition of a cationic surfactant to the aqueous dispersion. Adsorption of the cation in the manner described above results in a hydrophobic surface, and flushing then takes place readily.

Quaternary ammonium salts These are of the following general type:

$$\left[\begin{array}{c} R \\ | \\ CH_3—N—CH_3 \\ | \\ CH_3 \end{array}\right]^+ Br^- \qquad\qquad \left[\begin{array}{c} R \\ | \\ R'—N—CH_3 \\ | \\ CH_3 \end{array}\right]^+ Br^-$$

Alkyl trimethyl ammonium bromide

Dialkyl dimethyl ammonium bromide

The cations of these compounds are, like the ammonia type, strongly adsorbed on a number of surfaces, and their main use is in bituminous compositions and paints for application to wet surfaces.

Non-ionic surfactants

These do not dissociate in aqueous solution and their activity is due to polar groups such as hydroxyl or ether linkages within the molecule. Individually they are less hydrated than the ionized polar groups in anionics, but an equivalent total hydration can be achieved in the molecule if there is an adequate number of non-ionizing polar groups present. Non-ionic surfactants are made on this principle.

A number of types are available, the two most important being the polyglycerol esters of fatty acids and polyglycol ethers of polyethenoxy compounds. The latter are condensation products of compounds possessing an amido, hydroxyl or carboxyl group with a long-chain polyglycol ether radical, e.g.

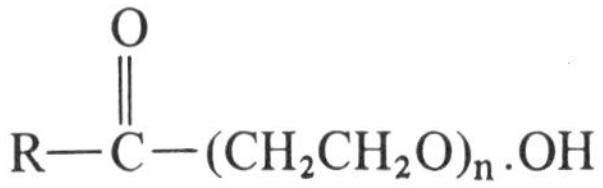

$$R{-}\overset{\overset{\displaystyle O}{\|}}{C}{-}(CH_2CH_2O)_n.OH$$

Ester linkage

$$R{-}\overset{\overset{\displaystyle O}{\|}}{C}{-}N\begin{cases}(CH_2CH_2O)_X.H\\(CH_2CH_2O)_Y.H\end{cases}$$

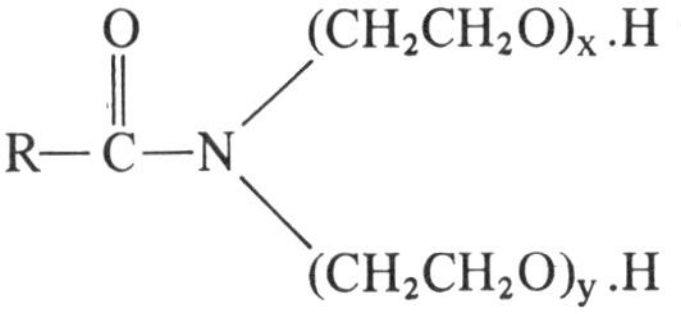

Amide linkage

These may be further esterified at the free hydroxyl groups to give polyoxyethyleneglycol diesters.

Properties of non-ionic surfactants

The water solubility of these non-ionics increases with increase in length of the polyoxyethylene chain. They can be subdivided on the basis of solubility into three types: (a) oil-soluble or slightly soluble in water, (b) self-dispersible in water, and (c) soluble in water. The water-soluble types can be used to replace anionics in most applications and possess the advantage of a lower tendency to promote foam formation. When used in aqueous paint systems they promote rapid wetting of the pigment when used at 0·5 to 1·0 percent (on the pigment). In emulsion paints they improve the brushability and levelling properties.

Non-ionics of the low ethylene oxide type are used to assist the dispersion of pigments in organic media. Small additions reduce the time required for dispersion and deflocculation. They show no preferential wetting in pigment mixtures. They are also used to assist the flushing of hydrophilic pigments.

Ampholytic surfactants

These are materials in which the polarity is internally neutralized. This type of structure occurs in the compound:

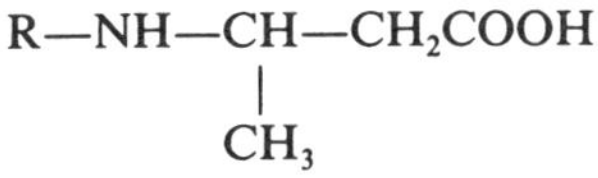

$$R{-}NH{-}\underset{\underset{\displaystyle CH_3}{|}}{CH}{-}CH_2COOH \text{ (Armeen Z) [8]}$$

and is also present in lecithin (*below*).

Lecithin This is a by-product obtained in the refining of natural drying oils such as linseed, soya, etc. The soya type is the more common and has been used as a surfactant on a wide scale for many years due to its versatility and relatively low cost. The material as marketed contains 35 to 45 percent of the parent oil together with a mixture of phosphatides. These contain a polyhydric alcohol

such as glycerol or other alcohol, esterified partly with fatty acids and partly with phosphoric acid. The phosphoric acid is further esterified with a hydroxyamino compound such as choline, ethanolamine or serine.
Each molecule contains lyophilic fatty acid groups and hydrophilic phosphoric acid and amino groups. The latter confer ampholytic properties on the molecule. In non-polar media and in aqueous dispersion the phosphatides assume the form of large micelles.

Uses Lecithin is used to assist pigment dispersion (0·5 to 1 percent on the pigment, but the amount may be critical for some systems) and generally as a modifier for pigment flocculation and dispersion.

Hyperdispersants

A range of specially designed dispersing agents which are claimed to raise dispersion efficiency as well as to improve the properties of the finished paint in respect of optical and flow properties [9]. They contain a synergist to enhance the attachment of the main surfactant chains to the pigment surface thereby promoting dispersion stability and the optimum effect is obtained by coating the pigment surface with a unimolecular layer of hyperdispersant. The amount required can therefore be calculated from a knowledge of the surface area of the pigment. Alternatively it can be determined practically by additions to a dispersed mill base. The type of hyperdispersant to be employed depends on the type of pigment as well as on the nature of the resin and solvent. Their use is claimed to effect appreciable savings in manufacturing costs as well as improvement in quality of product [9].

Stage 3: Stabilization of dispersions

A stable dispersion is one which, ideally, is free from any of the following defects.

1 Tendency of the pigment to settle to a dense mass.
2 Excessive flocculation of the pigment, a condition which can impair the rheological properties of the paint and influence the optical properties of the paint film [2].
3 Viscosity increase resulting from physical and/or chemical reaction between pigment and medium (as distinct from flocculation).

Pigment settlement [10]

The familiar form of the Stokes equation applies to spherical particles falling through a large mass of liquid:

$$V = \frac{2r^2(\rho_1 - \rho_2)g}{9\eta}$$

where V = velocity of fall of particle,
r = radius of particle,

ρ_1 = density of particle,
ρ_2 = density of liquid,
η = coefficient of viscosity of liquid,
g = gravitational constant.

Although a pigment dispersion does not satisfy the requirements for the application of this equation there is a degree of compliance, and it serves as a useful basis for our discussion.

V can be kept as low as possible (preferably zero) by reducing the radius of the particles; by equating, as far as possible, the densities of pigment and medium (ρ_1 and ρ_2); and by increasing the viscosity of the medium, preferably by controlled thixotropy.

If we consider a pigment particle with adsorbed low-molecular-weight polymer or surfactant, the effective radius will be increased, tending to more rapid settlement, but this will be negated by a reduction in overall density or ρ_1. The effect of the increased particle radius, caused by adsorbed long-chain molecules, will be a steric hindrance to free settlement. This will, nevertheless, take place slowly, but the settled layer of pigment will be more voluminous and will be more readily re-incorporated by agitation or stirring.

Prevention of dense pigment settlement is effected also by use of 'surface coated' precipitated calcium carbonate [11]. The coating consists of stearic acid (1·5–3·0 percent by mass on the calcium carbonate) which produces an enlarged hydrophobic particle. The presence of these particles introduces a steric hindrance to settlement of other pigment particles, but the quantity used should not exceed 10 percent of the total pigment. Above this figure, interference with flow properties can result.

The rate of pigment settlement is reduced by an increase in η, the coefficient of viscosity of the system. This can be achieved by introducing 'thixotropy' or structure into the paint. A thixotropic system is one which forms a gel when at rest but which becomes fluid and flows when a stress is applied and reverts to a gel as soon as the stress is removed. When a paint is in the gel condition, pigment settlement cannot take place. In practice, complete gelation is employed only in the 'non-drip' or thixotropic decorative paints. Generally a controlled degree of thixotropy or structure is built into the system and the following types of compounds are used.

1 Hydrogenated castor oils [12]. These require a temperature of 40°C to 60°C for gelling.
2 Organically reacted montmorillonite clays [13]. Types are available to suit solvents of different polarities. Some require to be pre-gelled but some newer types will gel spontaneously when mixed with paint.
3 Aluminium alcoholates [14].

Pigment flocculation

A flocculated system is one in which the pigment particles have formed clusters of 'flocculates' after they have been dispersed. The flocculates contain liquid and may be closely-knit or loose associations according to the forces acting between the particles; prevention of flocculation requires some means of opposing or neutralizing these attractive forces. In aqueous systems (p. 262) the particles can be given an ionic charge to render them mutually repellent, but this method cannot be employed in organic media. In these it is thought that adsorption of long chain molecules of polymer or surfactant tends to prevent the particles from approaching each other sufficiently closely for the short-range Van der Waals forces to operate [15].

In the more lightly pigmented paints such as gloss finishes, flocculates tend to settle rapidly, but their bulky nature results in a soft deposit of considerable volume which can readily be re-dispersed by stirring. The supernatant liquid is usually clear, whereas in non-flocculated or dispersed systems this layer is generally cloudy as a result of the very fine particles being kept in suspension by Brownian movement.

Flocculation in highly pigmented paints such as undercoats is usually accompanied by abnormal flow properties such as thixotropy and pseudo-plastic flow.

Generally a controlled degree of flocculation can be of advantage in preventing pigment settlement on storage.

The Flocculation Monitor [16]

An instrument developed by Tioxide UK Ltd, the uses of which are described by Gamon [17]. It is used to measure the degree of flocculation of pigment dispersions, especially titanium dioxide, in wet and dry paint films. In the latter it measures the amount of radiation of 2.5 μm wavelength which is reflected (back scattered) by dry paint films of differing thickness. A graph of reflectance plotted against film thickness gives a 'flocculation gradient'. High values indicate flocculation whilst low values indicate good dispersion. For liquid paints a sample is placed in a special cell but only one value is obtained.

Flocculation of stainers

In single pigment paints, flocculation generally presents no colour problems in the film. When, however, certain pale shades are produced by addition of dispersed stainer to a white base, for instance, difficulties can arise as a result of stainer flocculation. Pale blues, for example, made by addition of a dispersed phthalocyanine blue to a white base, have been known to flocculate on standing for a relatively short period. Many of the flocculates are re-dispersed on stirring but others can be more resistant. The latter are usually dispersed by shear, with the result that brushed films exhibit blue streaks whereas a mottled appearance results on spraying. The fault is reduced considerably by treatment of the phthalocyanine blue with surface-active agents to reduce the mutually attractive forces, and grades of 'flocculation-resisting' blues are available. The ideal situation is one in which co-flocculates are formed, involving both white pigment and stainer. This requires matching the relative free surface energies, a state difficult to achieve [18].

'Pigment shock' or 'dilution shock' is a severe form of flocculation which sometimes occurs during the 'letting down' of dispersed ball-mill bases. The latter consist of pigment dispersed in a dilute solution of resin in solvent (p. 270). A state of equilibrium exists, the resin molecules being subjected to the attractive forces of the pigment and the solubilizing characteristics of the solvent. This equilibrium can be disturbed very easily, and the sudden introduction of a large quantity of resin solution of high concentration can lead to severe flocculation of the pigment; further milling is usually required to disperse the flocculates. The added resin solution should be as dilute as possible and should be added in small portions with mill rotation between additions.

Effect of solvent on flocculation

A high proportion of inorganic pigments are hydrophilic and when dispersed in certain non-polar solvents yield 'buttery' pastes. For this reason these solvents have been described as 'poor wetters', but this may not be so. An example is afforded by the action of white spirit on anatase titanium dioxide, as quoted by Bowman and Hughes [19]. The 'buttery' dispersion, on treatment with a 2 percent resin solution, is converted to a free-flowing liquid. Actually the white spirit wets the anatase readily, but the dispersion is highly flocculated. The added resin is adsorbed on the anatase surface and probably inhibits the inter-particle attraction. Deflocculation results and the dispersion flows freely.

Flotation (or 'floating')

Some paint films can exhibit a mottled appearance shortly after application. It occurs when mixtures of pigments are used, the individual members of which differ appreciably in colour, particle size, specific gravity, and wetting characteristics. The cause is also associated with the type of solvent evaporation and the influence of convection currents in the wet film. Flotation is accentuated when one of the pigments has a tendency to flocculate. Systems which can exhibit troublesome flotation are titanium dioxide/phthalocyanine blue, titanium dioxide/carbon black and lead chrome/Prussian blue [20]. Flotation can be converted to the less objectionable 'flooding' by addition of certain silicone oils (1 part to 40 000 parts of paint) or proprietary materials [21].

(The causes of flotation and flooding have been examined in more detail in Chapter 5.)

Flocculation during drying

The loss of solvent from a freshly applied paint film results in film contraction and a reduction in distances between the pigment particles. As a result, flocculation can take place before the film has set, and the consequence is a loss of opacity and colour strength. Carr has presented evidence of the occurrence of flocculation during the drying of both decorative and industrial paints as well as printing inks [22]. It has also been observed during the cure of acid-cured alkyd/amino finishes pigmented with rutile titanium dioxide. Here the flocculation appears to be associated with reaction between the basic surface coating of the rutile titanium dioxide and the added acid [23].

Increase in viscosity (on storage)

This can result from reaction between pigment and medium and may be physical and/or chemical in nature. The most common causes are the following.

1 Adsorption by the pigment of certain low molecular weight fractions of the polymer binder, as discussed on p. 254.
2 Chemical reaction between an acidic medium and basic pigments such as zinc oxide or certain grades of coated rutile titanium dioxide. The increase in viscosity is progressive and can result in gelling. In the early stages the thickening can be checked by addition of small quantities of benzoic, salicylic or citric acids which act as peptizing agents.
3 Reaction between metallic aluminium pigment and an acidic medium can result in gassing accompanied by an increase in viscosity of the paint. This can lead to gelling.

Some other causes of thickening on storage are mentioned on p. 469.

WATER-SOLUBLE PAINTS

Paints in which the polymer binder has been rendered soluble in water are used mainly for application by electrodeposition (p. 363), although they are also applied by conventional spray and dipping. In each case the films are cured by stoving.

Nature of the binders

These are of two types and are described in Chapter 19. Briefly they are (a) alkali silicates, which are now used only in certain types of zinc-rich paints and in fire-resisting compositions for interior use, and (b) the water-soluble resins. Water-solubility has been conferred by the introduction of either acidic or basic groupings followed by neutralization by bases or acids respectively. The main use of the water-soluble resins is in primers for application by electrodeposition, but appreciable amounts are used in stoving primers and finishes.

Pigmentation of water soluble resin solutions for conventional application generally follows normal procedure, dispersion stability being achieved by the use of appropriate surfactants. Pigmentation for electrodeposition is considered in Chapter 20.

AQUEOUS EMULSION PAINTS

Polymer emulsion paints

These are usually three-phase systems consisting of aqueous (or continuous) phase, pigment and dispersed polymer. The manufacturing operation involves two operations; (a) dispersion of pigment in the aqueous phase, followed by (b) mixing of the pigmented aqueous phase with the polymer emulsion.

Pigment dispersion in the aqueous phase

A very high proportion of the emulsion paint used is white or tints prepared from white; the pigment employed is rutile titanium dioxide and the following discussion relates to the dispersion of this pigment. (The technique for dispersing full colours is a little different and is described later.) The aqueous medium consists, basically, of a dilute solution of either a cellulose ether or sodium carboxymethyl cellulose or an ammonium salt of a polyacrylic acid.

Rutile titanium dioxide is a hydrophilic pigment and usually has an adsorbed film of moisture on the surface of the particles. No great difficulty is encountered in wetting such a pigment by an aqueous medium, and the breakdown of the pigment aggregates is not taken to the degree required, for example, in gloss finishes. The main problem is the stabilization of the titanium dioxide dispersion. The particles must be prevented from flocculating and from any tendency to 'flush' to the organic polymer on storage. (With hydrophilic pigments, however, this tendency is very slight.)

The particles in the polymer emulsion are usually 'protected' by an anionic surface-active agent which gives them a negative charge. These emulsions are known as anionic and described in Chapter 19.

Stabilization of the pigment dispersion

The dispersion of titanium dioxide particles can be effected by the use of either anionic or long-chain non-ionic surfactants. Both these surfactants confer a negative charge on the particles as a result of adsorption. The negativity in anionics is due to the $-COO^-$ group, while in non-ionics it arises from the multiplicity of ether linkages. Non-ionics are usually preferred since the anionics have a great tendency to produce foam. In the presence of water-soluble thickners, foams can be difficult to break.

Adsorption of the surface-active agent results in an 'electric double layer' round the particle, the negative charges being accompanied or surrounded by a second layer of positive ions, as in Fig. 14.3. An electrical double layer to this type is also produced by the adsorption of electrolytes (e.g. sodium polyphosphates) or anionic polyelectrolytes (ammonium salts of polyacrylic acid) on the pigment surface. Whilst sodium hexametaphosphate [24] is used in emulsion paints as a sequestering agent for calcium and other polyvalent metal ions, reservations on the use of this type of material have been expressed by Capella and Bieleman [4]. They consider that polyphosphates can be hydrolysed to orthophosphates which not only act as pigment flocculants but also cause a loss of gloss and provide a nutrient for bacteria.

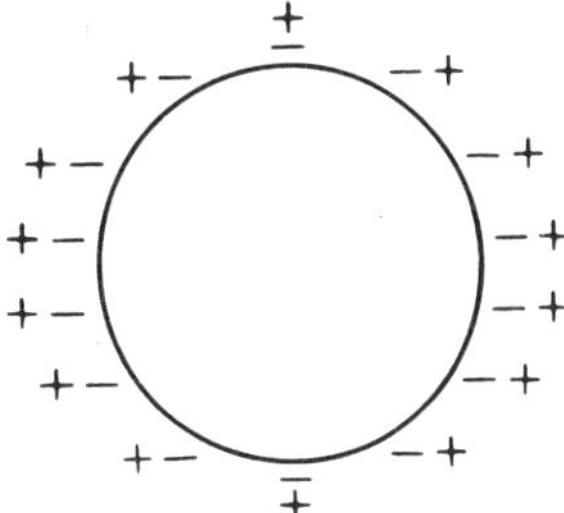

Fig 14.3

Further stabilization of the titanium dioxide dispersion, and of the emulsion paint, is effected by the use of water-soluble colloids or 'thickeners' which introduce structure in the system and so decrease the mobility of both pigment and polymer particles. Materials used for this purpose include methyl cellulose, hydroxyethyl- and hydroxypropyl-cellulose, ethylhydroxyethyl cellulose, and ammonium polyacrylates. These colloids have been described in Chapter 12.

Dispersion of hydrophobic pigments

Pale shades of emulsion paints are produced by addition of dispersions of coloured pigments (stainers) to the white emulsion paint. The stainers must be efficiently stabilized to prevent flocculation and changes of colour. The majority of inorganic pigments and extenders are hydrophilic and, in general, are readily dispersed and stabilized in aqueous media. Many organic pigments, particularly the pigment dyestuffs such as the arylamide yellows and toluidine reds, are hydrophobic and require treatment with anionic surfactants for both wetting and dispersion. The level of surfactant is relatively high to ensure the stability of these dispersions when used as stainers. Unless they are protected in this way, hydrophobic pigments will flush readily to the organic (polymer) phase, with serious effects on the colour of the emulsion paint. Similar remarks on the necessity for adequate protection apply to hydrophobic pigments used as the principal pigments in full-colour emulsion paints.

These usually require organic pigments, which are generally hydrophobic, to give the necessary brightness. They can be made in the following manner.

1 A concentrated dispersion of the organic pigment is prepared in an aqueous solution of a cellulose ether and a non-ionic surfactant.

2 An emulsion 'base' is made by dispersing the extenders in an aqueous solution of a cellulose ether together with the surfactants and other additives. This is then mixed with part of the polymer emulsion.

3 The required quantity of **(1)** is stirred into **(2)** and the remainder of the polymer emulsion is added.

References to Chapter 14

[1] Sheppard, I. R. and Cope, G., Methods available for dispersion testing. *JOCCA*, 1963, **46**, 220.

[2] Carr, W., The theory and practice of pigment dispersion. *JOCCA*, 1978, **61**, 397.

[3] Varley, D. M. and Bower, H. H., Dispersion of pigments: what is the ultimate? *JOCCA*, 1979, **62**, 401.

[4] Gans, D. M., Wetting, spreading and contact angles. *JPT*, 1966, **38**, No. 497, 322. *Idem*, Wetting and wetting. *JPT*, 1967, **39**, No. 511, 501. Capella, A. and Bieleman, J. H., The application of surface active agents in the manufacture of paints and varnishes today and in the future. *PPCJ*, 1979, **169**, N. 4007, 844; 1980, **170**, No. 4016, 65.

[5] Crowl, V. T., The interaction of the pigment with the medium. *JOCCA*, 1963, **46**, 169.

[6] *see also* Joppien G. R. and Hamann, K., The structure of layers of adsorbed polymers at pigment/solution interfaces and their influence on the dispersion stability of pigments in paints. *JOCCA*, 1977, **60**, 412.

[7] Stephen, H. G., Hydrogen bonding - key to dispersion? *JOCCA*, 1982, **65**, 191.
[8] Akzo Chemie UK Ltd, Wandsworth, London.
[9] Cowley, A. C. D., Improving dispersion of pigments with hyperdispersants. *JOCCA*, 1987, **70**, 207.
[10] *Hess's Paint Film Defects*, 3rd Edn, p. 8. Chapman & Hall, London, 1979.
[11] 'Calofort S' (J & E Sturge Ltd, Birmingham); 'Winnofil' (ICI).
[12] 'Thixcin' (NL Chemicals UK Ltd, Wilmslow, Cheshire); 'Perchem MF' (Perchem Ltd, Harlow, Essex).
[13] 'Bentones' (NL Chemicals UK Ltd); 'Perchem' range (Perchem Ltd).
[14] 'Alusec', 'Manalox' (Manchem Ltd, Manchester).
[15] Parfitt, G. D., *Dispersion of Powders in Liquids*, 2nd Edn. Applied Science Publishers, London.
[16] Pearson Panke Equipment Ltd, London, NW7.
[17] Gamon, M., *JOCCA*, 1986, **69**, 102.
[18] Patton, T. C., *Paint Flow and Pigment Dispersion*, 2nd Edn, Wiley-Interscience, New York, 1979.
[19] Bowman, A. and Hughes, W., Some observations on wetting and dispersion with particular reference to titanium oxide. *JOCCA*, 1951, **34**, 412.
[20] *Hess's Paint Film Defects*, 3rd Edn, p. 76. Chapman & Hall, London, 1979.
[21] 'Troykyd Anti-float' (Troy Chemicals Corp.) (Cole Chemicals Ltd, Purley, Surrey). 'Perchem' 37, 44, 48 (Perchem Ltd, Harlow, Essex).
[22] Carr, W., Optical properties of films. *JOCCA*, 1976, **59**, 443.
[23] *Hess's Paint Film Defects*, 3rd Edn, p. 118. Chapman & Hall, London, 1979.
[24] 'Calgon' (Albright & Wilson Ltd, London).

15

Methods of pigment dispersion

The breakdown of pigment aggregates and subsequent dispersion of the smaller aggregates or single particles is brought about by application of shear and/or attrition to a mixture of pigment and paint medium. The composition of the mixture in terms of the relative amounts of pigment, binder and thinner will be influenced by the type of machine to be used.

The methods employed can be classified under two general headings.

1 Plastic milling at high consistencies. Pastes are sheared either in horizontal or in vertical mixers. The Mastermix High Torque Mixer (Fig. 15.1) is a widely used machine of this latter class.
2 Fluid milling at lower consistencies and employing high rates of shear, as in the ball and pebble mills (p. 269) and the Attritor (p. 271), or high rates of shear and attrition as in sand and bead mills (Fig. 15.6 and 15.7), as well as in turbine dispersers of the Silverson type (Fig. 15.4).

Both shear and attrition are involved in the majority of machines, the relative amounts varying from one machine to another. The ultimate aim is the same in all cases, but the machines differ considerably in design, rate of output and applicability.

PLASTIC MILLING PUG MIXERS

These are heavy-duty machines used for high consistency milling; they may be of the horizontal or vertical type.

Horizontal pugs are used for the highest consistency pastes and consist of a horizontal U-shaped steel trough, which may be jacketed for water cooling and fitted with two horizontal paddle shafts. These may be fitted with Z-blades as in the Werner-Pfleiderer mixer, or the blades may be of special design. The bottom of the trough is designed to eliminate 'dead' spots. In addition to the manufacture of putties, stoppers and mastics, horizontal pugs are used for the dispersion of pigments into cellulose nitrate and other binders in the so-called 'chip' process (*below*).

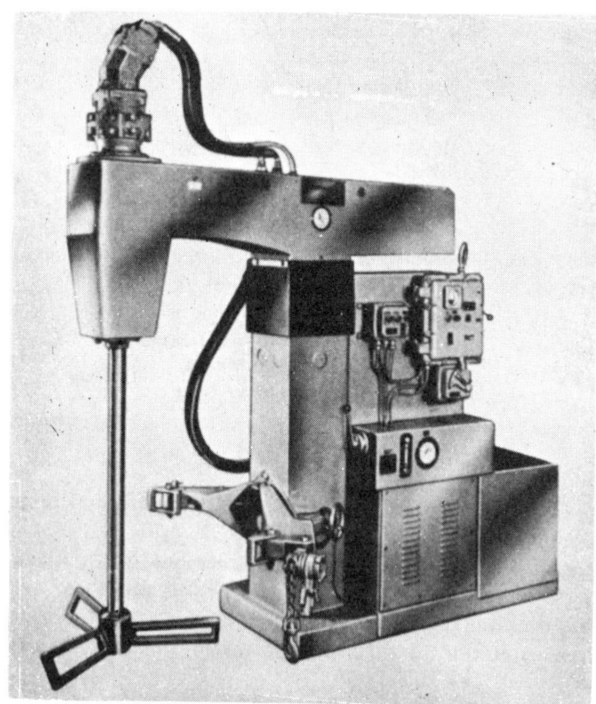

Fig 15.1 High torque mixer. Courtesy Masterix Eng. Co. Ltd.

Chip process Colour 'chips' consist of a high concentration of pigment dispersed in binder. When treated with solvent and additional binder the chips yield a dispersion of pigment which, in many cases, requires no further milling. The process of manufacture involves pigment and solvent-free binder. The heat produced in mixers of the Beken type is sufficient to maintain the mixture in a plastic state until dispersion is effected. The plastic mass is then discharged on to cooled rolls, from which it emerges as a solid sheet which is broken into chips.

The Bridge-Banbury mixer operates on a principle similar to that of the Beken but is totally enclosed. Pressure can be applied to the contents of the dispersing chamber and volatile solvents can be included.

Vertical mixers for heavy pastes comprise a fairly slow-moving stirrer which can be lowered into a mixing vessel (Fig. 15.1); mixing vessels of different sizes can be used and are securely attached to the mixer body when in use.

Paste mixers of the vertical change-pan type can be employed as pre-mixers for triple roll mills.

Triple roll mills are used for further processing the pastes produced in pre-mixers. The principle of operation is combined crushing and shear produced by passing the paste between rollers revolving at different speeds (Fig. 15.2). The rolls are of chilled iron and are hollow for water cooling.

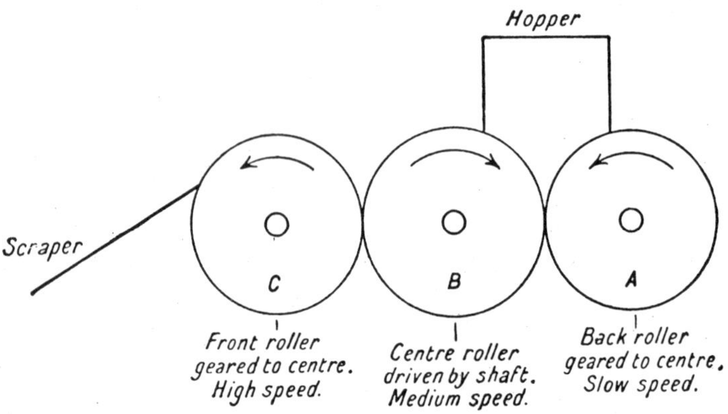

Fig 15.2 Principle of triple roll mill

The clearance between the rolls is of vital importance and a specific setting is required for each type of paste.

The gap between the back roll (A) and centre roll (B) determines the thickness of paste delivered to the nip between rolls B and C. Crushing and shear are applied in the A–B nip and the final dispersion is by shear alone in the B–C nip. On account of the higher speed in this nip the clearance can be considerably less than that between A and B.

Triple roll mills are effective in breaking down aggregates and dispersing the individual particles. No reduction in the ultimate particle size takes place. The capacity of these mills depends on (a) the rotational speed and speed differentials (the ratio is commonly 1:2:4), (b) the clearance between the rolls, (c) the physical dimensions of the rolls, and (d) the fluidity or stickiness of the paste. The latter is important as many media lack the necessary tack to adhere to the rolls, and the paste rides in the hopper. The addition of linseed stand oil introduces considerable tack, and this oil is widely used where circumstances permit.

Certain surface-active agents can be used for the same purpose and consequently increase the grinding efficiency.

Triple roll mills have a number of limitations which, with the introduction of more efficient mills, have resulted in a great decrease in their use in recent years. The following are the most important drawbacks.

1 They are very slow and need constant attention.
2 They are unsuitable for media containing very volatile solvents.
3 There is no means of preventing foreign bodies from entering the nips. The shock is taken by springs, but an emergency switch is located near the operator's hand.
4 Cleaning is very hazardous, the chief danger lying at the contact of the back and centre rolls.

Though now little used in the paint industry, triple roll mills continue to be used in the printing ink and other industries when materials of high consistency are processed.

FLUID MILLING

Several types of mill are available for pigment dispersion at fluid or 'slurry' consistencies. They may be divided into two general classes: (a) direct charge mills, in which the pigment and medium are fed directly into the dispersing chamber, and (b) mills which require the pigment and medium to be premixed.

Direct charge mills

Ball and pebble mills

These mills consist basically of a steel cylinder mounted horizontally and capable of being rotated about its axis. Ball mills are steel lined and contain steel balls to effect dispersion; pebble mills are lined with ceramic or silica blocks and are charged with balls of steatite, alumina or porcelain. (The terms 'ball mill' and 'pebble mill' are used rather loosely and the former is widely used for both types.)

Steel lined mills charged with steel balls can be used only for dark colours; erosion results in discoloration of whites and pale shades. These mills show a high dispersing efficiency, but the charges of steel balls are extremely expensive.

Pebble mills are used for whites and colours, but they are slower than the all-steel mills in reaching a desired degree of dispersion.

In operation, the mill is charged with pigment and medium and rotated at the appropriate speed. Dispersion takes place mainly by shear resulting from the relative movement of the balls, and to a lesser extent by impact.

It is of vital importance that these mills should be run at the correct speed. The optimum working speed has been found by experience to be 50 or 60 percent of the critical speed. The latter is the minimum speed at which the balls are carried round by centrifugal force and, can be calculated using an expression derived by Coghill and De Vancy [1]

$$\text{Critical speed (rev/min)} = \frac{54 \cdot 18}{\sqrt{r}} \quad \text{where } r = \text{internal radius of mill in feet}$$

The correct rotational speeds are, however, generally given by the manufacturers of the mills.

The efficiency of a ball or pebble mill is influenced by the following variable factors.

Load of steatite balls

The volume occupied should be 40 to 45 percent of the total volume of the mill; steel balls should occupy 35 to 40 percent. In a mill charged with steatite balls of 1″ (25mm) to 1½″ (38mm) diameter, the void volume is 23–25 percent of the total mill volume. Maximum grinding efficiency is achieved when the volume of mill base is equal to that of the void volume. High-density materials give more rapid dispersion, and ball mills are therefore more efficient than pebble mills for the darker pigments.

Size of steatite balls or pebbles

This should be as small as possible. Although Fischer [2] has found that a mixed-size charge has about the same efficiency as a single-sized charge of the same mean diameter, most manufacturers use a range of sizes. For example, for mills ranging from 450 litres (100 gallons) to 2250 litres (500 gallons) the charges usually consist of a mixture of 1" to 1½" steatite balls; steel balls are usually smaller, generally ½" to 1" diameter.

Volume of mill base

The optimum volume of mill base is equal to or slightly greater than the void space in the balls or pebbles.

Composition of mill base

In ball mills, as in all batch mills, the prime aim is the dispersion of the maximum amount of pigment per charge. It is known that low viscosities give most rapid dispersion, and the mill base should therefore contain the maximum amount of pigment at the lowest viscosity, i.e. in a medium containing a high ratio of solvent to binder. Complete elimination of binder leads to pigment flocculation difficulties as well as to severe wear on the inside of the mill.

In the early days of the ball mill, mill-base compositions were largely arrived at by empirical methods until Daniel [3] suggested a method whereby the optimum composition could be determined. The method, briefly, is as follows.

Daniel flow-paint method A number of solutions of the grinding medium are made up in the proposed solvent at concentrations ranging from 10 to 30 percent.

20 grams of the pigment are placed in a 100 ml beaker and titrated with the weakest of the resin solutions, using a thick glass rod to incorporate the pigment. The resin solution is added and thoroughly mixed in after each addition until the dispersion drips from the end of the rod and the residue snaps back like elastic. After a little practice this end-point can be repeated with a surprising degree of accuracy. The titration of 20 grams of pigment is repeated with each of the solutions in turn, and the volumes used are then plotted against concentration. For a medium of long-oil alkyd in white spirit a graph of the type shown in Fig. 15.3 is obtained. The minimum point on the curve represents the volume of solution as well as the concentration to be used in 20 grams of pigment.

The Daniel flow-point method forms the basis of the so-called 'slurry grinding' and has greatest application to paints where the pigmentation is low, i.e. gloss finishes. It enables sufficient pigment to be dispersed to give many times the mill volume in finished paint. It is not as readily applied to more highly pigmented systems.

When the pigment has been dispersed satisfactorily, the balance of the medium can be added to the mill, or the mill base can be dropped into mixers and let down in these. In either case care must be taken to add the medium

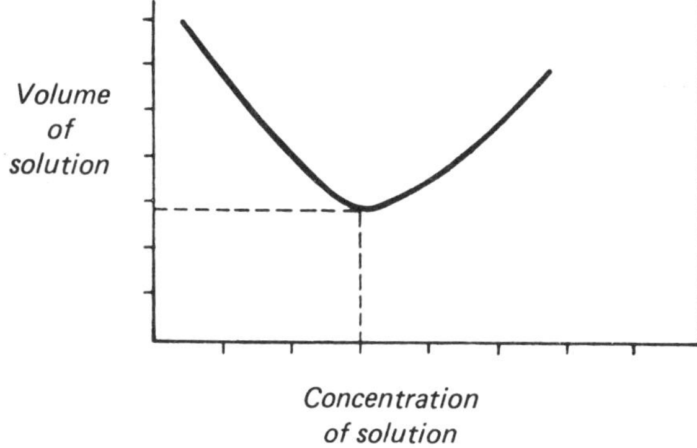

Fig 15.3 Daniel flow curve

slowly, preferably in portions with running or stirring for a short time after each addition. Rapid addition of a large quantity of medium can result in pigment shock, a severe form of flocculation.

Ball mills continue to be used extensively for batch production. Where long runs of an individual paint are involved it is customary to use the sand or bead mills (*below*).

The Szegvari Attritor can be regarded as a development of the ball mill in which the grinding chamber is fixed vertically and the balls are agitated by a vertical shaft fitted with arms. The grinding chamber is usually stainless steel and the steatite or stainless steel balls range in diameter from 3 mm to 12 mm, according to the type of material being processed. It has been stated by Wadham [4] that the attritor is, on the average, ten times as fast as the ball mill in producing a desired degree of dispersion. The mills range in size from 25 litres (5·5 gallons) to 400 litres (88 gallons) and can be jacketed for water-cooling. The pigment and medium can be charged directly into the machine or may be pre-mixed. Then, by means of an external pump system, the charge is re-circulated through the chamber until the required degree of dispersion is obtained.

Turbine dispersers embody a high-speed rotor which may be a propellor type enclosed in a form of cage (as in the Silverson machine) or it may consist of a disc with serrated edge of special design.

The Silverson mixer (Fig. 15.4) has a propellor enclosed in a circular cage, the walls of which are metal screens of differing design. The action of the propellor consists in drawing in the mixture of pigment and medium and then ejecting it at high velocity through the wire mesh. Dispersion is by impact, shear and attrition.

For operation, sufficient medium is placed in the pot to cover the dispersing head, and pigment is fed in gradually. More medium and pigment are fed in alternately and the machine run until the desired degree of dispersion is reached. Appreciable heat is developed in the process. The Silverson mixer is admirably suited to emulsion paint production, the pigment being dispersed in the aqueous phase before mixing with the polymer solution.

a

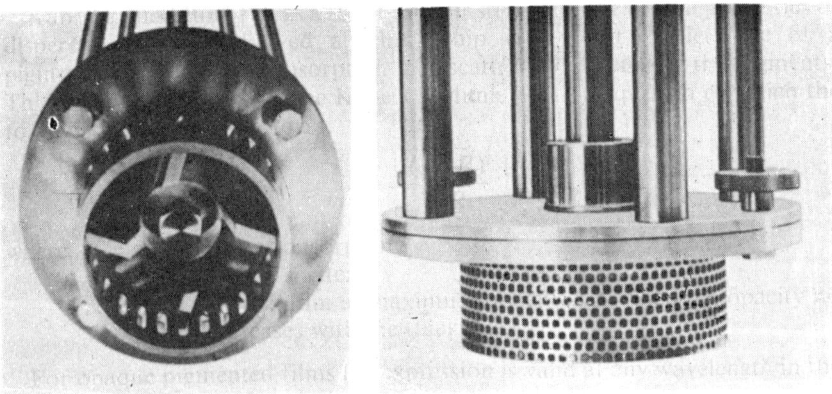

b c

Fig 15.4 Silverson high shear disperser; (a) general view; (b) high shear head; (c) high shear screen

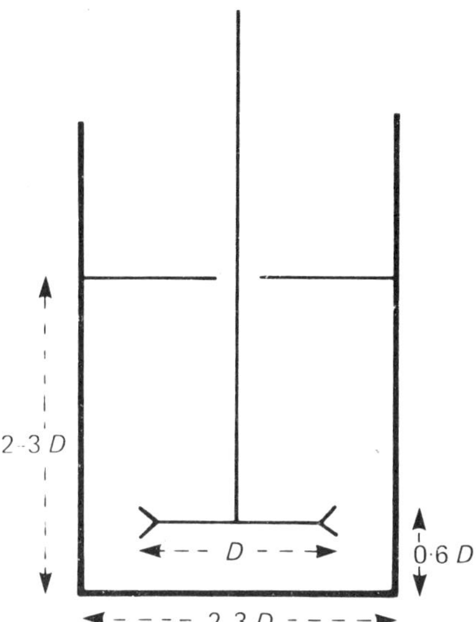

Fig 15.5 Dimensions for maximum efficiency high shear disperser

High-speed impellers are used as pre-mixers for other machines, as dispersers for 'soft' pigments, as thinning and tinting mixers, and as dissolvers. A circular steel plate is attached to the end of a steel shaft and the periphery of the disk is cut and then bent to form an irregular edge. The disk is suspended in a pot which is usually jacketed for water-cooling, and for maximum efficiency there is a definite relationship between the diameter of the disk and the dimensions of the pot (Fig. 15.5). When used as pre-mixers or dispersers, care must be taken to ensure the correct composition and consistency of mill base. The principles and operation of these mills have been discussed by Lester [5]. Examples of this type of mill are the Torrance Cavitation Mixer [6], the Cowles Dissolver [7] and the Molteni Disperser [8].

Batch bead mill

This mill functions in a similar way to the sand mill and the horizontal bead mill (*below*) in effecting dispersion by rapid agitation of ballotini in presence of the mill base. It differs from the two mills mentioned in requiring no premixer.

The machine consists of a vertical mixing vessel surrounded by a jacket for cooling or heating. The mixing vessel can be spun for discharge and is fitted with a rotor disc which produces the desired 'vortex' in operation. The rotor disc is underdriven in the Centrimill [9] and overdriven in the Torrance Batch Bead Mill [6].

Mills which require premixing

The sand mill

In mills in which pigment dispersion is effected by movement of a large number of balls, such as the attritor and ball mill, the efficiency, as measured by the rate of dispersion, increases with decreasing diameter of the balls. By using very small spheres and stirring these at high speed in the presence of the pigment slurry, dispersion is effected very rapidly.

The sand mill is a very simple machine, consisting of a vertical water-jacketed pot which contains graded sand (20 to 40 mesh) or glass ballotini of 0·5 to 1·0 mm diameter. This is agitated by a specially designed impeller comprising a vertical steel shaft to which steel disks are attached at intervals.

The pre-mixed pigment slurry is pumped in at the bottom of the container and rises through the sand which is kept in a 'fluid' condition by the rapidly rotating impeller. Dispersion of the pigment takes place as a result of shear as it rises through the chamber, the rate of throughput being regulated by the speed of the pump. The majority of pigments are satisfactorily dispersed when they reach the top of the chamber, and the dispersion passes out through a special mesh which retains the sand particles.

With an adequate supply of pre-mixed pigment slurry the process can be continuous, and the great virtue of sand-milling is the high output obtainable from a small mill.

The composition of the premixed slurry can be determined by the Daniel Flow Point method but the figures for different pigments may require adjustment in the light of experience with the mill.

The mill is easily cleaned by flushing through with solvent, but when changes to entirely different colours are involved it is usual to change the sand. For this reason the mill is most efficiently used on long runs of one type of paint.

The horizontal bead mill

This operates on the same principle as the sand mill, using glass ballotini or 'beads', but the dispersing cylinder is mounted horizontally (Fig. 15.6). Horizontal bead mills are available in a range of sizes, from small units with a one-litre dispersing cylinder giving a throughput of 4 to 20 litres per hour to large machines of 150 litres capacity giving a throughput of 600 to 3000 litres per hour. It is claimed that materials up to 50 poise viscosity can be processed by these mills. Cleaning is readily carried out by flushing through with solvent.

Paint re-conditioners

These machines employ rapid vibration to containers of paint in order to mix the contents thoroughly or to re-disperse settled pigment. Well-known examples of this type of machine are the 'Red Devil' [10] and the 'Vibro Shaker' [11]. Typically the machine consists of two recessed circular plates, A, between which the paint tin B is secured (Fig. 15.7). The plates are faced with polyurethane foam and are tightened by the wheel C. A steel shaft, D, is connected

Fig 15.6 Horizontal bead mill (courtesy of OBS Machines Ltd)

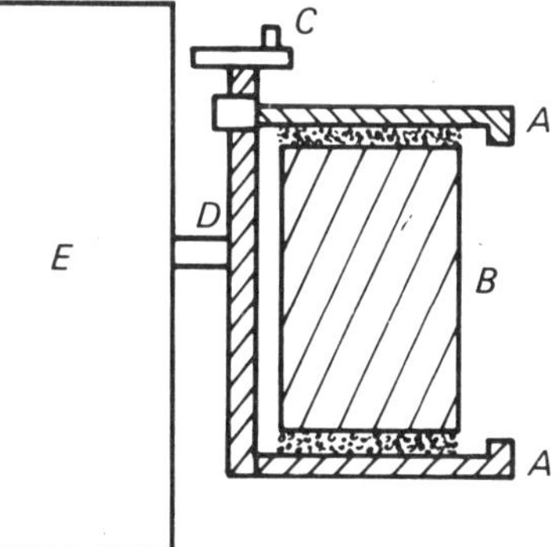

Fig 15.7 Paint reconditioner

to a gear and motor E so that the plates and paint are given a rapid oscillatory movement through a small arc. Settled pigments can be re-dispersed rapidly.

The machine is also useful for the rapid laboratory preparation of small samples of paint, often necessary in development work. Small glass bottles, charged with glass beads and the paint materials, are clamped between the plates. Dispersion is effected, in most cases, in 10 to 15 minutes.

Thinning mixers

This term covers the many types of general-purpose mixers used for thinning mill bases and for stirring during tinting. For small operations there are portable propellor stirrers which are clamped to the edge of the mixing vessel. Larger operations employ mixers in which the shaft and stirring head can be lowered into portable pots and raised on completion. The stirring head may be a slow paddle type or a high-speed rotating disc of the type used in high-speed dispersers. In fact, the same high-speed machine can be used for both dispersing and thinning.

In large-scale operations, paint is often pumped into large closed containers for thinning and tinting; stirrers of various types are incorporated.

STRAINING

Although manufacturers exercise great care to avoid contamination of products by extraneous matter such as dust, the paints may nevertheless contain particles of skin or gelled resin, fibres from bags, or the odd pigment aggregate. They therefore need straining to remove this matter. Strainers utilize metal or synthetic fibre gauge, the mesh being determined by the nature of the paint. Finishes, for example, require a much finer mesh than undercoats or primers.

The strainers in use are of four types.

Open gravity sieves

A strainer of this type is the Russel Finex (Fig. 15.8). The screen, which is detachable, forms the base of a circular tray which fits into a funnel-shaped hopper from which the paint passes to the filling container. The screen and hopper are vibrated to keep the meshes clear. The vibration also prevents 'set up' in thixotropic materials. The screens are either of metal (mesh sizes down to 64 micrometres) or supported nylon or other synthetic fibre gauge (mesh sizes down to 20 micrometres).

Rotary strainers

The Jenag [12] strainer illustrated in Fig. 15.9 is a widely used machine of this type. It consists of a vertical chamber containing two vertical perforated plates. Between these is sandwiched the straining gauge of nylon or other fibre with any desired aperture size down to 20 micrometres. The paint is fed by gravity or pump to the front of the plates and drawn through by vacuum. A small quantity

Fig 15.8 Heavy-duty vibrating paint strainer. Courtesy Russ & Finex Ltd.

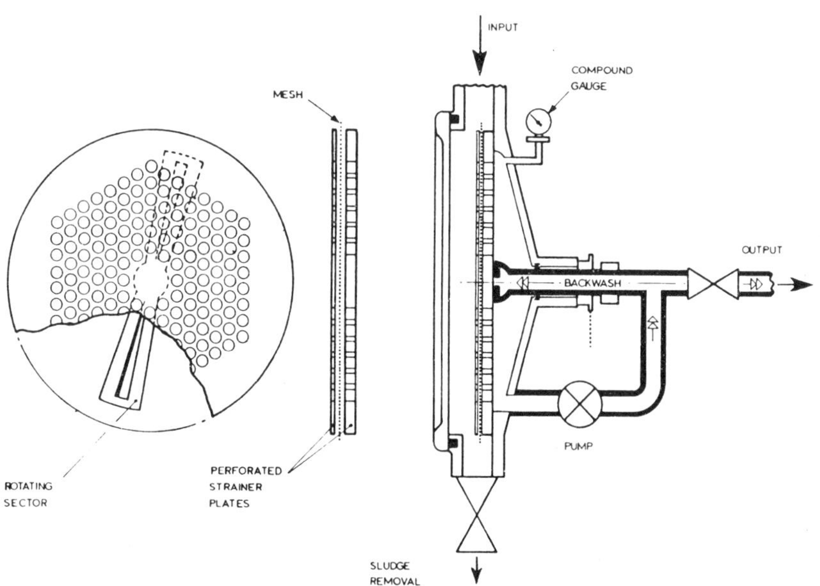

Courtesy: Jenag Equipment Ltd

Fig 15.9 Principle of Jenag paint strainer system

of the cleaned paint is by-passed back and applied to the rear plate by a rotary back-washing blade driven round the chamber. By means of this counter-current the straining gauge is kept free from clogging. Machines are available to handle materials of viscosities up to 200 poise, the rate of throughput depending on the viscosity.

Cartridge filters

These are employed for straining high-quality finishes such as those for cars and industrial products. The cylinders of the cartridges consist of wound polypropylene or other resin and are manufactured in different densities. The paint is pumped through the cartridge, and a special device is incorporated to indicate blockage of the filtering surface.

Bag-type strainers

These are filter bags made either from felts (rayon, polypropylene or nylon) or gauze (polypropylene, nylon or polyester) [13]. Mesh sizes available range from 3 to 100 micrometres. The bags are attached to the flanged end of a supply line and can be supported by a wire basket which, in turn, can be vibrated. Filter bags, which are washable, are generally used for small batches.

FILLING

Although the majority of paints are sold by volume, in practice it is found more convenient to fill by weight, making use of the specific gravity of the paint. Either manual or machine filling is employed, depending on the number and size of container to be filled. It has been stated [15] that a generally accepted minimum for semi-automic filling is 100 litres in small tins or 200 litres in 5-litre containers.

References to Chapter 15

[1] Coghill, W. H. and de Vancy, F. D., *U. S. Bur. Mines Tech. Paper No. 381* (1937).
[2] Fischer, E. K., *Colloidal Dispersions*. John Wiley, New York, 1950.
[3] Daniel, F. K., *Nat. Paint. Var. Lacq. Ass. Sci. Sect. Circ. 1950*, 744.
[4] Wadham, H., *JOCCA*, 1964. **47**, 669.
[5] Lester, G. R., *JOCCA*, 1964, **47**, 719.
[6] Stothert & Pitt Fluid Engineering Ltd, Bitton, Bristol.
[7] Moorhouse International Ltd (Russell Finex Ltd, London).
[8] D. H. Industries Ltd, Barking, Essex.
[9] OBS Machines Ltd, Milton Keynes.
[10] Red Devil Inc., Worcester.
[11] Westlairds Ltd, Slough, Bucks.
[12] Jenag Equipment Ltd, Waterlooville, Hants.
[13] 'Filbag' system (D. H. Industries Ltd, Barking, Essex).
[14] *OCCA Paint Technology Manuals*, Part 7. Works Practice, p. 845. Chapman & Hall, London, 1974.

16

The protection of metals

FERROUS METALS

With the exception of gold and silver, very few metals are found in nature in the free state; they exist in the form of compounds such as oxide, sulphide, or carbonate, which are more stable. The pure metals tend to revert to a more stable compound if exposed to a suitable environment, and an excellent example is afforded by iron. When exposed to air and moisture this will revert rapidly to the oxide, the familiar process of rusting. To prevent rusting or corrosion the metal must either be isolated from the environment or coated in such a way that the corrosion reactions are neutralized.

Corrosion takes place on both ferrous and non-ferrous metals, but the former present much the greater problem.

CORROSION OF FERROUS METALS

Ferrous metals comprise the various alloys of iron, and the following are commonly encountered.

Pig iron is the crude product from the blast furnace and consists of 92 to 95 percent iron, about 4 percent carbon, together with small amounts of other metals. When remelted and cast into moulds it forms *cast iron*, a hard and brittle material which is often used for heavy machine bases.

Wrought iron is produced by burning out the carbon and other impurities from pig iron. It consists of pure iron admixed with small quantities of slag and is more resistant to corrosion than is mild steel.

Steel is the general name given to alloys of iron containing small amounts of carbon, and capable of being hardened by heating followed by rapid cooling. It is produced from pig iron by burning out all carbon and other impurities, followed by addition of controlled amounts of carbon and other metals. The hardness and working properties of the steel are largely controlled by the carbon content, and the following figures indicate the percentage of carbon in common steels:

Structural (mild) steel	0·1 –0·25%
Sheet and strip steel	0·05–0·15%
Tool steels	0·8 –1·4%

Stainless steels contain appreciable quantities of chromium and nickel. The well-known 'Staybright' or '18/8' stainless steel contains 18 percent chromium and 8 percent nickel. The rust-resisting properties of the steel are attributed to a very thin film of oxides of chromium which forms over the surface.

Mild steel is the type most often encountered and is used in RSJs (rolled steel joists), girders, etc. in structural work; in ships' plates, and in the form of sheet or strip for many industrial purposes. Processing consists of hot-rolling the red-hot ingots until the desired shape and thickness is reached, and this rolling is accompanied by the formation of a considerable amount of oxide (millscale) on the surface.

Where a very smooth surface is required the hot-rolled sheets are descaled and then further reduced by cold-rolling.

Types of corrosion

Direct oxidation

With iron and steel this is a relatively slow process at ordinary temperatures. A piece of clean bright steel will rapidly become coated with a thin, invisible film of oxide. The rate at which this film grows decreases with increasing thickness, as oxygen must pass through the film for reaction at the metal surface.

The rate of oxidation increases rapidly with temperature, and when iron is heated to above 600°C a surface film is produced which consists of layers of the three oxides (A in Fig. 16.1). The layer of FeO is stable only at the high temperatures and on cooling decomposes, forming a dispersion of metallic iron in Fe_3O_4 (B in Fig. 16.1)

$$4FeO \rightarrow Fe_3O_4 + Fe$$

At lower temperatures FeO is not formed, but layers of Fe_3O_4 and Fe_2O_3 take its place, the proportions varying with the temperature.

The oxide layer illustrated in Fig. 16.1 is formed in the hot-rolling of mild steel and is known as *millscale*. Temperatures of rolling vary between 1300° and 1100°C and the layer of millscale produced is hard and very adherent.

The coefficient of expansion is less than that of the steel, and dimensional changes in the steel resulting from temperature fluctuations can result in cracking of the scale (Fig. 16.2). Moisture and oxygen enter the crack and cause

Fig 16.1 Millscale

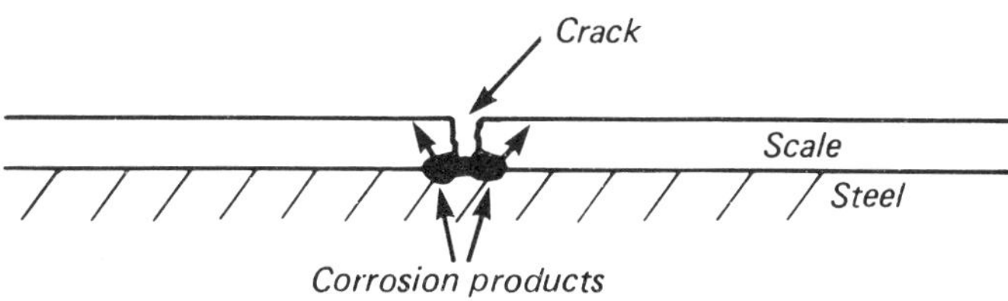

Fig 16.2

corrosion of the steel, the corrosion tending to 'creep' under the scale. Since the corrosion product (rust) occupies a larger volume than the steel, pressure is exerted on the scale and this ultimately flakes off. It is essential, therefore, to remove scale before painting. The application of paint to millscale will only delay the breakdown. Most paints are permeable to moisture and oxygen to some degree, and ultimate failure is inevitable.

Atmospheric corrosion

It was shown by Vernon that the tendency for iron to rust in the atmosphere is controlled by the relative humidity, and the rate of rusting is controlled by the degree of pollution, especially sulphur dioxide and solid particles. In the presence of sulphur dioxide the initial product of attack is ferrous sulphate, which reacts with water and oxygen to give ferrous hydroxide; this is further oxidized to hydrated ferric oxide (rust). Sulphuric acid is liberated and then reacts with the iron to regenerate ferrous sulphate. The reactions can be represented thus:

$$Fe + SO_2 + O_2 \rightarrow FeSO_4$$
$$4FeSO_4 + O_2 + 6H_2O \rightarrow 2Fe_2O_3H_2O(\text{rust}) + 4H_2SO_4$$
$$4H_2SO_4 + 4Fe + 2O_2 \rightarrow 4FeSO_4 + 4H_2O$$

Ferrous sulphate is usually present in the rust produced by exposure of steel to industrial or urban atmospheres and is very difficult to remove completely, even by the normal blast-cleaning process. This problem has been investigated by Bullett and Dasgupta [1] who state that blistering and eventual disruption of many paint films will occur if they are applied over pockets of ferrous sulphate.

The fact that chloride and sulphate ions on the surface of steel can cause premature breakdown of coatings is confirmed by Morcillo *et al* [2]. They state that amounts of the order of 100 mg/M^2 of sodium chloride or 1000 mg/M^2 of ferrous sulphate can shorten the life of many paint systems quite significantly.

Work by Harrison and Tickle [3] on the performance of paint applied to weathered steel has shown that contamination by atmospheric ammonium compounds plays a significant part. Ammonium sulphate was found to be retained on rusted steel after wire brushing, and this can become actively corrosive if moisture penetrates the protective paint film. The most satisfactory way of removing these ammonium compounds is by the use of water-blasting methods (see p. 291).

Electrochemical corrosion

It has been established that the three factors essential for the corrosion of iron and steel are moisture, oxygen, and a potential difference between the iron and other bodies with which it is in contact. These can be scale or rust on the surface or cabron (graphite) as a constituent of the steel. The presence of dissolved salts (chloride or sulphate) or carbon dioxide dissolved in the moisture will catalyse the corrosion reactions. The corrosion may be uniformly distributed over the surface (general rusting) or highly localized (pitting).

The basic mechanism in these types of corrosion is electrochemical and involves the passage of electrons from areas of high to lower electron density (or from negative to positive areas).

When a clean plate of mild steel is exposed to the atmosphere, a layer (usually non-continuous) of oxide is formed very rapidly. This oxide layer is, in the early stages, so thin as to be invisible to the eye. If such a plate is immersed in water or a neutral salt solution, differences of potential are set up between the oxide film and the metal.

At the base of a fissure in the oxide film (A in Fig. 16.3) there is iron, or a very thin oxide film. At this point iron will tend to go into solution (by virtue of its solution pressure) as ferrous ions, and electrons will be liberated on the metal. The reverse reaction will also take place, and in the absence of other factors a state of equilibrium will be reached:

$$4Fe \leftrightharpoons 4Fe^{++} + 8e' \text{ (anode reaction)}$$

In the present instance, the iron is in contact with oxide which exhibits no such tendency to dissolve and remains more positive. Electrons will therefore flow to the oxide film and there, in the presence of water and oxygen, are consumed in the formation of hydroxyl ions:

$$4H_2O + 2O_2 + 8e' \rightarrow 8OH' \text{ (cathode reaction)}$$

This flow of electrons away from the iron upsets the equilibrium and the metal

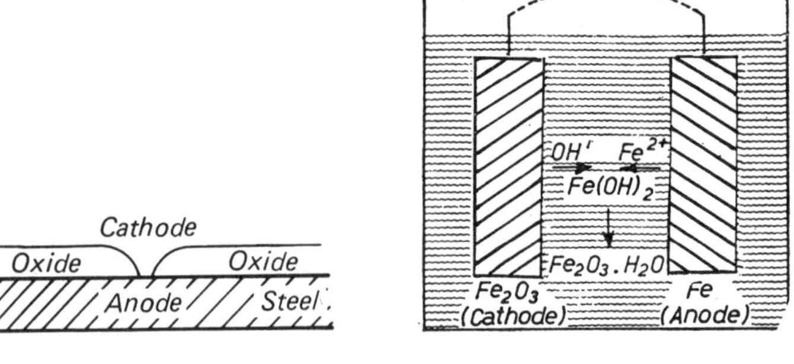

Fig 16.3 Corrosion cell

dissolves at the base of the fissure. The rate of solution of the iron anode will depend on the rate of removal of electrons and this will, in turn, depend on the rate of supply of oxygen at the oxide surface (cathode). If this is plentiful and the cathode is large, reacion will be rapid, and severe pitting of the steel will take place.

The ferrous and hydroxyl ions migrate towards the oppositely charged electrodes and ultimately react to form ferrous hydroxide, a process illustrated at B in Fig. 16.3.

$$4Fe^{++} + 8OH' \rightarrow 4Fe(OH)_2$$

This is then oxidized to hydrated ferric oxide (rust):

$$4Fe(OH)_2 + O_2 \rightarrow 2Fe_2O_3.H_2O + 2H_2O$$

The precipitation of ferrous hydroxide and subsequent oxidation takes place away from the metal surface and consequently the rust does not protect the metal from further attack.

In the presence of a neutral electrolyte such as sodium chloride, the rate of corrosion is considerably increased, but in the presence of oxygen the electrode reactions are substantially the same. Sodium ions are not discharged at the cathode since this is not sufficiently negative due to the cathode reaction with oxygen (*above*). However, in the absence of oxygen, hydrogen is discharged:

$$8e' + 8H^{+} \rightarrow 8[H] \rightarrow 4H_2$$

This can lead to accumulation of hydrogen on the cathode, and the rate of cell reaction will gradually decrease, a process known as polarization.

Methods for the prevention of corrosion are designed to suppress one or both of the electrode reactions, or to provide a complete barrier between the steel and the environment.

Galvanic corrosion

This occurs as a result of contact between metals, in a suitable environment, when there is an appreciable difference in the electrode potentials of the metals as set out in the Electrochemical Series (Table 16.1). The Series illustrated contains only the metals of concern with surface coatings, and these are arranged according to their electrode potentials compared with that of a hydrogen electrode.

Electrode potential Suppose that a metal M is immersed in a solution of its ions M^{2+} at a definite concentration. The solution pressure of the metal will cause a number of atoms to lose two electrons and to form soluble ions:

$$M - 2e' \rightarrow M^{2+}$$

The electrons liberated will confer a negative charge on the metal and this will attract the metal ions:

$$M^{2+} + 2e' \rightarrow M$$

Ultimately a state of equilibrium will be established when the rates of the two reactions are equal:

Table 16.1 Electrochemical Series

Normal electrode-potential (volts) with respect to hydrogen:

Silver	+0·80
Copper	+0·34
Hydrogen	0·00
Lead	−0·13
Tin	−0·14
Nickel	−0·25
Cadmium	−0·40
Iron	−0·44
Chromium	−0·56
Zinc	−0·76
Aluminium	−1·67
Magnesium	−2·38

Each metal exhibits a positive (less negative) potential toward each metal below it in the list. Thus, if copper and iron are in contact in a corrosive environment, electrons will flow from iron to copper and the iron will dissolve (corrode).

$$M \leftrightharpoons M^{2+} + 2e'$$

The concentration of electrons on the metal will give it a certain potential with regard to the solution. The potential of such an electrode can be measured against that of a hydrogen electrode (assigned a potential of 0·00 volts), and the values comprise the Electrochemical Series. The increasing negative potential as the series is descended means that a metal will act as an anode in contact with a metal above it. This can be illustrated by considering cells of iron-copper, and iron-zinc.

Iron-copper cell Suppose a cell (Fig. 16.4) contains separate electrodes of iron and copper in a neutral electrolyte. At the surface of each metal an equilibrium will ultimately be reached:

$$Fe \leftrightharpoons Fe^{2+} + 2e'$$
$$Cu \leftrightharpoons Cu^{2+} + 2e'$$

Iron has a much lower electrode potential than copper, i.e. it has a greater concentration of electrons on the metal. If now the two metals are connected by a conductor, electrons will flow from the iron to the less negative copper. The

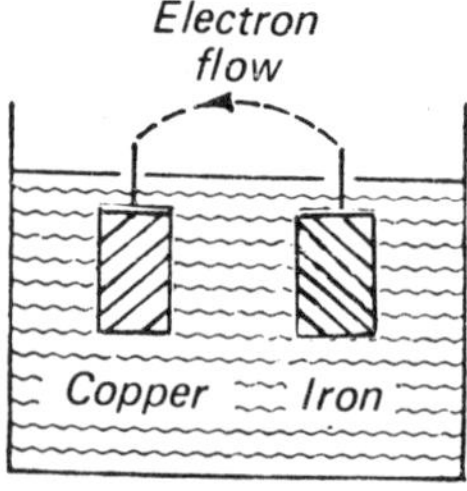

Fig 16.4 Iron-copper cell

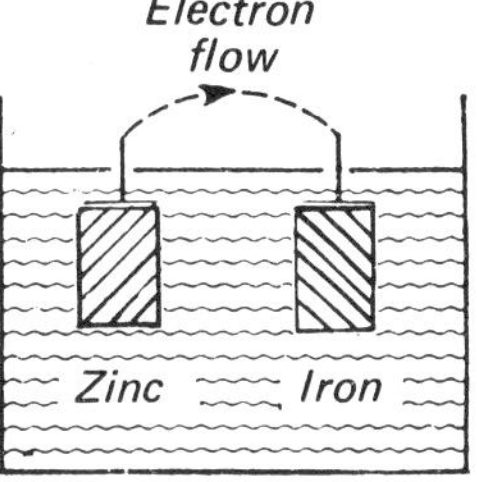

Fig 16.5 Iron-zinc cell

iron becomes the anode and the iron equilibrium is displaced, iron going into solution or being corroded. The reaction at the cathode (copper) is formation of hydroxyl ions as in other types of corrosion cell.

Contact between mild steel and copper results in rapid corrosion of the iron and such contact is rigorously avoided in all engineering operations where corrosive conditions may be encountered.

Iron-zinc cell (Fig. 16.5) When the two separate metals are immersed in a neutral electrolyte, the following equilibria are set up at their surfaces:

$$Fe \leftrightharpoons Fe^{2+} + 2e'$$
$$Zn \leftrightharpoons Zn^{2+} + 2e'$$

Zinc has the greater negative potential and when the two are connected by a conductor, there is a flow of electrons from zinc to iron, and zinc goes into solution. As the cathode, the iron is unattacked and this is known as *cathodic protection*. The zinc dissolves and is said to afford *sacrificial protection* for the iron. This principle is utilized in several methods of protection which are discussed later.

Cathodic protection by impressed current It has been seen from the foregoing examples that the metal receiving electrons in a cell is designated the cathode and is not corroded. If a steel structure, which has many anodic and cathodic areas on its surface, is connected to an external source and electrons are supplied, the whole surface becomes a cathode and corrosion is inhibited. The *impressed current* can be supplied either through existing sacrificial anodes (so augmenting them) or through non-consumable anodes such as graphite.

This type of protection is used to supplement paint coatings on exposed and buried pipe-lines. In the latter case iron or steel anodes are employed and are often packed round with coke breeze which effectively increases the anodic area [4]. It is also widely used on ships' hulls but here it can present a hazard in respect of paint adhesion over the protected areas. These negatively charged areas attract sodium ions which when discharged produce sodium hydroxide. This can saponify certain types of paint media at the metal-paint interface and adhesion fails. This problem is discussed further in Chapter 23.

Bacterial corrosion

Under anaerobic (oxygen free) conditions, certain bacteria can be responsible for rapid corrosion of cast iron or steel. It often occurs when the metal, usually in the form of pipe-lines, is buried in heavy or waterlogged soil devoid of oxygen, but it can occur also when steel supports are submerged in polluted waters of some river estuaries. Sulphate-reducing bacteria are usually present and these cause cathodic depolarization of the corrosion cells (Fe-C) on the metal surface.

Consider an iron-graphite cell (Fig. 16.6) in such an oxygen-free environment. The anodic reaction will take the normal form:

$$Fe - 2e^- \rightarrow Fe^{2+}$$

In the absence of oxygen, the cathode reaction will involve reduction of hydrogen ions, derived from water:

$$H_2O \rightleftharpoons H^+ + OH^-$$
$$2H^+ + 2e^- \rightarrow 2[H] \rightarrow H_2(\text{gas})$$

The hydrogen gas would be expected to polarize the cathode and consequently stop the corrosion reaction. It does not do so as a result of bacterial action, whereby the hydrogen is used to reduce soil sulphate to sulphide:

$$CaSO_4 + 8[H] \rightarrow CaS + 4H_2O$$

It is thought that the following series of reactions then ensues:

$$CaS \rightleftharpoons Ca^{2+} + S^{2-}$$
$$Fe^{2+} + S^{2-} \rightarrow FeS$$
$$Fe^{2+} + 2OH^- \rightarrow Fe(OH)_2$$
$$Ca^{2+} + 2OH^- \rightleftharpoons Ca(OH)_2$$

The ultimate result is conversion of the cast iron or steel to a mixture of ferrous sulphide, ferrous hydroxide, calcium hydroxide and graphite or carbon.

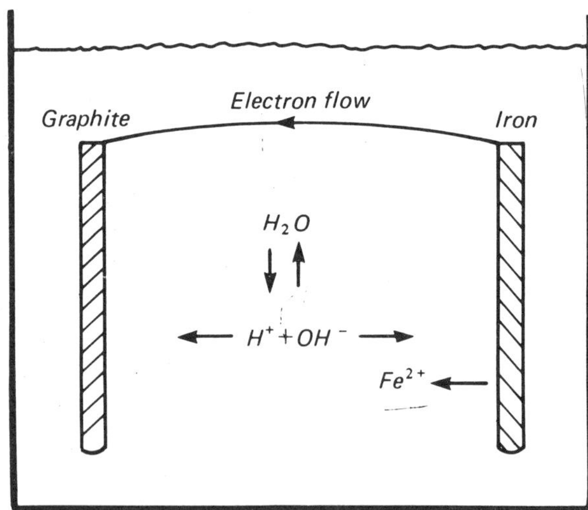

Fig 16.6 Iron-graphite cell

Corrosion of this nature is usually controlled by wrapping the pipes or steelwork in bitumen-impregnated tapes, and control is often augmented with cathodic protection.

Stress corrosion

Parts of steel members which are under stress in a corrosive environment tend to become anodic in relation to other parts. In consequence there is preferential corrosion of the stressed areas. As with other forms of corrosion, it is accelerated by traces of chloride or sulphate in the environment. Alloyed materials such as austenitic stainless steels are more sensitive than the mild steels.

An example of stress corrosion is afforded by the reaction of a common wire nail when placed in a dilute sodium chloride solution. The head and point of the nail are under stress as a result of the manufacturing operation. If small quantities of potassium hexacyanoferrate and phenolphthalein are added to the salt solution, a blue coloration (Prussian blue) will form rapidly round the two ends of the nail, indicating dissolution of iron. A pink coloration along the body of the nail indicates alkali (hydroxyl ions) from the cathodic reaction.

Corrosion resulting from differential aeration (or oxygen deficiency)

This type of corrosion can occur at crevices, joints or other locations where moisture can accumulate and a difference in dissolved oxygen concentration can arise between the surface and the deeper parts. Differences in potential on the metal surface result, the parts in contact with the lower oxygen concentration becoming anodic and corroding. Other circumstances where this type of corrosion can occur are under lifting paint, under dirt deposits, and under bacterial or fungal growths on steel surfaces.

A simple experiment can be performed to illustrate this form of corrosion. A heap of moist sand is placed on a clean steel panel (Fig. 16.7), kept moist for a few days, and then removed. Corrosion will be found under the centre of the heap of sand due to oxygen deficiency in this area.

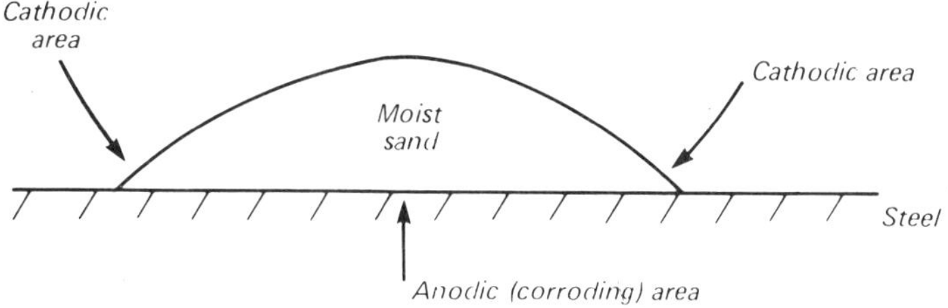

Fig 16.7 Corrosion of steel from moisture deficiency

PREVENTION OF CORROSION OF FERROUS METALS

Corrosion can be prevented (or suppressed) by a number of methods, of which one or more may operate at the same time. The following are the important methods.

1 **Modification of the composition of the steel** Stainless steel, for example, contains an appreciable amount of chromium and nickel (a common type of stainless steel contains 18 percent and 8 percent respectively of these metals). Such steels are expensive and used only for special purposes.

2 **Chemical reaction** of the steel surface to form a 'skin' of compound more resistant to corrosion than the original metal. These are known as 'conversion coatings'. The best-known is Bonderite [5] which converts the surface iron to a micro-crystalline layer of phosphate. This is discussed in more detail later.

3 **Cathodic protection** using anodes of either zinc or magnesium, or by the use of impressed current. In the latter method, a battery or D. C. generator is used to supply electrons so that the structure becomes cathodic. Cathodic protection is usually augmented by painting, but care must be taken in the choice of paint since alkali (hydroxyl ions) is produced at cathodes or on the surface of a cathodic structure. Under these conditions, paint binders based on alkyds or oleoresinous varnishes undergo saponification.

Metallic barrier coats

Barrier coats isolate the steel from both oxygen and moisture.

Metallic coats have been used for many years and the following metals are the most commonly used.

Zinc In addition to acting as a barrier coat, zinc will provide cathodic protection if the surface is damaged and the steel exposed. It is applied by several processes.

Hot-dip galvanizing The thin steel sheet is 'pickled' in sulphuric acid, washed and dried. It is then dipped in a bath of molten zinc. On a good quality sheet the weight of zinc should be 305 to 610 g per sq. metre. The addition of 0·2 percent of aluminium to the zinc gives more flexible and thinner coatings and limits the formation of the brittle iron-zinc alloys at the interface.

Sherardizing [6] This is a cementation process suited to small articles especially of intricate shape. These are cleaned by shot-blasting, packed with zinc powder in steel drums and heated to 375°C (707°F). The film of zinc is very adherent and uniform.

Spraying Metal in the form of wire or powder is fed through a blowpipe flame in a special type of spray gun. The molten metal is atomized by a stream of compressed air and solidifies on the surface. By this method practically any thickness of zinc can be deposited, but the film is porous. In addition to zinc, other metals, including aluminium, tin, lead, solder, cadmium, silver, gold, and stainless steel can be deposited by this method.

Electroplating is used to give dense, smooth deposits of zinc.

Tin Thin sheet-steel coated with tin is familiar as tin-plate. It has been manufactured for many years by a hot-dip process similar to that used for zinc, but with tin the protection is entirely physical. The quality of tin-plate is assessed by the weight, i.e. thickness, of tin, and if this film is broken rapid corrosion of the iron can occur since iron will form the anode in the corrosion cell set-up.

Very uniform films of tin are deposited by electroplating.

Chromizing The steel articles are heated with powdered chromium at 1300° to 1400°C (2372° to 2552°F). Chromium is also applied by electroplating and this gives a smoother, glossier finish than chromizing.

Calorizing is a process similar to sherardizing but using aluminium in place of zinc. The temperature employed is much higher.

Nickel and cadmium are also used for the protection of ferrous metals and are usually applied by electroplating.

Organic barrier coats

Organic protective coatings can be classified as 'barrier coats' which provide a purely physical barrier to the ingress of moisture and oxygen and paints (including primers) in which the barrier effect is augmented by corrosion-inhibiting pigments. The line of demarcation is often indistinct.

Organic barrier coats can be used for either temporary or 'permanent' protection.

Temporary coats - also known as 'temporary rust preventives' - are used to protect steel while in stock. They are based on materials which remain soluble in the parent solvent and so can be removed by wiping or washing. Materials used for this purpose include petroleum jellies and greases, wool grease and wool grease/rosin mixtures. The admixture with rosin gives films hard enough to be handled without removal. These coatings are frequently pigmented.

Permanent barrier coats are required to resist penetration by moisture and oxygen and are applied in thicker films than normal paints. The following types are used.

Bitumen solutions consist of asphaltic bitumens such as Gilsonite mixed with petroleum bitumens, oils or resins. The solvents used are generally aliphatic such as white spirit or distillate. Bitumen films oxidize slowly on weathering and the surface ultimately shows a characteristic crazing.

Pitch solutions These consist of solutions of pitch in aromatic solvents such as the naphthas. They give harder and more glossy films than bitumen. Although bitumen and pitch solutions are black, they can be pigmented to give dark greys, dark greens and browns. They soften with heat and cannot be overpainted with lighter shades as they 'bleed' and cause discoloration.

Pitch-epoxy compositions are prepared using special grades of pitch. These are blended with the epoxy resin and mixed with the hardener - usually an amine adduct or a polyamide resin - immediately before use. The pitch/epoxy mixture and the hardener are kept in separate packs (a 'two-pack' system) and sufficient is mixed for the working period only as the mixture has a limited pot-life. Isocyanates also can be used to cure the pitch/epoxy mixture and have the advantage that cure will take place (albeit slowly) down to 0°C.

Pitch-urethane compositions also are two-pack systems. The pitch is mixed with the polyester/polyol and cured by isocyanate which is added immediately before use.

Both pitch/epoxy and pitch/urethane coatings adhere well to metals (as well as to a number of other surfaces); the films are impermeable to moisture, oxygen and many chemicals. They are used extensively on ships' bottoms and other surfaces subject to complete immersion for long periods. Very thick films of up to 250 μm (10 mils) can be applied per coat.

Chlorinated rubber, suitably plasticized, yields films which are among the most resistant to water and most chemicals. They can be pigmented and, by the incorporation of thixotropic agents such as organo-clays [7] or hydrogenated castor oil [8], high build coatings can be produced. The films also resist petroleum solvents and lubricating oils, but are softened or dissolved by aromatic hydro-carbons and some esters and ketones. To ensure good adhesion the surfaces to be painted must be clean and dry. (Chlorinated rubber paints are discussed in more detail in Chapter 21.)

Powder coatings Protective 'plastic' coatings can be produced by the application of a powdered resin to a metal surface, followed by heating to fuse the resin into a uniform film. A number of different types of resin are employed. Some are applied directly to metal, whereas with others a primer is required. Powder coatings are considered in more detail in Chapter 22.

Protection by primers

The use of corrosion-suppressing primers is by far the most widely used method for protecting steel, and it is now universally accepted that the protection afforded by any primer is influenced by the condition of the surface to which it is applied. Good protection is associated with good adhesion, and this requires a clean surface. Without this no paint system will protect for more than a very short time.

A great deal of attention has therefore been given to cleaning and preparing the steelwork, and the important methods are described below.

Cleaning and pre-treatment processes

The basic principle of pre-treatment processes for steel is chemical reaction of the metal surface to produce a compound which is more resistant to corrosion than the metal itself and prevents 'creep' of corrosion under the paint film in the event of damage. At the same time it provides a good key for the adhesion of subsequent coats of paint. These 'chemical conversion' coatings are not weather-resistant in themselves but require protection by overpainting before exposure to weather.

Pre-treatment processes are necessarily preceded by very thorough cleaning of the steel surface. The cleaning methods and subsequent treatment depend on the type of steel being processed. Hot-rolled mild steel forms the bulk of the steel used in constructional work. It is usually coated with millscale, and the most efficient method of scale removal is by blast cleaning.

Blast cleaning

This term is used to cover methods in which a coarse abrasive powder is directed at a surface at high velocity and removes scale and rust by abrasion.

Abrasives The abrasives employed are classified under the general headings 'grit' and 'shot'. They include siliceous materials such as iron slag, copper slag and silicon carbide (carborundum). Sand (silicon dioxide) is used in water-blasting but the use of dry sand is banned in the UK on health grounds. Among the metallic abrasives are steel and chilled-iron grit and shot [9]. Non-metallic abrasives are generally used on structural steelwork where recovery of the grit is very difficult or impossible. Metallic grits or shot are used when recovery of the grit or shot is possible as, for example, in blast cleaning units at works.

Compressed air blasting The abrasive is mixed with air and directed at the surface at about 100 psi (7 kg/cm^2) pressure. Dust is a major hazard and operators must be adequately protected. In vacuum blasting the dust, grit and scale are removed by a vacuum attachment.

Air/water blasting Water is injected into the abrasive/air-stream near the nozzle and serves to damp down the dust. To prevent corrosion of the steel a corrosion inhibitor [10] is dissolved in the water.

Water-blasting is the most effective method of removing soluble salts from pitted and rusty surfaces. Non-metallic grit is mixed with water containing a corrosion inhibitor and directed at the work under pressure. Pressures employed are of the order of 7000 to 8000 psi (490–560 kg/cm^2) for high-pressure systems and 80 to 90 psi (5·6–6·3 kg/cm^2) for low-pressure methods. The latter is the safer and more widely used.

Impellor or wheelabrator blasting The dry abrasive is hurled at the surface at high velocity. This method is generally employed only in workshops where steel is cleaned before fabrication.

'Soft impact' blasting is used in precision work where tolerances are critical. Granulated plastic is used as abrasive and it is possible to remove paint coatings, one layer at a time, without disturbing the underlying substrate [11].

Nature of the blast-cleaned surface

The degree of cleaning, i.e. the amount of scale removed, can of course vary considerably, but for optimum paint performance the metal must be cleaned to a high standard. BS 4232 [12] sets out figures and contains line diagrams showing the recommended degrees of cleanliness. 'First quality' refers to 100 percent scale or rust removal; 'second quality' allows 5 percent residual scale, but the specification sets out the permissible distribution; 'third quality' allows 20 percent (maximum) residual scale.

These degrees of cleanliness correspond to the designations SA3, SA2½ and SA2 which are photographs of actual blasted steel contained in Swedish specification SIS 05 59 00 [13].

A very important feature of a blast-cleaned surface is the amplitude of the profile, i.e. the height from trough to peak. This is determined largely by the size of the grit and to some degree by the pressure employed. An amplitude of 100 μm (4 mils) is generally agreed to be the maximum acceptable figure, but 'rogue' peaks often occur which are appreciably higher and pose a problem in protective painting. Measurement of amplitude or surface roughness can be made with a Talysurf. A fine-pointed stylus is drawn in a straight line across peaks and valleys and records a trace of the surface. An alternative method for determining overall roughness is provided by the PRA Roughness Meter [14]. The end of a specially designed cylinder is pressed on the surface and air pressure is applied. The rate at which air leaks between the cylinder end and the surface is a measure of the roughness.

Hand cleaning

Hand cleaning is considerably less effective than blast cleaning and is employed only in situations where the latter is impracticable. The methods can be divided into two classes: (a) *abrasion* which employs hand or mechanical wire brushes and grinding wheels, and (b) *impact tools* such as chipping hammers and needle guns. Wire brushes remove loose scale and rust but not embedded corrosion products and they tend to polish adherent scale or rust. This can lead to poor adhesion of primer. Impact tools are more effective but, unless used with care, can damage the metal surface.

Other methods for removal of scale and rust

Flame cleaning [15]

A very hot flame (oxy-acetylene or oxy-propane) is played on the surface. The adhesive bond between scale and steel is broken by differential expansion, and rust is dehydrated. The surface is scraped immediately and wire brushed; it is left dry and warm and so in a good condition to receive the primer. Flame cleaning is used in maintenance painting, particularly in cold or damp weather.

Acid pickling

This is often used for descaling small articles and gives a much smoother surface than is obtained by blast cleaning. The articles are immersed in dilute mineral acid containing an inhibitor to minimize attack on the metal after loosening the scale. Among inhibitors in use are the tolylthioureas and *N*-ethylquinoline [16]. The widely used Footner process employs inhibited 5 percent sulphuric acid solution. This is followed by a hot-water rinse and then immersion in 2 percent phosphoric acid solution containing a small amount of iron. The surface of the steel is coated with a thin film of phosphate which acts as a temporary rust inhibitor.

Cold-rolled sheet steel

This is not usually coated with scale, although scale can be present as a result of rolling hot-rolled steel plate to thinner sheets. More frequently the sheets have been lubricated with oil during rolling or may have acquired a degree of rust.

Rusted sheet

This is either shot-blasted or derusted by treatment with a phosphoric acid solution of the following general type:

Orthophosphoric acid	10 parts by volume

n-Butanol	40 parts by volume
isopropanol	30 parts by volume
Water	20 parts by volume

Oiled or greasy sheets

The cleaning of these sheets can be effected by several processes.

Solvent wipe This is used for small or difficult articles and is the least efficient method.

Vapour degreasing The articles are suspended in the vapour of the boiling solvent. Condensation will take place as long as the temperature of the article is below that of the vapour, and the time during which effective washing takes place is therefore limited. This method does not remove adherent stains.

The solvents used are trichlorethylene (b.p. 86·6°C, 187·7°F) and perchlorethylene (b.p. 121·1°C, 248°F) both of which are non-inflammable but highly toxic. Water-cooling coils are fitted near the tops of the tanks and the heavy vapours are thereby confined.

Modifications of the process include spraying first with solvent or dipping in stirred solvent, followed by immersion in the vapour. It is very effective, but articles must not be handled subsequently with the bare hands.

Solvent degreasing is also used as a system comprising water over a layer of trichlorethylene or methylene chloride. This method is useful if the contaminants include water-soluble materials which are insoluble in the organic solvent. It is operated at room temperature.

Alkali and emulsion cleaning Soiled metals can be cleaned with hot aqueous alkali, and the following are commonly used either alone or in admixture: sodium hydroxide, sodium carbonate, sodium phosphate, sodium metasilicate, and borax. Increased efficiency is obtained by the use of non-ionic or anionic wetting agents. If the latter are used with hard water, Calgon [17] should be added to prevent precipitation.

The solution may also contain emulsified petroleum solvents, and such an emulsion will remove oils and greases by combined solution, saponification and emulsification. The articles may be washed with a petroleum emulsion alone or before alkali treatment.

Alkali degreasing must be followed by efficient rinsing, or undesirable residues would be left. Several water rinses are required to prevent carry-over of alkali salts. The final rinse is often made with water containing a small amount of chromic acid. This facilitates drainage from the surface.

Chemical conversion coatings

It has been stated earlier (p. 288) that the tendency of steel to corrode can be reduced considerably if the steel surface is converted, chemically, to a compound which does not corrode. Such a compound or 'coating' is formed when steel is treated with dilute phosphoric acid. A microcrystalline layer of complex iron phosphate is formed which forms a good substrate for paint and is very adherent, thereby preventing the creep of rust under the film.

The efficiency of this phosphating process is improved very considerably if certain metal ions are present in the phosphating bath and this forms the basis of the Bonderite Processes [5]. Among the metals used are manganese, copper and zinc. Several Bonderite processes are available, designed to suit different alloys and different methods of application. They have been proved to be very effective and are used throughout the metal finishing industry.

Electrogranadising is deposition of a coating of zinc phosphate from a zinc hydrogen phosphate electrolyte. The articles must be scrupulously clean before treatment.

Treatment of non-ferrous metals

The non-ferrous metals most comonly treated with surface coatings are aluminium and its alloys, zinc, magnesium, and cadmium. These metals corrode in air but the rate is very much lower than that of steels, and in some cases, such as aluminium, the oxide film exerts a protective action.

The main problem with non-ferrous metals is poor adhesion of paint coatings, and pre-treatment processes are designed, in the main, to provide surfaces with greater adhesion.

Aluminium

The metal has probably been lubricated in the forming operations and any oil or grease must be removed by suitable treatment before any pre-treatment of the metal is carried out. Care must be exercised if alkali or emulsion cleaners are used. The following pre-treatments are used.

Light mechanical abrasion This is used for small areas–especially where the work has been assembled.

Chemical etching As an amphoteric metal, aluminium is attacked by both acid and alkali.

Acid etch The article is immersed for 20 minutes at 53° to 55°C (127° to 131°F) in a solution of the following general composition:

Sodium dichromate	7·5%
Sulphuric acid (conc.)	22·0%
Water	70·5%

After treatment the article is rinsed thoroughly in cold followed by hot water, dried rapidly, and painted.

Alkaline etch (MBV Process) The article is immersed for 4 to 5 minutes at 95° to 100°C (203° to 212°F) in a solution containing 4·7 percent sodium carbonate (Na_2CO_3) and 1·5 percent sodium chromate (Na_2CrO_4). The article is then washed thoroughly with water, dried and painted.

Both these processes are also known as *chromatizing* and result in improved adhesion.

Phosphoric acid wash gives a fine crystalline deposit of acid aluminium phosphate.

Anodizing The oxide film on aluminium is protective and the degree of protection depends on the thickness of the coating. Anodizing is an electrolytic process which increases the thickness of the oxide layer. It is carried out in the following way.

The object to be treated is made the anode in a bath containing sulphuric, chromic, or oxalic acid. Sulphuric is most commonly used (Fig. 16.8). The cathode can be any convenient material such as carbon. When potential is applied, the following reactions take place at the anode:

$$SO_4'' - 2e' \rightarrow [SO_4] + H_2O \rightarrow H_2SO_4 + [O]$$

$$\downarrow$$

$$2H^+ + SO_4''$$

$$2Al + 3[O] \rightarrow Al_2O_3$$

The oxide film has a high electrical resistance, so that at constant voltage the current decreases with increase in thickness of the oxide film. At the same time outer layers of the film are subjected to attack by the acid, and as the rate of growth of the film decreases with decreasing current, a point will be reached when the rate of growth is balanced by the rate of solution in the acid. Under a given set of conditions, therefore, there is a limit to the thickness of film which can be produced.

Anodized films are very porous when freshly produced and will absorb dyestuffs to give the familiar anodized coloured surfaces. They also afford good adhesion for paint films.

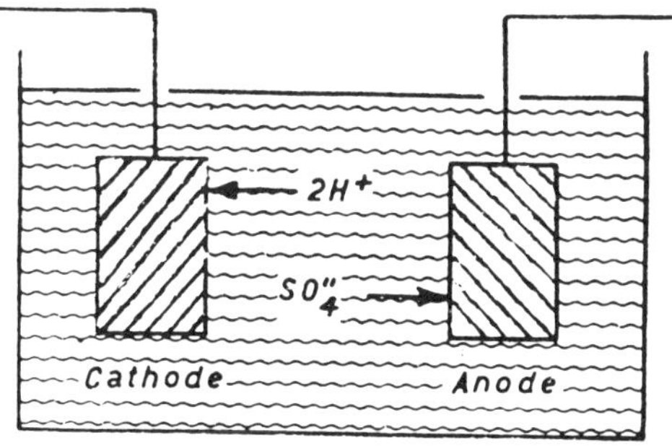

Fig 16.8 Anodizing bath

Etch primer is very frequently used as a pre-treatment process on aluminium and aluminium alloys and gives very good results. The reactions involved are discussed on p. 310. This primer is also used on sprayed aluminium and aluminium alloys as well as on zinc surfaces (*below*).

Magnesium alloys

These are cast, rolled or extruded and are frequently coated with a thin film of lubricant. This is removed by solvent, alkali, or emulsion cleaner, washed and dried. The surface can then be given one of the following pre-treatments:

Acid chromate } as for aluminium
Phosphoric acid wash } as for aluminium
Etch primer

Cadmium

Cadmium is usually encountered in the form of plating. The most successful pre-treatment is the use of etch primer.

Zinc

The commonest form in which zinc is encountered is hot-dipped galvanized iron, but zinc-coated steels are used in some industrial work. According to the thickness of the zinc coating, galvanized iron has a life of 5 to 15 years in the unprotected state under normal atmospheric exposure. Painting is often required for both aesthetic and protective reasons, but on the new surface most primers of the oil or oleoresinous type fail by breakdown in adhesion. A study of the factors which affect paint adhesion to galvanised steel surfaces has been made by van Ooij and Groot [18] who conclude that the binder, the corrosion inhibiting pigment and the surface composition of the zinc all affect paint performance. The following pre-treatments are used.

Weathering is used for buildings. The time required is 3 to 6 months, when the surface acquires a very thin coating of oxide and carbonate.

Treatment with mordant solution The zinc surface is treated with a dilute solution of phosphoric acid containing copper phosphate. A solution of this type was evolved in the laboratories of British Rail and designated 'T-wash'. It is made with the following ingredients:

	% by weight
Orthophosphoric acid (SG 1·70)	9·0
Ethyl cellosolve [15]	16·5
Industrial methylated spirit	16·5
Water	57·0
Copper carbonate	1·0

When this solution is applied to a clean zinc surface the latter turns black rapidly, but the black appearance is not obtained if the surface is contaminated. The galvanized surface must therefore be cleaned thoroughly with either solvent or detergent solution until the T-wash gives a uniformly black colour. The surface is primed as soon as possible after the T-wash is dry (the drying time will depend on the ambient conditions). T-wash is effective only on new galvanized iron; it does not react with a weathered surface.

Etch primer also is used on new galvanized iron. It gives satisfactory results provided the surface is clean, but unfortunately there is no rapid method for indicating the degree of cleanliness, as with T-wash.

Sprayed zinc (and aluminium) coatings are very porous and require sealing before painting or other form of coating. Several types of sealer – clear or pigmented – are suggested in BS 5493 [19] but before sealing, the sprayed metal surface must be treated with etch primer.

References to Chapter 16

[1] Bullett, T. R. and Dasgupta, D., Characterisation of blast-cleaned surfaces and its relation to paint performance. *JOCCA*, 1969, **52**, 1099.
[2] Morcillo, M, Fellu, S., Galvan, J. C. and Bastidas, J. M., The effect of water soluble contaminants at the steel/paint interface on the durability of the paint coating. *JOCCA*, 1988, **71**, 1.
[3] Harrison, J. B. and Tickle, T. C. K., New aspects of the atmospheric corrosion of steel and their implications. *JOCCA*, 1962, **45**, 571.
[4] Cutler, A. J. B. *et al.*, Corrosion. *Chem. Brit.*, December 1986, p. 1109.
[5] The Pyrene Company Ltd, Iver, Bucks.
[6] Zinc Alloy Rustproofing Co. Ltd, Wolverhampton.
[7] 'Bentones' (NL Chemicals Ltd, Wilmslow, Cheshire); 'Perchem' (Perchem Ltd, Harlow, Essex).
[8] 'Thixcin' (NL Chemicals Ltd); Perchem Ltd, Harlow Essex.
[9] BS 2451. Chilled iron shot and grit. British Standards Institution.
[10] 'Nalfloc' (Nalfloc Ltd, Northwich, Cheshire).
[11] Guyson International Ltd, Otley, W. Yorks.
[12] BS 4232. Surface finish for blast-cleaned steel for painting. British Standards Institution.
[13] Standardiseringskommissionen i Sverige (SIS), Stockholm.
[14] Paint Research Association, Teddington, Middx.
[15] British Oxygen Co. Ltd, Chester-le-Street, Co. Durham.
[16] *Hess's Paint Film Defects*, 3rd edn, p. 225. Chapman & Hall, London, 1979.
[17] Albright & Wilson Ltd, London.
[18] van Ooij, W. J. and Groot, R. C., Paint adhesion to galvanised steel surfaces. Part 1: effects of binder and pigments. *JOCCA*, 1986, **69**, 62.
[19] BS 5493. Protection of iron and steel structures against corrosion. British Standards Institution.

17

Primers for metals

IRON AND STEEL

The principal function of a primer for iron and steel is prevention of corrosion, and the primer should ideally possess the following properties.

1 Be suitable for application by the specified method under the required conditions.
2 Effective solvent release, enabling it to dry within the specified time.
3 Good adhesion to the metal surface under both dry and wet conditions.
4 High scratch and impact resistance (not unconnected with **3**).
5 Sufficient elasticity to accommodate dimensional changes in the metal.
6 The surface must provide a good 'key' for subsequent coats.
7 When a corrosion-inhibiting pigment is used it must be non-toxic and the amount must be adequate.
8 A high degree of impermeability to moisture and oxygen.

The components must not present a toxic hazard during either application or subsequent removal. Many of these properties are self-evident, but others require some comment.

Adhesion depends on both the nature of the binder and the cleanliness of the surface. The presence of polar groups in the binder as, for example, in alkyds and epoxies, confers greater adhesion than is shown by non-polar or weakly polar binders. Adhesion is discussed more fully on p. 437.

Mechanical properties (elasticity, abrasion and impact resistance) depend on the type of binder as well as the pigment content. For a given binder, pigmentation above the critical pigment volume concentration (CPVC) will result in an underbound film with poor mechanical properties.

Permeability to moisture and oxygen In the UK exposed paint films are wet with water for an appreciable part of their lives. Primers are protected by differing thicknesses of undercoats and finishes but no paint films are completely impervious to moisture. Furthermore, all paint films are permeable to oxygen also in varying degree [1], [2] and the presence of both moisture and oxygen at the metal surface can result in corrosion unless a corrosion-inhibiting pigment is present.

The degree to which paint films are permeable depends on several factors including the nature of the binder, thickness of coating, pigment characteristics and degree of pigmentation (PVC).

Formulation of undercoats and finishes is discussed in Chapter 18. In this chapter we are concerned with primers, the performance of which can influence that of the complete paint system.

Classification of primers

When binders were limited to oils and oleoresinous varnishes it was usual to describe primers by the major pigment used; for example, red lead primer, red oxide primer, and so on. The introduction of synthetic binders led to primers possessing individual characteristics, and it became necessary to use the binder name in the description. In the following pages this dual classification, i.e. pigment and binders, is employed. The formulation is influenced also by the method of application and this is included in the description. A full description of a primer could take the form:

Zinc phosphate/epoxy primer for steelwork. Airless spray.

(Methods of application are discussed in Chapter 20.)

Formulation of primers for iron and steel

The factors to be taken into consideration would include the following.

1 Nature of the exposure environment.
2 Nature of the steel surface, i.e. degree of blast cleaning, wire brushing, etc.
3 Method of application, i.e. brush, airless spray, dipping, etc.
4 Drying and overcoating times.
5 Nature of the overcoating system. There must be no incompatibility between coats.
6 Any flash point or other restrictions.
7 Pot life (for two-pack systems) in relation to application conditions.

Blast primers

Steel which has been blast cleaned rusts very rapidly and it is essential to protect the surface as soon as possible after cleaning. It is recommended in BS 5493 [3] that priming should be carried out within 4 hours if inside a building, but if exposed to weather, the steel must be coated immediately after blasting.

Blast primers are quick-drying materials designed to protect the steel until further coats of paint are applied. This inverval may be of short duration at works, but occasions arise when protection is required for longer periods – of up to six months or more – on site: in these circumstances the blast primer is followed by a thicker coat of 'holding primer', sometimes referred to as a 'travel coat'. In spite of this, it is difficult to protect all rogue peaks, and the application of one or more undercoats before delivery of steel to the site is a safeguard against premature rust-spotting.

The following are among the blast primers in current use.

Zinc chrome in a polyvinylbutyral/phenolic medium

In a study of the relative merits of different blast primers Bayliss and Wall [4] obtained favourable all-round results with this type. They used a primer of the following composition:

	% by weight
Polyvinylbutyral resin	5·2
Phenolic resin	5·2
Zinc tetroxychromate	9·7
Asbestine	1·4
IMS 64 O. P.	26·3
Methylethylketone	19·4
Toluene	2·4
Orthophosphoric acid	5·4
n-Butanol	25·0

This type has been used satisfactorily as a blast/holding primer under conventional oxidizing systems. Often part of the zinc textroxychromate is replaced by red oxide to give greater opacity. Whilst this type of primer has been found satisfactory under oxidizing systems, failures in adhesion have occurred when overcoated with chlorinated rubber systems. (See chlorinated rubber/alkyd primer, *below*.)

Zinc phosphate in a polyvinylbutyral/phenolic medium

This is sometimes used as a blast primer where white or pale colours are desirable, but more often as a 'patch primer' under oleoresinous systems in maintenance work.

Zinc phosphate/chlorinated rubber/alkyd blast primer

This type of primer was evolved for use on structural steelwork where difficulties were experienced when chlorinated rubber paints were applied over the PVB/phenolic type of blast primer. Adhesion failures occurred in the early life of the system. By using a binder consisting of two parts chlorinated rubber and one part long-oil alkyd (by weight), pigmented with zinc phosphate, a blast primer is produced which gives satisfactory results under chlorinated rubber systems. A higher consistency modification can be used as a first undercoat. The drying time of the blast primer is, however, longer than that of the PVB/phenolic type.

Zinc phosphate/epoxy blast primer

This primer is recommended for use where steelwork is likely to be exposed for appreciable periods before being overcoated. It is a two-pack system in which the curing agent can be polyamide, amine adduct or isocyanate. The latter gives the fastest cure at low temperatures but requires care in use (particularly in mixing) in view of the toxic nature of isocyanates.

Formulation of this primer follows conventional lines for epoxy systems, but the surface drying time can be controlled to some degree by selection of the appropriate solvents. For fast drying, a mixture of methylethylketone (3 parts), xylene (1 part), toluene (1 part) and isopropanol (1 part) is a useful starting point, the proportions being varied according to the nature of the resin.

This primer is slower drying than the PVB/phenolic type but is more suitable for 'on site' application. In common with epoxy coatings generally, overcoating should be carried out within 48 hours or left for 7 days when the coating has cured completely. It is then very hard and requires abrasion to ensure good adhesion.

Metallic zinc ('zinc-rich') primers

It has been noted earlier that contact between metallic zinc and steel in a corrosive environment results in the steel being protected by sacrificial action of the zinc. In other words, the zinc becomes the anode in the corrosion cell and the steel is cathodically protected.

The utilization of this principle in a primer requires a zinc-pigmented film in which the metal particles are in actual contact with the steel surface and with each other. To achieve this, the film must contain 92 to 95 percent by weight of metallic zinc powder. A number of binders have been employed, but the two most widely used are polyamide-cured epoxies and chlorinated rubber. The following formulations are indicative of the two types which are covered by BS 4652 [5]:

Epoxy-base, 2-pack

	% by weight	
Zinc dust	83·0	
Bentone 38[6]	0·8	
Epikote 1001[7]	3·5	
Solvent	8·9	*Solvent*
Polyamide resin (Versamid 115)[8]	1·9	Equal volumes Methoxypropanol/MIBK/Xylol
Solvent	1·9	

This material can be used as a blast primer at a dry film thickness of about 25 μm, or, for prolonged exposure, a further 'full' coat of 50 to 75 μm can be applied to give protection for 6 months or more.

Chlorinated rubber base (One pack) [9]

	% by weight
Chlorinated rubber	3·9
Cereclor 48 [10]	2·6
Zinc dust	74·8
Hydrogenated castor oil	0·2
Epoxidized oil	1·0
Aromosol H [10] or Solvesso 100 [11]	17·5

Materials Hydrogenated castor oil is used to provide structure and so check pigment settlement in the can. Epoxidized oil absorbs traces of hydrochloric acid which may originate in the chlorinated rubber. All materials must be dry, since moisture can result in gassing in the tin, even in the absence of free acidity.

This one-pack primer is used where exposure conditions are not severe and do not justify the two-pack epoxy type.

Properties and uses of zinc-rich primers

The freshly dried film of zinc-rich primer is porous and pinholing or blisters can result in applied films of some types if these are applied early. High build expoxies can be affected in this way. The defect can usually be avoided if the zinc rich primer film is first given a coat of blast primer of the two-pack epoxy zinc phosphate or red oxide type. If the film of zinc-rich primer is exposed to the weather for a prolonged period (3 to 6 months, depending on location), white corrosion products - mainly basic zinc carbonate - are formed on the surface and must be removed before paint is applied. This is best effected with a stiff brush and clean water. The surface must then be allowed to dry before painting.

It is generally accepted that zinc-rich primers exert cathodic protection in early life. On exposure to air, corrosion products of zinc are formed in the film and the latter gradually becomes impervious and acts as a barrier coat. Sealing the film by etch primer has a similar effect. If the system is scratched or cut, the zinc-rich film exerts cathodic protection until the exposed zinc and the adjacent bare steel are coated with a protective layer of zinc corrosion compound.

The two-pack epoxy zinc-rich primer is used extensively on structural steelwork. It also adheres well to new galvanized iron where it can augment the protection given by the zinc coating.

Zinc dust/alkali silicate primers

When zinc dust is mixed with an alkali silicate solution and the mixture is cast as a film, reaction between the two leads to the formation of a complex zinc silicate in which particles of metallic zinc are embedded. The two components are supplied in separate packages and mixed immediately before use.

In the early stages the film on steel will provide cathodic protection, but as the setting reaction proceeds, the film gradually becomes impervious and acts as a barrier coat. Cathodic protection, however, comes into play if the film is scratched or cut through. The completely cured film is adherent, very hard and withstands temperatures up to 500°C.

Sodium silicate was the first binder employed, but it was found that the ratio Na_2O/SiO_2 was critical. A study of the effects of variation in silicate composition on film properties has been made by Pass and Meason [12]. Lithium silicate was found to give improved film properties, but this, in turn, has been superseded by the pre-hydrolysed ethyl silicate (*below*).

Zinc dust/alkyl silicate primer

The medium employed is a partially hydrolysed alkyl silicate solution with which zinc dust is mixed immediately before use. The degree of hydrolysis of the ester is of importance, and 80 percent is usually taken as the optimum. Unfortunately these solutions have a limited life, generally of the order of six months.

These zinc silicate primers require very clean surfaces for satisfactory adhesion. Steel should be blast cleaned to First Quality or SA3. In the film, reaction between the zinc and the alkyl silicate results in an impervious matrix of zinc particles in zinc silicate. Protection is essentially by barrrier effect but, in the event of damage, cathodic protection comes into play.

It has been widely accepted, and there is a considerable amount of supporting evidence derived from practical experience, that curing of these films is catalysed by moisture. Under very dry conditions that cure is assisted by applying a fine spray of water to the primer surface. Nevertheless, Ginsberg, Merriam and Robeson [13] have concluded from their work on the mechanism of cure that in the relative humidity range of 7 to 93 percent the rate of film formation is independent of humidity and takes place by a solvent-evaporation mechanism.

Properties and uses of zinc dust/alkyl silicate primer

These zinc silicate primers can be overcoated with most one-pack and two-pack systems. The films are impervious to moisture, resistant to solvents, and will withstand dry heat up to 400°C. They are used for severe exposure conditions in marine and industrial atmospheres.

General primers for steel

It is generally conceded that blast cleaning is the most efficient method of cleaning steel and that for maximum paint performance, particularly under severe exposure conditions, it is absolutely essential.

Very many occasions arise, however, when blast cleaning is not practicable or when circumstances do not justify the expense. A range of primers exists which performs satisfactorily on steel which has been flame-cleaned or abraded either manually or using mechanical tools. The most widely used of the air-drying primers are described below.

Zinc phosphate primers

Since the introduction of zinc phosphate, described by Barraclough and

Harrison [14], the pigment has become very widely used for the protection of steelwork. It possesses several attractive properties – it is white, non-toxic, compatible with most media and has good corrosion-inhibiting properties. The mechanism of corrosion inhibition has been the subject of considerable interest. Barraclough and Harrison claimed that under damp conditions zinc phosphate primers supply phosphate ions which slowly cause phosphating of the steel surface. The phosphate ions also stifle corrosion caused by residues of ammonium salts in badly cleaned steel.

On the other hand, Clay and Cox [15], in comparing zinc phosphate with zinc chromate, found little evidence of solubility of the phosphate.

The fact that inclusion of zinc phosphate in the undercoat, as well as in the primer, improves the protective properties of the system points to a physical barrier or absorption mechanism. The barrier effect can result from a possibly dense form of pigment packing; the evidence in favour of a moisture-absorption mechanism is very scant.

Primers utilizing zinc phosphate in two-pack epoxy and chlorinated rubber/alkyd media have been mentioned above. Other media in which it is used include alkyds, epoxy esters, one and two-pack polyurethanes, chlorinated rubber, certain vinyls and tung/linseed/phenolics. The last-named are suitable for use where non-toxic brushing type primers are required to replace red lead primers. It is recommended in BS 5493 [3] that in zinc phosphate primers the pigment should contain not less than 40 percent of zinc phosphate.

Red oxide primers

Primers based on red iron oxide as the major pigment component have been used on iron and steel for many years, but no conclusive evidence of any positive corrosion-inhibitive properties has been advanced. The pigment is inert and the protection afforded by these primers is due mainly to the binder, although the packing of the pigment doubtless reduces the moisture permeability of the film. This, however, may be offset to some degree by the hydrophilic nature of the red oxide pigment.

Red oxide primers often contain a proportion (frequently very low) of zinc chrome to provide some positive corrosion resistance but the use of zinc chrome is being discontinued (*see below*).

Red oxide/zinc chrome primers

Primers based on a mixture of red iron oxide and zinc chrome [16] have been used extensively for a number of years but, on account of the hazardous nature of zinc chrome, these primers are being phased out. The mechanism by which they are thought to protect steel is of interest and justifies their inclusion. To provide effective corrosion inhibition, the pigment should contain 20 percent and upwards of zinc chrome. The following formulation illustrates the general type:

	% by weight
Red iron oxide	35·9
Zinc chrome [16]	10·0
Micro talc	3·0
Whiting	1·5
Lecithin	0·4
Medium oil alkyd (70% solids)	36·4
White spirit	11·4
Lead naphthenate (24%)	1·0
Cobalt naphthenate (6%)	0·4

Both the red oxide of iron and the zinc chrome must be free from soluble salts since chloride and sulphate ions are active accelerators of corrosion.

This type of primer can be made on a variety of air-drying and stoving media, but the stoved films generally possess the greater corrosion resistance.

Properties and uses of red oxide/zinc chrome primers

Red oxide/zinc chrome primers possess good flow and levelling properties. The red oxide makes no positive contribution to the corrosion-inhibiting properties but confers good opacity. These primers are widely used in industrial work on articles which are not likely to encounter severe corrosive conditions.

Zinc chrome contains a proportion of water-soluble chromate which is responsible for its corrosion-inhibiting properties. The powerful oxidizing action of the chromate ion is thought to oxidize ferrous ions to ferric oxides on the anode surfaces, and at the same time combination takes place to form an adherent ferric/chromate complex. This seals the anodic areas and so inhibits the corrosion.

Note: Etch primers and blast primers of the PVB type contain zinc tetroxychromate which does not release soluble chromate to the same extent. The mechanism of action of these primers is quite different (see Etch primers, p. 310).

Lead-based primers

Certain lead pigments have, in the past, been found to be very effective corrosion inhibitors but, for various reasons, not the least being toxicity, their use has declined considerably. Red lead, for example, when dispersed in linseed oil, has not been surpassed as a protective for steelwork, but red lead primers are highly toxic and possess other drawbacks. They are now very little used.

The mechanism of protection by lead primers has attracted considerable attention and for this reason a brief account of lead-based primers is retained.

Red lead It is generally accepted that red lead gives best results when dispersed in linseed oil. This type of primer has been used on structural steelwork for many years. It is however very slow in drying and hardening.

Red lead primers are usually manufactured to BS 2523 [17] which covers three types known as A, B, and C. The requirements of these three are shown in Table 17.1.

Table 17.1

	Type A	Type B	Type C	
Non-setting red lead	78–82%	60% minimum	2 parts	77–82% (Type C: red lead, white lead and mineral suspending agent combined)
White lead	–	–	2 parts	
Mineral suspending agent	max. 4%	10–15%	1 part	
White spirit	max. 6%	max. 6%	max. 8%	
Linseed oil	remainder	remainder	remainder	

Type A, which contained only 4% of mineral suspending agent was prone to rapid and very hard settlement. The protective qualities were very little better than those of Type B and the latter was the more generally used.

Type C is a pink colour and is used as the second coat when two coats of primer are required, the difference in colour being of considerable help to the painter.

Properties and mechanism of red lead primers

Red lead/linseed oil primers must be applied by brush (or roller) but show poor levelling properties. As oil paints they require 24 hours to dry and some 2 to 3 days before they are hard enough to be over-painted. Faster-drying primers can be made by replacing part of the oil by alkyd resin. Red lead primers can also be made on other media, including plasticized chlorinated rubber, two-pack epoxy, and epoxy ester.

When immersed in water, red lead/linseed oil films absorb considerable quantities resulting in swelling and, ultimately, blistering. They are therefore unsuitable for use in submerged situations. For this use, media such as plasticized chlorinated rubber or two-pack epoxy are used.

The mechanism of protection by red lead has for some time been considered to be associated with its oxidizing properties. This causes the formation of oxides in intimate contact with the anodic areas, and these cease to function as a result of precipitation inhibition.

Mayne and van Rooyen [18] have shown that red lead reacts with acidic scission products formed in the drying oil film to produce lead soaps, some of which are soluble in water. These are adsorbed on the steel surface and prevent the functioning of the corrosion cells – a process known as 'adsorption polarization'.

Whilst this mechanism is possible with drying-oil media it cannot apply to binders such as chlorinated rubber where scission products are not formed. In this case the protection is afforded largely by the barrier effect of the chlorinated rubber. This binder is not, however, completely impermeable to moisture and oxygen [1] and the red lead may possibly act as a depolarizer of corrosion cells set up by these penetrants.

Calcium plumbate

The structure of this pigment is similar to that of red lead and, like the latter, it performs best when dispersed in linseed oil. BS 3698 [19] recommends a minimum of 50 percent calcium plumbate in the primer pigment. The following formulation illustrates the general composition of calcium plumbate oil primers:

	% by weight
Calcium plumbate	50·9
Barytes	8·7
Talc	8·7
Pale boiled oil	8·6
40-poise stand oil	9·9
Lead naphthenate (24%)	0·7
Cobalt naphthenate (6%)	0·4
White spirit	12·1

Linseed oil may be replaced wholly or in part by other compatible media to give faster-drying and harder films. The media used include alkyds and epoxy esters. The pigments used should be as dry as possible since moisture causes the paint to thicken on storage.

Properties and uses of calcium plumbate

Calcium plumbate primers are superior to red lead in flow and levelling properties. They also dry and harden more rapidly. The oil-type primer has been used on structural steelwork both as a primer and as a second coat over red lead. Prolonged immersion in water will lead to softening and blistering of the film. It has also been used as a primer for woodwork. The harder type is one of the few primers which adhere satisfactorily to new galvanized iron and it is used extensively for this purpose. It would appear likely that acidic scission products produced in the oxidation of the oil or other oxidizing binder are neutralized by the basic pigment. Consequently the zinc surface is not attacked and the adhesion remains intact.

The mechanism of the protection of steel by calcium plumbate primers has been discussed by Read [20]. Calcium compounds formed from scission products in the film can precipitate iron oxides in close proximity to the anodes. This, together with the oxidizing action of the PbO_2 radicle, is thought to cause precipitation inhibition at these points.

Metallic lead

It has been found that finely divided metallic lead, when dispersed in linseed oil or other medium and applied to clean steel, gives excellent protection to the surface. Originally linseed oil was used, but this was displaced by oleoresinous varnishes and by quick-drying binders such as chlorinated rubber. The formulations in Table 17.2 illustrate the general compositions employed.

The corrosion-inhibiting properties of metallic lead are not impaired when mixed with limited quantities of other pigments, and good results have been obtained with proportions down to 50 percent of the total pigment.

Properties and uses of metallic lead

Metallic lead primers have been used extensively for the protection of structural and other types of steelwork. On occasion it has been used as a first

Table 17.2 Metallic lead primer

Oleoresinous type	% by weight	Chlorinated rubber type (ICI)	% by weight
Metallic lead in oil [21]	24·0	Alloprene R20 [10]	16·2
Basic lead sulphate	11·3	Cereclor 42 [10]	7·0
Rutile titanium dioxide	0·5	Metallic lead in	
Barytes	18·8	Cereclor 42 (91 %)	30·0
Talc	12·0	Silica graphite	13·3
Soya lecithin soln (25 %)	0·3	Hydrogenated castor oil	0·5
Varnish (3/1 oil length)	24·4	Epoxidized soya bean oil	1·0
Linseed/tung/phenolic		White spirit	8·0
(56–59 % solids)		Aromosol H [10]	24·0
White spirit	8·6		
Lead/manganese driers	0·3		

undercoat over another primer. Metallic lead shows a high soluble lead figure and consequently brushing is the usual method of application.

The mechanism by which metallic lead protects steel has been the subject of great interest. From the electrochemical series it would appear that lead should be slightly cathodic to iron when the two metals are in contact under corrosive conditions, but Whitby [22] has shown from potential measurements on coated steel that metallic lead pigment is actually anodic to steel. The true mechanism of its action is not, however, galvanic. When used in oil or oleoresinous media, acid scission products may give lead soaps which are adsorbed on the surface [18]. In non-oxidizing media such as chlorinated rubber these scission products are not produced, and Wild [23] has suggested that the great tendency to oxidation of finely divided lead results in its acting as an oxygen absorber. In this way oxygen is prevented from reaching the surface and corrosion is inhibited. This would explain the efficiency of lead in media other than oxidizing types.

Lead cyanamide is widely used in primers for steel on the Continent of Europe but less so in the UK. It performs most efficiently in linseed oil medium but is claimed to give satisfactory protection in other media such as chlorinated rubber and oleoresinous varnishes. In the presence of moisture a lead cyanamide film shows high alkalinity, in which respect it resembles calcium plumbate. High alkalinity favours precipitation on anodic areas, and both pigments therefore provide anodic protection. Lead cyanamide also forms lead soaps which are probably adsorbed on the metal surface in a manner akin to that postulated for red lead.

Lead silicochromate [24] consists of silica particles coated with lead chromate. The corrosion-inhibiting properties are not of a high order, but it has been incorporated in undercoats and finishes to augment the protection afforded by the primer.

Leadless primers

In addition to the types already discussed (metallic zinc, red oxide/zinc chrome and zinc phosphate), the following leadless pigments are on the market.

Barium metaborate [25] is claimed to possess corrosion-resisting properties but is more widely known as an ingredient of fire-resisting and mould-resisting paints. See also Chapter 19.

Molybdates of zinc and calcium-zinc are white non-toxic pigments claimed to give good protection in solvent type and water-borne media respectively [26].

Other corrosion inhibitors for iron and steel

Shieldex [27]

Shieldex is a unique type of material produced by ion-exchanging calcium ions on to a finely divided, porous, amorphous silica support. It is a white free-flowing powder, non-toxic and is marketed in two grades of average particle size 3 and 5 μm. It is claimed that aggressive corrosive ions penetrating the paint film are preferentially exchanged at the silica surface and thereby prevented from reaching the metal surface. Whatever the precise mechanism, both accelerated and outdoor tests have shown Shieldex to function very satisfactorily in both solvent and water-borne (Haloflex - ICI) media and that it appears to be a suitable replacement for lead and chromate primers for steelwork.

Properties of Shieldex

Shieldex has the following physical characteristics:

Specific gravity 1·8 pH 9·0 Calcium content approx. 6%

Water - borne primers for steel

The use of water as solvent in air-drying corrosion-inhibiting primers (as well as finishes) has been a desirable aim but has proved to be a very intractable problem. Much work has been done using conventional emulsion type systems but the very appreciable amount of water solubles has proved a stumbling block. (Water-soluble types have been used for some years but these are cured by stoving.)

A breakthrough in water-borne air-drying primers was the introduction of Haloflex [28], an acidic emulsion of a vinyl/acrylic copolymer. Extensive tests on this material when pigmented with a number of corrosion-inhibiting pigments including zinc phosphate, K-White [29] and Shieldex [27] have given excellent results on steelwork with no accompanying rust problems. It can be overcoated with pigmented forms of the same material or with conventional air-drying systems.

Rust converters or stabilisers

These materials react with firm, adherent rust to form compounds and so stabilise the surface so that further rusting is inhibited. They are used in situations where blast cleaning is impracticable.

Tannins

It has long been known that certain tannin extracts have the property of preventing rusting of iron and steel and of reacting with existing rust to form blue/black compounds (chelates) [30]. These tannin extracts are the basis of many commercial rust converters. The tannins can also be blended with Haloflex and this provides some protection for the iron chelates until further coats of paint are applied. These are generally of the conventional air drying types. A dispersion of tannin in a chlorinated rubber medium has been used in a similar manner for a number of years but in this case the choice of overcoating paints is limited to those compatible with chlorinated rubber. These include short oil alkyds, vinyls and, of course, chlorinated rubber.

Corroless [31]

This is a proprietary rust stabiliser which can be added to primers for steel. It is claimed to convert hydrated ferric oxide to magnetite, a process which takes about one year. Corroless is useful when, for some reason, thorough surface preparation is impossible.

NON-FERROUS METALS

New galvanized surfaces

Most new galvanized surfaces are coated with a very thin greasy film which must be removed by solvent or detergent before painting. However, in spite of cleaning, primers based on oil or oleoresinous media have been found to fail through breakdown in adhesion. It has been suggested that water-soluble acids, produced as scission products during oxidation, react with the zinc surface to form zinc soaps; breakdown in adhesion at the zinc/primer interface ensues.

The two treatments most widely used on new galvanized surfaces are mordant solution (T-wash) and etch primer. Zinc-rich and calcium plumbate primers also adhere satisfactorily as also do primers based on the water-borne Haloflex [28].

Mordant solution (T-wash)

The application of this material has been discussed on p. 296. Mordant solution is an etching process and not a primer *per se*. It is usually followed by zinc chrome primer (*below*).

Etch primers

Etch or wash primers contain zinc tetroxychromate Type 3, BS 282 [16], as the major pigment component. They are supplied as both two-pack and single-pack systems, but the former is the more efficient. The classic 'WP1' formula is an example of the two-pack system:

Etch primer, WP1 Type

Base component	*Parts by weight*
Polyvinyl butyral resin	7·2
Zinc tetroxychromate	6·9
Talc	1·1
Isopropanol or 95% ethanol	48·7
n-Butanol	16·1
	80·0
Acid component	
Orthophosphoric acid 85% (S.G. 1·7)	3·6
Water	3·2
Isopropanol or 95% ethanol	13·2
	20·0

The two components are mixed shortly before use in the proportions shown above.

Etch primers are classified as *pretreatment* primers as they are always applied to clean metal. The mechanism of action is complicated in that there would appear to be several reactions taking place more or less simultaneously. According to Prosser [32] the reactions appear to be as follows.

1 Reaction between phosphoric acid and the pigment to give a chromium/phosphate/zinc complex which forms a corrosion-inhibiting film.
2 Cross-linking of the polymer by the phosphoric acid.
3 Etching of the surface by the phosphoric acid to give the adhesive bond.

The water content of the primer must be rigidly controlled to ensure satisfactory pot-life, but certain other variations in composition are possible. A proportion of the zinc tetroxychromate can be replaced by other pigment, and the film properties can be varied by replacing part of the polymer by certain other resins. No other acid has been found to be a suitable substitute for phosphoric.

Properties and uses of etch primers

Etch primers are applied by spray, dip or brush. They dry rapidly to a matt finish which is an ideal surface for the following coat of paint. They are not weather-resistant in the unprotected state but give good corrosion protection when suitably overpainted. They are used on a variety of non-metals.

One-pack etch primers

When a less reactive inhibiting pigment such as chromium phosphate is used to replace the zinc tetroxychromate in the two-pack formulation the pot-life is very greatly extended. This forms the basis of the one-pack etch primers. In the above instance the pot-life can be extended to 12 to 18 months while still retaining the etching properties. The following formulation illustrates the general composition of a one-pack etch primer suitable for conventional spray application (Rhône-Poulenc) [33]:

	% by weight
Polyvinyl butyral resin	5·9
Chromium phosphate	5·8
Talc	1·5
Isopropanol	56·3
n-Butanol	18·7
Bentone 34 [6]	1·8
Orthophosphoric acid	1·8
Water	0·9
Isopropanol	7·3

It has been stated [33] that the use of equal weights of chromium phosphate and zinc tetroxychromate results in a pot-life of 6 to 9 months and yields a product with improved adhesion to steel and greater resistance to sea-water than the above formulation.

Zinc chrome primers

Primers based on zinc chrome ($4ZnCrO_4.K_2O.H_2O$) [16] as the major ingredient are widely used on non-ferrous metals. They adhere well to etched surfaces and augment the protective qualities of etch primer. These primers are formulated on the following general lines.

	% by weight
Zinc chrome [16]	17·0
Zinc oxide	3·7
Micronized barytes	8·5
Water-ground mica	12·8
VT alkyd (70% solids)	23·5
Solvent naphtha	15·3
White spirit	18·6
Lead naphthenate (24%)	0·4
Cobalt naphthenate (60%)	0·2

This primer will air-dry in 2 to 3 hours or can be stoved satisfactorily at 120°C (248°F) for ½ hour. In the latter event the driers are omitted.

Zinc chrome primers with excellent adhesion can be made on a number of media including epoxy/polyamide, epoxy/amine, epoxy/isocyanate, polyurethane and alkyd/amino for stoving applications. In view of the carcinogenic properties of chromates, the use of zinc chrome is being phased out.

Calcium plumbate primer

It was noted earlier that many primers based on oil or oleoresinous binders fail on new zinc surfaces through breakdown in adhesion. Calcium plumbate primers (p. 306) are an exception in that they adhere well to galvanized iron at all stages, and it is likely that the alkaline environment characteristic of this pigment neutralizes any acidic scission products produced in the film. There is then no attack on the zinc and adhesion is maintained.

Zinc-rich primers

Zinc-rich primers (p. 301) based on a two-pack epoxy binder adhere well to the zinc surface and serve to increase the effective thickness of the zinc coating.

Weathered galvanized surfaces

Galvanized iron which has been exposed to the weather for some years - the actual time depending on the location–acquires a dull appearance due to formation of basic zinc compounds (mainly basic carbonate). Before painting it should be cleaned down with clean water and a brush and then allowed to dry. Most conventional and two-pack primers will adhere satisfactorily. T-wash, however, will not develop the characteristic black colour on weathered zinc. In fact, etching of the surface to promote primer adhesion is unnecessary.

Sprayed zinc

These coatings are porous, and on exposure to weather the particles become coated with a basic carbonate which fills the voids and produces an effective barrier coating. The white zinc reaction products are plainly visible, and if painting has been delayed until their formation they must be removed from the surface. The use of clean water and a stiff brush is usually sufficient. The surface must be allowed to dry before painting.

The two main processes in use for the protection of sprayed zinc coatings are (a) application of etch primer followed by a sealer, generally a pigmented or non-pigmented lacquer type, as suggested in BS 5493 [3]; materials used include chlorinated rubber, vinyls and polyurethanes: (b) application of etch primer followed by a zinc chrome primer and the desired paint system.

Aluminium and aluminium alloy sheeting

Aluminium, when exposed to air, forms a thin dense oxide film which protects the metal from further oxidation. The main problem in painting aluminium sheet is adhesion rather than prevention of corrosion, and the preparation of the surface has been discussed. Etch primer is frequently used, followed by zinc chrome primer. Where the metal has been etched chemically, zinc chrome primer can be applied to the metal. It shows good adhesion and gives good protection. The soluble chromate serves to maintain the protective film of oxide, especially if the latter is damaged.

Sprayed aluminium: treatment

Sprayed aluminium is being used on an increasing scale for the protection of steelwork as an alternative to sprayed zinc. On exposure, the particles form an oxide film, but this is self-sealing and the film remains porous. The treatments for sprayed aluminium are essentially the same as those for sprayed zinc (*above*), namely, sealing by a lacquer or a paint system comprising etch primer, zinc chrome primer, undercoat and finish.

References to Chapter 17

[1] Guruviah, S., The relationship between the permeation of oxygen and water through paint films and corrosion of painted mild steel. *JOCCA*, 1970, **53**, 669.

[2] Haagen, H. and Funke, W., Prediction of the corrosion protective properties of paint films by permeation data. *JOCCA*, 1975, **58**, 359.

[3] BS 5493: 1977. Code of Practice for the protection of iron and steel structures from corrosion. British Standards Institution, 2 Park Street, London, W1A 2BS.

[4] Bayliss, D. A. and Wall, D. C., Prefabrication primers for structural steelwork. Part I, *JOCCA*, 1966, **49**, 770; Part II, *JOCCA*, 1968, **51**, 792.

[5] BS 4652: 1979. Metallic zinc-rich priming paint (organic media). British Standards Institution.

[6] NL Chemicals (UK) Ltd, Wilmslow, Cheshire.

[7] Shell Chemicals Ltd, London.

[8] Cray Valley Products Ltd, Farnborough, Kent.

[9] Parker, H. E., in *Treatise on Coatings*, ed. Myers, R. R. and Long, J. S., Vol. 1, Part 1, p. 169. Edward Arnold, London, 1967.

[10] ICI (Mond Division), Runcorn, Cheshire.

[11] Esso Petroleum Company Ltd, London.

[12] Pass, A., and Meason, M. J. F., Alkali silicates in surface coatings. *JOCCA*, 1965, **48**, 897.

[13] Ginsberg, T., Merriam, C. N. and Robeson, L. M., Kinetics of film formation of alkyl silicate zinc-rich coatings. *JOCCA*, 1976, **59**, 315.

[14] Barraclough, J. and Harrison, J. B., New leadless anti-corrosive primers. *JOCCA*, 1965, **48**, 341.

[15] Clay, H. F. and Cox, J. H., Chromate and phosphate pigments in anti-corrosive primers. *JOCCA*, 1973, **56**, 13.

[16] Zinc chrome Type 2. BS 282, 389: 1963, British Standards Institution.

[17] BS 2523: 1966(1983). Lead based priming paints. British Standards Institution.

[18] Mayne, J. E. O. and Van Rooyen, D., *J. Appl. Chem.*, 1954, **4**, 384.

[19] BS 3698: 1964(1979). Calcium plumbate priming paints. British Standards Institution.

[20] Read, N. J., *JOCCA*, 1958, **41**, 352.

[21] Associated Lead Manufacturers Ltd, Newcastle.

[22] Whitby, L., PRS Technical Paper No. 96, Paint Research Association, Teddington.

[23] Wild, G. L. E., Metallic lead pigment for anti-corrosive protection. *JOCCA*, 1965, **48**, 43.

[24] 'Oncor M 50' (NL Chemicals (UK) Ltd, Wilmslow, Cheshire).

[25] 'Busan 11-M1' (Buckman Laboratories Inc., Memphis, Tenn.).

[26] Sherwin-Williams Chemicals, New York (Kingsley & Keith Ltd, Croydon).

[27] W. R. Grace Ltd, Northdale House, London NW10 7UH.

[28] ICI Chemicals & Polymers Group, Runcorn, Cheshire.

[29] Bromhead & Denison Ltd, Welwyn Garden City, Herts.
[30] Knowles, E. and White, T., *JOCCA*, 1958, **41**, 10.
[31] Lawrence Industries Ltd, Mitcham, Surrey.
[32] Prosser, J. L., *PRA Technical Paper No. 205*. Paint Research Association, Teddington, Middx.
[33] Vinyl polymers for the protection of metals. Rhone-Poulenc (UK) Ltd, 161, Fleet Street, London, E. C. 4.

18

Decorative and building paints

For many years linseed oil in one or other of its various forms was the standard binder for practically all paints. These oil paints were slow-drying, possessed poor initial gloss and gloss retention, and showed poor levelling. Nevertheless, they afforded reasonably good protection. The introduction of oleoresinous varnishes led to shorter drying times and harder films with, in the case of finishes, greater gloss and gloss retention. They gave rise to the so-called *hard gloss* paints. In gloss finishes and undercoats for decorative work oleoresinous varnishes have been replaced by alkyds but they continue to be used in some undercoats and primers for structural painting.

Paints for decorative and building work are generally applied by brush or hand-roller, but on new work, and especially on prefabricated sections, spraying is used. Generally, decorative paints are slow drying (4 to 16 hours according to type) and contain long media and slow-evaporating solvents. This combination allows large areas to be painted in sections without difficulties in joining up.

The following discussion follows the sequence of application, namely sealers, primers, fillers and stoppers, undercoats and finishes. Special products receive separate treatment.

SEALERS

Wood often contains knots which slowly exude resinous matter which will soften and discolour paint. These knots can be sealed by the application of *knotting* which is a 25 to 30 percent solution of orange shellac in IMS.

Creosoted wood may also be sealed with knotting to prevent the creosote from bleeding into and discolouring subsequent coats of paint, provided the creosote has been on for some years. Newly creosoted wood cannot be sealed in this way. PVA emulsion sealer is also effective on weathered creosoted wood and is appreciably cheaper than knotting.

Plasters, cement render, etc. These surfaces can be highly alkaline as well as porous and require an alkali-resistant sealer with low penetration properties, i.e. high molecular weight. Polyvinyl acetate emulsion is widely used, but dilute solutions of chlorinated rubber as well as oleoresinous sealers based on tung oil/coumarone mixtures are also employed. PVA emulsions have an advantage for interior work as both aqueous and non-aqueous paints will adhere to the surfaces.

Plaster surfaces must be thoroughly dry and free from any tendency to efflorescence. Otherwise the coating of paint will be lifted by moisture or by crystallization of soluble salts beneath the film.

Hardboards and building boards can be sealed with PVA emulsion sealer. For hardboard the latter can be made by addition of water to normal emulsion paint. Where surfaces are more porous, an effective sealer can be produced by mixing emulsion paint (2 volumes) with unpigmented emulsion (1 volume) and water (1 volume).

Preservatives for wood

Nature of wood A cross-section of a tree trunk shows a series of annular rings which comprise the sheaths of new wood formed each year. These completely cover the old wood and enclose any lateral branches. The heartwood, which occupies the central part of the trunk, has changed with age and is darker in colour, contains less water, and is less permeable to liquids than the more recent sapwood which surrounds it. Sapwood consists of alternate layers (grain) of spring wood, formed in the early part of the growing season, and summer wood which is formed later.

The grain consists of interconnected elongated cells with rigid cellulose walls. Spring cells are thin-walled and large, but as the season advances the cells decrease in size and the walls thicken. There is a sudden change in cell structure from one year's summer wood to the following year's spring wood, and this is very evident in softwoods such as pine, spruce, larch and Douglas fir. A high proportion of the external joinery used in the United Kingdom is Scots pine or 'Baltic redwood' (*Pinus sylvestris*). The sapwood of this tree is very porous and susceptible to attack by fungi [1] and in the UK climate this presents a major problem [2]. Greater stability is shown by some of the South American hardwoods now widely used for window frames, etc., which usually are only varnished. Hardwoods are more irregular in structure, and the wood is frequently oily; common British hardwoods are oak, birch, chestnut, and walnut.

Types of preservative

Creosote is the time-honoured preservative for exterior joinery where preservation is more important than decoration as, for example, wooden fences or sheds. The preservative qualities are contributed by the phenolic compounds present. It is sometimes claimed that creosoted timber which has been 'weathered' can be overpainted but the results with paints other than black are seldom satisfactory. 'Bleeding' takes place into the paint film resulting in patchy discolouration. Application of a sealer such as knotting can prevent bleeding on smooth timber but the treatment can be expensive and is not recommended on the rough-sawn timber of many fences.

A very large number of wood preservatives is available. These include the naphthenates and 8-hydroxyquinolinates of copper [3], pentachlorophenol, zinc naphthenate and tributyl tin compounds. Busan 1009 [4] is a mixture of methylene bis (thiocyanate) and 2-(thiocyanomethylthio) benzothiazole. Tests have indicated a high efficiency of this material against a wide range of fungi.

The Madison formula, a pigmented linseed oil/wax mixture, was designed for the restoration of weathered cedarwood.

Application of wood preservatives

Water uptake along the grain is very much greater than across and so it is essential to seal the ends of timber effectively before units are assembled.

The efficiency of a preservative increases with the depth of penetration and whilst woodwork 'in situ' must necessarily be treated by brush or spray, manufactured pieces can be dipped before assembly. Dipping is more effective than brush or spray for end grain but the deepest penetration of the entire members is achieved by immersion in the preservative solution using a vacuum/pressure technique.

Preservative stains

In place of paint, exterior woodwork is now frequently treated with a preservative stain which can be either water-borne or solvent based.

The water-borne types are based on a mixture of alkyd and acrylic emulsions pigmented with transparent UV-absorbing oxides of iron or other light-fast pigments. The stains may be transparent, translucent or opaque and exhibit very good durability. They are discussed also in Chapter 19.

The exterior solvent-based stains are best suited to hardwoods. They are based on a specially designed flexible alkyd [5] and pigmented with transparent UV-absorbing oxides of iron. Their performance is far superior to that of conventional alkyds.

The painting of wood

Behaviour of paint on wood

When paint is applied to wood it will fill the superficial cavities in the rough surface, and the medium will tend to enter the cell spaces, passing from one cell to another. The degree of penetration will depend on the molecular size and will be deeper for raw than for bodied oils. The paint medium does not enter the cell walls. Spring cells are larger than summer cells and so tend to absorb more medium, but pigment cannot pass into the cells. The result is a variation in pigment/binder ratio in the film over the spring and summer wood. This will be different for each type of wood.

The moisture in wood is in equilibrium with that in the atmosphere. It enters wood mainly along the grain, i.e. from one cell to another, and travels through the cell walls, which expand in the process. Lateral penetration also takes place but to a much lesser degree. Expansion is greater in the thicker summer-cell walls, and there will therefore be greatest movement of the wood at the junction of spring and summer cells. This movement, together with the difference in pigment/binder ratio, causes considerable strain along lines in the film and failure usually commences along these lines.

In order to maintain a constant pigment/binder ratio in the dry primer film it is common practice to incorporate a proportion of polymerized oil or varnish which will show less tendency to penetrate the wood.

It was formerly thought that deep penetration into the wood was necessary for satisfactory adhesion, but from recent work it appears that unbodied oils penetrate only to a depth of 2 to 3 cells. The adhesion of paint films to wood depends, as with other substrates, on good wetting of the surface by the paint. Hence the advisability of ensuring that the wood is dry when priming is carried out. Rothwell [6] has pointed out that 'the method of drying timber, direction of grain, proportions of oil and resin in the timber (particularly in hardwoods) were all important in determining the degree of penetration and paint adhesion'.

The durability of a paint system on wood will be influenced by the ability of the system - particularly the primer - to accommodate the dimensional changes in the wood caused by absorption and loss of moisture. Two essential properties of the primer are adhesion and elasticity, but certain other properties, including a finite degree of moisture permeability, are also important.

Preservatives for softwoods

Primers for softwoods

The traditional 'pink primer', based on a mixture of white and red leads in linseed oil, gave excellent protection to exterior woodwork but has now been phased out for reasons of high toxicity.

Calcium plumbate primers also have given good results on softwoods and these have been marketed as 'universal' primers for use on wood, steel and galvanized iron. However, as lead compounds, calcium plumbate primers are also being phased out.

Leadless primers are manufactured on a number of media including oleoresinous varnishes, alkyds, and urethane oils. Non-toxic pigments and extenders are employed to give primers suitable for both interior and exterior joinery.

Most prefabricated timber units are primed at works and should of course be primed before assembly to ensure sealing of end-grain in joints. The primers employed are quick-drying types applied by spray or dip. The formulation below illustrates the general type:

	% by weight
Titanium dioxide	17·0
Paris white	11·8

Red oxide of iron	2·9
*Medium (50% solids)	67·7
Driers	0·6

*A variety of media can be employed, the type chosen being often influenced by the drying time required.

Prefabricated joinery units such as windows and frames can be exposed to weather on building sites for considerable periods; it is essential, therefore, that good quality primers are used and that close attention is paid to application.

Primers for brush application can be made on the formulation set out above using a medium/long alkyd binder. Where highly porous sapwoods are encountered, the addition of 10–20 percent of light stand oil (with the appropriate driers) has been found advantageous.

Aluminium primer Lamellar pigments such as aluminium are thought to even the strain at points where it is greatest in the film, i.e. at the junction of spring and summer woods. They also yield films resistant to moisture penetration. The following is a typical example:

	% by weight
Non-leafing aluminium paste	19·0
Solvent naphtha	8·0
Medium (60% solids)	72·8
Cobalt naphthenate (6%)	0·2

(Medium: 2/1 oil length tung oil/modified phenolic. 2 poise)

A leafing type of aluminium would give a superior moisture barrier, but the non-leafing type provides a more suitable surface for the adhesion of subsequent coats. This primer is suitable for use on both interior and exterior woodwork and dries in 4 to 6 hours.

Emulsion primers Suitably pigmented acrylic emulsions are used successfully on softwoods. A typical formulation is given on p. 348.

Acrylic emulsion primers are non-toxic, flexible and adhere well to wood surfaces. They tend to dry slowly at low temperatures and at high humidities. Difficulties have arisen in respect of the long-term adhesion of glazing putties over these primers [6].

Painting of plywood

The use of plywood for exterior cladding has increased in recent years. It is more dimensionally stable than timber but is susceptible to surface cracking along the grain on weathering. Protective systems discussed by Gunn [7] comprise textured coatings on oleoresinous media and heavily textured coatings based on vinyl and acrylic emulsions. The latter are stated to be the more satisfactory.

Primers for hardwoods

Oil-type primers are unsatisfactory on hardwoods, particularly if these are inclined to be oily. Adhesion is poor and flaking usually occurs in 12 to 18 months in the UK. More satisfactory results are obtained using the aluminium primer described above. Other aluminium primers which have given very promising results are based on epoxy esters and on urethane oils.

Finishing systems for wood

Interior The conventional alkyd-type primer, undercoat and finish gives satisfactory results and is the commonly used system for interior decoration.

Exterior Several systems are in use.

1 The conventional alkyd-based primer, undercoat and finish can give satisfactory performance provided the wood has been effectively sealed against moisture penetration [2]. Unfortunately this is often not done and the progressive embrittlement of the coating system leads, ultimately, to cracking and flaking. These defects are discussed in Chapter 27.

2 The most interesting development in the painting of wood has been the use of the acrylic emulsions. An acrylic primer and undercoat under an alkyd finish gives a fairly good life by reducing stress cracking and providing a degree of permeability [2]. However, the same authors consider that an all-aqueous gloss finishing can have a life expectancy in excess of ten years. Further evidence of the excellent performance of all-acrylic systems has been provided by Feist [8] and by Miller and Boxall [9].

It is interesting to note the trend to water-borne systems either as all-acrylic paints or as acrylic/alkyd preservative stains.

Primers for plaster and other wall surfaces

Plasters used for the surfacing of internal walls are based on calcium sulphate and formed by the action of heat on gypsum. When the latter is heated to 120°–130°C (248°–266°F), partial dehydration occurs with formation of the hemihydrate plaster of Paris. When the latter is mixed with water, gypsum is formed. The two reactions can be represented in the following way:

$$\underset{\text{Gypsum}}{2CaSO_4.2H_2O} \underset{\text{Water}}{\overset{\text{Heat}}{\rightleftharpoons}} \underset{\text{Plaster of Paris}}{[CaSO_4]_2.H_2O} + 3H_2O$$

The reaction between plaster of Paris and water takes place too rapidly for application to large areas and it is therefore retarded by addition of alum or borax. The well-known 'Parian', 'Thistle' and 'Sirapite' plasters are retarded hemihydrates of this type.

These plasters are neutral and set to a porous mass. In practice a proportion of lime is added to facilitate spreading, and the dry plaster is therefore highly alkaline as well as porous. In the presence of potassium sulphate (from alum), potassium hydroxide can be formed:

$$Ca(OH)_2 + K_2SO_4 \rightarrow CaSO_4 + 2KOH$$

and could cause saponification of certain types of media.

The type of treatment to be applied is determined largely by the degree of porosity. Highly porous surfaces require application of a sealer before priming, but on less porous surfaces the primer is used also as the sealer. Even so, it is usually necessary to apply two coats for satisfactory results.

Types of primer for internal walls

Oleoresinous type These are based on alkali-resisting highly polymerized media. Tung oil/phenolic and tung oil/coumarone have been found suitable. The following formulation illustrates the general type of oleoresinous primer:

	% by weight
Anatase titanium dioxide	9·3
Gypsum	18·0
Talc	5·5
Water-ground mica	3·7
Medium (60% solids)	52·0
White spirit	11·5

(Medium: 2/1 oil length tung oil/Coumarone. 3 poise)

This primer dries in 6 to 8 hours. The alkali resistance can be upgraded by the incorporation of isomerized rubber. The finish is glossy and can be overpainted with non-aqueous systems.

Emulsion type This is a lightly pigmented vinyl emulsion applied at 25 to 30 percent solids. It has the necessary alkali-resistance and sealing properties and, in addition, shows the following advantages over the oleoresinous type.

1 The drying time is approximately one hour.
2 It is much cheaper.
3 It can be overpainted with emulsion, or solvent type paint or can be papered.

Primer for hardboard Hardboard is best primed with a vinyl emulsion paint which has been thinned with water. This acts also as a sealer and gives a finish suitable for application for either aqueous or non-aqueous paints.

Primer for chip-board Chip-board can be primed with vinyl or acrylic type emulsion paint which has been thinned a little with water.

Primers for exterior surfaces

The types of surface likely to be encountered are roughcast, stucco, concrete, asbestos sheeting, and brickwork. All these, when new, are highly alkaline (brickwork through the pointing) and porous, but the degree of alkalinity decreases on exposure to weather. Painting can be required on new surfaces or on old surfaces the condition of which may vary considerably. The latter is not difficult (provided the surface is properly prepared) but the former needs alkali resistance and sealing properties of a high order. The treatment will be determined by the nature of the following coats and the following primers are in use.

1 Vinyl co-polymer and acrylic emulsions. These are used under emulsion finishes and some of the proprietary masonry finishes.
2 Solution types based on styrene-acrylic type co-polymers. These are employed in some masonry paints and are alkali-resistant.
3 Chlorinated rubber is highly alkali-resistant and is used on very alkaline surfaces such as new asbestos sheeting and concrete. The application of chlorinated rubber paint is the usual treatment for concrete swimming pools where high resistance to water and alkali is required.

STOPPERS AND FILLERS

These are highly pigmented compositions used to fill deep or shallow surface indentations. They are always applied over primers.

Stoppers are paste materials used for deep indentations such as nail holes or cracks and are applied by knife. They are sometimes described as paste fillers. Pigmentation is kept very high to minimize shrinkage and to enable the dry material to be rubbed down to a smooth finish without clogging the paper. The resulting surface is porous and must be sealed before painting.

A white stopper for woodwork can be made on the following formulation:

	parts by weight
Titanium dioxide	20
Barytes	130
Whiting	120
Medium (60% solids)	60

(Medium: short-oil alkyd or oleoresinous varnish)

The hardness can be controlled by the extenders used, and this stopper could be made harder by replacing the whiting by ground natural silica or by slate powder, on a volume basis. The latter would give a grey stopper.

Glazing putty is a stopper based on linseed oil, whiting and soft soap in the following approximate proportions: putty whiting 85·8 percent, raw linseed oil 13·8 percent, soft soap 0·4 percent. Putty is a flocculated system formulated at approximately the critical pigment volume concentration and for this reason does not stick to the hands. A slight increase in oil content would deflocculate the system and it would become sticky.

This type of putty is used in glazing wooden frames and slowly forms a skin on the surface, the underlying mass remaining plastic for a considerable time.

Metal casement putty is generally similar to glazing putty but contains a small amount of fatty acid or sulphonated oil to improve the wetting characteristics and consequently the adhesion to the metal surface.

Fillers are highly pigmented liquid materials used to fill shallow indentations and irregularities in surfaces. They are applied by brush or spray, and dry to a hard surface which can be rubbed down. The stopper (formulation above) when thinned further with white spirit would make a suitable filler.

Coloured stoppers or fillers for wood These are used on natural coloured woods when the latter are to be polished or varnished. Semi-transparent pigments are used and the stainers can be varied to match the different types of wood. They are made on the following general lines:

	parts by weight
Burnt umber	20
Barytes	25
Whiting	25
Short oil alkyd	30

UNDERCOATS

Undercoats are highly pigmented paints which follow the primer and any filler or stopper and provide the necessary ground for the finishing coat.

An ideal undercoat would possess the following properties.

1 Good adhesion to the primer.
2 Satisfactory opacity
3 Good flow and levelling, i.e. the dry film should show no brushmarks.
4 Fairly rapid set to avoid sagging and pulling away from sharp edges, yet remain open long enough to be joined up to other areas.
5 An eggshell finish to the dry film so that the finishing coat can adhere satisfactorily.
6 A colour similar to, but preferably a little lighter than the finishing coat.

It is extremely difficult to embody all these desirable properties in any one undercoat and it is usually necessary, as with most paints, to produce the best compromise. The main difficulty is to produce the desired flow properties with the high pigmentation required to produce an eggshell finish. Pigmentation and flow properties are closely interrelated and merit brief discussion before considering undercoat formulation.

Pigmentation

For many years the relationship of pigment to binder was expressed as a simple ratio by weight, but this was of limited value in view of differing densities and therefore of differing volume of pigments. It is now usual to express the pigment content of a paint film in terms of the pigment percentage by volume. This is defined as the Pigment Volume Concentration (PVC) and has been discussed in Chapter 1. It is essential that paints be pigmented below the CPVC and this is particularly important with undercoats. These are made just below the CPVC to ensure absence of porosity (which would lead to sinkage of the finish) and failure to produce the desired slight eggshell finish to the film.

The high pigment content of undercoats tends to reduce the overall mobility of the pigment particles, and as a result abnormal flow properties can be encountered.

Flow properties of paints

When a liquid such as water or raw linseed oil is stirred, the rate at which it flows is directly proportional to the force applied.

Force applied (shear stress) = constant × rate of flow (rate of shear)

or

$$\frac{\text{Shear stress}}{\text{Rate of shear}} = \text{constant } (\eta)$$

η is defined as the *coefficient of viscosity* of the liquid and is expressed in poise (P) – the unit of dynamic viscosity. If ρ is the density of the liquid, η/ρ = stokes (v) – the unit of kinematic viscosity.

Liquids which behave in the manner of water or raw linseed oil are known as Newtonian liquids and are said to exhibit Newtonian flow (after Newton, whose Hypothesis of Viscous Flow forms the basis of viscosity measurements).

If the results are plotted on a graph, a straight line is obtained (Fig. 18.1) the slope of which gives the viscosity of the liquid.

Non-Newtonian flow

When a Newtonian liquid is pigmented, the solid particles increase the viscosity, but at low concentrations the effect is small. When the volume of pigment present is large, interference with the flow properties can take place, and often a certain stress (force) has to be applied before the system will flow. This stress is known as the *Bingham yield value (f)* and is indicated graphically in Figs 18.2 and 18.3.

The yield value was postulated by Bingham to be the force expended in overcoming the internal stresses.

A paint which shows a yield value will, after this is exceeded, possibly give a straight line when shear stress is plotted against rate of shear. This line will not give the viscosity as this is not constant. Such behaviour is known as *plastic flow*.

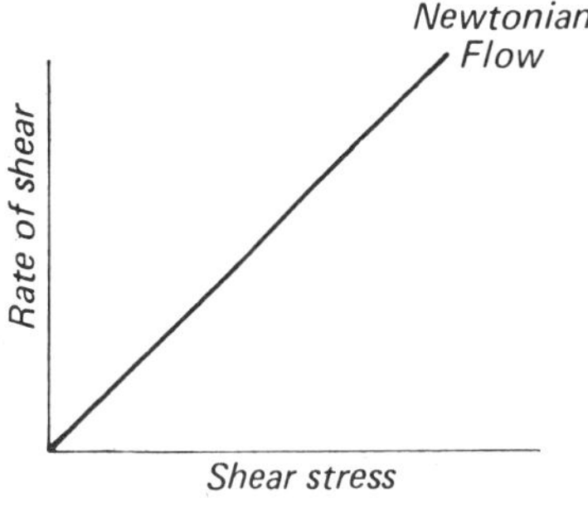

Fig 18.1 Newtonian flow

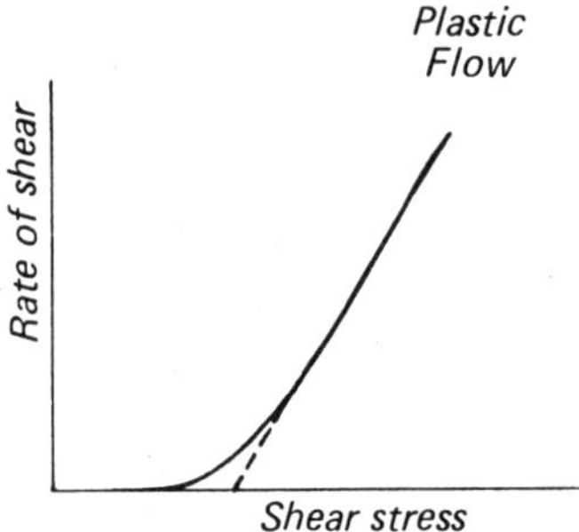

Fig 18.2 Yield value

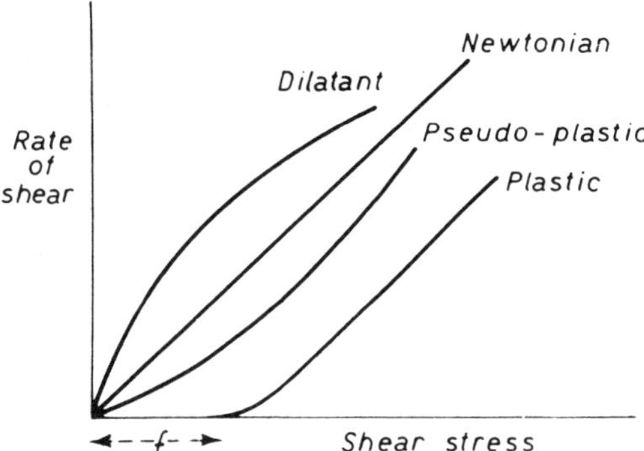

Fig 18.3 Flow curves

The presence of a yield value in a paint indicates the existence of structure, and when such a paint, which has become fluid on stirring, is allowed to remain at rest, the original structure is recovered. This behaviour is known as *thixotropy* and the paint is said to be thixotropic (*below*). This has become a familiar phenomenon since the introduction of the thixotropic or jelly paints.

Some paints appear to be heavy but break down immediately on application of stress, i.e. when stirred. They show no yield value and recover their structure immediately the stress is removed. These paints are said to exhibit *pseudoplastic flow*.

Occasionally highly pigmented systems are encountered in which the resistance to flow increases with increasing shear stress. In these systems there is little free liquid and movement of particles is restricted. At high shear rates they tend to pack before the blade of the stirrer. This behaviour is known as *dilatancy* and is rare in paint systems.

The various types of flow are illustrated in Fig. 18.3.

Thixotropy

When the flow properties of a thixotropic paint are measured on a rotational viscometer (Chapter 25), two values are obtained for the shearing stress at the same rate of shear. One value is obtained when the rate of shear is increasing and the other when it is decreasing (Fig. 18.4). The area of the hysteresis loop is sometimes taken as an indication of the degree of thixotropy in the paint.

A paint which exhibits thixotropy shows the following advantages: (i) it does not settle in the tin; (ii) provided the recovery rate is not too low, it does not sag on vertical surfaces nor pull away from sharp edges; (iii) heavier coats can be applied, so economizing in labour.

Application to flow properties of undercoats

Ideally, an undercoat should possess a degree of thixotropy, i.e. a yield value together with plastic flow. This ideal is extremely difficult to achieve since more than one type of flow is generally present. Pigment particles tend to flocculate (form clusters), and the degree of flocculation influences the type of flow property. Pseudo-plastic flow is often present, resulting in poor levelling properties. Brushmarks persist in the dry film and can mar the uniform gloss of the finishing coat.

Undercoat formulation consists largely of choosing pigments, and more particularly extenders, with the correct physical characteristics to give the desired flow and levelling. This calls for experience and a thorough knowledge of pigment properties in relation to the media being used. Further complications arise as a result of the little-understood inter-particle reactions, with the result that few undercoats approach the ideal type.

Undercoats generally are formulated at a PVC of 40 to 55 percent, depending on the type of pigment and medium used. The solid matter consists of true pigment (which is the more expensive solid and so used in minimum quantity to give the necessary colour and opacity) and extenders. The latter are used in sufficient volume to give the necessary degree of sheen, but care has to be exercised in their choice. There is little point, for example, in using large quantities of extenders of high specific gravity, such as barytes, which will give settlement troubles. Lower specific gravity pigments make a larger volume contribution for a given weight and are often cheaper.

Certain extenders, e.g. types of precipitated silica, possess very high oil

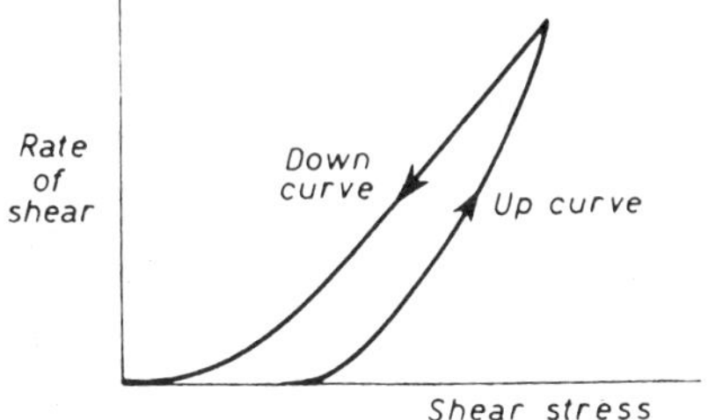

Fig 18.4 Thixotropy hysteresis loop

absorptions and have a pronounced flatting effect on paint. These are generally used in small quantities. If too much is used they induce pseudo-plastic flow and hence destroy the levelling properties of the paint.

Formulation of undercoats

The properties required in the ideal undercoat have been listed on p. 324. Properties which are amenable to control are opacity, adhesion, degree of sheen, ease of brushing, and drying characteristics. Flow properties present the most difficult problem. Ideally the paint should possess sufficient flow to eliminate brushmarks but, unfortunately, this degree of flow is accompanied by sagging on vertical surfaces and pulling away from corners and sharp edges. There is no ready solution to this problem and a compromise is necessary – as occurs so often in paint formulation. In order to minimize sagging on vertical areas and pulling away from edges, a degree of brush-marking has to be tolerated.

In structural maintenance work, application is largely by airless spray, brushing being used only for small areas. Sagging is eliminated by the use of thixotropic or high-build paints whilst edges are given a 'stripe coat' to provide the necessary film build.

Media for undercoats

Experience has shown that the best performance of a paint system is achieved when undercoat and finish are based on the same general type of medium. Decorative finishes are usually formulated on long-oil alkyds, either straight or modified. The use of a long-oil alkyd in a highly pigmented undercoat usually results in a system showing poor flow and levelling properties. On the other hand, a highly polymerized alkyd provides excellent flow and levelling properties, but the cost is high and the enamel 'hold-out' is sometimes suspect. A mixture of the two will provide a reasonable compromise which satisfies most requirements. Alternatively, the long-oil alkyd can be mixed with a thixotropic alkyd to provide a thixotropic undercoat. The degree of thixotropy can be controlled by the relative proportions of the two resins.

Pigments for undercoats

Whites and tints account for a very high proportion of decorative undercoats sold. Titanium dioxide is the white pigment used, the rutile type being preferred for the superior flow properties. A wide range of extenders is employed and the choice and quantity can be critical. Opinions differ on the best combination of extenders to provide the necessary film build with minimum interference with flow properties. Maximum 'flatting' is obtained with extenders of high oil-absorption (fine particle size) but the amounts used are usually critical and flow can be impaired. Extenders of low oil-absorption (usually modular types) can be used in greater quantity to provide film build. A compromise is usually reached in which both types are employed.

For coloured undercoats the pigments must possess good opacity, but the brightness and lightfastness are not necessarily of the same standard as required in the finish. For this reason, inorganic pigments are used in undercoats whenever possible, but the more expensive organic pigments have to be used on occasions with finishes which are very bright and/or lacking in opacity.

Replacement of one pigment by another in a formulation is sometimes required, without alteration of the essential characteristics such as gloss and consistency. Replacement must be on a volume basis, and this is satisfactory provided the two materials in question possess similar oil-absorptions. If these differ widely then they must be taken into account in the substitution. Often this means reformulation.

Examples of decorative undercoats

These formulations are intended to give the reader a general idea of the composition of undercoats. Appreciable latitude is possible in the choice of extenders and resins. The final formulation is often a matter of personal preference.

White: alkyd base	*% by weight*
Rutile titanium dioxide	21·8
White barytes(micronized)	17·4
Fine silica	5·5
Coated calcium carbonate	5·5
Long-oil alkyd (75% solids)	12·5
Polymerized alkyd (40% solids)	21·5
White spirit	12·4
Dipentene	2·5
Lecithin	0·3
Driers (Co/Zr octoates)	0·5
Methylethylketoxime	0·1

Specific gravity: 1·46
Solids by volume: 45·3%
PVC: 43·8%

(The figures for solvents may vary with the types of alkyd resin used and the solids content.)

Note on materials

Titanium dioxide the rutile type confers better flow properties than anatase.

Barytes and silica confer film build and strength. Silica provides ‘tooth’ and aids the adhesion of the finish.

Coated calcium carbonate prevents hard settlement. The effect is augmented by the lecithin.

Alkyds The flow properties of the undercoat are influenced by the relative proportions of the two alkyds. An increase in the polymerized alkyd content gives improved flow, whereas the long-oil alkyd confers a controllable degree of stickiness which assists film build by preventing the application of excessively thin coats.

Dipentene is an excellent solvent for most alkyds and possesses some anti-skinning properties.

Methylethylketoxime is a powerful anti-skinning agent and also extends the wet-edge time. Both functions are a consequence of complex formation between methylethylketoxime and cobalt which decomposes when the film is exposed to air. The freed cobalt then functions normally.

White Undercoat: alkyd base, thixotropic	*% by weight*
Rutile titanium dioxide	20·3
Micronized barytes	33·9
Talc	10·5
Thixotropic alkyd (50% solids)[10]	21·4
Pale boiled oil	2·5
Co/Zr drier	0·3
White spirit	11·0
Methylethylketoxime	0·1

Specific gravity: 1·68
Solids by volume: 53·4%
PVC: 53·8%

FINISHING COATS

Oil-based paints

The oldest types of gloss paint are those based on linseed oil in one or other of its forms. They are highly pigmented and contain only a small quantity of thinner. Consequently they give films of high build with a moderate degree of gloss.

Paint properties such as flow and levelling, as well as film properties such as elasticity, hardness and durability are dependent largely on the nature and degree of pigmentation. In this way oil finishes differ from the modern alkyd types in which the film properties are controlled to a large extent by the resin binder.

The demand for oil-based paints is now very small.

Modern gloss finishes

A considerable improvement was effected in gloss finishes by the replacement of oil by oleoresinous varnishes. These gave harder films with improved gloss and gloss retention. The earlier types were known as varnish paints and were heavily pigmented, but reduction in extender content, together with improved oleoresinous media, led to the superior hard gloss paints that were in general use until the Second World War. These paints were based on long-oil media produced from linseed oil with fused congo copal or copal ester, and gave finishes with high gloss, but the gloss retention was not as good as with modern alkyds. The durability was, by modern standards, reasonably good.

The first step in the introduction of alkyds was admixture with oleoresinous media, and such mixtures containing 50 percent and upwards of long-oil alkyd produced finishes with excellent gloss, gloss retention, and durability.

The earlier alkyds used were of the linseed/glycerol, and later linseed/pentaerythritol, types which showed poor brushability on large areas. This brushability and the wet-edge time were improved considerably by admixture with oleoresinous media or enamel oils. These mixtures tended to yellow on exposure, a fault which was very apparent in white and pale shades.

The change to the modern all-alkyd type of finish became feasible with the introduction of the long-oil pentaerythritol alkyds based on linoleic acid (or 'linoleic rich' acids) derived from soya bean and other semi-drying oils. These alkyds show excellent brushing properties and very little tendency to yellowing on exposure.

Formulation of alkyd gloss paints

The demand for whites and light colours is such that a high percentage of the alkyd gloss paints produced are pigmented with rutile titanium dioxide, with possibly a small quantity of anti-settling agent such as a coated calcium carbonate or a Bentone [11]. These are pigmented at a PVC of 15–19 percent, the lower figures being used when stainers are to be added. A typical formulation for a white alkyd gloss is set out below at (A). Black finishes require only 2·5–3·0% of a good quality carbon black (Formulation B). In bright colours the pigment usually contains a high percentage of light-fast organic pigments, and examples of these are illustrated at (C) and (D) below.

(A) **White alkyd gloss**	*% by weight*
Rutile titanium dioxide	29·3
Coated calcium carbonate	1·4
Long-oil alkyd (70% solids)	52·2
White spirit	13·3
Dipentene	2·1
Methylethylketoxime	0·1
Cobalt/zirconium octoate (6%Co, 9%Zr)	0·4
Calcium octoate (4%)	1·2

Specific gravity: 1·20
Solids by volume: 50·0%
PVC: 18·6%

(B) **Black alkyd gloss**	*% by weight*
Carbon black	2·7
Long-oil alkyd (70% solids)	71·2
Nuact [12]	0·1
White spirit	20·9
Dipentene	3·0

Methylethylketoxime	0·1
Cobalt/zirconium octoate (6%Co, 9%Zr)	0·5
Calcium octoate (4%)	1·5

Specific gravity: 0·88
Solids by volume: 41·8%
PVC: 3·4%

(C) **Signal Red alkyd gloss**	*% by weight*
Toluidine Red	6·0
Permanent Red 2G	7·3
Rutile titanium dioxide	1·9
Long-oil alkyd (70% solids)	59·4
White spirit	20·4
Dipentene	3·0
Methylethylketoxime	0·1
Cobalt/zirconium octoate (6%Co, 9%Zr)	0·4
Calcium octoate (4%)	1·5

Specific gravity: 0·98
Solids by volume: 46·7%
PVC: 19·7%

(D) **Light Green alkyd gloss**	*% by weight*
Phthalocyanine Blue	1·0
Hansa Yellow G	10·0
Ferrite Yellow	3·5
Long-oil alkyd (70%solids)	59·9
White spirit	20·6
Dipentene	3·0
Methylethylketoxime	0·1
Cobalt/zirconium octoate (6%Co, 9%Zr)	0·4
Calcium octoate (4%)	1·5

Specific gravity: 0·97
Solids by volume: 50·3%
PVC: 25·5%

Note on materials

The ratio of thinners to alkyd resin will depend on both the nature of the alkyd and the viscosity. Alkyds based on the linoleic-rich oils, such as those derived from soya bean and safflower seeds, generally require less than the linseed type. The final paint viscosity is adjusted to 3 to 4 poise at 20°C (68°F).

Methylethylketoxime functions as an anti-skinning agent and also prolongs the wet-edge time. In warm weather, when wet-edge times are inclined to be short, the amount of methylethylketoxime can be increased without detrimental effects on other properties.

Decorative paints are now made on a leadless basis, and a mixture of cobalt and zirconium drier metals has been found to be the most satisfactory replacement for the traditional lead/cobalt mixture. Octoates (ethylhexoates) are preferred as the acid radicles to avoid the 'after odour' associated with the use of naphthenates.

In the black gloss a small quantity of Nuact [12] is incorporated to prevent absorption of driers by carbon black on storage. Loss of driers leads to prolonged drying times.

The calcium naphthenate plays a dual role. It prevents precipitation of the drier metals through reaction with phthalic acid and also reduces the incidence of 'crystalline bloom' on the dry film.

Crystalline bloom consists of a deposit of fine crystals of ammonium sulphate and occurs under certain conditions in urban atmospheres containing ammonia and sulphur dioxide. It can be removed readily with water.

According to Holbrow [13], condensed water droplets on the film dissolve ammonia and sulphur dioxide which react slowly to produce ammonium sulphate. This remains as a bloom when the droplets evaporate. Calcium naphthenate induces condensation as a film rather than as droplets. Evaporation is more rapid, and the time available for the formation of ammonium sulphate is reduced. It was shown, however, by Hill [14] that moisture is not essential for the formation of crystalline bloom. He produced the defect on painted panels by thermal precipitation of particles of ammonium sulphate.

General properties of alkyd enamels

The important attributes of alkyd enamels are good gloss retention and good durability. Application is usually by brush or roller and, under normal conditions, they dry in 4 to 8 hours.

Alkyds are more sensitive than oleoresinous media to adverse conditions during drying, but there is an appreciable difference between types. Pentaerythritol alkyds are more sensitive than the glycerol types, but greater tolerance is claimed for the vinyl toluene types.

Low temperatures increase the drying time of alkyds considerably and they are very sensitive to moisture during the drying process. When alkyd enamels are applied on exterior surfaces during periods of heavy dew formation, application late in the day can result in moisture deposition during the drying process. Some moisture is absorbed by the surface layers of paint, resulting in swelling. Micro-rivelling occurs when the moisture dries out and the result is loss of gloss.

The sensitivity to moisture renders alkyds unsuitable for use on surfaces subject to immersion in water. Poor alkali resistance renders them unsuitable for use on alkaline surfaces such as plaster or stucco.

Thixotropic paints

Decorative thixotropic paints are based on thixotropic alkyds and are also known as non-drip paints. They are made in the form of gloss, semi-gloss and eggshell finishes as well as undercoats.

The great advantage of thixotropic paints lies in the saving of time and labour. Thick coats can be applied which remain open long enough for levelling but set before sagging can set in. They can be made solely on thixotropic alkyds or on blends with conventional types. In the latter case the degree of thixotropy is reduced, i.e. the gels are less firm and the recovery time is greater.

The behaviour of thixotropic paint films towards water and chemicals follows a similar pattern to that of conventional alkyds.

Flat finishes, also known as *flat wall paints*, are used as interior wall finishes, but their use has declined considerably since the introduction of latex emulsion paints. The older types were based on stand oils or oleoresinous media heavily pigmented with lithopone and containing aluminium stearate as an anti-settling agent. They possessed poor levelling properties and were stippled to remove brush-marks.

Modern types are based on highly bodied alkyds of the non-yellowing linoleic-rich type pigmented with titanium dioxide and extenders at a PVC of 55 to 60 percent. These modern types are more sensitive than the older oil finishes to variations in the porosity of the wall. It is important, therefore, that the wall should be adequately sealed, otherwise variations in sheen (flashing) will occur.

Eggshell and semi-gloss finishes are used to a limited extent as wall finishes and are made on conventional or thixotropic alkyds. Rigid control of PVC is essential for control of gloss.

Anti-condensation paints are used to provide a heat insulating barrier between the cold walls and the warm moist air of a room. Heat losses to the wall are reduced and condensation is minimised. They usually consist of an emulsion or flat oil paint to which is added a proportion of heat insulating material in the form of fine particles. Cork powder was formerly used but is no longer available. Very small polymer or glass spheres, such as 'Microsphere' [15] are now commonly used.

Heat-resisting paints. The formulation of these paints will naturally be influenced by both the temperature to which the films are likely to be exposed and the exposure-time which is often for prolonged periods.

Considerable variations occur between the different types of organic binders and pigments (both inorganic and organic) in respect of resistance to heat. The number which will withstand prolonged exposure to temperatures above 120° to 130°C (248° to 266°F) is relatively small. The most usual effects of heat are darkening, which can be due to both pigment and binder, and embrittlement [16]. The latter is a consequence of the acceleration of the normal ageing process brought about by the high temperature.

An investigation into the long-term heat stability of finishes was carried out by Berger [17] who pointed out that temperatures of 90° to 120°C (194° to 248°F) were most frequently encountered in practice in central heating systems, chemical equipment, motors and so on. He found that at these temperatures the only finishes unaffected were silicones; polyurethanes cured by aliphatic isocyanate showed slight discoloration. With alkyds and with polyurethanes cured by aromatic isocyanates the protective value was retained in spite of discoloration.

Domestic radiators and heating panels are usually coated with a stoved finish at works. Repainting in light colours requires air-drying finishes in which colour stability of binder (as well as the pigments) is as high as possible.

By basing the paints on a linoleic type alkyd, reinforced with a melamine/formaldehyde resin, the discoloration can be minimized. Linseed type alkyd can be used for darker colours where darkening is not so apparent.

Krauss [18] has shown that the greatest changes in film properties occur in the first 1000 hours of exposure and that little change takes place after 10 000 hours.

In a number of industrial applications much higher temperatures are encountered. Oelsner [19] studied the resistance of finishes at 150°C (302°F) and above and found that only silicones remained unchanged for 500 hours at these temperatures.

Temperatures of 600° to 700°C are reached in the lower sections of some steel chimney stacks, while many exhaust manifolds are only a little lower. These temperatures are above the limits of stability of most organic binders. For these surfaces aluminium paints based on silicone resins are usually employed. The silicone resins decompose at about 350°C (482°F), but the aluminium becomes sintered to the steel surface and provides protection.

Flame-retardant paints are designed to play a dual role. Firstly they must check or eliminate the spread of flame across the surface (flame will spread rapidly across most conventional paint films) and, secondly, they must in most cases be decorative. In this way they can reduce the fire hazard of combustible substrates and not detract from the flame-resisting properties of non-combustible materials.

The earlier types of paint were based on alkali silicate solutions. Although they would not completely prevent damage or charring of combustible substrates such as wood as a result of excessive heat, they were very effective in preventing the spread of flame [20]. Sodium silicate was the cheapest of the alkali silicates, but the paints developed an efflorescence which spoiled the colour. Potassium silicate paints were free from this defect and were used for stage scenery for many years. Very little silicate paint is used for flame-retardant purposes today.

Modern flame-retardant paints are produced on both solvent and emulsion bases. They include the following types.

1 Oxygen-barrier coatings, often containing borates. On heating, these form a carbonaceous mass or a glassy barrier and evolve CO_2 and water rather than flammable products. Oxygen is excluded from the surface.

2 Media containing chlorinated compounds (such as chlorinated wax) and a pigment containing a high proportion of antimony oxide. Application of heat results in the formation of a blanket of volatile antimony chloride which prevents oxygen from reaching the surface. This type of paint, is protected by patents [21].

3 Intumescent coatings which produce a substantial layer of foam when exposed to fire. These paints often contain starch, ammonium phosphate and a blowing agent to produce the foam.

The testing of flame-retardant paints is carried out on the desired substrate by the methods set out in BS 476 [22]; these include tests for ignitability, fire propagation, and surface spread of flame. For spread of flame the paint/substrate combinations are assigned to one of four classes: (a) very low rate, (b) low rate, (c) medium rate, (d) rapid.

Supplies of fire-retardant paint must be accompanied by a Certificate of Compliance with BS 476 and issued by the Fire Research Station, Boreham Wood.

Aluminium paints for decorative and structural work are required to show maximum retention of brilliance over long periods. A high degree of leafing is essential and this requires a medium of low viscosity, low acid value and high solids. The most favoured media are tung oil/coumarone, tung oil/phenolic or urethane oil. Pigmentation is by aluminium paste to give about 130 to 150 grams of metal per litre. The final viscosity of the paint should be about 2 poise at 20°C (68°F).

Aluminium paints are used on metal lamp-standards, roof trusses and other structural work. They are not recommended in buildings where inflammable solvents are used since a spark can be generated if the surface of the paint is struck with a steel object.

Bronze paints are made in a range of colours from Pale Gold to Antique Bronze. The metallic pigments used are copper alloys in flake form; pigmentation is usually at the rate of 400 to 500 grams per litre in a neutral medium such as a polymer solution. Polyvinyl acetate, acrylics and polystyrene are used. Bronze paints are more sensitive than aluminium paints to acidic media and weathering; they are used only for interior decorative work.

Masonry finishes 'Masonry finish' is a general term applied to coatings used on exterior walls, excluding wood. Surfaces involved include stucco, cement render, brickwork, roughcast, and stonework.

Important reasons for painting exterior wall surfaces are (a) improvement in appearance, (b) prevention of water penetration, particularly into stucco and brickwork, (c) prevention of spalling and disintegration resulting from frost, (d) reduction of dirt retention. Complete sealing of the surfaces is not desirable [23] and masonry paints are formulated with a microporous structure which provides a degree of permeability and so enables the wall to 'breathe'.

A number of binder types are in use but, in view of the alkaline nature of many of the materials, it is usual to employ alkali-resistant types. The following are among the types of binder most widely used. The finishes dry either to a smooth film or can be textured to exhibit an imitation stone effect.

1 '100%' acrylics or co-polymers of styrene/acrylate or vinyltoluene/acrylate. These are generally in solution form and, when suitably pigmented, will produce films which adhere well and show very good durability.

2 Acrylic emulsions, in both flat and textured finishes, also show very good durability.

3 Emulsions of vinyl co-polymers in which the second monomer can be either ethylene or an acrylic ester. Terpolymers produced from vinyl chloride, vinyl acetate and ethylene are also used and are claimed to possess very good weather resistance.

4 Chlorinated rubber, suitably plasticised. Lower-viscosity grades are recommended to assist penetration into the surface. Chlorinated rubber paints are highly resistant to alkali and to atmospheric contaminants.

5 Proprietary cement paints supplied in the form of dry powder and mixed with water immediately before use.
6 Long oil alkyd types which are usually applied as a two-coat system comprising sealer/undercoat and finish. They usually dry to a semi-gloss finish which can be smooth or textured.

Floor paints Floors are painted for the following reasons, (a) to improve appearance, (b) to protect the floor surface from abrasion, (c) to protect the surface from attack by spillages of various kinds and so present a surface from which they can be cleaned off without damage to the floor.

There is a wide selection of binder types which makes it possible to select the type best suited to the type of floor and the conditions of use. The following list sets out the most commonly used types and their main characteristics.

1 Medium-oil alkyd Used for general purpose floor paints where traffic is mainly pedestrian. Suitable for wood, quarry tiles, composites and mature concrete. Not resistant to alkali and so unsuitable for new concrete or cement screeds. (The preparation and painting of concrete floors is discussed in Chapter 21.)
2 Epoxy ester Resembles alkyd insofar as usage is concerned but harder and more wear resistant.
3 Moisture-cured polyurethane Dries very hard and is very wear resistant. Used in clear form on parquet and other floors where appearance is to be retained. Used in pigmented form on other types of floor. Since this material cures by reaction with atmospheric moisture, ullage cans must be kept securely closed or the contents are likely to 'gel' and become useless.

Other floor paints are considered in Chapter 21.

Fungistatic paints are used in locations favourable to the growth of moulds and fungi. They are usually conventional paints to which has been added one to two percent of a fungistat. Among the latter are chlorinated phenols and dodecylamine salicylate but a number of proprietary materials are available. Before application of fungistatic paint it is usual to neutralize any growths in the surface by application of a dilute solution of fungicide.

Multicolour paints produce films exhibiting two or more separate colours when applied as a single coat by spraying. They are polyphase systems containing one or more pigmented organic phases dispersed in a non-pigmented aqueous phase. The latter contains a thickener or stabilizer such as a cellulose ether and possibly an emulsifying agent.

The most widely used multicolour paints are based on cellulose nitrate as the medium for the dispersed organic phase. This is pigmented in the normal way and the lacquer stirred into an aqueous solution of stabilizer. More than one coloured lacquer can be used to produce the multicolour effect, but each is stirred into a separate portion of stabilizer solution and the individual dispersions are then mixed carefully.

The size of the coloured droplets varies from 1 to 8 mils in diameter, depending on the speed of stirring, and the shape depends on the viscosity of the lacquer. Low viscosity gives spherical and high viscosity elongated shapes.

Can stability of these paints depends on the densities of the aqueous and organic phases being as close as possible, and on the absence of water-miscible solvents in the organic phase.

Multicolour paints are used mainly for interior wall decoration in public buildings, schools and offices.

References to Chapter 18

[1] Froggatt, J. J., Preservation, priming, putty and paint. *JOCCA*, 1980, **63**, 237.
[2] Hilditch, E. A. and Woodbridge, A. J., Progress in timber finishing in Great Britain. *JOCCA*, 1985, **68**, 217.
[3] Cuprinol Ltd, Frome, Somerset.
[4] Buckman Laboratories.
[5] Synolac 6005W. Cray Valley Products Ltd, Farnborough, Hants. *see also*, The formulation of high durability varnish stains. O'Hara, K. and Newton L. *JOCCA*, 1987, **40**, 44.
[6] Rothwell, G. W., The problems of painting timber. *JOCCA*, 1971, **54**, 658.
[7] Gunn, D. J., The coating of external plywood cladding panels. *JOCCA*, 1980, **63**, 28.
[8] Feist, W. C., *J. Coatings Tech.*, 1982, **54**, 43.
[9] Miller, E. R and Boxall, J., Waterborne paints for exterior wood. *Information Paper IP 9/84*, BRE, Princes Risborough, 1984.
[10] Gelkyd 331W (Cray Valley Products Ltd).
[11] Bentone (NL Chemicals (UK) Ltd, Wilmslow, Cheshire).
[12] Nuact (Durham Chemicals Ltd, Chester-le-Street, Co. Durham).
[13] Holbrow, G. L., Atmospheric pollution, its measurement and some effects on paint. *JOCCA*, 1962, **45**, 701.
[14] Hill, G. V. G., The mechanism of the formation of crystalline bloom on paint films. *JOCCA*, 1974, **57**, 342.
[15] Metaspheres (Microfine Minerals & Chemicals Ltd, Raynesway, Derby).
[16] *Hess's Paint Film Defects*, 3rd edn, p. 285. Chapman & Hall, London, 1979.
[17] Berger, W., Studies in the long term heat stability of finishes. *JOCCA*, 1967, **50**, 244.
[18] Krauss, W., Farbe und Lack, 1958, **64**, 309.
[19] Oelsner E., Farbe und Lack, 1959, **65**, 625.
[20] *Leaflet No. SS2-14. Silicate paints*. Crosfield Chemicals Ltd, Warrington.
[21] Cookson Ceramics & Antimony Ltd, Stoke on Trent, Staffordshire.
[22] BS. 476. Part 6, 1981. Method of test for fire propagation of products. Part 7, 1981. Method for classification of the surface spread of flame of products. British Standards Institution.
[23] Price, C. A., Stone – decay and preservation. *Chem. Brit.*, Oct. 1975, p. 350.

19

Water-borne coatings

The use of water as a paint solvent has always been attractive for reasons of low cost and absence of fire risk and toxicity. It was probably the first solvent used by man in his primitive paints and has, of course, been used in artists' water colours for many centuries. A number of coating materials which are now virtually obsolete, among which were distemper and limewash, were water borne.

An impetus to the extended use of water as a paint solvent has been provided by the spiralling costs of solvents and considerations of environmental pollution.

The water-borne media in use today [1] can be divided into two main classes, (a) polymer emulsions, which are used chiefly in building paints and (b) water-soluble or water-dispersible binders, used mainly for industrial paints especially metal primers applied by electrodeposition. An obviously attractive application of water-borne binders is in air-drying decorative finishes and undercoats and considerable work has been done in attempts to produce suitable emulsions. It is now claimed that paint resins can be emulsified satisfactorily using a tailor-made emulsifier [2] possessing an appropriate HLB value (*below*).

POLYMER EMULSIONS

Nature of emulsions

An emulsion is a two-phase system consisting of a liquid, in the form of very fine droplets (the dispersed phase), suspended or dispersed in a second liquid (the continuous phase) in which it is insoluble. The commonest form of emulsion contains water as the continuous phase; polymer emulsions are of this type. When the dispersed phase is a solid the system is termed a 'dispersion', but although the polymer is usually more solid than liquid, the term 'emulsion' is always used for the polymer systems. (Emulsions exist in which water is dispersed in an organic continuous phase, but these are rarely encountered in surface coatings.)

Formation of emulsions

Suppose that it is desired to produce an emulsion of linseed oil in water. Rapid stirring of the two liquids together will produce a suspension of oil droplets, but when stirring ceases, the two liquids will separate into clearly defined layers (Fig. 19.1). If, however, the water contains a small quantity of a soap or a surface-active agent (or 'surfactant') the oil droplets remain dispersed in the water and an emulsion is formed. When this emulsion is allowed to stand for some time, the dispersed oil droplets will tend to rise (the specific gravity of linseed oil being lower than that of water). This is known as 'creaming' of emulsions, by analogy with the separation of cream from milk. The 'cream' is readily redispersed on stirring, but the formation of 'cream' can be prevented by increasing the viscosity of the water phase.

Among materials used for this purpose are cellulose ethers, alkali salts of polyacrylic acid, alginates and alkali caseinate. Such thickening agents are described also as 'emulsion stabilisers' or 'protective colloids'.

Surfactants

These are compounds which, when dissolved in a liquid, reduce the surface tension of that liquid or the interfacial tension between two immiscible liquids or between the liquid and a solid. Their effectiveness is related to their chemical constitution and a very large number of these compounds are known. When a surfactant is used to produce an emulsion, as in the example given above, it is classed as an 'emulsifying agent', and the suitability of a surfactant for this purpose is related to the HLB value.

HLB value A surfactant molecule contains both hydrophilic and lipophilic (hydrophobic) groups and it is the relationship between these groups which controls the emulsifying properties of the surfactant. Empirical values were assigned to these groups by Davies and Rideal [3]. Some of these values are quoted in Table 19.1.

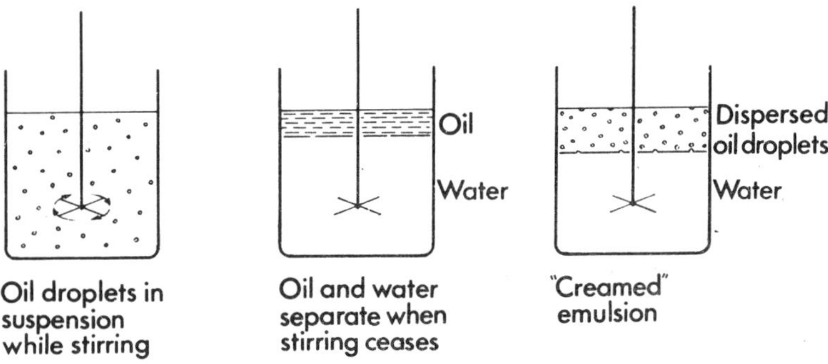

Fig 19.1

Table 19.1

Hydrophilic groups			Lipophilic groups	
$-SO_4^{2-}$	Na^+	38·7	$-CH_3$	0.475
$-COO^-$	Na^+	19·1	$-CH_2-$	
$-COOH$		2·1	$-CH-$ (with vertical bond)	
$-OH$		1·9		
$-O-$		1·3		

Using these figures the HLB or Hydrophilic/Lipophilic Balance can be calculated using the expression:

$$\text{HLB} = \Sigma(\text{Hydrophilic groups}) - \Sigma(\text{Lipophilic groups}) + 7$$

Very good agreement was obtained between calculated and experimental values.

The correlation between HLB values and practical application of detergents is shown in Table 19.2 [4].

Table 19.2

HLB range	Application
3·5– 6	W/O emulsifier
7 – 9	Wetting agent
8 –18	O/W emulsifier
13 –15	Detergent
15 –18	Solubilising agent

In practice the HLB value is useful in the choice of emulsifiers but is less important in detergents.

An interesting physico/chemical relationship between HLB value and emulsion stability was pointed out by Davies and Rideal [3]

$$\text{HLB} = \log \frac{\text{coalescence rate W/O}}{\text{coalescence rate O/W}}$$

where W/O refers to the water in oil emulsion and O/W to the oil in water emulsion.

Surfactants are classified into four main groups designated anionic, non-ionic, cationic, and ampholytic. The anionics are the more commonly used as emulsifying agents. The other classes are used in pigment wetting and dispersion and for special purposes.

Anionic surfactants

The most common types are the alkali metal salts of straight chain carboxylic or sulphonic acids of 11 to 17 carbon atoms. Sodium laurate is a typical example. When dissovled in water it ionizes:

$$CH_3(CH_2)_{10}COONa \rightleftharpoons CH_3(CH_2)_{10}COO^- + Na^+$$

The surface activity is due to the polar nature of the anion which consists of the hydrophilic (water-attracting) COO^- group and the hydrophobic or lipophilic

Fig 19.2

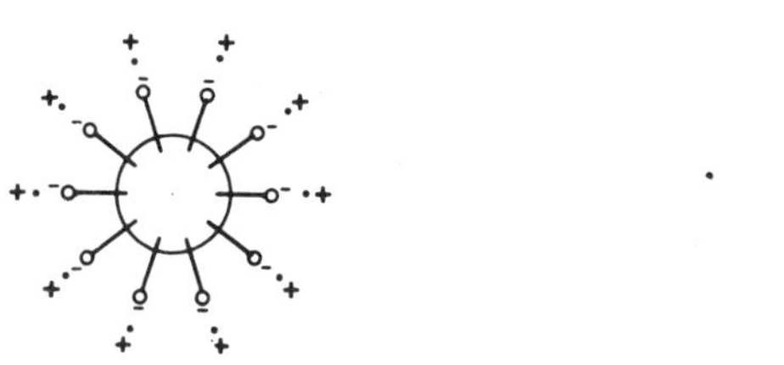

Fig 19.3

(water-repelling or oil-attracting) hydrocarbon chain. Except at very low concentrations the anions form micelles (Fig. 19.2, where the anions are represented as—$\bar{o}$) in which the hydrocarbon ends are directed inwards and the carboxyl groups outwards into the water.

When oil droplets are introduced (as in the introductory example) the hydrocarbon ends are attracted to the oil and the carboxyl groups project into the water (Fig. 19.3).

The oil droplets thereby acquire an overall negative charge and are mutually repellent. Coalescence of the droplets is prevented and an anionic emulsion results. The anions tend to concentrate at the surface of the solution and lower the surface tension by an amount proportional to the concentration. In a similar way they are adsorbed at the interface between the solution and a solid and so reduce the contact angle. Wetting of the surface is thereby facilitated.

Other types of compounds used as anionic surfactants are water-soluble salts of amines and fatty acids, e.g. triethanolamine oleate; sulphated higher alcohols, e.g. sodium lauryl sulphate; sulphonated compounds, e.g. sulphonated castor oil (Turkey Red Oil), and aromatic petroleum sulphonates.

Formation of polymer emulsions [5]

When a liquid monomer, such as vinyl acetate, is added to an aqueous solution of an anionic surfactant under rapid stirring, droplets of various sizes are produced and these become coated with the anions. The smaller droplets become enclosed in micelles and are rapidly 'solubilized'. Addition of the polymerization catalyst results in rapid and vigorous reaction in which the solubilized monomer is converted to the polymer (polyvinyl acetate). The large droplets are solubilized more slowly, but ultimately the whole of the monomer is converted to polymer. The polymer particles are coated with anions, so acquiring a negative charge. The final system is an anionic emulsion. Emulsions produced in this way contain 45 to 55 percent of polymer by mass.

The polymerization reactions have been described in Chapter 11.

Homo- and co-polymers

Homopolymers The product obtained by polymerization of a single monomer is known as a 'homopolymer'. A large number of homopolymers have been produced and they show wide variation in physical properties, particularly hardness. For example, polyvinyl acetate is a hard, brittle solid, whereas polybutyl acrylate is soft and flexible in consequence of the different groups in the chains. Polyvinyl acetate therefore does not form continuous films when the emulsion is applied to a surface and allowed to dry. If, however, a suitable plasticizer (up to 20 percent by mass on the polymer) is stirred into the emulsion, migration into the polymer takes place, resulting in softer particles which are capable of coalescing to form a film.

Ester-type plasticizers are the most suitable but, unfortunately, they possess drawbacks. On prolonged exposure of the plasticized film, loss of plasticizer takes place gradually, either by evaporation or by migration into the substrate. The result is progressive embrittlement, leading to failure.

Co-polymers If vinyl acetate is co-polymerized with a second monomer, which itself gives a soft, flexible polymer, then by adjustment of the relative quantities, products can be produced with any desired degree of hardness and which are capable of yielding continuous films without addition of plasticizer. The proportion and type of second monomer can be varied, and the types used include ethylene and other olefines, vinyl chloride, VeoVa [6] (a vinyl ester of a branched chain carboxylic acid), and acrylic esters. Other vinyl co-polymers with more specific applications include vinyl chloride/VeoVa, styrene/acrylic and vinyl toluene/acrylic.

A series of vinyl acetate/VeoVa 10 co-polymers has been designed [7] to accept higher pigment loadings than normal without appreciable loss of scrub resistance. Matt finishes with a PVC of 70% are possible giving opacity by 'dry hiding'. Economies can be effected by replacement of a proportion of the titanium dioxide by extender. Co-polymers of this type are known as 'high binders'.

Vinyl co-polymers, particularly of the vinyl acetate/olefin type, are the standard media for the majority of emulsion paints for interior and, but to a lesser extent, exterior application.

A range of co-polymers of vinyl chloride and vinylidene chloride modified by inclusion of a proportion of acrylic ester has been produced by ICI and marketed under the trade name 'Haloflex' [8]. These anionic emulsions differ in several respects from the conventional types. A low level of stabiliser is used in manufacture and so it is necessary to add stabiliser to the paint formulation. A block co-polymer of ethylene oxide/propylene oxide (Synperonic PE 39/70—ICI) has been found very suitable.

'Haloflex' is marketed at a very low pH (1·0 – 2·0) since this type of polymer tends to undergo dehydrochlorination at higher pH levels. Emulsion paints based on 'Haloflex' are pigmented at a pH of 4 – 5 and this imposes some restrictions on the types of pigments which can be used. Zinc phosphate, modified aluminium triphosphate [9] and 'Shieldex' [10] have given satisfactory results when used in primers for steel.

Applications of the 'Haloflex' co-polymers include primers for blast cleaned steel (at 50 – 80 μm dry film thickness) and they show good adhesion to new galvanized steel, concrete, cement render, wood and many plastic surfaces.

The 'Haloflex' range includes co-polymers suited to finishes as well as to primers.

VINYL EMULSION PAINTS

Emulsion paints contain three distinct phases, (a) the aqueous or continuous phase in which the thickener, preservative and other additives are dissolved, (b) the dispersed polymer phase, and (c) the dispersed pigment phase. Table 19.3 illustrates typical compositions of flat and semi-gloss vinyl emulsion paints. They are followed by more detailed discussion of the components and of the finished paints.

Table 19.3 Examples of vinyl emulsion paints

	Percentage composition	
	Flat finish	Semi-gloss finish
Rutile titanium dioxide	20·00	20·00
Whiting	13·50	–
Blanc fixe	9·40	–
Hydroxyethylcellulose	0·45	0·40
Water	29·45	13·70
Non-ionic surfactant	0·70	0·50
Sequestering agent	0·20	0·20
Water	1·30	1·30
Preservative	0·20	0·20
Anti-foam	0·15	0·15
Coalescing solvent	1·20	1·10
Co-polymer emulsion (50% solids vinylacetate/ethylene)	23·45	62·45
Specific gravity	1·48	1·22
Solids by volume	34·15%	40·70%
PVC	52·00%	15·0%
Water (by weight)	42·5%	46·2%

Materials and manufacture of vinyl emulsion paints

Pigmentation

The pigments are dispersed in the aqueous phase, usually with the aid of a surfactant, preferably of the non-ionic type. This type is preferred to the anionic which has a great tendency to cause foaming (see 'Anti-foam' below). The greatest demand in emulsion paints is for whites and pale shades in which the major pigment is titanium dioxide. The rutile grade is preferred for chalk resistance and better flow properties.

Dispersion of the titanium dioxide is conveniently effected by a high-speed disperser of the Silverson type (p. 272). The pigment (and extender, if used) is added to the aqueous solution of colloid, surfactant and other ingredients under the mixer. The degree of dispersion required is not as high as for gloss finishes. The pigment dispersion is then mixed with the emulsion, tinted if necessary, and pumped to the straining and filling area.

Where deep colours are required the strong pigments are generally used in the form of concentrated aqueous dispersions. Dry pigments can be used, provided care is taken to ensure that the dispersed pigment is in a completely hydrophilic condition (this applies particularly to many organic pigments which are hydrophobic) by using sufficient surfactant. Extenders are used to make up the PVC to the required figure, the extenders being dispersed in the aqueous phase before addition of the concentrated pigment dispersion.

Protective colloids (also known as 'thickeners')

These are used to provide a degree of 'structure', so reducing the mobility of the dispersed pigment and polymer particles and improving stability. In addition they are thought to be adsorbed on the dispersed pigment particles. Whilst the primary function of protective colloids is stabilization of the paint, they also exert a significant effect on flow properties (particularly levelling) and film formation (*below*).

In vinyl emulsion paints the types used are the non-ionic cellulose derivatives such as methyl-, hydroxyethyl- and ethylhydroxyethyl-cellulose, the anionic sodium salt of carboxymethyl cellulose and the anionic ammonium polyacrylate. Mixtures are frequently required to meet formulation requirements.

Other types of thickening agents are the hydrophilic (montmorillonite) clays [11] which absorb considerable amounts of water and disperse to form firm gels. They are more resistant to bacterial and fungal attack than are the cellulose ethers.

Preservatives

Protective colloids of the types enumerated above are prone to attack by bacteria in the can, the consequence being a drop in viscosity, often accompanied by gassing. Again, in damp situations the dried film can be attacked by moulds and fungi and it is necessary, therefore, to include a preservative in the formulation. Sufficient may be present to protect the liquid paint, but a higher concentration is necessary if the film is also to be protected. Paints of the latter type are marketed under the description 'Fungistatic Emulsion Paint'. (The adjective 'fungicidal' is often used but would be difficult to substantiate.) A fairly wide choice of preservatives is available and includes the chlorinated phenols, certain organotin compounds, chloroacetamide and a number of proprietary materials [12].

Sequestering agents

These are used in emulsion paints where hard waters or waters containing iron compounds are used. It is well known that polyvalent metal ions (cations) will precipitate anionic dispersions. The function of the sequestering agent is removal of these metal ions. Calgon [24], which is commonly used, is a sodium hexametaphosphate and functions by ion exchange.

Anti-foam

Foam can be produced when dry pigment is added rapidly under vigorous stirring to an aqueous solution of cellulose ether or other colloid. The structure of the medium retards movement of the dispersed air-bubbles and the foams can be very persistent.

A diversity of 'anti-foams' or 'defoamers' is available but in some cases the mode of action is far from clear. It has been stated [14] that the mechanism of one type [15] consists in the rapid spread of a low tension insoluble oil over the surface of a system causing weaknesses in the bubble walls leading to their collapse. The oil used contains additives such as spreading agents and 'entering agents' to carry the defoamer into the inter-bubble spaces.

Film formation and coalescing solvent

When an emulsion paint is cast as a film, evaporation of water leaves a matrix of pigment and polymer particles interspersed with the minor components. If a continuous film is to be produced, the polymer particles must be sufficiently fluid to coalesce and to embody the particles of pigment. To reach this condition the temperature of the film must be at or above the 'minimum film-forming temperature' of the polymer. This is generally above the 'Glass Transition temperature' (Tg) which is the temperature at which a polymer or resin changes from a brittle solid to a plastic condition, or vice versa. The minimum film-forming temperature is a characteristic of a particular polymer or co-polymer. The relationship between minimum film forming temperature and Tg has been examined by Nicholson [16]. It can be reduced to assist film formation at lower temperatures by addition of small quantities of certain solvents to the emulsion. These 'coalescing solvents' are absorbed by the polymer particles, resulting in softening and so facilitating coalescence. Among materials used as coalescing solvents are ethers of propylene glycol, Texanol [17] and Dalpad A [18].

Whilst coalescing solvents are effective aids to film formation at lower temperatures they detract from the resistance of the paint to freezing. Soft polymer particles are more prone to coagulation as a result of freezing than the harder types.

In a study of factors affecting the optical properties of emulsion paint films,

Williams and Fairless [19] discuss film formation in some detail, particularly the part played by coalescing solvents.

Bondy and Coleman [20] have studied film formation and the factors which influence tensile strength and water permeability. They consider that coalescence of polymer particles is influenced by the presence of colloid stabilizers and the (usually uneven) distribution of pigment particles in the film.

The drying of an emulsion paint film is accompanied by the development of stresses before the film ultimately reaches a stable condition and this can result in detachment of the film from an unstable substrate. These stresses have been measured by Perera [21] who has also shown how they are affected by the type and quantity of coalescing solvent.

Stainers for emulsion paints can be aqueous dispersions or the universal type. Inorganic pigments present little trouble, but hydrophobic organic pigment dyestuffs in aqueous dispersion must be treated with anionic or (preferably) non-ionic surfactant to avoid the risk of flushing to the polymer phase.

Properties and uses of vinyl emulsion paints

These paints have been very popular with both professional and amateur painters for a number of years. They are characterized by the following properties.

1 Absence of fire risk and toxic solvents.
2 Ease of application by either brush or roller.
3 Speed of drying, enabling a further coat to be applied within a few hours.
4 Tools and containers can be cleaned with water.

Their main application is on interior walls and ceilings where several coats can, if necessary, be applied in a working day. On absorbent surfaces such as new plaster it is advisable to thin the first coat with water (instructions are usually printed on the can), but no thinning is necessary for non-absorbent surfaces such as old paintwork. This should, of course, be thoroughly rubbed down. Vinyl emulsion paints (thinned if necessary) can be used also as primers for hardboard and many types of building board. They can be overpainted with conventional air-drying undercoat and finishing systems.

Some vinyl co-polymer emulsions are used on exterior surfaces. These usually contain mineral grit or mica, and the dry film resembles stonework.

ACRYLIC EMULSIONS AND PAINTS

The polymers employed, although frequently referred to as '100 percent acrylics' are, in effect, co-polymers of acrylic esters. Variations in properties can be obtained by varying the types and relative proportions of the esters. For example, polymerisation of ethyl acrylate yields an extremely soft polymer of little practical use; methyl methacrylate, on the other hand, yields a hard brittle product. By co-polymerising the two esters in different proportions polymers differing in hardness and flexibility can be produced. Other esters co-polymerised with methyl methacrylate include butyl acrylate, and butyl methacrylate; styrene and acrylonitrile also are used.

Acrylic emulsions being marketed today are characterised by fine particle size, good freeze-thaw stability and a high pH (9·5 – 10·0).

Formulation is, in general, somewhat similar to that of the vinyls except for the appreciably higher pH (about 9·5) which requires fairly close control. Film formation takes place in a manner similar to that in vinyl emulsion paints, and the factors which influence the coalescence or fusion of the polymer droplets have been studied by van der Wiel and Zom [22]. The films are characterised by good adhesion to a wide range of surfaces, lasting flexibility and good durability. Among their applications are primers for wood and finishes for exterior surfaces such as cement render, stucco, roughcast and concrete.

The formulation below illustrates the general composition of an acrylic primer for softwoods.

	% by weight
Titanium dioxide	14·0
Calcium carbonate	11·4
Fine natural silica	4·5
Ammonium polyacrylate 2% solution	8·6
Preservative	0·4
Non-ionic surfactant	0·6
Anti-foam	0·2
Propylene glycol	2·3
Acrylic emulsion (46% solids)	58·0
Ammonia solution	q.s.

Specific gravity, 1·26; solids by volume, 44%.

When dry this primer can be overcoated with conventional oxidising systems or with acrylic emulsion paint (below).

The following formulation for an exterior quality acrylic emulsion paint illustrates the general composition of this type.

Acrylic emulsion paint: exterior, white	*% by weight*
Water	5·2
Non-ionic surfactant	1·1
Anti-foam	0·2
Ammonium polyacrylate solution (2%)	9·6
Rutile titanium dioxide	20·5
Talc	8·4
Paris white	9·2
Proplene glycol	2·1
Pine oil	0·3
Preservative	0·2
Acrylic emulsion (46%)	43·0
Ammonia (0·880)	0·2

Specific gravity, 1·40; solids by volume, 41·0%; PVC, 38·7%; pH 9·5–9·6.

Properties and uses

Some of these have been mentioned above. In addition, the acrylic emulsion paints are more freeze-thaw resistant and less permeable to moisture than are the vinyls. Besides their use as wood primers and exterior finishes they are used on interior walls and ceilings, and, by incorporating the harder type of polymer, as floor paints. They have also been used successfully as primers for new galvanized iron.

Exterior joinery, such as window frames, is now frequently treated with a preservative stain instead of paint. A widely used stain is water-borne and consists of a mixture of alkyd and acrylic emulsions, together with light-fast pigments. Transparent, translucent and opaque stains are used. It is considered that the good durability of these stains is contributed by the acrylic component whilst the alkyd is responsible for penetration and adhesion. Low opacity red iron oxides, which are effective UV absorbers, are often used for pigmentation.

WATER-SOLUBLE BINDERS

Early types

Among the earliest binders for pigments used by man were the vegetable gums, some of which have been employed for special purposes down to the present day. Examples are gum tragacanth and gum arabic which are still used in limited quantities. Egg yolk is another material used by medieval painters and which continues to be used as a component of artists' 'tempera' medium.

Animal glues, obtained from hides and bones, were in general use until the middle of the present century as binders for 'soft' distempers (e.g. ceiling whites) and as protective colloids in oil-bound or washable distempers. These materials are now virtually obsolete.

Casein, a protein obtained from milk, is soluble in dilute alkali and was used as a binder *per se* in 'oil-free' distempers and as a protective colloid in the oil-bound or washable types. Casein possessed one advantage over the animal glues. A solution in dilute ammonia gave a water-insoluble film by loss of the alkali. Alkali caseinate solutions continue to be used on a limited scale as a protective colloid in certain types of emulsion paint.

Alginates, obtained from certain species of seaweeds, form colloidal solutions of high viscosity. They have been used as binders for soft distempers.

Modern types

These can be divided into two classes: (a) alkali silicates, and (b) water-soluble resins. The latter are the more important.

Alkali silicates

Sodium silicate was used as the binder in fire-retarding paints (e.g. for fire curtains and stage scenery) for a number of years. However, the films tended to whiten due to formation of sodium carbonate, and for this purpose sodium silicate was later replaced by potassium silicate. Sodium silicate was also used in paints required to resist moderately high temperatures, but the films were not sufficiently weather-resistant for outside work.

An interesting use of sodium silicate has been in the so-called 'inorganic zinc silicate' paints in which zinc dust is mixed with the sodium silicate solution shortly before use. The film consists of zinc silicate containing unreacted zinc and is an effective primer for steel. It is essential, however, that the SiO_2/Na_2O ratio in the sodium silicate be kept within fairly narrow limits.

Lithium silicate was later used to replace sodium silicate in zinc silicate paints, but the use of alkali silicates for this purpose has dropped considerably since the introduction of the prehydrolysed ethyl silicates.

Water-soluble resins

A number of synthetic resins can be produced in water-soluble form by incorporating acidic groups, e.g. carboxyl, in the molecular structure. These groups are then neutralized by bases such as ammonia or amines to give products soluble in water.

Water-soluble resins have been used as binders in conventional type stoving enamels for a number of years, but surfaces to be coated must be very clean. The slightest traces of oil or grease can impair wetting of the surface and result in serious cissing or contraction of the film [10].

In 1963, metal primers based on water-soluble resins were introduced in the car industry and applied by electrodeposition. This has probably been the most important application of these resins to date. The process of electrodeposition involves several simultaneous reactions and is considered in Chapter 20, pp. 000.

Types of resin used

Alkyds

These are produced by selecting the appropriate polyol and acid to give the desired free acid groups in the molecule. A wide range of suitable acids is available among which are maleic and isophthalic and trimellitic anhydride. Caron *et al.* [23] describe a method for introducing carboxyl groups during chain formation by using 2:2-bishydroxymethyl propanoic acid or by modification of the preformed chains using various aldehydes. The products are claimed to exhibit air drying times and application properties similar to those of conventional alkyd resins. A comparison of the performance of water-soluble alkyds with conventional solvent-based systems has been published by Engelhardt [24]. Burckhardt and Luthardt [25] have examined water-soluble short-oil alkyds, particularly for use by conventional and electrostatic spray. They conclude that their behaviour is determined by the functional groups causing the solubility. Water-soluble methoxylated melamines are often added to crosslink with the alkyd in stoving finishes.

Solubilizing groups can be introduced by the use of appropriate fatty acids or by maleinization [26]. Certain solubilized epoxy esters can be cured by reaction with water-soluble melamines or phenolics followed by stoving.

Other resins

Other water-solubilized resins of interest, particularly for electrodeposition, are polyesters, acrylics, aminos and phenol/formaldehyde. Appreciable amounts of water-soluble resins are used in paints for application by coil-coating. Polyesters and modified acrylics are widely used and are cross linked with water-soluble melamine during stoving.

References to Chapter 19

[1] *see also* Nicholson, J. W., *OCCA Monograph No. 2. Water borne coatings.* Oil & Colour Chemists Association.

[2] 'Atlas G. 4809' emulsifier (ICI Speciality Chemicals).

[3] Davies, J. T. and Rideal, E. K., *Interfacial phenomena.* New York, 1961.

[4] Hellsten, M., in 'Surface Chemistry'. *Proceedings of 2nd Scandinavian Symposium on Surface Activity*, Munksgaard, Copenhagen, 1965.

[5] *see also* Fries, N. and Nylagen, I., Kinetic study of the emulsion polymerisation of vinyl acetate. *J. Appl. Polymer Sci.*, 1973, **17**, No. 8, 2311.

[6] Shell Chemicals Ltd.

[7] 'Viking' series. Kirkless Chemicals, Batley, W. Yorks.

[8] ICI (Mond Division). *see also* Humphries, R. G., The versatile paint resin—Haloflex. *JOCCA*, 1987, **70**, 150.

[9] 'K White' (Bromhead & Denison Ltd, Welwyn Garden City, Herts).

[10] W. R. Grace & Co. Ltd, London.

[11] 'Perchem' range (Perchem Ltd, Harlow, Essex).

[12] 'Busan 11-M1' : 'Busan 1030' (Buckman Laboratories).

[13] Albright & Wilson Ltd, London.

[14] Priestman, K. M., Foam control in the surface coatings industry. *JOCCA*, 1982, **65**, 21.

[15] Bevaloid Ltd, Beverley, Yorks.

[16] Nicholson, J. W., Water borne coatings: A review of basic principles. *JOCCA*, 1987, **70**, No. 1, 1.

[17] 'Texanol' (Eastman Chemical International A. G., Hemel Hempstead, Herts).

[18] 'Dalpad A' (Dow Chemical Co. Ltd, Staines, Middx).

[19] Williams, J. M. and Fairless, T., Some factors affecting the brightness and opacity of emulsion paint films. *JOCCA*, 1970, **53**, 629.

[20] Bondy, C and Coleman, M. M., Film formation and film properties obtained with acrylic, styrene/acrylic and vinyl acetate/VeoVa co-polymer emulsions. *JOCCA*, 1970, **53**, 555.

[21] Perera, D. Y., Internal stresses and film formation in emulsion paints. *JOCCA*, 1985, **68**, 275.

[22] van der Wiel, H. and Zom, W., Water-borne acrylics and urethanes for the coating industry. *JOCCA*, 1981, **64**, 273.

[23] Caron, D., Fradet, A and Marechal, E., Synthesis and study of water-soluble alkyd resins with high long-chain fatty acid contents. *JOCCA*, 1985, **68**, 44.

[24] Engelhardt, R. R. Water reducible industrial coatings. *Pig. & Resin Tech.*, 1979, **8**, No. 8, 5.

[25] Burckhardt, W. and Luthardt, H. J., Water-borne industrial thermosetting systems; physical chemical behaviour of their components and aspects of application technology. *JOCCA*, 1979, **62**, 375.

[26] Bijleveld, J. and Krak, H., Some aspects of spray applied water-borne paint binders. *JOCCA*, 1975, **58**, 279.

20

Methods of application and cure

The transfer of a liquid paint from container to surface can be effected in a number of ways. Decorative paints are applied by brush or hand roller, but spraying is used on many large exterior surfaces. Both spray and brush application are employed in industrial maintenance painting as, for example, on bridges and other complex structures. A variety of methods are in use in industrial paint application. In the majority of cases these are followed by an accelerated curing operation.

The principal methods of paint application are listed below and later outlined in the text:

Brush	Spray – conventional (air-assisted), hot, airless, electrostatic, and combinations
Hand roller	Tumbling or barrelling
Dip coating	Silk screen
Flow coating	Electrodeposition
Curtain coating	
Roller coating	

Brush application

Brushes are made in many shapes and sizes, ranging from the small round 'pencils' to large flat types. The quality of the bristle is important, and hog bristle has long been regarded as the finest type. The bristles subdivide and so produce a number of fine ends which help to minimize brush-marking in the film. Nylon bristles also are used; these are more wear-resistant than the animal bristles.

New brushes are usually *broken in* by use with primers. Brush application is the most satisfactory method of applying primers to most surfaces as the paint can be 'scrubbed' in, giving good wetting and adhesion. The scrubbing action, however, must be avoided in the application of undercoats and finishes. With these, the paint is applied more liberally and with longer strokes, cross-brushed, and finally *laid off* with light strokes across the surface.

After use, paint brushes should be cleaned in white spirit or other suitable solvent and stored suspended by the handle so that they do not rest on the bristles. Brushes used for decorative work can be kept in a 'soft' condition by immersing the suspended bristles in a mixture of white spirit (1 volume) and raw linseed oil (1 volume). They must, of course, be washed out in white spirit before further use.

Paints suited to brush application are the slow-drying types thinned with white spirit or other slow-evaporating solvent. By this means large areas can be painted in sections and the latter joined up smoothly before the film has dried. The period during which the paint film remains 'open' to enable joining up to be done is known as the *wet-edge time.*

Defects which can occur in brush painting include the following [1].

Brush drag This can be due to excessively high viscosity, in which case the remedy is thinning, but this may be accompanied by loss of opacity and film build. Another cause is suction from an underbound priming coat or from a substrate which has been inadequately sealed.

Poor levelling or persistent brushmarks. This usually occurs in low-sheen paints and can result from excessive pigmentation, flocculation, or incorrect choice and quantity of extenders. Addition of a surfactant may be effective in inducing flow but, if not, reformulation of the paint is necessary.

Sagging (or 'curtaining') on vertical surfaces, accompanied by pulling away from sharp edges, indicates excessive flow or excessively heavy application. The former can be checked by introducing a limited degree of thixotropy by the use of a Bentone [2], hydrogenated castor oil or a thixotropic alkyd.

Too short wet-edge time Wet-edge time has been defined above. It depends on the nature of the resin (linoleic-rich alkyds set up more slowly than the corresponding linseed types), the volatility of the solvents, and the temperature. Remedies include addition of an anti-skinning agent such as methylethyl-ketoxime to delay the initial oxidation of the film, or (if possible) an increase in the proportion of high-boiling solvent.

Hand rollers

The roller is made of lamb's-wool or soft plastic foam and is charged with paint by rolling to and fro in a shallow tray with a small quantity of paint. These rollers are very useful for painting large flat areas such as walls, and are popular with home decorators. The soft rollers can be removed easily for cleaning and must be washed out very thoroughly after use, or they will harden and become useless.

Pressure-fed rollers

For large-scale work the use of the tray can be eliminated by conveying the paint to the roller through the hollow handle. The paint is pumped at low pressure and the flow can be controlled by a trigger mechanism. The use of an extension handle allows large areas to be painted with minimum use of scaffolding [3].

Dip coating

This is the simplest method of applying paint and provides a cheap and quick method of coating large numbers of articles.

The uniformity of the coating depends on the rate of withdrawal, and a rate of 2 to 4 inches (50 to 100 mm) per minute is usually satisfactory. It is necessary to adjust the viscosity of the paint and the rate of solvent evaporation, so that the film sets fast enough to ensure adequate thickness at the top of the article but slow enough for flow to ensure a uniform coating. Paint viscosities are low and many pigments tend to settle. Uniformity of the paint is maintained by pumping round an external circuit.

After removal from the tank, articles are allowed to drain and the excess paint is returned to the tank. Loss of solvent, and the fact that oxidizing types of paint are often employed, results in an increase in viscosity in the tank. The latter has to be checked daily for viscosity and solids content, and rectified as necessary.

A disadvantage of dipping is the quantity of paint required and, in large installations, the attendant fire risk. The risk is, of course, absent when the process is used for the application of water-soluble primers.

Hand-dipping allows rapid immersion and withdrawal, but bubbles and 'tears' can be troublesome. The article is then allowed to drain for about 15 minutes before stoving. Advantages of hand-dipping are speed, low paint loss, and low cost.

Slow dipping involves withdrawal at about 1 inch per minute and gives superior results. The paints used can be of higher viscosity than in other forms of dipping and produce very uniform films. This method is used for brush and tool handles which are dipped in high-viscosity cellulose nitrate lacquer. The articles are fixed in a suitable jig.

Automatic dipping The articles are suspended from a conveyor which dips over the tank and later rises. The excess paint drains on to a chute and is returned to the tank while the articles, after a 'flash off' period on the conveyor, pass through a camel-back (p. 367) or other type of oven. The viscosities of the paints used are of the order 0.7–1.0 poise at 25°C (77°F) so that bath contents must be kept in motion to avoid pigment settlement. This is achieved by pumping through an external circuit.

With articles which tend to retain drops of paint on the lower edges, the excess paint can be removed by the Ransburg electrostatic de-tearing process. The article is earthed (through the conveyor) and, after initial drainage, is suspended for a short time (20 to 100 seconds) over a negatively charged grid at 90 000 to 100 000 volts. Induced positive charges on the drops cause them to fly to the grid, leaving a uniform coating of paint on the article. With this process it is possible to use paints of slightly higher viscosity.

Rotational dipping, in which the article is rotated while immersed in the paint and while draining, is used for articles of complicated design or containing pockets from which the paint cannot otherwise drain.

Slipper-dip is used to apply an extra coat of paint to the lower half of car bodies. The latter are passed through the tank with only the lower half immersed.

Defects A number of defects can appear in dip coating; these have been summarized elsewhere [4]. The two most common are the following.

Fat edge and 'tears' These can be caused by poor flow properties, indicating an incorrect solvent balance. Alternatively, removal from the dipping tank may have been too rapid.

Excessive drainage from the top of the article due to excessive flow. A slight degree of thixotropy can be introduced by means of a Bentone [2] or hydrogenated castor oil.

Types of paint employed Conventional stoving enamels of the medium/short alkyd and alkyd/amino types are widely used. Paints thinned with trichlorethylene offer certain advantages in non-inflammability, excellent solvent power, and high evaporation rate. The latter allows higher rates of production with fewer fat edges and drips. Although trichlorethylene is highly toxic, this hazard has been overcome by design of suitable plant.

Items coated by the dipping process include parts for agricultural and other implements, tools, and electrical components (with impregnating varnish). Dipping is also employed for the phosphate pretreatment of steel, particularly in the car industry.

Flow coating

This can be regarded as a variation of the dip process. The paint is either flowed over the objects and the excess returned for re-use, or the paint is directed at the objects from a number of strategically placed nozzles. In the latter case the objects are conveyed through a tunnel in which the paint application is carried out. Any drops which collect on the lower edges of the objects are removed by electrostatic de-tearing.

As a result of aeration the films may contain air-bubbles and other imperfections, and the objects are therefore passed through a chamber of solvent vapour, when the film flows out to a uniform coating. The articles are allowed to flash off, and are then passed into the stoving oven.

Flow coating is used for objects of intricate shape, or which are large in comparison with the area to be painted. For this reason it is used for agricultural machinery. The jets can be so positioned that every part of the object is coated.

Compared with the dip process relatively small amounts of paint are required, but solvent losses are much greater.

Curtain coating

This is, in effect, a refined form of flow coating. It is used for large flat panels, which may have raised mouldings, of wood, plywood, hardboard and similar materials. The panels are carried on a conveyor through a curtain of paint which is formed when the paint, from a pressure head, flows through a V-shaped slot which runs the width of the machine. The flow is controlled by the width of the slot, which can be varied from 0·1 to 5 mm and is adjusted to give the required film thickness.

Two-component systems can be applied by the use of two curtains about 3 inches apart. The film remains sufficiently fluid for the two components to mix.

Roller coating

Application of paint is by transfer from rollers to one or both sides of flat materials such as strip metal, tin-plate sheets, and boards of various types. The paint is transferred through systems of rollers to the application rolls (Fig. 20.1). The thickness of the film is controlled by the gap between the feed and application rolls.

The basic requirements of the paint are high opacity, for one-coat coverage, good flow, and correct solvent balance. The solvent must not evaporate too slowly or the high production rates for which this process is well suited will not be achieved.

A wide variety of paints are used, including short-oil alkyds, alkyd/amino combinations, vinyl co-polymers, and modified epoxies. A high proportion of high-boiling aromatic is included in the solvent mixture.

In 'reverse' roller coating the application roll rotates in the opposite direction to the movement of the sheets. The shearing action at the line contact results in a more uniform film. It is particularly useful in applying filler or primer to absorbent boards such as plywood and chipboard, where the filler is forced into the surface and a smooth finish is obtained.

Losses and wastage in roller coating are extremely low.

Coil coating

The coating of metals in the form of continuous strip is now well established. The strip, received in the form of a coil, is subjected to a sequence of operations, namely uncoiling, degreasing, pretreatment, roller coating with paint, cure by stoving, cooling and re-coiling. Aluminium and steel strip up to 5 ft (1·52 metres) wide are coated at speeds up to 600 ft (213·4 metres) per minute [5]. The dry-film thicknesses of the paint are of the order of 25 to 35 micrometres.

Obviously a very high performance standard is required of the paints used. The major requirements are as follows.

1 Outstanding flexibility and adhesion to withstand forming operations.
2 High opacity, since one-coat systems are often employed.
3 Satisfactory inter-coat adhesion if more than one coat is applied.
4 A rapid stoving schedule (usually less than 60 seconds) without impairment of properties.

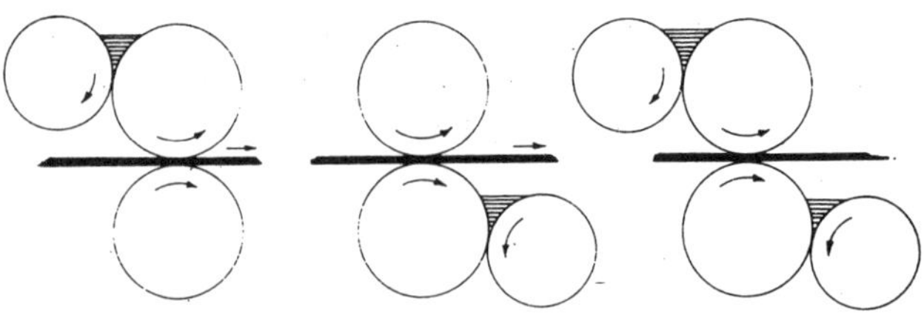

Fig 20.1 Roller coaters

Types of paint used include alkyd/amino combinations, modified alkyds, plastisols, organosols and polyvinyl fluorides. The use of aqueous coil coatings offers certain attractions, not the least being freedom from toxic and inflammable solvents. Their merits are discussed by Percy and Nouwens [6].

Coated metal strip has a wide field of application, ranging from cladding for buildings to office filing cabinets and bottle tops.

Tumbling or barrel coating

This method is used to coat large numbers of small and irregular objects such as screws, nuts, clips, etc. These are placed in a barrel inclined at about 30 degrees to the vertical, and the precise amount of paint to coat the articles is added. The amount of paint required is determined accurately on a pilot run with a smaller number of objects.

After tumbling, the objects are tipped into a wire basket to flash off. If lacquered, they will dry rapidly, but oleoresinous or japanned finishes are stoved. The conventional alkyd/amino stoving enamels are unsuitable for this method of application.

Centrifuging is used for small objects such as artificial jewellery which would suffer damage in a barrel. They are placed in a wire basket, dipped, and finally centrifuged to remove surplus paint.

SPRAY METHODS

Conventional spray application

The general principle in spray application is atomization of the paint into a fine spray which is directed on to the object. A number of methods have been devised to perform this, and the earliest was the use of compressed air with a spray gun.

Spray guns

For small-scale work the paint is contained either in a cup fitted on top of the spray gun so that the paint feed is by gravity, or in a cup below the gun, from which the paint is drawn by suction. These guns are very useful if frequent colour changes are involved.

For large-scale and production work a continuous supply of paint to the gun is required. The paint is contained in a pressure vessel connected to compressed air and fed to the gun through pressure hose. A cut-away section of this type of gun is illustrated in Fig. 20.2. The paint leaves the needle valve and is mixed with the compressed air from jets surrounding the valve. Mixing of the paint and air can take place either outside or inside the tip of the gun, and this is determined by choice of gun fittings appropriate to the type of paint to be sprayed. A trigger mechanism controls both paint flow and compressed air and is arranged so that the air flow precedes the paint.

Operation Apart from choice of the correct gun fittings, important factors are paint viscosity and air pressure. The viscosity of the paint influences the degree of atomization and the film properties. If too high, a coarse spray and

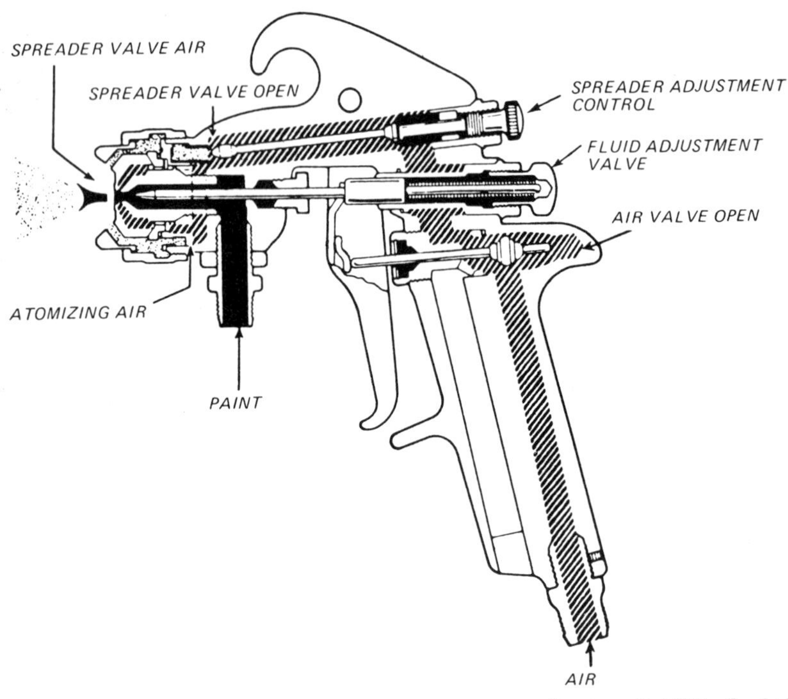

Courtesy: DeVilbiss Co. Ltd

Fig 20.2 Section through air-assisted spray gun

orange peel film will result. If too low, it can give dry spray and a thin unsatisfactory film. The optimum spraying viscosity of a paint is characteristic of the type and can vary appreciably from one type to another.

The best air pressure is usually determined by preliminary tests, and it is essential to use the lowest pressure that will give adequate atomization. Pressures higher than this can lead to defects (*below*). The majority of paints can be sprayed at 40 to 60 lb/sq. in, but certain types require higher values.

The nature of the paint film is also determined by the distance of the gun from the surface. This should be about 9 to 12 inches and the gun should be held at right angles to the surface; movement should be in parallel lines, each overlapping the preceding one by 25 to 30 percent.

Spray booths A large volume of air passes through a conventional spray gun and there is, inevitably, a certain amount of overspray (the amount depending on the geometry of the article) together with rebound from the sprayed surface, which gives rise to 'spray fog'. To remove this paint-laden air the spraying operation may be carried out in a booth with open front and exhaust at the rear, so that the overspray is carried away as soon as it is produced. There are two general types of spray booth – 'dry' and 'water-washed'.

Dry booths The contaminated air is removed by a fan and then scrubbed with water before release to the atmosphere. Baffles are usually fitted in front of the fan, but these, as well as the walls and roof, become coated with paint very rapidly and a fire hazard is created. To eliminate this risk it is usual to line the booth with paper or other strippable coating.

Water-washed booths In the smaller booths the extracted air is drawn through a cascade of slightly alkaline water at the back of the booth. The water/paint mixture is treated to agglomerate the paint, filtered, and the water recirculated. Booths in which large items are sprayed have floor gratings through which the air is exhausted. Filtered air enters through the roof. The exhausted air is scrubbed by passage through a series of water jets before release to the atmosphere. The water/paint mixture is treated as described above.

Spray booths are available, some incorporating special features, for handling articles of differing sizes and shapes [7].

Defects The most common defects occurring in spray painting are as follows.

'Orange peel' appearance of the film. This can be due to poor technique, incorrect solvent balance, or too high paint viscosity.

Sagging indicating an excessively thick film. Causes are the gun being held too close to the work, or the use of excess high-boiling solvent, leading to slow set.

Dry spray The surface is rough and the paint droplets have not coalesced to a smooth film. This can be due to excess of very volatile solvents, to too high pressure, or to holding the gun too far from the surface.

Hot spray application

When a paint is heated the viscosity is reduced, and in this state it can be sprayed at higher solids content, i.e. containing less solvent. It is not necessary to alter the basic composition of conventional spray types except to reduce the thinner content. The viscosity, measured at 25°C (77°F), will be influenced by a number of factors, but generally it is in the region of 2 to 3 poise.

The paint is heated in a special container and supplied to the gun at 60° to 71°C (140° to 160°F). Coatings of 2 to 4 mils dry-film thickness can be applied in one operation and this results in considerable savings in labour cost.

The Bede Airless Hot Spray is widely used and embodies the advantages of both systems. The pressures employed (400 to 600 lb/sq.in) are considerably lower than those required by the cold airless method (*below*).

Airless spray application

In this process no air passes through the spray gun but pressure is applied to the paint. The fluid pressure at the gun orifice is of the order of 1800 to 3000 lb/sq.in. (depending on the properties of the paint) and, on release, the sudden drop in pressure 'atomizes' the paint, the droplets of which are carried to the surface by their momentum. The orifice size is chosen to suit the consistency of the paint. Diameters range from 11 mils (for lower consistencies) to 21 or 23 for high consistency materials. Pressure is applied to the paint by a pump which may be operated by electricity or by compressed air. In the latter case the fluid (paint) pressure can be calculated from the expression

$$\text{Fluid pressure} = \text{air pressure} \times \text{pump ratio.}$$

Heating the paint (see Bede Airless Hot Spray process, *above*) assists the atomization and produces a finer spray. It also reduces the setting time of the film.

The use of high pressure enables high-build or thixotropic paints to be sprayed to give considerably thicker coats without sagging. The types of paint best suited to use as high-build coatings are the non-convertible (chlorinated rubber and vinyls) and chemically cured (two-pack, e.g. epoxy) types. It should be borne in mind that solvent release will be slower from the thicker films and overcoating intervals are, therefore, longer. This does not apply, of course, to the solvent-free chemically cured (two-pack) coatings, and very thick coatings of these can be applied, the limiting factor being any tendency to sag.

A modified airless spray system is marketed under the name 'Airmix' [8]. By introducing a proportion of air the atomising pressure can be reduced from 1000–3000 psi to 200–500 psi. The slower speed of the particles results in reduced overspray and rebound and, it is claimed, produces a finer finish.

Oxidizing media, such as alkyds and oleoresinous binders, depend on absorption of oxygen for cure, and in consequence there is a limit to the thickness of film which will cure satisfactorily. With the high-build types, sufficient time must be allowed for solvent release and cure by oxidation before the film is overcoated.

The advantages of airless spray are summarized below.

1 Reduced overspray and rebound, making the process particularly suited to large areas.
2 High application rates.
3 Permits the use of high-build coatings with the result that fewer coats are required to achieve a specified film thickness, with consequent savings in labour costs.

Electrostatic spray

This process utilizes the fact that if atomized paint droplets are given a negative charge in the vicinity of a positively charged object, the droplets are attracted to, and form a uniform coating on, the exterior surfaces of the object.

This process was patented by Ransburg in the USA and named the 'Ransburg No. 1 Process'. The metal (or other conducting) articles to be painted are usually attached to a conveyor which is earthed. A system of electrodes or wires, well insulated and carrying a high negative potential of 150 000 to 200 000 volts, is fixed in a position to create an electrostatic field between the electrodes and the article. Paint droplets, produced by a conventional automatic gun, are injected into the field, whereupon they acquire a negative charge and are attracted to the earthed articles. Areas facing the electrodes are coated first with an insulating layer of paint. The charged droplets are then deposited on areas further away and, with many articles comprising cylindrical or rounded shapes, a 'wrap-round' effect results in the entire surface being coated from one side. Very few paint droplets are lost and the operating costs are low. On the other hand, this process involves a high initial outlay and requires very careful control of paint viscosity and spray gun setting.

In the 'Ransburg No. 2 Process' an electrostatic field is created between the earthed articles and the gun. The paint is distributed from the edge of a disc or bell rotating at about 90 rpm and which is given a negative potential of about 90 000 volts. The paint, at a viscosity of 0·5 to 0·8 poise at 25°C (77°F) and with a resistivity within the range 400 to 500 kΩ, is fed to the centre of the disc by a pump and is thrown outwards and atomized as it leaves the edge. The charged droplets are attracted to the earthed objects on which a uniform wrapround coating is deposited. Many objects, except those which are large or flat, can be completely coated from one side.

In production units (Fig. 20.3) a horizontal disc is made to reciprocate vertically to cover large areas; large or difficult objects are taken round the disc and also rotate on their own axes to expose all surfaces to the gun.

A number of variations of this basic process have been produced. In some versions, low-pressure air is used to assist the atomization, and under these conditions the voltage can be reduced. This air-assisted method has been found very suitable for electrostatic spraying of water-thinned paints.

Electrostatic spraying shows very considerable savings in paint compared with conventional spray processes, since there is virtually no overspray. The automatic nature of the process leads to very uniform results. Disadvantages

Fig 20.3 Electrostatic disc unit. Courtesy Ransburg UK Ltd.

are its inability to coat the insides of articles and a tendency, unless conditions are very rigidly controlled, to give hazy or orange-peel effects.

For smaller or special types of work an electrostatic hand gun can be used, a modern form of which is illustrated in Fig. 20.4.

Paints for electrostatic spray

A number of types can be applied by this method, but a correct solvent balance is highly important for the following reasons.

1 The evaporation rate must be slow enough for the charged droplets to reach the surface in a sufficiently fluid condition to flow out to a smooth film, but fast enough to avoid sagging. Furthermore, dry deposits must not form on the rotating discs or bells.

2 The resistivity of the paint must be sufficiently low to enable the paint droplets to acquire the maximum charge. A mixture of non-polar and polar solvents is usually employed to give a resistivity of the order of 400–500 kΩ at 20°C (68°F). Electrostatic spray is used also for the application of powder coatings (pp. 365)..

Silk screen coating

This technique is used for the application of lettering and decorative patterns to surfaces where sharp definition is required and repetitive work is involved.

The silk or other mesh is stretched tightly over a wooden frame and the stencil

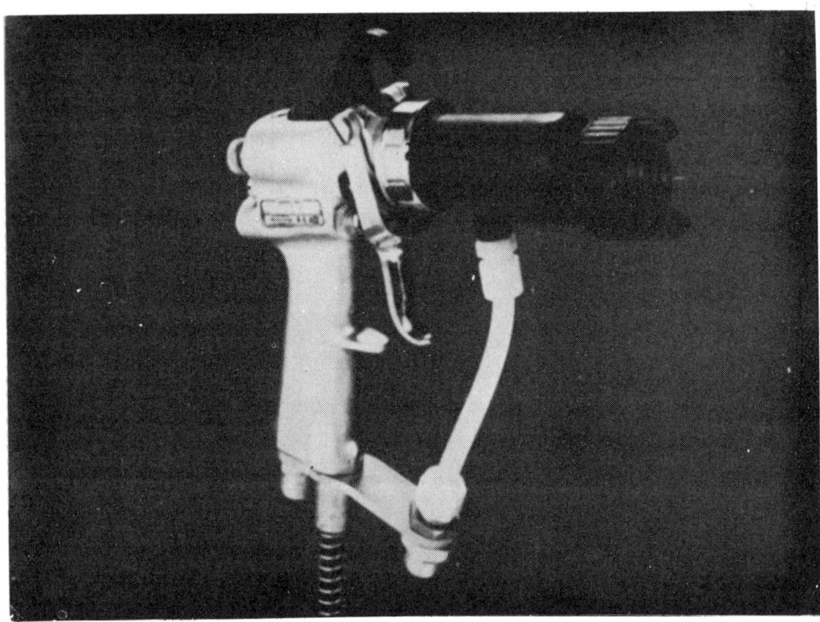

Fig. 20.4 Electrostatic hand gun

applied to the underside with water-soluble adhesive. The screen is placed on the surface to be treated and the ink or paint applied with a rubber or plastic squeegee.

The flow properties of the inks or paints must be controlled carefully and the types used include cellulose nitrate, drying-oil alkyds, and oleoresinous media. A thin coat is applied so pigmentation has to be high. In recent years thixotropic paints have been used. They work easily and give very sharp definition.

ELECTRODEPOSITION (also known as 'Electro-coating')

General principles

The majority of emulsions consist of a negatively charged dispersed phase, often stabilized by a water-soluble colloid. If two electrodes are immersed in an emulsion and potential difference applied, the negatively charged dispersed species will migrate to the anode (a process known as electrophoresis) where the charges are neutralized, and deposition or coagulation of the dispersed species takes place.

In the early part of this century this process was applied to the coagulation and deposition of crude rubber from the natural latices.

Later, in the 1930s, the deposition of can coatings from emulsified oleoresinous varnishes was investigated, but the process proved to be too slow for production lines.

The rapid development of electrodeposition as a viable commercial process followed the introduction of the water-soluble and the less soluble water-dispersible resins.

Until relatively recently electrodeposition was applied exclusively to the coating of metallic anodes, a process not without difficulties, as will be discussed later. Research on water-soluble resins has resulted in the production of types which can be deposited on the cathode and this process has proved to be so much more efficient than the anodic process that it has virtually replaced the latter. The anodic process is now largely of historic interest but still merits a brief description.

Anodic deposition

A range of resins, which includes modified alkyds, epoxy esters and maleinized oils, can be made water-soluble by introducing carboxyl groups followed by neutralization with ammonia or other base. In dilute aqueous solution these water-soluble resins ionize, giving a large anion embodying the resin complex (denoted by R) and the ammonium (or other) cation:

$$R.COONH_4 \rightleftharpoons R.COO^- + NH_4^+$$

When electrodes are immersed in the solution and a potential difference applied, electrophoresis takes place, the large anions migrating to the anode, where the following reactions are considered to take place.

1 Decomposition of water, yielding hydrogen ions and oxygen gas:

$$\begin{aligned} 2H_2O - 2e^- &\rightarrow 2OH^- + 2H^+ \\ 2OH^- &\rightarrow H_2O + [O] \\ 2[O] &\rightarrow O_2 \uparrow \text{ (gas evolved)} \end{aligned}$$

2 Neutralization of the charged anion by hydrogen ion:

$$R.COO^- + H^+ \rightarrow R.COOH$$

The spongy deposit of resin is then consolidated by electro-osmosis (movement of water outwards from the mass).

Some anode metals tend to dissolve during the process, giving cations which react with the resinous anions. For example, a bare steel anode will yield ferrous ions and these react to give a coloured product:

$$2\,RCOO^- + Fe^{2+} \rightarrow (R.COO)_2Fe$$

This would discolour a white coating. Phosphated steel and zinc-coated steel give no discoloration. The former yields no ions, whereas those of zinc (Zn^{2+}) are colourless.

Discolouration of product due to dissolution of steel anodes was one of the main drawbacks of the anodic process. Evolution of oxygen gas also caused problems. On the other hand, the water-soluble resins were cheaper to produce than those for the cathodic process.

Cathodic deposition

The article to be coated forms the cathode. The bath solution contains pigmented resinous cations or dispersed polymer. Types of binder containing amino/amide groups which are solubilised by acid or dispersed non-ionic resins stabilised by cationic surfactants are described by Majia [9]. Precipitation or coagulation of the resin at the cathode results from the alkaline conditions at the surface.

Reactions at the cathode surface

1 Electrolysis of water, producing hydroxyl ions and hydrogen gas:

$$\begin{aligned} &2H_2O \rightarrow 2H^+ + 2OH^- \\ &2H^+ + 2e^- \rightarrow 2[H] \rightarrow H_2 \text{ (gas)} \end{aligned}$$

2 Neutralization of charged cation (or cationic surfactant) by hydroxyl ions:

$$R - \underset{\substack{|\\ H}}{\overset{+}{N}} - R_2 + OH^- \rightarrow R - NR_2 + H_2O$$

The range of resins employed in cationic depositions is more restricted than that used in the anionic process. Self-cross-linking properties can be built on to 'backbones' of epoxy/amine, epoxy/acrylic and epoxy/isocyanate resins, but cross-linking agents may be included in the bath solution.

The advantages of cathodic electrodeposition are summarized below.

1 No dissolution of electrode.
2 No staining of pale shades.
3 Gives a hard dense coating with excellent corrosion resistance. Superior to anodically deposited films.
4 High utilisation of paint materials and no environmental contamination.
5 Excellent throwing power and hence ability to coat recessed areas difficult to treat by other methods.

POWDER COATINGS

Metal articles can be coated with certain resins by applying the powdered resin to the surface, followed by heating to melt the resin so that it flows and forms a uniform coating on the article. The method eliminates the fire and toxicity hazards associated with the use of solvent-based paints and also shows savings in cost.

There are three methods in use for application of the powders, namely (a) fluidized bed, (b) electrostatic spray, and (c) flame spraying.

Fluidized bed

The equipment is illustrated diagramatically in Fig. 20.5.

The rectangular container A is fitted with a porous plate B and inlet C for low-pressure air. The powdered resin rests on plate B. Compressed air is admitted at C, diffuses through porous plate B and then rises through the resin. The mass of powdered resin now behaves as a fluid and is said to be in a fluidized condition.

The metal article to be coated is heated to a temperature above the melting-point of the resin and is then dipped in the fluidized bed. Resin particles in contact with the metal become attached by fusion, and the article is removed and passed through a stoving oven, when complete fusion of the resin takes place and a uniform coating is formed. The fluidized bed tends to give thicker coatings than the other methods.

The electrostatic fluidized bed eliminates the need to dip articles in the

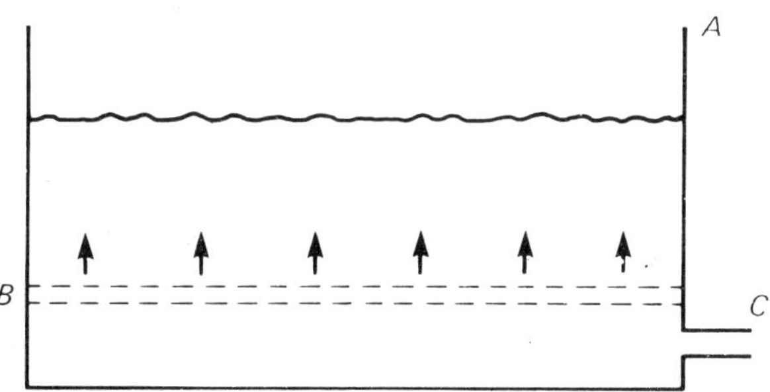

Fig 20.5 Fluidized bed for powder coating of metals

fluidized powder. The resin particles are given a high negative charge by means of electrodes in the container, whilst the articles are earthed. These pass over the bed of powder (or through a cloud of powder) when resin particles fly to the surface of the earthed article and are held by electrostatic attraction. The article then passes through a stoving oven to melt the resin into a uniform film. This method is suitable for continuous coating of small articles of simple shape.

Electrostatic spray

The resin powder is blown through a special type of gun in which the particles receive a very high negative charge. The article to be coated is earthed, so that the resin particles are attracted to the surface and adhere by electrostatic attraction. The charges are lost only slowly and the article, with adherent resin, is then stoved to melt the resin into a continuous coating.

In electrostatic spraying, resin dust in the air at certain concentrations can create an explosion hazard, and precautions must be taken to ensure that no source of ignition is present. The principal safety precautions to be taken in the application of powder coatings by electrostatic spray are set out in a publication for the Paintmakers' Association [10].

The toxic hazard of resin dust is mainly irritation.

Flame spraying

The resin powder is blown through the gun by compressed air. The particles are melted in a high-temperature flame and adhere well to the surface of the article. The process has been used widely with epoxy powders. Materials employed in powder coatings are discussed in Chapter 22.

METHODS OF CURE

The term 'cure' refers to the conversion of a fluid or resinous paint binder into a hard, tough and adherent film by intermolecular reaction. This reaction, which can take one or more of several forms according to the nature of the binder and the temperature, occurs in convertible coatings only. Non-convertible coatings, of which chlorinated rubber and cellulose nitrate are examples, dry by solvent evaporation only and no cure or intermolecular reaction takes place.

Types of cure

1 **Auto-oxidation** at ambient temperature. Drying alkyds, oleoresinous binders, epoxy esters and urethane alkyds contain unsaturated linkages and undergo oxidation at these points. Free radicals are produced which promote cross-linking in the film.

2 **Chemical reaction** between two components (two-pack materials) at ambient temperature or, more rapidly, by application of mild heat. Types of binders employed include epoxies cured by polyamides, polyamines or isocyanates and polyesters cured by isocyanates (polyurethanes) or styrene.

3 Application of heat (stoving) The object of stoving is to bring about the cure or hardening of a paint film in a relatively short time. The cure consists generally of cross-linking of the resins, but oxidation also occurs when medium or short-oil oleoresinous or drying alkyd films are stoved. Stoving schedules can vary appreciably; for example, alkyd/amino systems are stoved at 120°C (248°F) for ½ to ¾ hour, whereas silicone resins require 200°–230°C (392°–446°F) for 1 to 2 hours. The main types of oven in use are as follows.

Box oven This is the simplest form and can be heated indirectly or directly by gas, electricity or steam. It requires adequate ventilation and is suited to batch processes.

Conveyor ovens are used for continuous production. They can be heated by convection or employ radiant heat (infra-red).

Convector ovens The most familiar convection oven is the 'camel-back' (Fig. 20.6) which has open ends somewhat lower than the central section which is at the highest temperature. It is heated indirectly by hot air. In this oven, heating and cooling of articles takes place more gradually than in other types.

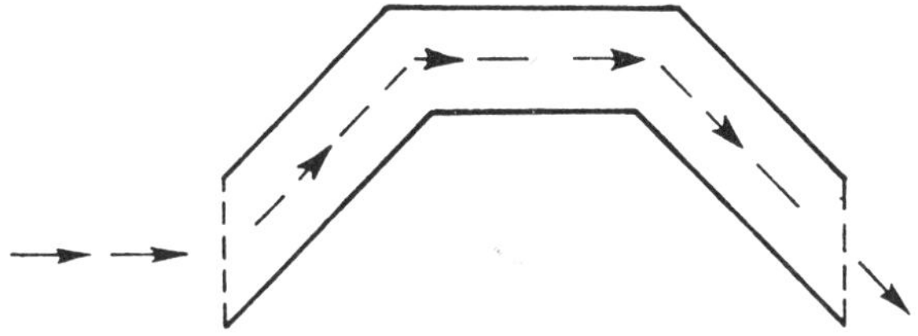

Fig 20.6 Camel-back oven

Radiant heat (infra-red) ovens may be heated by gas or electricity. Gas type ovens are built from a series of flat or curved panels which are heated by gas flames impinging on the backs. The panels act as dull emitters of radiant heat, and the surfaces of objects passing between them reach a high temperature very quickly. The time required for stoving is very much lower than in the convection type ovens, but the radiant heat may not reach screened areas of articles of complicated shape. These are best stoved in convection ovens.

Electric infra-red ovens are composed of a series of heating lamps backed by reflectors and often arranged to form a tunnel. The principle of operation is the same as that of the gas types.

Radiation curing

Where the cure of paint films is effected by a 'free radical' mechanism it can be accelerated by exposure of the wet film to either a beam of electrons or to ultraviolet radiation. In both cases cure is effected in a matter of seconds but, as described below, there are important differences between the two. Both methods show considerable energy savings over conventional methods of cure [11].

Electron beam curing

The fact that polymerization of certain types of monomer could be achieved by radiation has been known for many years, but it was only in the 1960 to 1970 decade that the use of a beam of electrons became a commercial possibility.

This period was one of great activity in the field and work in later years has been concerned mainly with improvements in plant and a study of the polymerisation reactions involved. Briefly the plant consists of a thermionic cathode or other source of electrons, a tube through which the electrons are accelerated and a thin titanium foil 'window' through which they emerge. They then impinge on the paint film but it is important that the space above the film be oxygen-free since this gas is known to be an effective scavenger of free radicals.

Using voltages of 300 to 500 kV, the emergent electrons possess sufficient energy to penetrate both clear and pigmented coatings of conventional thickness and to liberate free radicals which promote cross-linking and cure. Types of media which have been cured satisfactorily include unsaturated polyesters, acrylics, epoxy acrylics, and PVC plastisols.

Electron beam curing is rapid and low in energy demand. A drawback is the high cost of installation. X-rays are produced at the same time as the electrons, so that the equipment must be adequately screened. The application of electron beam and ultra-violet curing to inks has been described by Pelling [12].

Ultraviolet curing

Ultra-violet radiation of the desired wavelength is produced by mercury vapour lamps of which there are several types. Medium pressure mercury arc lamps are widely used and a comparison with other types is made by Pelling [12].

UV radiation cures, in a similar manner to electron beam, by the release of free radicals and the process has worked satisfactorily with clear coatings. Pigmentation, however, has presented problems and the efficiency of cure appears to be related to the refractive indices of the pigments. A high refractive index means poor penetration by the UV radiation. (Electron beams are not affected in this way.) The difficulty has been overcome by the use of photoinitiators which furnish the necessary free radicals when irradiated. A number of compounds has been used for this purpose among which are the aminobenzoates [13], thioxanthones, benzoin ethers, benzophenone and acetophenone derivatives. The effect of various pigment types on the efficiency of UV curing and an explanation of the mechanism by which photoinitiators work is contained in a paper by Kaden and Eckert [14].

Ultraviolet curing is now used extensively, the fields of use including wood finishes, metal decorating, and inks for various purposes. The most commonly used vehicles are unsaturated polyesters, acrylates either alone or in combination, and modified epoxies.

Initial outlay is lower than for electron beam curing; energy consumption also is low.

Radiation methods of cure are attractive in the following respects.

1 Films are solvent-free, hence no pollution problems.
2 Cure is effected in seconds.

3 Energy consumption is low.
4 Sensitive substrates can be treated.

Nevertheless, they have certain limitations. They are suited only to flat surfaces and, particularly in the case of the electron beam, expensive to instal.

References to Chapter 20

[1] *Hess's Paint Film Defects*, 3rd edn, pp. 46, 50, 76. Chapman & Hall, London, 1979.
[2] NL Chemicals (UK) Ltd, Wilmslow, Cheshire.
[3] Hurst, E. A. and Goodier, J. H., *Painting and Decorating*, 9th edn, p. 89. Charles Griffin, 1980.
[4] *Hess* (Ref. 1), p. 30.
[5] Tatton, W. H. and Drew, E. W., *Industrial Paint Application*, 2nd edn, p. 62. Newnes-Butterworth, London, 1971.
[6] Percy, E. J. and Nouwens, F., Aqueous coil coatings. *JOCCA*, 1979, **62**, 392.
[7] Binks-Bullows Ltd, Walsall, Staffs. Berridge Engineering Ltd, Beeston, Notts.
[8] 'Airmix' system (Kremlin Spray Painting Equipment Ltd, Slough, Berks).
[9] Majia, J., *Metal Finishing*, 1980, **78**, 52.
[10] Paintmakers' Association of Great Britain Ltd, *Code of Practice-Application of powder coatings by electrostatic spraying*, 3rd imp., 1975.
[11] O'Hara, R., Ultra-violet and electron beams; energy efficient methods of curing. *PPCJ*, 1981, **171**, No. 4038, 11.
[12] Pelling, D., Accelerated drying techniques. *JOCCA*, 1987, **70**, 31.
[13] Ethyl 4-(dimethylamino) benzoate (Lambson Ltd, Castleford, W. Yorks).
[14] Kaden, V. and Eckert, P., Colouration of radiation curing systems. *JOCCA*., 1985, **68**, 251.

21

Paints for industrial structures and industrial buildings

This chapter contains descriptions of most of the general types of paints used on industrial structures and buildings. Although protection is generally the first priority and the majority of the paints are functional rather than decorative, the appearance of the finished structure is also of importance.

Many industrial structures are difficult to paint, often as a consequence of difficult access or the fact that continuous processes are taking place. The paints employed, therefore, are designed to give maximum performance and durability, so that intervals between repaints are as long as possible.

Primers

Primers for metals have been described in Chapter 17; those for woodwork, plaster, exterior masonry, render, etc., in Chapter 18.

UNDERCOATS FOR STRUCTURAL STEELWORK

The type of undercoat depends on the nature of the paint system chosen and this will be determined by the exposure environment and the life required. Often several coats are superimposed in one system to achieve the desired coating thickness.

Oleoresinous undercoats are employed to a limited extent, pigmented mainly with micaceous iron oxide (MIO), in systems for the protection of bridges and other steel structures in moderate exposure conditions. They are generally applied over the PVB type of blast primer and can be self-overcoated or treated with decorative gloss finish. A typical oleoresinous/MIO undercoat, based on a tung oil/phenolic or tung oil/modified phenolic varnish, is given below.

Oleoresinous/MIO undercoat

	% by weight
Micaceous iron oxide	48·0
Wet ground mica	7·0

Talc	5·0	
Bentone B34 [1]	1·4	pregelled
IPA 99%	0·5	pregelled
White spirit	2·0	pregelled
Tung oil/phenolic varnish (60% solids)	22·4	
White spirit	13·0	
24% Lead naphthenate	0·3	
6% Cobalt naphthenate	0·2	
Methylethylketoxime	0·2	

Specific gravity, 1·65; solids by volume, 46%; PVC, 52%.

Undercoats of this type, containing a high percentage of lamellar micaceous iron oxide, are highly resistant to penetration by moisture and are sufficiently weather-resistant to be self-overcoated and used as finishes. A brighter and more decorative finish is obtained by the addition of a small quantity (1 to 2 percent) of non-leafing aluminium pigment.

Greys and other light colours can be produced by the introduction of rutile titanium dioxide and other pigments. If a glossy decorative finish is required, the MIO undercoat can be overcoated with alkyd-type undercoat and finish or with alkyd finish only.

Other undercoats are designed for use with specific finishing systems and are described under the relevant headings.

FINISHING PAINTS: (1) NON-CONVERTIBLE COATINGS

Chlorinated rubber

The general properties of chlorinated rubber and the plasticizers employed have been described in Chapter 10.

The use of chlorinated rubber paints for the protection of both structural steelwork and concrete structures has increased considerably in recent years, mainly on account of the following attractive properties.

1 Fast drying. The film is ready for overcoating as soon as the solvents have evaporated. This can be 2 to 6 hours according to the thickness of coating (high-build types take appreciably longer), the volatility of the solvents, and the ambient temperature.
2 Good inter-coat adhesion. As a non-convertible coating, there is physical fusion between coats, and no difficulties are encountered in inter-coat adhesion. Furthermore there is no upper limit to overcoating intervals.
3 High resistance to moisture, alkalis, acids, a wide range of chemicals, and many solvents.
4 Good durability, resulting in maintenance intervals which are considerably longer than those with most conventional coatings.
5 One-pack systems with an indefinite storage life.

Chlorinated rubber films have the following weaknesses.

1 Difficulty with brush application. The application of chlorinated rubber primers to uneven or rough surfaces such as concrete is best carried out by brush, but brush application of further coats can result in 'pick up' due to softening of the primer. These further coats can be laid on with a full brush with no attempt to lay off, but this is wasteful and generally results in sagging on vertical surfaces. The difficulty can be mitigated to some degree by inclusion of a proportion of non-solvent, such as white spirit, which reduces the solvent power of the blend and increases the time required for softening of the primer coat.
2 Chlorinated rubber films are softened by animal fats and oils. In addition, they are dissolved by certain strong solvents, notably aromatic hydrocarbons and some esters and ketones.

Primers for chlorinated rubber systems

Alkaline surfaces The surfaces of new concrete, cement screeds and cement render are highly alkaline as well as absorbent. An effective primer/sealer for such surfaces can be prepared on the following lines:

Chlorinated rubber primer/sealer (for alkaline surfaces)

	Parts by weight
Alloprene R 20 [3]	23
Cereclor 52 [3]	12
Cereclor 48 [3]	4
High aromatic solvent (e.g. Aromasol H – ICI)	50
Methoxypropyl acetate	1
White spirit	10

Specific gravity, 1·02; volume solids, 28%.

Brushing is the preferred method of application so that 'bridging' of cavities is avoided.

Steel Whilst chlorinated rubber primers containing conventional plasticizers adhere to two-pack etch primer films, the adhesion to one-pack etch primer and to bare steel tends to be erratic. A range of chlorinated rubber primers containing rust-inhibitive pigments are available for steelwork, but they are not used as widely as some other types.

The adhesion of chlorinated rubber primers to steel and non-ferrous metals is improved by the addition of resin, but the nature of the resin and the proportion used are of vital importance. A list of the resins which give good adhesion to steel when incorporated into chlorinated rubber primers has been compiled by Parker [2]. Certain long-oil alkyds have been found to give very satisfactory results; consequently primers for chlorinated rubber systems on steelwork are now frequently based on this mixture. The optimum ratio would appear to be two parts chlorinated rubber to one part long-oil alkyd (by weight). This type is illustrated by the following formulation:

White chlorinated rubber/alkyd primer (for airless spray application)

	% by weight	
Zinc phosphate	20·3	
Titanium dioxide	3·5	
Barytes	4·0	
Talc	5·3	
Alloprene R5 chlorinated rubber [3]	11·8	
Long-oil alkyd (65% solids)	9·1	
Cereclor 48 (plasticizer) [3]	2·8	
Solvesso 100 [4]	12·8	
Xylol	27·2	
Bentone 38 [1]	2·3	Pregel with
IPA 99%	0·6	proportion of xylol
24% Lead naphthenate	0·2	
6% Cobalt naphthenate	0·1	

Specific gravity, 1·29; solids by volume, 34%; PVC, 37%.

Although formulated for airless spray application, this primer can be brushed on small areas and can be thinned for use as a blast primer. A brushing type can be prepared by increasing the proportion of high boiling aromatic solvent.

Other primers which can be used under chlorinated rubber systems are epoxy/zinc phosphate (p. 301), epoxy zinc-rich (p. 301), and inorganic zinc silicate (p. 303).

Chlorinated rubber/MIO paints

Pigmentation of chlorinated rubber with lamellar micaceous iron oxide has been found to augment the already high moisture-resistance of the film and to improve the mechanical properties. The presence of the MIO also results in a reduced plasticizer demand. The following formulation for a high-build paint, which can be used both as undercoat and finish, has been suggested as a basis by Carter [5].

Chlorinated rubber/MIO, high build

	% by weight
Micaceous iron oxide	28·0
Barytes	6·3
Colloidal zinc oxide	0·7
Alloprene R10 [3]	14·5
Cereclor 52 [3]	5·7
Hydrogenated castor oil	1·2
Epoxidized oil	0·6
Xylol	34·0
Aromosol H [6]	9·0

Specific gravity, 1·38; solids by volume, 32%; PVC, 33%.

Paints of this type are widely used in the preservation of bridges and other steel structures which are exposed to severe environmental conditions.

Chlorinated rubber finishes

The formulation of finishes is, in the main, quite straightforward, but modifications can be made to meet special requirements. Table 21.1 illustrates two brushing-type gloss finishes and a high-build quality.

Table 21.1 Chlorinated rubber finishes: percentage composition

	White gloss	Black gloss	White high-build
Titanium dioxide, rutile	18·8	–	15·1
Carbon black	–	2·6	–
Micronized barytes	–	–	11·5
Alloprene R20 [3]	20·0	27·5	–
Alloprene R10	–	–	19·3
Cereclor 48 [3]	13·0	11·0	–
Cereclor 70	–	–	12·8
Cereclor 42	–	–	6·5
Hydrogenated castor oil	0·5	0·5	2·3
Xylol	–	3·0	27·2
High aromatic solvent [7]	39·6	47·5	5·3
White spirit	8·1	7·9	–
Specific gravity	1·19	1·04	1·36
Solids by volume	34%	29%	50%
PVC	16·6%	5·5%	17·6%

Formulations of this type allow appreciable latitude in the choice and amount of plasticizers (depending on the end use of the paint) and in the solvent blend. A range of solvent combinations for chlorinated rubber paints has been published by Parker [2].

Airless spray application The white high-build formulation above yields a paint of which the consistency is too high for airless spray application. The addition of thinners to achieve spraying consistency will result in a porous film. For airless spray the solvent should contain a proportion of a higher boiling ingredient such as Methoxy propanol, and the PVC must be increased. The following formulation, quoted by Parker [2], illustrates the type suited to airless spray.

	% by weight
Rutile titanium dioxide	12·5
Micronized barytes	12·5
Alloprene R10 [3]	12·5
Cereclor 70 [3]	7·4
Cereclor 42	4·1
Hydrogenated castor oil (modified)	1·9
Methoxy propanol	9·7
Xylol	39·4

Specific gravity, 1·24; solids by volume, 31%; PVC, 24%.

The use of modified hydrogenated castor oil requires heat (42° to 45°C) (107° to 113°F) for gelation. It can be gelled separately in a portion of solvent, but if the paint is made in a high-speed disperser, the heat produced can be utilized. The rise in temperature occurring in ball mills is usually insufficient to gel a modified hydrogenated castor oil.

Other chlorinated rubber paints

Paints for concrete swimming pools These pools may be constructed with or without a waterproof membrane.

Pools with waterproof membrane on the outer surface of the concrete. This can take the form of bitumen, a layer of which is laid down before the bottom of the pool is cast and a thick coating is applied to the outer sides before the 'in-filling' is done. Alternatively, the membrane may be thick plastic sheeting, care being taken to ensure that the joins of sheets are effectively sealed.

The purpose of this membrane is prevention of seepage of water outwards from the pool in periods of drought or into the pool if the water table rises above the bottom of the pool.

The concrete in these pools can be allowed to dry out thoroughly before painting. The surface of new concrete is highly alkaline (pH 11–12) and chlorinated rubber paints are used for their alkali resistance and impermeability to water. The concrete is given one or two coats (depending on the density) of chlorinated rubber primer/sealer, as described above, followed by one or two coats of chlorinated rubber finish. The following formulation is indicative of the type of finish employed.

	% by weight	
Rutile titanium dioxide	20·0	
Micronized barytes	9·8	
Alloprene R20 [3]	16·8	
Cereclor 48 [3]	11·2	
Bentone 34 [1]	0·5	Pregelled with proportion of solvent
IPA 99%	0·2	
High aromatic solvent [7]	32·0	
White spirit	9·5	

Specific gravity, 1·32; solids by volume, 36%; PVC, 27%.

Pools without waterproof membrane When a membrane is not used, moisture is able to diffuse through the concrete from either side. An impermeable paint film, such as that above, could be forced off by hydrostatic pressure from the outside, particularly when the pool is empty. A permeable paint film, however, will allow the moisture to pass through without becoming detached. Such a paint can be based on the above formulation but the PVC is increased to a little above the critical point by the addition of extenders.

Chlorinated rubber is used also in the types of paint listed below; these are described on the pages indicated.

Floor paints (p. 388)
Road-marking paints (p. 413)
Masonry finishes (p. 336)
Fire-retardant paints (p. 335)
Marine paints (p. 403)

Vinyl paints

The chemistry of the vinyl resins has been discussed on p.

Etch primers both one- and two-pack, are generally based on polyvinyl butyral resin and are described on p.

Vinyl coatings (primers, intermediate coats and finishes) are based on copolymers of vinyl chloride and vinyl acetate. Among the resins in common use are those manufactured by Union Carbide Ltd, and designated Vinylite VYHH, VYHD, VMCH, etc. The most widely used are the following.

VYHH The ratio of monomers, by weight, is approximately 87 parts vinyl chloride to 13 parts vinyl acetate. The solubility, even in strong solvent mixtures, is low; a mixture of equal weights of ketone and aromatic hydrocarbon dissolves only 15 to 20 percent (by weight) of the resin. Adhesion of the air-dried film to metal is poor, but the film exhibits high resistance to chemical attack.

VMCH is similar in composition to VYHH but is modified with one percent dibasic acid. The presence of the carboxyl groups results in good adhesion to bare metal but can result in 'feeding' with basic pigments. This resin is recommended for air-drying maintenance paints, underwater coatings and finishes for light alloys.

VMCC is specially modified to give good adhesion to clean metal surfaces.

VAGH is partially hydrolysed and, compared with VYHH, shows wider compatibility and improved adhesion. Cross-linking properties result from the presence of hydroxyl groups. It is compatible with various types of long-oil and short-oil drying alkyds, non-drying alkyds and certain urea- and melamine-formaldehyde resins with which it will cross-link. VAGH is claimed to be the only vinyl resin which develops good adhesion on air-drying on WP1 and other etch primers.

Formulation of vinyl paints

The vinyl co-polymers dissolve in ketones and certain esters such as propylene glycol derivatives to give solutions of low concentration. Aromatic hydrocarbons also can be incorporated. The paints usually contain a proportion of plasticizer such as tricresyl phosphate or Cereclor [3]. The volume of solids is generally low (20 to 22 percent) so that a thixotropic agent or a thickener is required to produce a reasonable film build. Intercoat adhesion of vinyl paints is usually satisfactory, but care must be taken in the choice of resins for coatings to be applied to bare metal. For this purpose VMCH is used alone or in admixture with VYHH.

Typical formulations

Vinyl finish for metal

	% by weight
Vinylite VMCH	16·5
Tricresyl phosphate	3·3

Rutile titanium dioxide	13·2
Methylisobutyl ketone	33·0
Xylol	33·0
'Anti-settle' [9]	1·0

Specific gravity, 1·02; solids by volume, 19%; PVC, 17·4%.

This formulation yields a paint which dries to a gloss finish; the gloss can be reduced, if desired, by the incorporation of neutral extenders such as silica or talc. The low volume of solids results in a thin film; but a high-build paint, giving greater film build, can be produced by the use of Bentone [1] or other thickening agent.

Intermediate coat (for use over etch primer)

	% by weight
Vinylite VAGH	17·4
Zinc tetroxychromate	9·0
Talc	2·7
Tricresyl phosphate	1·8
Methylisobutyl ketone	41·5
Toluene	27·6

Specific gravity, 1·0; solids by volume, 17%; PVC, 19·6%.

Properties and uses of vinyl coatings

They exhibit excellent resistance to chemicals, moisture and weather. Their solubility characteristics, however, result in paints of low solids content by volume and consequent low film built. As mentioned earlier, greater film thickness can be obtained by the incorporation of gelling agents or thickeners.

The adhesion of the resins on different substrates varies considerably, and the following guide is taken from the Union Carbide [8] literature.

Substrate	*Most satisfactory resin*
Clean smooth metal	VMCH, VMCC
Phosphated metal	VMCH, VMCC
Alkyd resin paint	VAGH
Vinyl butyral resin (etch primer)	VAGH
Concrete	VMCH, VMCC
Plaster	VMCH, VMCC

Vinyl coatings are used for the protection of steel, non-ferrous metals, concrete, and other surfaces exposed to very corrosive conditions. They are used also in certain types of marine paint and anti-fouling compositions.

Bituminous paints

Bitumens and bituminous paints have been considered in Chapter 16. Bituminous paints made on straight bitumen solutions give water-resistant films and provide excellent protection for steelwork under certain conditions. Their principal use is protection of underground and submerged steelwork where high-build types are often an advantage.

On exposure to weather, bitumen films undergo surface oxidation and hardening, while the underlying material remains soft. The surface then develops the familiar crazed pattern. The weather resistance of bitumen is improved by the introduction of drying oil, and still further improvement is conferred by long-oil alkyds. The following formulations illustrate the three types:

Bituminous black

	% by weight
80/90 Bitumen	60·0
High aromatic solvent [7]	20·0
White spirit	20·0

Specific gravity, 0·97
Solids by volume, 53%

Bituminous black, oil modified

	% by weight
Gilsonite	30·7
Mexphalt	7·3
Boiled linseed oil	9·7
High aromatic solvent [7]	18·7
White spirit	33·3
24% Lead naphthenate	0·2
6% Cobalt naphthenate	0·1

Specific gravity, 0·92
Solids by volume, 41%

Bituminous black, alkyd modified

	% by weight
80/90 Bitumen	26·3
Long-oil alkyd solids	18·4
White spirit	36·6
High aromatic solvent [7]	18·0
24% Lead naphthenate	0·5
6% Cobalt naphthenate	0·2

Specific gravity, 0·9; solids by volume, 37%.

Bituminous aluminium paint (see Chapter 10) is used widely on roofs. The leafing of the aluminium provides a heat- and light-reflective surface and also improves the durability of the bitumen. The paint is used also as a barrier coat in the protection of ships' bottoms.

(2) CONVERTIBLE COATINGS

Alkyd base

Paint systems based on medium- and long-oil alkyd resins are used extensively in general building and decorative work. Similar systems are used in industrial maintenance work where exposure conditions are relatively mild and decorative finish is required. Here the resins employed are usually of the medium-oil length to give shorter drying times and greater hardness. In many cases these medium-oil length alkyds have shown better durability than the long-oil counterparts.

Formulation of alkyd undercoats and finishes for maintenance work follows lines similar to those for the decorative types (p.   ).

Silicone/alkyd co-polymers (cf. p. 196). These are used in place of unmodified alkyds in situations where a decorative system combined with long-term durability is required. The co-polymer must contain 30 percent silicone to be effective. Formulation follows lines similar to those for the decorative types (p. 331) but they are more expensive.

Alkyd/chlorinated rubber The addition of a proportion of chlorinated rubber to certain long-oil alkyds results in an upgrading of both chemical and moisture resistance. The optimum effect is obtained from a mixture of three parts alkyd solids to one part chlorinated rubber (cf. chlorinated rubber/alkyd primers, p. 300). This blend is used in quick-drying primers for alkyd systems and in marine finishes. A gloss finish is illustrated by the following formulation:

White alkyd/chlorinated rubber finish

	% by weight
Rutile titanium dioxide	18·4
Long-oil alkyd (70% solids)	36·4
Chlorinated rubber	8·5
High aromatic solvent [7]	35·9
24% Lead naphthenate	0·5
6% Cobalt naphthenate	0·2
Methylethylketoxime	0·1

Specific gravity, 1·1; solids by volume, 36%; PVC, 14%.

Epoxy esters

The nature of these resins has been discussed (p. 204). They are used as binders for primers, undercoats and finishes and can be blended with alkyds of the appropriate oil length to upgrade the chemical resistance and to shorten the drying time.

The air-dried epoxy ester films show good adhesion to most common substrates, good flexibility, and good resistance to chemicals and many solvents. Their resistance properties lie between those of alkyds and those of the two-pack epoxy systems. Complete epoxy ester systems have shown very good durability on storage tanks in both marine and industrial environments. The following formulations indicate the general composition of an epoxy ester undercoat and finish:

Epoxy ester undercoat (brushing)

	% by weight
Rutile titanium dioxide	21·1
Micronized barytes	11·3
China clay	6·2
Epoxy ester (60% solids) [10]	37·7
Aromatic solvent [7]	2·9
White spirit	20·6
6% Cobalt drier	0·2

Specific gravity, 1·22
Solids by volume, 37·5%
PVC, 33·1%

Epoxy ester finish (brushing)

	% by weight
Rutile titanium dioxide	26·2
Micronized talc	1·2
Epoxy ester (60% solids) [10]	47·9
Aromatic solvent [7]	3·4
White spirit	21·1
6% Cobalt drier	0·2

Specific gravity, 1·10
Solids by volume, 34%
PVC, 22·6%

Two-pack systems

Film formation in two-pack systems results from chemical reaction between the two components which are mixed immediately before use. The reaction commences in the container and the mixture, in consequence, possesses a limited pot-life. Reaction in the film takes place more rapidly as solvent evaporation results in an increasing concentration of reacting molecules.

Epoxies

The chemistry of epoxy resins has been discussed on p. 201. Briefly, an epoxy resin is produced by reaction between epichlorhydrin and diphenylol propane; the curing or cross-linking agent (sometimes referred to as 'hardener') can be a polyamine, amine adduct, polyamide or isocyanate. The pot-life of the mixed components depends on the type of resin modification and curing agent. The amount mixed should be capable of being used within the pot-life period; any material remaining unused at the end of the period should be discarded and not added to a new mix.

Since the curing of an epoxy resin film is a chemical reaction between resin and hardener, atmospheric oxygen is not involved, and there is, in theory, no upper limit to the thickness of coating which can be cured. The thickness is, however, controlled by other factors, the most important of which are the tendency to sag by thick films on a vertical surface and solvent retention by the film. Complete loss of solvent from a thick film can take a very considerable time. Nevertheless, the high-build types of two-pack epoxy paints are used extensively. Solvent-free coatings are produced by the use of low molecular-weight liquid resins and liquid hardeners.

When applied at 20°C (68°F) epoxies can be overcoated at not less than four hours and not more than 48 hours after application. If the overcoating is delayed beyond this time the film will have hardened to a degree when intercoat adhesion is likely to be poor, unless the first coat is abraded to provide a key.

A curing period of up to 7 days is usually required for the development of the maximum chemical and solvent resistance. The rate of cure is markedly influenced by temperature, and when amines or polyamides are used as curing agents, the material should not be applied when the ambient temperature is below 7°C (44°F). Actually curing becomes progressively slower as the temperature falls below 10°C (50°F). The use of isocyanate as curing agent allows epoxies to be applied and cured at temperatures down to 0°C.

Typical formulations

Primers for metal The formulation and use of zinc-rich primers based on epoxy resins are discussed in Chapter 17. Other epoxy-based primers include red lead (now being phased out on toxicity grounds), zinc phosphate and red oxide/zinc chrome. The following formulations illustrate the general composition:

Zinc phosphate/epoxy primer

Base	*% by weight*	
Zinc phosphate	38·4	
Titanium dioxide	8·5	
Talc	6·9	
Bentone B27 [1]	0·3	Pregelled with proportion of toluol
n-Butanol	1·7	
Epoxy resin [11] 70% solution in xylol	16·7	
Methylethylketone	7·8	
Toluol	7·8	
Hardener		
Polyamide solution [12]	8·4	
Methylethylketone	1·8	
Toluol	1·7	

Specific gravity (of mixed components), 1·53; solids by volume (of mixed components), 49%; PVC (of mixed components), 50%.

Red oxide/zinc chrome/epoxy primer

Base	*% by weight*	
Red oxide of iron	31·7	
Zinc chrome [13]	9·0	
Talc	5·9	
Bentone B27 [1]	0·8	Pregelled with proportion of toluol
n-Butanol	1·8	
Epoxy resin [11] 70% solution in xylol	16·6	
Methylethylketone	10·0	
Toluol	10·0	
Hardener		
Polyamide solution [12]	8·4	
Methylethylketone	3·0	
Toluol	2·8	

Specific gravity (of mixed components), 1·44; solids by volume (of mixed components), 41%; PVC (of mixed components), 43%.

Both these materials are used extensively as blast primers. They can be applied by airless spray, dry rapidly, and adhere tenaciously to blast-cleaned steel.

Epoxy finishes

These are produced in white, black, and a full range of colours, the pigments being dispersed in the resin phase. The curing agents can be polyamines, amine adducts, polyamides or isocyanates. Polyamines and amine adducts give the greatest chemical resistance; the amine adducts are the less toxic and the less unpleasant in use. Polyamides give greater film flexibility and water resistance. There is also more latitude in the ratio of curing agent to resin than exists with the polyamines.

The following formulations illustrate the general composition of standard and high-build epoxy finishes.

Standard epoxy finish, white

Base	*% by weight*	
Rutile titanium dioxide	27·0	
Epikote 1001 [14]	23·2	
Xylol	14·6	
Methoxy propanol	10·1	
isoButanol	1·7	Pregelled with proportion of xylol
Bentone B34 [1]	0·4	
Hardener		
Versamid 115X [12]	16·4	
Xylol	3·3	
Methoxy propanol	3·3	

Base and Hardener are mixed in the proportions shown.

Specific gravity (of mixed components), 1·22; solids by volume (of mixed components), 46%; PVC (of mixed components), 18%.

High build epoxy finish (for airless spray)

Base	*% by weight*	
Rutile titanium dioxide	23·0	
Micronized barytes	16·8	
Talc	4·9	
Epikote 1001 [14] 75% solution in xylol	26·9	
Bentone B38 [1]	1·6	Pregelled with proportion of xylol
isoButanol	1·3	
Xylol	5·6	
Methoxy propanol	1·0	
Hardener		
Versamid 115X [12]	13·5	
Xylol	3·8	
isoButanol	1·6	

Specific gravity (of mixed components), 1·5; solids by volume (of mixed components), 57%; PVC (of mixed components), 29%.

Solventless coatings By the use of low-viscosity liquid epoxy resins and low-viscosity hardeners, it is possible to dispense with the use of solvents. The fire hazard is virtually eliminated and the paints can be applied in thick films which do not shrink on drying; a high degree of protection is thereby achieved. Solventless coatings are used also for underwater application to steelwork and for the interiors of storage tanks and other vessels.

General properties of two-pack epoxy systems

Overcoating times Epoxy resins cure to extremely hard films, and overcoating times are restricted. They can be overcoated after minimum periods of 4 to 16 hours from time of application, the period depending on the resin and type of paint. Overcoating must not be delayed for more than 48 hours, since the film will then be very hard and abrasion will be necessary to promote adhesion.

Resistance The films show outstanding resistance to chemicals, solvents and abrasion. They tend to chalk on exterior exposure.

Adhesion to the majority of surfaces is outstanding, provided they are clean.

Uses Blast and holding primers for steelwork; finishes for structural steelwork and concrete where high resistance to chemicals and solvents is required: marine paints; flooring compositions; adhesives.

Epoxy/pitch coatings

Epoxy resins can be diluted with certain grades of coal-tar pitch and the mixture can be cured by the normal curing agents. Extenders are usually incorporated to give high-solids products which are appreciably cheaper than the straight epoxies. They are applied by brush or airless spray to give thick films which are usually sag-resistant up to 190°C (374°F). The films are tough and show excellent resistance to dilute acids, water and alkalis, but the resistance to solvents is lower than that of the undiluted resins. The following formulation is presented to show the general composition:

Epoxy/pitch coating

Base	*% by weight*	
Special pitch No. 4 [15]	33·6	
Epoxy resin solution [16]	19·4	
Silica	1·9	
Talc	23·1	
Xylol	3·4	
Bentone B34 [1]	2·4	Pregelled with xylol
isoButanol	3·1	
Hardener		
Versamid 115X [12]	10·0	
isoButanol	1·2	
Xylol	1·9	

Specific gravity (of mixed components), 1·27; solids by volume (of mixed components), 68%.

The relative proportions of pitch, resin/hardener and extender can be varied, but any appreciable variation will be reflected in the properties.

Uses of epoxy/pitch coatings

They are used on steel and concrete surfaces which are exposed to severe conditions where protection rather than decoration is the prime consideration.

These include ships' bottoms, piers and pilings, lock gates, barrages, piping, dams and sewers.

Polyurethane coatings

These are two-pack materials consisting of a pigmented polyester or polyester/polyol mixture cured by reaction with an isocyanate (p.). A wide choice of polyesters and polyols is available to produce films varying in elasticity and hardness from coatings suitable for soft rubber articles to tough, abrasion-resistant floor finishes. The use of aromatic di-isocyanates as second component produces films which yellow on exterior exposure. This defect can be avoided by the use of aliphatic di-isocyanates.

Polyurethane films show good weather-resistance and high resistance to abrasion. They show, also, outstanding resistance to chemicals and solvents. The fact that any desired degree of flexibility can be built into the film, coupled with the intrinsically high durability, has resulted in the widespread use of polyurethanes on exterior woodwork as well as metals. For use on wood they can incorporate the flexibility required to accommodate the dimensional changes which take place. Advantage is taken of their excellent durability in marine work where they are used as finishes on both steel and aluminium superstructures (see p. 403). The primer employed for aluminium is frequently a two-pack etch followed by zinc chrome primer, whilst for steel a two-pack epoxy primer is commonly used.

The following formulation illustrates the general composition of a polyurethane finish which exhibits good non-yellowing properties, gloss retention and weather resistance:

White polyurethane finish

Part A	*% by weight*
Rutile titanium dioxide	16·0
Desmophen 650 [17]	35·3
Cellulose acetobutyrate, 10% solution in Methoxypropyl acetate	8·3
Methoxypropyl acetate	8·2
Xylol	3·7
Part B	
Desmodur N75 [17]	28·5

Specific-gravity (of mixed components), 1·19; solids by volume (of mixed components), 49%; PVC (of mixed components), 10%.

PAINTING SPECIFICATIONS

It is generally agreed that the average life of a paint system on exposed iron and steel structures is considerably less than the potential life and this is frequently attributed, in the majority of cases quite erroneously, to deficiencies in the paint. An analysis of the causes of film breakdown has shown, however, that whilst the paint may occasionally be at fault, other factors also exert a major influence on the life of a paint system. Among these are the following.

1 Condition of the surface. Before painting, steel surfaces must be clean and free from scale, rust, grease and general dirt. Application of paint to a dirty or partly cleaned surface will result in poor adhesion and early failure.
2 The weather conditions under which application and drying take place. Painting at site is often dogged by adverse weather conditions such as low temperatures, mist, fog or high humidity. In consequence, items of new work are blast-cleaned and then painted, as far as possible, at the fabricator's works.
3 The paint film must be of uniform and adequate thickness. The protection afforded by a paint film is, within limits, proportional to the film thickness, and the maximum figure depends on the type of paint and exposure conditions. For exposed steelwork this can range from 150 to 300 micrometres (6 to 12 mils).
4 The location of the structure, i.e. industrial, marine or rural.
5 The design of the structure. It should be free from pockets or grooves which can retain dirt and moisture and from sharp edges which are difficult to coat satisfactorily.
6 The paint may, for some reason, be unsuitable.

In order to achieve maximum protection of exposed steelwork it is essential to prepare a detailed painting specification and, in major works, to employ independent painting inspectors to supervise all stages of the work [18].

The specification

New work. The specification is best prepared at the design stage [19] in order to avoid pockets which retain water and dirt and to ensure, as far as possible, that all areas will be accessible for cleaning and painting when maintenance is carried out. It should, preferably, be prepared by persons with a sound knowledge of painting materials and procedures. Co-operation between the engineer and the paint technologist can be beneficial since the consequences of using an inadequate specification can be serious [20].

The specification should state clearly the requirements in respect of surface preparation, the type of paint system, number of coats and their thickness (maximum as well as minimum), the method of application, and the standard of finish required. Other factors to be considered in preparing the specification are the location of the structure (industrial, marine or rural) and the required period before minor and major maintenance become necessary. Useful guidance on the preparation of painting specifications is contained in BS 5493: 1977 (Code of practice for protective coating of iron and steel structures against corrosion) [21].

For new work it is a great advantage if, as mentioned earlier, the steel members can be cleaned and painted, as far as possible, under controlled conditions at the fabricator's works. They are delivered to the site in 'holding primer' or undercoat. Any damage from handling is then made good and the remaining coat(s) applied after erection.

The following skeletal example illustrates a painting specification for a steel structure to be exposed to a severe environment.

At works Blast clean to specified standard (SA2½ or 3) [22]
Apply chlorinated rubber/alkyd blast primer at 25 μm dry-film thickness.
Apply chlorinated rubber/alkyd first build coat at 35 μm dry-film thickness.
Apply stripe coat to edges.
Apply one coat chlorinated rubber, high build, at 75 μm dry-film thickness.

At site Make good any damage caused in transit.
Apply one coat chlorinated rubber paint, high build, at 75 μm dry-film thickness.
Apply one coat chlorinated rubber gloss finish at 30 μm dry-film thickness.
Total film thickness : 240 μm.

MAINTENANCE PAINTING

Usually this is delayed until the paintwork on a structure shows serious breakdown which may be general or in isolated areas. Before a re-painting specification can be prepared, a technical inspection of the structure is essential so that the correct treatment can be applied. The main points to be considered are the following.

1 The overall condition of the surface. General breakdown, usually accompanied by widespread rusting, will require complete removal of the existing system. If the breakdown is in isolated areas and the remainder of the paintwork is sound, then, naturally, only the affected areas are removed.
2 In the event of isolated areas of breakdown, a knowledge of the type of paint of the structure is useful. In some instances, this is on record.
3 The location of the structure and the life expectation of the new system.

Cleaning

When general breakdown has occurred, blast-cleaning is the most efficient method of removing the existing paint system and the underlying rust. If the structure is exposed to severe environmental conditons, the painting specification outlined under 'new work' would be appropriate. For less severe conditions the specification could embody a PVB type of blast primer followed by oleoresinous/MIO undercoats and alkyd finish. The life of this system can be extended by using a finish based on silicone alkyd, but this is appreciably more expensive than the conventional alkyd type.

In localized breakdown, the affected areas can be either blast-cleaned or abraded. In difficult situations, a needle gun or mechanical wirebrush may be required. The edges of sound paintwork should be bevelled and the surface given a light blast (preferably by water blast) to assist adhesion of the new paint.

Painting

The cleaned areas are 'patch primed' and then build coats are applied to bring the level up to that of the sound paint. The entire areas are then given undercoat and finish to the desired thickness.

In maintenance painting it is essential to ensure good adhesion between old and new paint films, and to this end it has long been considered good practice to use the same type of paint, i.e. alkyd on alkyd or oleoresinous; oleoresinous on oleoresinous. These systems, however, can become very hard with the passage of time and satisfactory adhesion of the new paint is not always achieved.

Chlorinated rubber paints have shown good adhesion in such cases, presumably as a result of attack on the old paint film by the strong solvents present. In consequence the use of chlorinated rubber paints in maintenance re-painting is now widespread. Where any doubt exists in respect of the age and hardness of the existing paint it is advisable to treat a small area and to note any tendency to 'lift' before large areas are painted.

PAINTING OF CONCRETE

Concrete consists of a mixture of cement, sand and aggregate, the proportions of which can be varied to suit the end use. Cement is largely a mixture of tricalcium silicate, dicalcium silicate and tricalcium aluminate all of which hydrate at different rates when mixed with water. The result is a mass of inter-locking crystals of hydrated silicates and aluminates which, in admixture with sand and aggregate, produces concrete. At the same time some calcium hydroxide is formed [23] and this is responsible for the high alkalinity (pH 11–12) of new concrete.

On exposure to air the outermost layers react slowly with atmospheric carbon dioxide to produce calcium carbonate (a reaction known as 'carbonation') and this results in a fall in pH value to about 8–9. Concrete is attacked also by soluble sulphates and atmospheric sulphur dioxide with formation of the sparingly soluble calcium sulphate whilst chlorides (often in the form of sodium chloride) produce the soluble calcium chloride which is leached from the film.

The 'carbonation' process extends gradually into the concrete mass by penetration of the micropores by carbon dioxide and can exert a damaging effect when the concrete contains steel re-inforcing rods. These rods are protected from corrosion by the highly alkaline concrete surrounding them. If the process of carbonation reaches the concrete immediately surrounding the rods the pH will fall to a point when corrosion can result. Then the more voluminous rust exerts pressure on the concrete and cracking or spalling can result.

Anti – carbonation coatings

The object of these coatings is provision of a barrier to carbon dioxide penetration whilst, at the same time, allowing the passage of moisture. Furthermore, such coatings must retain these properties as the film weathers.

This aspect has been investigated by Robinson [24] who measured the carbon dioxide permeability of a number of coatings of different types before and after accelerated weathering for 3000 hours. He points out the importance of critical pigment volume concentration (CPVC) in formulation of the coatings. Further work is in hand in correlating natural with the accelerated weathering results.

Work by Klopfer [25] in Germany has led to the establishment of 'minimum acceptance criteria' for anti-carbonation coatings and these are now accepted in Germany and some other European countries. These criteria involve a 'carbon dioxide diffusion resistance coefficient' and coating thickness. It is claimed that a dry film thickness of 150 micrometres is essential in an efficient anti-carbonation coating. Among coatings which have shown good test results are high build and textured forms of acrylic and vinyl copolymer emulsion paints.

Painting of non-reinforced concrete, cement render, etc.

Although the process of carbonation takes place with these materials no structural damage results and painting is generally for aesthetic reasons. The following materials are commonly used.

1 Emulsion paints, flat or textured, based on acrylic or vinyl copolymer resins.
2 Solvent type masonry paints based on acrylic or acrylic copolymer resins .
3 Proprietary cement paints.

The three types above can be applied to alkaline surfaces.

4 Alkyd based finishes can be used on mature (neutral) surfaces or over a non-saponifiable sealer.

Painting concrete floors

1 New floors Before any form of treatment is applied, concrete floors must be allowed to dry out thoroughly. The time necessary will depend on several factors including the nature of the mix, thickness, temperature and relative humidity. As an approximate 'rule of thumb' guide, one month should be allowed for each inch in thickness.

The surface of the concrete should first be examined carefully for the presence of laitence. This is a mechanically weak surface layer, rich in cement, formed as a result of excessive tamping or the use of too much water. It forms a weak substrate for paint and should be removed. This can be done by one of three methods.

(a) *Vacuum blast* This is the cleanest method and is essential if the floor is not fitted with drainage channels.
(b) *Acid etching with dilute hydrochloric acid* This solution is made by adding (carefully) one volume of commerical hydrochloric acid (also known as 'Spirits of salts' or 'Muriatic acid') to two volumes of water. The water should be placed in a plastic container and the acid added slowly. The operator must be protected by suitable clothing and boots and must wear an eyeshield. The amount of dilute acid required can be calculated from the fact that 6 litres (2

litres concentrated acid + 4 litres water) is sufficient for 24 square metres. The acid is spread with a stiff broom. When the reaction (evolution of gas) has ceased – this usually takes about 15 minutes – the floor is rinsed well with clean water and must then be allowed to dry.
(c) *Acid etching with sodium bisulphate* ($NaHSO_4$). The floor is wetted thoroughly with water and sodium bisulphate powder is sprinkled uniformly over the surface at the rate of one Kilo per 25 square metres. When the reaction (evolution of gas) has ceased the floor is rinsed thoroughly with water and must then be allowed to dry.

The following paints are among those commonly used on new concrete.
Chlorinated rubber, for resistance to alkali and spillages of dilute acids, chemicals, etc. It is not resistant to ketones, esters or animal fats.
Two-pack epoxy for resistance to alkali, solvents, animal fats. It is very hard and resistant to abrasion.
Moisture-cured polyurethane for resistance to alkali and solvents. It is very hard and resistant to abrasion.
2 **Mature (neutral) concrete floors** These are often contaminated with oil and/or grease. These are removed by scrubbing with detergent solution but, in difficult cases, steam or pressurised water may be required. (Solvent should not be used to remove oil or grease from concrete. The oil or grease is carried further into the floor.) The floor must be allowed to dry.

Since the surface is neutral a wider selection of paints is available.
Alkyd based a general purpose paint which is resistant to pedestrian and light traffic.
Epoxy ester, for heavier traffic. It is more resistant to chemicals than alkyd.
Chlorinated rubber for resistance to alkali, acid and other chemicals.
Two-pack epoxy for heavy traffic. It is resistant to chemicals and solvents.
Two-pack polyurethane for resistance to chemicals, solvents and abrasion.

Immersed or buried concrete

Here appearance is irrelevant but protection from attack by soluble salts, such as chloride and sulphate, can be vital. The following types of protective coating are in use: bitumen, chlorinated rubber/pitch; vinyl resin/pitch; epoxy/pitch; polyurethane/pitch.

Thermal insulation

Heat losses through asbestos sheeting, particularly if weathered and water-absorbent, are very high and a system of insulation has been developed which reduces these losses very considerably. Briefly, the process consists in cleaning the surface followed by application of a moisture-impermeable adhesive/sealer. Polyurethane foam, often 4 to 5 cm in thickness, is then applied by spray. This foam is highly water-absorbent and is also subject to rapid deterioration by ultraviolet light. It must therefore be protected as soon as possible after application. A number of paint types have been used for this purpose, including synthetic and chlorinated rubbers, vinyls and acrylic emulsions. Adhesion to foam coupled with elasticity and low moisture permeability are the principal requirements.

References to Chapter 21

[1] NL Chemicals (UK) Ltd, Wilmslow, Cheshire.
[2] Parker, H. E. in *Treatise on Coatings*, ed. R. R. Myers and J. S. Long, vol. 1, Part 1, p. 154. Edward Arnold, 1967.
[3] ICI Ltd (Mond Division), Runcorn.
[4] Esso Chemicals Ltd, Southampton.
[5] Carter, E., Micaceous iron oxide paints. *PPCJ*, 1981, **171**, No. 4053, 506.
[6] ICI Ltd, London.
[7] 'Aromosol H' (ICI) or 'Solvesso 100' (Esso).
[8] Union Carbide (UK) Ltd, London.
[9] 'Anti-settle' (Cray Valley Products Ltd, Farnborough, Kent).
[10] 'Synolac 400W' (Cray Valley Products Ltd).
[11] 'Araldite 6100' (Ciba-Geigy Ltd, Duxford Cambridge).
[12] 'Versamid 115X' (Cray Valley Products Ltd).
[13] BS 282/389: 1963, Type 2. British Standards Institution, 2 Park Street, London, W1A 2BS.
[14] Shell Chemicals Ltd, London.
[15] British Steel Corporation (Chemicals) Ltd, Chesterfield, Derby.
[16] 'Grilonit 1201' 70% solution in xylol/Cellosolve acetate (Grilon UK Ltd, Stafford).
[17] Bayer UK Ltd, Richmond, Surrey.
[18] Dunkley, F. G., Quality control of painting in the construction industry. *JOCCA*, 1976, **59**, 39.
[19] Duligal, E. A., Consultation before painting avoids litigation after. *JOCCA*, 1979, **62**, 303.
Goodman, D., Problems facing a designer when specifying surface coatings. *JOCCA*, 1968, **51**, 686.
[20] Failures due to inadequate specifications. *Hess's Paint Film Defects*, 3rd edn, p. 324. Chapman & Hall, London, 1979.
[21] British Standards Institution.
[22] Swedish Standard SIS 05 59 00. Standardiseringskommissionen i Sverige, Stockholm.
[23] Gay, P. J., *JOCCA*, 1974, **58**, 252.
[24] Robinson, H. L., Durability of anti-carbonation coatings. *JOCCA*, 1987, **70**, 193.
[25] Klopfer, K., The carbonation of external concrete and how to combat it. *Bautenschutz und Bausanterung*, No. 3, 1978.

22

Industrial finishes

The term 'industrial finish' covers a broad spectrum of coatings which are usually factory-applied to a wide variety of articles. Many of the paints discussed in Chapter 21 can be classified as 'industrial' since they are used also for purposes other than on buildings and structures. Chlorinated rubber paints, for example, are used also to protect plant and machinery likely to be exposed to severely corrosive environments.

Industrial finishes comprise both convertible and non-convertible coatings and differ from the decorative types in the following particulars.

1 The drying and curing times are very much shorter to satisfy mass production schedules. Methods used to attain this include the use of shorter oil media, more rapidly evaporating solvents, and the use of accelerated curing methods such as stoving and irradiation.
2 The resistance to impact, abrasion and general mechanical damage is greater.
3 The resistance to chemicals and solvents is of a higher order.
4 In many instances, the weather resistance is not of paramount importance.
5 They often have to withstand considerable temperature fluctuations or very high temperatures.
6 A high proportion of industrial finishes are applied to metal, either directly or over a primer.

For the purpose of the present discussion it is convenient to make a broad division into air-drying and stoving types, the former including lacquers, conventional air-drying, and chemically cured types; the last-mentioned can also be stoved.

AIR-DRYING TYPES

Lacquers and non-convertible coatings

Cellulose nitrate lacquers The formulation and pigmentation of these lacquers has been discussed in Chapter 12. The principal use of pigmented cellulose nitrate lacquers is now in car refinishing and reproduction furniture. The resistance to heat and solvents of these lacquers can be improved by the use of modifiers and these improved lacquers are used on domestic furniture made from wood or wood-veneered chip-board [1].

Chlorinated rubber paints The formulation and properties of chlorinated rubber paints has been discussed on p. 371. In addition to the uses mentioned in Chapter 21, chlorinated rubber paints are used for the protection of steel and other vulnerable materials which are continuously exposed to water and/or chemicals. Examples are plant in plating shops, chemical plant, sewerage and water works.

Acrylic lacquers The acrylic resins used in solutions are polymers of methacrylic acid esters or co-polymers of, for example, methyl methacrylate and an acrylic ester such as methyl acrylate. A number of esters are available and, by judicious selection, it is possible to produce polymers with any desired degree of hardness and flexibility.

The resins are usually colourless and are soluble in esters, ketones, etheralcohols and aromatic hydrocarbons.

Drying of the films is by solvent evaporation only. The films show good gloss retention, good durability (in this respect they are superior to cellulose nitrate) and good resistance to lubricating oils and greases. Resistance to water is only fair. The films are resistant to heat (with only slight discoloration) up to 180°C (356°F).

A white acrylic gloss finish can be made on the following formulation:

	% by weight
Rutile titanium dioxide	16·5
Acrylic solution (55%) [2]	43·3
Xylol	40·2

Specific gravity, 1·05; solids by volume, 28%; PVC, 15%.

Uses of acrylic lacquers These lacquers have a wide field of application which includes media for metallic and fluorescent paints, clear lacquers for metal and sealers for sprayed metal, aircraft components, protective lacquer for vacuum-deposited metals, and pigmented finishes for kitchen cabinets and washing machines. Acrylic resins also form the basis of finishes for certain types of plastic articles.

Vinyl lacquers have been discussed on p. 376. They are based on co-polymers of vinyl chloride and vinyl acetate and are used where a high degree of resistance to chemical attack is required.

Cellulose acetate is used in strippable coatings which may be clear or pigmented. The following formulation gives a clear coating of this type:

	% by weight
Cellulose acetate	16·6
Diacetin	12·0
Acetone	66·0
IMS 64 O.P.	5·4

Shellac solutions in methylated spirit are pigmented, usually with bright non-toxic pigments, and used on wooden playthings such as children's bricks. Pigmentation follows normal procedure and for articles of uniform shape application is often by tumbling.

Shellac varnish provides a rapid drying film with high build and gloss. It is often used on timber to improve appearance and to reduce water penetration for limited periods.

The following is a typical formula:

	% by weight
Orange lac	42·5
Raw castor oil	2·5
Ethyl cellulose	2·1
IMS 64 O.P.	53·1

Convertible coatings

Oxidizing types

Alkyd resins are probably the most widely used binders for industrial air-drying paints, especially in the medium- and short-oil types. They are used alone, in modified form (VT or acrylic modifications), or in admixture with other resins such as chlorinated rubber, acrylic or expoxy ester. Tung oil and phenolic modified types possess excellent resistance to lubricating and cutting oils and are therefore suited to machinery finishes. Rapid air-drying primers and finishes can be based on short VT modified alkyds. For example, a grey oil-resisting machinery enamel can be based on the following general lines:

	% by weight
Rutile titanium dioxide	19·2
Carbon black	0·7
Coated calcium carbonate [3]	1·9
VT Alkyd solution (50% solids)	56·7
Xylol	11·2
Solvesso 100 [4]	10·0
6% Cobalt naphthenate	0·5

Specific gravity, 1·4; solids by volume, 35%; PVC, 19%; drying time, 1–2 hours: this can be reduced by replacement of Solvesso 100 by xylol.

Epoxy esters are used to upgrade the chemical and moisture resistance of long-oil alkyds. They are also used alone, when they show very good adhesion to many surfaces and moderate to good resistance to chemicals. The linseed types exhibit good durability and have given satisfactory service as media for paints for steel storage tanks and gas holders. The DCO types give very hard and abrasion resistant films and are used in floor paints.

Straight epoxy esters are compatible with white spirit and dipentene and so can be applied by brush. This may be the most convenient method of application on some types of plant.

The following is a typical formulation for a brushing type of white base machinery enamel:

	% by weight
Rutile titanium dioxide	26·0
Coated calcium carbonate	1·9
Epoxy ester solution	47·6
Dipentene	8·1
White spirit	16·2
Cobalt naphthenate (6%)	0·2

One-pack urethane alkyds The properties and uses of these materials are discussed in Chapter 11.

Chemically cured coatings

The acid-cured alkyd/amino and the hydroxylated acrylic/isocyanate systems are described below. The other important members of this class are the polyesters, polyurethanes and epoxies the properties and uses of which are described in Chapter 11.

Acid-cured alkyd/amino systems

The use of certain acids or acidic materials as catalysts enables cure of alkyd/amino systems to be effected at normal temperatures.

The ratio of non-drying alkyd to amino resin is usually between 1·5: 1 and 1:1 (by weight). Cure is effected by addition of controlled amounts of hydrochloric acid or acid butyl phosphate. Other acids, chiefly sulphuric and *p*-toluenesulphonic, are sometimes used.

Hydrochloric acid is very efficient and is added, in solution in ethanol or butanol, at the rate of 2%–5% (calculated as concentrated acid) on the weight of solid amino resin. At the higher concentration the pot-life of the mixture is very short. A drawback to the use of hydrochloric acid is the corrosion which can take place on metallic substrates.

Corrosion of metallic substrates can be avoided by the use of acid butyl phosphate as catalyst. It has a longer pot-life than systems using hydrochloric acid and is used at about 10 percent on the amino resin solids.

The acid-cured alkyd/amino systems yield films which are very tough, adherent and resistant to many solvents and other liquids. They are used, *inter alia*, for bar and counter tops.

Acrylic/isocyanate systems

Reaction between an hydroxylated acrylic resin and an isocyanate produces a binder possessing good durability and gloss retention. It is used in fast drying furniture finishes and in coatings for motor car refinishing.

STOVING TYPES

The application of heat to wet films of certain types of resin or resin mixtures accelerates the cross-linking reactions, and cure of the film takes place rapidly, often in a matter of minutes. This enables mass-produced articles, from small components to motor car bodies, to be painted and cured on a production line basis.

Stoving cannot be applied to the long-oil types of resin used in air-drying decorative and maintenance paints.

Primers

Industrial stoving primers are, to a large degree, based on short/medium oil-length straight or modified alkyds. As an example, the widely used red oxide/zinc chrome primer can be produced on a formulation of the following type:

	% by weight
Red iron oxide	29·0
Zinc chrome, Type 2 BS 282, 389 [5]	8·2
Zinc oxide	1·9
Micronized talc	2·4
Whiting	1·9
VT Alkyd (50% solids)	31·5
Aromosol H [6]	16·6
Xylol	8·5

Specific gravity, 1·39; solids by volume, 35%; PVC, 42·7%.

This primer is used on steel articles and gives a tough, adherent film when stoved (convention oven) at 120°C (248°F) for ½ to ¾ hour.

Since zinc chrome is being phased out on toxicity grounds the coloured pigments in the above formula can be replaced by zinc phosphate. In many painting systems the use of a pale shade primer is an advantage. If corrosion is not likely to be a major problem, titanium dioxide can be used as the opacifying pigment. Modified alkyd binders are likely to give some discolouration but this can be avoided by the use of an alkyd/amino combination.

Alternatively, if stringent conditions such as hot detergents are likely to be met, a melamine/epoxy combination is preferable. The following primer is based on this type of medium and produces a tough, adherent and resistant film when stoved at 140°C (284°F) for about 10 minutes.

	% by weight
Titanium dioxide	14·8
Blanc fixe	13·6
Micronized talc	13·6
Beetle BE 649 [7]	29·2
Xylol	25·3
n-Butanol	3·5

Specific gravity, 1·33; solids by volume, 36·7%; PVC, 42·4%.

Primers of this type are used on certain types of domestic equipment such as washing machines and dish washers where detergent resistance is of paramount importance.

Finishes

Stoving finishes can be produced in a large number of possible resin combinations according to the degree of gloss, flexibility, adhesion, chemical resistance, etc. required.

Alkyd/amino resins Combinations of non-drying alkyds and ureaformaldehyde resins are the most widely used. Pigmentation is on conventional lines, the pigment usually being ball-milled into the alkyd solution. In whites where titanium dioxide is the standard pigment, the pigment/binder ratio is 0·6 to 0·8/1. Other pigments are used at concentrations similar to those used in decorative enamels.

The gloss and mar resistance of these finishes can be improved, if necessary, by replacement of part of the mixed resins by a medium/short drying alkyd based on dehydrated castor oil.

The following formulation illustrates a typical alkyd-U/F medium suitable for stoving at 120°C (248°F) for 30 minutes or at 240°C (464°F) for 3½ minutes:

	% by weight
Rutile titanium dioxide	23·1
U/F resin solution (53%)	25·7
Non-drying alkyd solution (50%)	28·2
Aromosol H [6]	4·7
Xylol	16·3
n-Butanol	1·2
Ethyl lactate	0·8

Specific gravity, 1·14; solids by volume, 35%; PVC, 19·0%.

Ethyl lactate is added to promote secondary flow in the stoving oven and so eliminates surface imperfections.

Mixtures of non-drying alkyds and melamine-formaldehyde resins give finishes with high gloss and a high degree of resistance to heat and chemicals. The ratio of alkyd to amino resin is higher than in the alkyd/urea combinations. The following formulation illustrates an alkyd/melamine finish:

	% by weight
Rutile titanium dioxide	22·5
Beetle BE 615 [7]	15·0
Synolac 9063X [8]	45·0
Xylol	17·5

Specific gravity, 1·16; solids by volume, 38·3%; PVC, 17·0%.

The alkyd/amino stoving types formed the basis of finishes for refrigerators, washing machines and other kitchen equipment for some time, but in recent years an appreciable amount of epoxy resin has been incorporated. This has the effect of upgrading the adhesion and general resistance of the finishes. Additions of melamine/epoxy resins can be made to the alkyd/amino combinations, but for high chemical and solvent resistance the former can be used alone.

The following formulation illustrates an epoxy-reinforced type, though the proportion of epoxy to melamine resin can be varied to suit circumstances:

	% by weight
Titanium dioxide	22·9
Uramex P138 [9]	12·9
Uralac 1308X [9]	38·2
Beetle 649 [7]	7·7
Xylene	18·3

Specific gravity, 1·16; solids by volume, 38·4%; PVC, 17·4%.

High solids coatings based on blends of melamine (HMMM) and polyesters are stated [10] to possess outstanding durability and chemical resistance. They are claimed to be superior to conventional alkyd systems and are suggested for use as automobile finishing coats.

Epoxy resins can be reacted with amino (chiefly melamine/formaldehyde) and with phenolic resins to give products which cure on stoving. With melamine/formaldehyde, the product is a very pale resin suitable for white finishes which possess very high resistance to wear and chemicals. An example is Beetle BE 649 [7] mentioned above. The reaction with phenolic resins yields films with high resistance and adhesion but possessing poor colour. These are used as coatings for the interior of cans and drums.

The cure of epoxy resins by polyamines or polyamides, a reaction which is relatively slow at ambient temperature, is accelerated considerably by the application of heat.

Vinyl resins This general term covers a number of resins which have been discussed in Chapter 11. Probably the most versatile of these are the copolymers of vinyl acetate and vinyl chloride which produce films with excellent toughness and flexibility but which are not resistant to temperatures above 150°C (302°F). The films are resistant to oils and greases, water, many chemicals and many solvents. They are used for container linings, chemical-resistant finishes, coatings for collapsible tubes, and in ships' bottom compositions.

The protective value is a function of the coating thickness, and hot spray application has been used to produce thicker coats. Considerably greater thickness of coating has been achieved by the use of plasticized polyvinyl chloride.

Plastisols are dispersions of polyvinyl chloride in plasticizer. They are of high consistency and are frequently referred to as 'PVC pastes'. These materials can be pigmented, but the pigments used must be resistant to the stoving temperature of 171° to 193°C (340° to 380°F).

Organosols are dispersions of polyvinyl chloride in plasticizer and solvent. The polyvinyl chloride is not soluble in the plasticizer at normal temperatures, but on stoving, the plasticizer is absorbed by and softens the PVC so that the particles sinter or fuse, forming a very hard and highly resistant coating. The stoving temperatures are in the region of 171° to 193°C (340° to 380°F). Under these conditions relatively thick coatings can be applied in a textured or smooth finish according to the type of plasticizer and the stoving temperature.

Unfortunately these films have shown very poor adhesion to metal and have to be applied over a primer which is usually based on two-pack epoxy resin. Plastisols and organosols are used where hard wear is encountered, such as heavy test equipment and office furniture. They are frequently applied by the coil coating process.

Water-borne coatings Most polymer emulsions are air-drying but stoving is the normal method of cure for the majority of water-soluble and water-dispersible systems which are applied by either conventional methods or electrodeposition. They are described in Chapter 19.

Silicone resins require temperatures in the region of 200° to 230°C (392° to 446°F) for satisfactory cure. They are used when resistance to high temperatures is required, and when pigmented with aluminium will give protection up to 600°C (1112°F). It is probable that before this temperature is reached the silicone resin has decomposed and the aluminium has been sintered to the surface. These paints are used on exhaust manifolds and steel chimney stacks. Their general composition is illustrated by the following formulation:

	% by weight
Silicone resin (50% in xylol)	17·6
Ethyl cellulose	1·8
Aromosol H [6]	55·9
Polished aluminium paste (78%)	24·6
Manganese naphthenate solution	0·1

Specific gravity, 1·07; solids by volume, 17%.

The production of high temperature resisting blacks using polyalkoxysilane binders and a range of metal oxide pigments has been investigated by Bhatia *et al* [11]. They found that cobalt oxide, manganese dioxide or a 1:1 mixture showed the best heat resistance (330°–500°C) and film properties.

Flamboyant finishes are two-coat systems in which a coloured transparent lacquer is applied over an aluminium base coat. The base coat contains non-leafing aluminium in a conventional alkyd/amino medium and the latter is also used for the coloured lacquer. The colour in the lacquer is provided by a soluble dyestuff which is fast to light and stoving. The general effect is that of a brilliant coloured metal with a deep gloss and it is popular on cycles. Formulations can be based on the following:

Aluminium base for flamboyant finish

	% by weight
Synolac 9063X [8]	43·0
Uramex P138 [9]	29·8
Xylene	21·1
n-Butanol	2·6
Fine non-leafing aluminium paste	3·5

Specific gravity, 0·98.

Flamboyant finish

	% by weight
Synolac 9063X [8]	38·9
Uramex P138 [9]	34·7

Xylene	14·4
Ethoxy propanol	5·0
n-Butanol	3·3
Dyestuff	3·7

Specific gravity, 0·98.

Polychromatic finishes are pigmented or dyed stoving finishes to which is added a small quantity of non-leafing aluminium paste. The dyes or pigments must be fast to light and stoving. The general effect is a coloured metallic sheen; it is used for cars as well as cycles.

Wrinkle finishes Early types were made on tung oil oleoresinous varnishes to which a proportion of raw tung oil was sometimes added, and the pattern was dependent on the gas-checking of the tung oil medium in a foul oven in the presence of cobalt and manganese driers. The texture of the wrinkle depended on the thickness of the coat.

Modern types are based on short-oil alkyds of the linseed/tung oil or linseed/DCO types. With these media the size and pattern of the wrinkle can be controlled to a greater degree than with oleoresinous media. The nature of the wrinkle is influenced by the pigmentation and the type of extenders used. Surface-coated calcium carbonate enables great control to be exercised over the pattern. The texture of the wrinkle is also influenced by temperature, becoming finer as the temperature rises.

These finishes are applied by spraying and can be stoved in any of the conventional type ovens. The following formulation is typical of a wrinkle finish:

	% by weight
Short linseed/tung alkyd (60% solids)	46·9
Carbon black	1·9
Diatomaceous silica	4·7
Xylol	46·5

Cobalt and manganese naphthenates are added to give 0·3 percent cobalt metal and 0·1 percent manganese metal on the resin solids.

Wrinkle finishes of this type are stoved for about 1 hour at 120°C (248°F). They are used mainly on instruments of various types and on typewriters, but have been displaced to a considerable degree by hammer finishes. A serious drawback of wrinkle finishes is the tendency of the pattern to collect dust which cannot readily be removed.

Flock finishes are produced by application of a fairly heavy coat of oleoresinous varnish designed to remain very tacky for some time after loss of solvent. The nylon or silk flock is then sprayed through a special type of spray gun.

Polyurethanes The general properties of these materials have been discussed in Chapter 11. They can be cold-cured, though the curing process can be accelerated by stoving. This also increases the hardness and general resistance of the film but also causes some yellowing. Generally, polyurethanes are stable only up to 150°C (302°F).

The stoved films show a high degree of adhesion, and resistance to chemicals and a number of solvents. Among their uses are corrosion-inhibiting primers for steel, interior coatings for tanks and drums, and finishes for chemical plant.

Stoving phenolic resins These are heat-reactive, alkali-catalysed resins in which the polymerization reaction has been terminated before the resin becomes insoluble in solvents. The resins are supplied as solutions in an alcohol, a solvent of the Cellosolve [10] type, or a mixture. They are usually applied in non-pigmented form, and the article then being stoved when the polymerization reaction is completed. The films show excellent solvent- and corrosion-resistance and are employed as linings for cans and drums.

Hammer finishes These are so-called because they give the surface the general appearance of hammered metal. In this way they mask minor irregularities in the surface and at the same time produce a pleasing appearance.

The hammer pattern can be produced either by using a carefully selected blend of solvents with a rapid drying medium such as a short-oil alkyd or by addition of certain silicone oils to a conventional type of stoving medium. The latter gives the more satisfactory and consistent result and is generally employed. Pigmentation is by non-leafing aluminium at low concentration.

The following formulation illustrates the base paint for hammer finishes:

	% by weight
Butylated urea-formaldehyde resin (50% solids)	22·7
Short-oil DCO alkyd (60% solids)	68·1
Non-leafing aluminium paste	1·9
Xylol	4·9
n-Butanol	2·4

Silicone oils designed specifically to give the hammer effect are added in small quantity. The amount is critical, the size of the pattern varying inversely with the amount of silicone added. Too much silicone leads to *cissing*.

Application of hammer finishes is by spray at higher viscosities than used with conventional paints so as to minimize sagging on vertical surfaces. The viscosity range is 1–2 poise at 20°C (68°F). The articles are then stoved at 220°F (104°C) for about half an hour.

Coloured hammer finishes are produced by addition of conventional stainers to the above base paint. The quantity of stainers required is very small.

Uses Hammer finishes are used extensively on instruments, typewriters, office furniture such as filing cabinets, and sheet-metal work generally. They wear well and do not collect dirt to the same degree as wrinkle finishes.

Thermosetting acrylic systems

The film properties of these systems are controlled by the nature of the comonomers, the molecular weight, and the type of cross-linking mechanism.

One of the most widely used systems consists of reaction between an hydroxylated acrylic and a butylated melamine/formaldehyde resin. The combination cures in 30 minutes at 120°C and is used as a motor car finish. The film shows excellent durability and gloss retention.

Hydroxylated acrylics react also with epoxy and with polyester resins to give coatings with a very wide field of application.

Coatings for wood, suitable for UV curing, are produced from hydroxylated acrylics modified with epoxy, isocyanate, polyester or silicone.

Carboxylated acrylics can be reacted with the co-polymers mentioned above to give hard glossy films with outstanding resistance to chemicals, soaps and detergents. They are used on hospital equipment and laundry machines.

Powder coatings

The basic principles and methods of application of powder coatings have been described in Chapter 20.

Materials employed

The resins employed are drawn from both thermoplastic and thermosetting types.

Thermoplastic resins include the following.

Polyethylene (polythene) is of low density, with a softening point of 105°–114°C. It is used for wirework and articles such as supermarket baskets and trollies.

Polypropylene has a melting point of about 168°C and is tougher and more abrasion resistant than polyethylene.

Nylon (a polyamide) shows outstanding resistance to abrasion and is self-lubricating. It is used for moving surfaces in contact such as sliding racks, ball joints and so on.

PVC (polyvinyl chloride) is resistant to water and impact and has a wide range of uses such as out-door furniture, playground equipment and fencing.

Generally these thermoplastic resins require a primer, for the following reasons.

1 Their adhesion to metal is of a low order. Adhesion to primers is usually satisfactory but the primer must withstand the temperature of fusion of the resin.
2 They form a protective 'envelope' round the article. If this envelope is pierced or broken in any way, moisture and air will enter allowing corrosion to take place unless an efficient primer has been used.

Thermosetting resins The main thermosetting resins used are epoxies, saturated polyesters and acrylics. Other resins are used but in smaller quantities.

Epoxies are usually polyamide cured. They show outstanding adhesion, electrical insulation and resistance to chemicals and solvents. Their usage covers a wide field.

Polyesters are often blended with 'blocked' isocyanates (isocyanate reacted with a material such as caprolactone. The compound decomposes on stoving to release the isocyanate). Polyester films, in general, exhibit excellent durability, electrical properties and gloss. They are often used without primer.

Acrylics can be reacted with epoxies, isocyanates and polyesters to yield films of outstanding durability and mechanical properties.

Liquid powders are aqueous dispersions of powder coatings in which stainers, flow promoters, etc., can also be dispersed in the water. The advantages and drawbacks compared with dry powders have been summarized by Ramsay [12].

Advantages include: (1) conventional equipment, with minor modifications, can be used for application; (2) stainers can be added to the dispersion, so facilitating colour-matching; (3) no explosion hazard; (4) finer particle-size means smoother and thinner films.

Disadvantages are instability to shear (care must be exercised in the choice of pumps) and to temperatures above 80°C (194°F).

Liquid powders are being applied to a variety of steel articles, including tractors, scaffolding and car wheels.

References to Chapter 22

[1] Sparkes, A. J., New materials for the furniture industry. *JOCCA*, 1986, **69**, 5.
[2] 'Synacryl 823S' (Cray Valley Products Ltd, Farnborough, Kent).
[3] 'Calofort S' (J. & E. Sturge Ltd, Birmingham).
[4] Esso Chemicals Ltd, Southampton.
[5] British Standards Institution, 2 Park Street, London W1A 2BS.
[6] ICI Ltd.
[7] BIP Chemicals Ltd, Oldbury.
[8] Cray Valley Products Ltd.
[9] Synthetic Resins Ltd, Speke, Liverpool.
[10] Buter, R. and Broomgaard, R. E., Durable high solids polyester topcoats. *JOCCA*, 1986, **69**, 320.
[11] Bhatia, S. S. *et al.*, Poly (alkoxysiloxane) carrier vehicles for high temperature coatings. *JOCCA*, 1986, **69**, 216.
[12] Ramsay, R. G., Liquid powders—a new development in surface coatings. *PPCJ*, 1981, **171**, No. 4038, 17.

23

Marine paints and compositions

Paints for use under marine conditions are conveniently divided into those for (a) ships, (b) dock and harbour installations, and (c) off-shore structures.

SHIPS' PAINTS

The properties required in paints for the various parts of a ship depend on the location of the areas to be protected, and it is usual to designate the paints by these areas, i.e. superstructure, topsides, boot-top, and so on.

The superstructure includes fixtures above deck level, i.e. deck houses and fittings, funnels, etc. which may be constructed of steel or, in larger vessels such as liners, of light alloy. Conditions of exposure are severe. Temperature fluctuations can be considerable, and rain, salt spray and, at times, intense sunlight are encountered.

Primers for steel superstructures are now generally based on zinc phosphate in medium-oil alkyd, oleoresinous or epoxy ester media. The use of red lead has been discontinued and zinc chromate is being phased out since both pigments present a health hazard. Formulation of the primers is on similar lines to those described on p. 303.

Finishes, generally described as 'Marine Gloss', are usually white or pale colours, although dark colours and black are used on some merchant ships. They are based on medium/long alkyds of the non-yellowing linoleic-rich type. These are similar in composition to the gloss finishes used on land and are described on p.

The trend towards larger vessels and smaller crews results in less maintenance in service; shipowners therefore require paints with longer life than the conventional types. Among the 'long-life' coatings in use are chlorinated rubber, which is being used to an increasing extent [1], vinyls, two-pack epoxies and two-pack polyurethanes. The latter are based on aliphatic isocyanates to minimize yellowing. Both vinyls and polyurethanes are usually applied over two-pack epoxy primer since their adhesion to metals is often poor. The two-pack epoxies, although offering excellent protection, are used only when their chalking tendency is not of vital importance.

Light alloy superstructures are primed with zinc chromate in alkyd, epoxy ester or two-pack epoxy media. The metals are sometimes protected with an etch primer (p. 310) to improve the adhesion of the primer, as well as to give added protection. The finishes employed are similar to those used on steel, namely medium/long linoleic type alkyd, two-pack polyurethane, chlorinated rubber or vinyl.

Steel decks

The paint must exhibit maximum adhesion and abrasion resistance as well as corrosion inhibition. Additionally, it must resist sea-water and materials such as mineral or vegetable oils which may be carried as cargo. Non-slip properties, which must be unaffected by water or oil, are of vital importance. Two-pack systems – epoxies and polyurethanes – have proved to be the most satisfactory.

Steel decks are often coated with inorganic zinc silicate primer (p. 303). This dries to a very hard, durable and non-slip film and is often left uncoated.

Hull paints

The outer surface of the hull is divided into three areas.

1 Topsides, extending from the water-line (when laden) to the deck.
2 Boot-top – the area which is exposed when the ship is empty but immersed when laden.
3 Bottom, which is permanently submerged.

Steel hulls are fabricated from mild steel plate which is produced by hot rolling and is therefore coated with millscale. As was pointed out in Chapter 16, removal of millscale is essential for satisfactory paint performance, and it is usual to blast-clean the plates and to prime them immediately afterwards. In more leisurely times the hull was fabricated from the untreated plates and left exposed to the weather while the interior was fitted out. Most of the millscale was loosened, enabling the hull to be scraped and wire-brushed before painting. The method was unsatisfactory since areas of millscale remained intact. Sooner or later this scale would flake off, together with the superimposed paint film. Apart from the time involved, this method of cleaning would be grossly inadequate for modern types of paint.

In shipyards today, the plates are blast-cleaned, using impellor-type machines (wheelabrators) to a standard of SA 2½ (Swedish Standard 05 59 00) [2] or 2nd quality, BS 4232 [3]. This is followed within minutes by application of a blast primer. The latter is required to protect the steel during storage and fabrication of the hull, after which the finishing systems can be applied.

Blast primers

During fabrication of the hull the primed plates are welded and some are cut. In these processes the blast primer must not emit objectionable or toxic fumes, it must not impair the quality of the welds, and it must be heat-resistant. Bending of the plates requires a primer with good flexibility and adhesion.

A further problem is presented by the use of cathodic protection (p. 285) to control corrosion of the hull. This may take the form of sacrificial anodes or impressed current to render the hull surface cathodic. The cathodic reaction (p. 285) produces alkali (hydroxyl ions), the concentration of which may be high in the vicinity of the anodes. Any primer used in conjunction with cathodic protection must therefore be alkali resisting. Although it may be overcoated with a highly impermeable paint system, damage to the latter is always possible, and the cathodic reaction will then produce alkali, resulting in saponification and loss of adhesion.

Types of blast primers

In addition to the other desirable properties, blast primers must dry rapidly and so require the use of fast-evaporating solvents so that the plates can be handled in a matter of minutes after painting. The following types are in use.

1 Phenolic-modified PVB type, pigmented with red oxide of iron and zinc tetroxychromate [4] and thinned with ethanol and/or isopropanol. This type of primer has been described on p. 300.

2 Polyamide-cured epoxies pigmented with zinc phosphate, red oxide of iron, or metallic zinc (zinc-rich primer) (p. 301). The latter, whilst affording good protection by sacrificial action, does nevertheless have certain drawbacks as a 'holding' primer. (a) On exposure, white zinc corrosion products are formed on the surface. Before painting these must be removed with water and a stiff brush and the surface must be allowed to dry. (b) In welding and cutting operations, zinc oxide ('zinc fume') is produced. Some welders regard this as objectionable since it can cause temporary sickness, and this primer is therefore not popular in shipyards. (c) In view of the chemical reaction between metallic zinc and alkali, zinc-rich primers are not recommended for use on vessels when cathodic protection also is used.

3 Inorganic zinc silicate primers are sometimes employed and give good protection. They do, however, present problems similar to those encountered with zinc-rich primers (*above*) on plates which are to be flame-cut or welded.

4 Epoxy-coal tar pitch is employed in some cases in maintenance painting. It has good adhesion to steel and provides an excellent barrier to moisture penetration.

Finishing systems

Topsides

Alkyds have largely replaced oleoresinous media in topside paints, although the oleoresinous types continue to be used on some smaller vessels. Requirements are good adhesion to the primer (or to the undercoat, if used), good weather-resistance, and ability to withstand intermittent soaking and buffeting by waves. Pigmentation of these finishes follows established practice as discussed in Chapter 18. Light colours are based on the non-yellowing linoleic-rich types of alkyd, but black and dark colours can be made on linseed/tung alkyds or on linseed/tung/modified phenolic varnishes.

'Long life' finishes are based on chlorinated rubber or vinyls.

Boot-topping paints

The boot-top area of the hull is that area which is exposed when the ship is empty but immersed when laden. It is bounded by the topsides paint above and the bottom composition below. This area is the most difficult to maintain in good condition. When at sea it can be completely immersed in slightly alkaline sea-water (pH 8·0 to 8·2) or exposed to sun, wind, and salt spray. It is constantly buffeted by a moving stream of fully aerated sea-water, and when in port is subject to the rubbing of fenders and to attack by floating debris.

The requirements of the paint are therefore (i) good adhesion, (ii) maximum salt-water resistance, (iii) good abrasion resistance under wet conditions, and (iv) good weather-resistance. Systems which have given satisfactory service are epoxy, chlorinated rubber, polyurethane and vinyls.

The extent of the boot-top area varies considerably from one type of ship to another. In passenger liners it is a comparatively narrow band round the vessel, but in oil tankers it extends to almost deck level.

Toxins are sometimes incorporated into boot-topping paints to check the growth of fouling.

Bottom compositions

These are applied to the primed hull areas which are permanently submerged and they usually comprise two coats.

1 A barrier coat, the function of which is to isolate the primer from the anti-fouling composition and to provide extra protection to the hull since the anti-fouling composition can be regarded, to some degree, as expendable.
2 An anti-fouling composition whose function is prevention of attachment of marine growths such as barnacles, weed, etc., which can slow down a ship very considerably.

It has not been found possible to combine the two functions in one paint [5].

Barrier coats

These function as additional primers (as No. 655, *below*), so augmenting the blast primer, or as a physical barrier to moisture penetration (see Bituminous aluminium paint, *below*).

Primer No. 655 was evolved as a result of work by BISRA and has the following composition:

Pigment

Basic lead sulphate	2 parts by weight
Aluminium powder	1 part by weight
Barytes	1 part by weight
Burnt Island Red	1 part by weight

Medium

Linseed/tung/modified phenolic varnish, 2/1 oil length

Coatings of this type are widely used as primers and barrier coats in merchant shipping.

Bituminous aluminium paint is probably the most widely used barrier coat and consists of non-leafing aluminium pigment dispersed in a mixture of oleoresinous varnish and bitumen. The following composition is typical but there are many possible variations.

	% by weight
Aluminium N/L pigment	12·5
Basic lead sulphate	10·0
Gilsonite	22·3
Linseed/mod. phenolic varnish (solids)	7·5
Naphtha	43·0
White spirit	4·5
Cobalt drier	0·2

Apart from its use as a barrier coat between primer and anti-fouling composition this material is also applied to old but sound anti-fouling before application of a fresh coat of the composition.

Epoxy/coal-tar pitch is used both as a barrier coat and as a primer. It provides excellent resistance to moisture penetration and, by the use of airless spray, can be applied at 100 to 150 μm (dry-film thickness) per coat. Further coats can be applied to give total dry-film thickness of 300 to 400 μm. If multiple coats are applied, recoating times, both minimum and maximum, are of vital importance to ensure good intercoat adhesion. These times are influenced by temperature, and the following figures are suggested by Banfield [11].

10°C	Minimum 22 hours	Maximum 7 days
25°C	Minimum 8 hours	Maximum 2½ days

Epoxy/coal tar pitch is frequently used in repainting smaller vessels after hull cleaning. It is followed by bituminous aluminium, anti-corrosive composition (possibly) and anti-fouling composition.

Anti-corrosive compositions are not corrosion-resisting paints in the strict sense but are sealer coats. They were formerly used in repaint work to seal old anti-fouling compositions before the new compositions were applied. They have now been largely replaced by the bituminous aluminium barrier coats.

Anti-fouling compositions

Types of fouling

'Fouling' is the term applied to the growths, either animal (shellfish) or vegetable (weed) which can infest a ship's bottom. Fouling takes place mainly when the ship is at rest, and its presence can seriously reduce the speed of a vessel or result in increased fuel consumption to maintain the scheduled speed.

Fouling is usually preceded by the deposition of a film of 'slime' (containing bacteria, diatoms and protozoa) on which the larvae and spores feed in their early existence. Some modern anti-fouling compositions contain an anti-slime biocide.

Animal fouling involves barnacles, tubeworms, mussels, hydroids and others. Barnacles are the most common. The shells adhere strongly (even when the animal is dead) and give a very rough surface, causing high friction. Heavy chiselling or blast cleaning is necessary to remove them. Tubeworms can cause serious damage to the blades of screws, but they are readily killed by the toxins in anti-fouling compositions. Mussels also are easily killed by anti-fouling toxins.

Green weed ('grass') fouls the sides of hulls near the water-line, but brown and red species occur when the light is less intense.

Degree of fouling

The degree of fouling depends on several factors: (a) season of year – around the British Isles fouling takes place from April to September, in tropical waters, all the year round; (b) temperature of water; (c) geographical location; (d) the amount of light reaching the surface.

Types of toxins (poisons) used

Copper compounds have been the most widely used toxins since the time when copper sheathing was used to protect wooden vessels. The red cuprous oxide, the white thiocyanate and the black sulphide are used as also is the electrolytic copper powder.

The range of colours available with these toxins is limited and a demand for brightly coloured anti-fouling compositions by owners of small and racing craft provided impetus to the search for alternatives. Over the years a large number of organic compounds have been evaluated as toxins but the majority were useless. However, certain organo-tin compounds, particularly the tributyl- and triphenyl-tin oxides were found to be more effective than the copper compounds. These toxins were colourless and enabled the demand for brightly coloured compositions to be met.

A further development was co-polymerisation of an organo-tin compound (such as tributyltin methacrylate) with acrylates or styrene [6], [7]. The tin then became a component of the anti-fouling binder and a constant supply of toxin was available at the surface as the film slowly eroded. The erosion or dissolution of the binder was aided by the use of hydroxylated acrylic resin as the co-polymer. This type of binder forms the basis of the so-called 'self polishing' anti-fouling compositions.

It was found, however, that the widespread use of tin based anti-foulings had serious environmental repercussions. When numbers of small craft are moored in restricted waters such as estuaries and marinas of various types the concentration of tin in the water becomes sufficient to interfere with animal and vegetable marine growth. As a result, Legislation is now in force restricting the amount of tin in anti-foulings generally and virtually imposing a ban on the use of tin compounds on craft of 12 metres (39·37 feet) and under. The Regulations are the following:

Control of Pollution (Anti-Fouling Paints) Regulations, 1985.
Statutory Instrument, 1985 No. 2011
Control of Pollution (Anti-Fouling Paints) (Amendment) Regulations 1986
Statutory Instrument, 1986, No. 2300

The difficulty of producing coloured anti-fouling compositions using a copper toxin has been alleviated by the introduction of the 'Low-tint' cuprous oxide. Although the colours (and white) appear somewhat 'dirty' when freshly applied, they appear clean and bright after immersion for a short time.

On ocean going vessels cuprous oxide is not entirely satisfactory as a toxin for the green seaweed 'Enteromorphia'. When this growth is very troublesome, as it can be occasionally on the sides of tankers, the organo-tin compounds are used [8]. Tankers and other ocean going vessels are not in port long enough for serious pollution of the water to occur.

Types of composition

Soluble matrix These are based on a solution of rosin and oleoresinous medium. They function by slow dissolution of the acidic rosin in the slightly alkaline sea-water (pH 8·0–8·2) so that fresh poison is being exposed continuously. The solution rate can be controlled by inclusion of water-insoluble substances. Materials employed in the medium include rosin, linseed/tung/modified phenolic varnish, pitch, Stockholm tar, waxes, etc. In addition, the quantity and nature of the pigments and extenders present influence the solution rate.

The service life of this type of composition is influenced by the film thickness, and by the application of multiple coats at least two years' life is possible. The necessary film thickness can be obtained also by the use of high-build coatings or of high solids coatings which are applied hot.

Chlorinated rubber/WW rosin matrix The solution rate of rosin in sea water must be controlled or the film would rapidly be eroded with loss of expensive toxin. Where normal cuprous oxide is used, bitumen or oleoresinous varnish is used as control agent. When the bright colours are required plasticised chlorinated rubber is used as control agent. For optimum anti-fouling efficiency the ratio of chlorinated rubber to WW rosin is critical, a fact confirmed by the work of del Amo *et al.* [9].

Insoluble matrix In this type of composition the toxin is leached out slowly while the binder remains intact and unaffected by the sea-water. The concentration of toxin is sufficiently high to give particle-to-particle contact, so as each particle is dissolved by the sea-water, another is exposed. Concentrations of toxin are of the order of 0·8 to 1·0 kg of cuprous oxide per litre or mixed cuprous oxide and copper powder. Types of binder employed include chlorinated rubber and vinyl resins.

This type of composition is claimed to give longer periods at sea between repaints. Still longer periods are claimed [10] for compositions based on isobutylene and containing about 90 percent of cuprous oxide in the dry film.

Interior surfaces

Accommodation

Good quality decorative materials are employed, but the use of paint in these areas has decreased in recent years. Melamine-faced laminated panels are now widely used as linings for accommodation spaces [8]. Anti-condensation paint is used where steel bulkheads form walls of cabins or assembly rooms. These are frequently of a flame-retardant type (see Chapter 18).

Dry holds

In addition to protection of the steel, abrasion resistance is required, together with maximum light reflectance. A typical painting scheme consists of a red oxide/zinc chrome primer on a tung oil/phenolic medium followed by an aluminium paint on the same type of medium. Harder films are provided by epoxy esters, but when very high abrasion resistance is required, two-pack epoxies are used.

Oil tanks

These may contain cargoes of mineral, animal or vegetable oils and may often hold sea-water as ballast. Paints must of course be resistant to these materials. Two-pack epoxies are employed for refined products, but for crude oils, when slight discoloration would not be important, the cheaper epoxy/coal tar pitch coatings are used.

DOCK AND HARBOUR INSTALLATIONS

The surfaces to be protected fall into three classes: (i) those above the waterline such as warehouses, cranes, and pier superstructures; (ii) surfaces intermittently immersed in water, such as pier and jetty supports and the seaward side of dock-gates; and (iii) surfaces permanently immersed such as the dock side of lock gates and the bottoms of navigation buoys and lightships.

Surfaces above the water line

Conditions of exposure are severe and, in addition to maximum protection, painting schedules must frequently also provide decoration. Conventional decorative practice is followed, i.e. primers, undercoats and finishes, the latter being based on medium/long-oil alkyds. Epoxy ester systems have performed well on tankage in marine environments, whilst chlorinated rubber is finding widespread use on exposed steelwork generally.

Surfaces intermittently immersed and in splash zones

A considerable amount of hardwood has been used in jetty construction in the past and has been difficult to protect. Impregnation with creosote or tar before installation gives a measure of protection, but once installed, the time between tides is insufficient for drying out and consequently the wood remains unpainted.

Steel supports of piers and the seaward side of dock gates are exposed to the corrosive action of sea-water and the abrasive action of sand. They have to be repainted between tides when the surfaces are often difficult to dry. The types of coatings used in the past have been bitumen solutions and tar paints, but in their early life these are soft and readily abraded by sand. Much better results have been obtained by the use of epoxy/coal-tar pitch compositions, particularly the high-solids and solvent-free types. By incorporating certain cationic surfactants these can be applied to a wet surface, and the curing reaction will proceed when they are immersed in water. The cured films are tough and very water-resistant.

Surfaces permanently immersed

The bottoms of navigation buoys and lightships are subject to heavy fouling which progressively increases the weight. In the case of buoys, the increase in weight reduces the exposed area and hence the visibility. Heavy coats of anti-fouling compositions are applied, but when in due course this becomes exhausted and fouling commences, the ship or buoy is dry-docked, cleaned and repainted.

OFF-SHORE STRUCTURES [11]

These are complicated steel structures used by the oil companies (and known as 'rigs') or by the gas industry for the extraction and collection of natural gas (these are known as 'gas platforms'). Both types of structure are situated in the open sea and are therefore exposed to an extremely aggressive environment. They are designed for a life of 20 to 25 years and so require the most efficient form of protection available.

Totally submerged sections are given heavy protection at the fabrication stage. Several systems are employed, one of the most favoured consisting of blast cleaning to first-quality bright metal [3] or SA3 [2]. This is followed by zinc silicate primer and then epoxy/coal-tar pitch. If sacrificial anodes are used, the high-build type of epoxy/coal-tar pitch is employed.

In maintenance work, the underwater structures are cleaned by water jet or high-pressure air and then coated with a 'glovable mastic' applied by hand (by divers). The mastic is often given further support by wrapping in polythene sheet.

Tidal and splash zones also require maximum protection. Blast cleaning to SA3 is followed by zinc silicate primer. Finally epoxy/coal-tar pitch coatings to a total thickness of 600 to 700 μm are commonly employed.

Superstructure and deck areas Efficient blast-cleaning to SA3 is followed by either zinc silicate or two-pack epoxy primer. Several types of finishing coat are available. The zinc silicate primer can be overcoated with either two-pack epoxy, chlorinated rubber or vinyl finishes. (If vinyl finishes are to be used on zinc silicate primer, a 'tie-coat' is advisable.) Vinyl and chlorinated rubber finishes show good durability and present fewest problems in recoating.

References to Chapter 23

[1] Ford, K. S., Developments in marine paints. *JOCCA*, 1972, **55**, 584.
[2] Swedish Standard SIS 05 59 00. Standardiseringkommissionen i Sverige, Stockholm.
[3] BS 4232: 1967. Surface finish of blast-cleaned steel for painting. British Standards Institution, London, W1A 2BS.
[4] BS 282, 389: 1963 (1987). Zinc chrome, Type 3. British Standards Institution.
[5] van Louden, A. M., Testing and investigation of ship bottom paints. *JOCCA*, 1969; **52**, 141.
[6] Evans, C. J. and Smith, P. J., Organo-tin based anti-fouling systems. *JOCCA*, 1975, **58**, 160.
[7] Sghibartz, C. M., New developments in self-polishing anti-foulings. *JOCCA*, 1985, **68**, 233.
[8] Banfield, T. A., Marine finishes. Part 2. *JOCCA*, 1980, **63**, 93.
[9] del Amo, B. *et al.*, Anti-fouling paints based on WW rosin and chlorinated rubber: influence of binder composition and content. *JOCCA*, 1986, **69**, 178.
[10] *Marine Fouling and its Prevention.* Woods Hole Oceanographic Institute, U.S. Naval Institute, Annapolis, Md. 1952.
[11] *see also* Atherton, D., Original and maintenance painting systems for North Sea oil and gas platforms. *JOCCA*, 1979, **62**, 351.

24

Miscellaneous coatings and ancillary materials

This chapter contains coatings and relevant products which are difficult to classify under other chapter headings. Also, paints which have received but brief mention in other parts of the book are described more fully.

Paints for special floors

Asphalt floors are encountered in factories, warehouses, car parks, playgrounds and so on. They can be subjected to considerable pedestrian and vehicular traffic. Paints applied must not soften or dissolve the bitumen (or tar if tarmacadam has been used), a restriction which precludes the use of materials thinned with aliphatic or aromatic hydrocarbons. The traditional type of paint for these floors has been based on resins soluble in alcoholic solvents (industrial methylated spirit or isopropanol), particular attention being paid to pigmentation to give maximum hardness and wear resistance. Even so, areas exposed to constant traffic require repainting at frequent intervals.

Films of greater hardness and abrasion-resistance are produced by the use of water-borne expoxies. These are solvent-free materials in which the resin/pigment and hardener components are dispersed in water and supplied in two packs which are mixed immediately before use.

'Self-levelling' and 'seamless' floors are based on two-pack epoxies, two-pack polyurethanes or catalysed (benzoyl peroxide) polymethacrylates. The resin is mixed with sand or other aggregate before laying. Finishing paints can be based on the same general type of resin. With the very hard-drying epoxy systems the paint should be applied within 48 hours from setting of the floor. After this time the hardness of the floor surfaces affords poor adhesion for the paint coating, unless it is abraded.

Road marking paints are used on asphalt and concrete roads, surfaces of car parks, aerodrome runways and factories. As mentioned, paints for asphalt surfaces must be based on resins soluble in alcoholic solvents. Manila resin dissolved in industrial methylated spirit is commonly used, suitably pigmented to give maximum abrasion resistance. The following formulations illustrate two possible pigmentations for white line paints. Yellow line paints can be pigmented with lead chrome or, if required for use on concrete, with an alkali-resistant pigment of the arylamide type.

White line paint

	% by weight			*% by weight*
Lithopone	36·5		Rutile titanium dioxide	20·5
Fine silica	12·7		Barytes	16·0
Manila resin	25·4	or	Fine silica	12·7
IMS 64 O.P.	25·4		Manila resin	25·4
			IMS 64 O.P.	25·4

This type of paint dries rapidly and, depending on conditions, takes traffic in ½–¾ hour. It can be used also on concrete, but here other paints with a higher degree of alkali resistance are preferable. These are based on chlorinated rubber, vinyl toluene/acrylic co-polymer and thermoplastic acrylics.

Paints for aircraft

Subsonic types of aircraft have a basic structure of aluminium alloy with high-tensile steel and magnesium alloys for certain specialized parts.

For structural aluminium alloys, epoxy systems are favoured. Anodizing is followed by a primer based on epoxy/polyamide and pigmented with strontium chromate. Epoxy/polyamide finish is then applied. Painting is carried out before assembly. Steel parts are phosphated and the absorbent phosphate coating is impregnated with low-viscosity high-bake epoxy lacquer. This is followed by epoxy primer and finish [1].

The outside skin of the aircraft presents a smooth aluminium alloy surface and the finish must be decorative as well as protective. Protection is required against wide fluctuations of temperature, intense ultraviolet radiation at high altitudes, and impact with raindrops at high velocity. Paints must be resistant also to fuels and to hydraulic, brake and de-icing fluids. Temperature fluctuations can result in condensation of appreciable amounts of water inside the outer skin. This can boil at the temperature reached by the skin of some aircraft and is a potent source of corrosion.

The surface of the aircraft becomes increasingly hot as the speed increases; for subsonic civilian and transport aircraft this can be of the order of 100°C.

Priming and any pretreatment is carried out before assembly, and this usually follows the pattern outlined above for structural aluminium alloys. The metal is anodized and primed with an epoxy/polyamide resin paint. After assembly a second coat of primer is applied, followed by the finish and all decorative work. The finish which has been increasing in popularity in recent years is polyurethane based on aliphatic isocyanate. This provides good gloss-retention, colour and chemical resistance.

With high-speed aircraft, temperatures are higher. According to Hoey [2] a Mach 2·2 aircraft flying at 60 000 ft is likely to develop a surface temperature of about 120°C (248°F) but this figure is likely to be modified to some degree by the reflectivity and emissivity of the paint coating.

The exterior finishing system employed is somewhat similar to that described above, namely pretreatment of the aluminium alloy followed by epoxy/polyamide/strontium chromate primer and polyurethane finish based on aliphatic isocyanate. Acrylics are sometimes used: they show good colour stability at the temperatures reached, but their resistance to the hydraulic fluids is not as good as that of the polyurethanes. Two-pack epoxies are not favoured for the exterior of modern aircraft. They chalk rapidly at high altitudes where the ultraviolet radiation is more intense than at lower levels.

The interior alloy surfaces are given the same pretreatment and primer as the exterior surfaces, but they are often finished in two-pack epoxy enamel.

Clear wood finishes

Interior The factors of prime importance are preservation of the appearance, resistance to heat and fluids of various types, and resistance to mechanical damage. Resistance to weather is not of importance. The finishes in use include the following.

Copal varnish This is now seldom made on copal resin which has become too expensive. Usually it is a linseed/tung/modified phenolic varnish which dries to a hard film that retains its gloss for a considerable period. It does, however, tend to 'yellow' on ageing.

One-pack polyurethane varnish gives very hard and abrasion-resistant films. It is resistant to heat, alcohol (beverages), water and detergents. Applications include table tops, bar counters, floors and interior woodwork generally.

Acid-cured alkyd/amino clear finishes are used for similar purposes.

Exterior Clear finishes on exterior woodwork suffer rapid breakdown as a result of ultraviolet penetration. This can, of course, be reduced by the incorporation of UV absorbers (p. 55).

Protection of the wood is largely dependent on film thickness of the varnish so that 4 or 5 coats are applied to give protection for 2 to 3 years. Woods vary in their absorption characteristics [3] and the multicoat application ensures a uniform finish.

The use of 'water-repellent' finishes has become more general in recent years. These are clear or lightly pigmented materials which impregnate the surface layers of the wood with a water-repelling material. It is claimed that this allows the wood to 'breathe' more than paint does, but it does not bring out the grain as well as conventional varnishes. A fungicide is also incorporated. Such a material was developed in the USA and is known as the 'Madison formula'. The original formula contained boiled linseed oil, wax, zinc stearate, a fungicide (pentachlorophenol) and white spirit. It can be tinted to match the colour of woods. Application is by brush, the excess being wiped off after about two hours. In the UK it dries slowly and the film tends to hold dirt. Breakdown is by gradual erosion, the life being about three years. It can then be cleaned down and a further coat(s) applied.

A water-based clear finish based on a vinylidene co-polymer is claimed to accommodate the movement of wood without cracking, to possess high moisture-resistance, and to provide excellent grain enhancement [4]. It gives excellent results when used over a clear varnish (where it acts as an oxygen barrier) or as a four-coat system. The life is claimed to be at least two years.

Stains for wood consist usually of a solution of a dye in water, spirit or oil, but semi-transparent pigments dispersed in oil or varnish are sometimes employed.

Water stains are the least popular as they tend to raise the grain of the wood. They frequently contain natural colouring-matter such as logwood [5]. A dilute solution of potassium permanganate confers a brown stain on softwood.

Spirit stains consist of dyes dissolved in a solution of shellac or other resin in industrial methylated spirit. They penetrate well and dry rapidly. Generally the stained wood is varnished.

Oil stains are either solutions of dye in linseed oil and white spirit or dispersions, in the same medium, of semi-transparent pigments. After staining, the wood is varnished. Staining and varnishing can be carried out in one operation, but the result is influenced to a marked degree by the nature of the wood.

Scumble stain (or 'graining' colour). Scumbling is a process which was formerly used extensively but is now used to a very limited extent. It was designed to enable a decorator to imitate the grain of wood or to produce special grain effects. The surface is first given a flat ground coat or undercoat in a white or pale colour. Over this is applied a coat of scumble stain which consists of semi-transparent pigment dispersed in a varnish medium containing a gelling agent (usually a metallic stearate). A graining comb or other suitable tool is used to produce the desired pattern which dries without flowing out. Finally the surface is varnished or glazed [6]. Scumbling was used mainly on outside doors.

Exterior woodwork can be treated with oil or varnish stains or with one of the many water-repellent preservative stains on the market. The latter are often based on oil or varnish media containing fungicide and are lightly pigmented with semi-transparent pigment (see Madison formula, *above*). Over-varnishing is generally required.

Motor-car finishes

The basic requirements of paint systems for cars are protection of the metal, maximum durability, good polishing properties, and resistance to chipping.

The application of primers to the phosphated steelwork by electrodeposition (p. 363) has resulted in a very considerable improvement in protection, particularly in box sections and other parts which were not adequately coated by the conventional dipping methods.

Underbody protection, in addition to the electrodeposited primer, is generally provided by bituminous type coatings which exhibit good water-resistance. It is difficult to understand why this very necessary additional underbody protection or 'underseal' has for so long been an 'extra' for the purchaser and not a standard feature applied by the manufacturer. In recent years, however, PVC plastisols have been applied by some manufacturers to certain underbody areas. They are more resistant to petrol and oils than are the bituminous types and exhibit better durability.

For bodywork on new cars, the alkyd/amino type of stoving finish has been replaced very largely by hydroxy functional (thermosetting) acrylics, cross-linked with melamine. These show very good durability, particularly in metallic finishes [7], and can be cured at 120°C (248°F) for 30 minutes. Thermosetting acrylics, in the form of polymer dispersions in aliphatic solvents ('non-aqueous dispersions' or NAD) (ICI), give high-build finishes with good solvent release.

In 're-finishing' operations, stoving is not possible and so finishes must dry or cure at ambient temperatures. Finishes in use include modified alkyds, cellulose nitrate, and thermoplastic acrylics. Two-pack systems in which hydroxylated acrylic resins are cross-linked with isocyanate have shown excellent performance.

Paint removers

Alkaline type the oldest and cheapest type of paint remover is based on sodium hydroxide and relies on saponification to destroy the paint film. For this reason it is ineffective against many modern coatings and furthermore is not recommended for use on woodwork, zinc, aluminium, or porous surfaces. It is, in fact, suitable for use only on ironwork where other methods are too costly. Alkaline paint removers can be based on the following general type of formulation, but the figures quoted can be varied widely.

	% by weight
Sodium hydroxide	5·7
Water	28·3
Non-ionic surfactant	0·3
Glycerin	0·7
Whiting	58·0
China clay	7·0

Operatives using this material should wear face shields, gloves and protective clothing. The cleaned surfaces should be washed thoroughly with water and dried before repainting.

Solvent type This employs strong solvents to penetrate and swell the paint coating, so destroying the adhesion to the surface. The solvents are kept in contact with the film by the use of cellulose acetate and wax, which delay evaporation. An emulsifying agent is often added to permit the surface to be cleaned with water. These are described as 'water-rinsible'.

It is often stated that several applications are necessary to remove several coats in a system, but according to Hurst and Goodier [8] this is unnecessarily expensive. They suggest that a thin coating of the paint remover be applied and allowed to remain for 20–30 minutes. A second coating is then applied as thickly as possible and left undisturbed until the paint film has softened throughout. This may take several hours but the coating would then be removed completely to the substrate. The surface is then cleaned thoroughly with white spirit or water.

Solvent-type paint removers may be either flammable or non-flammable, the latter being the more common. In each case gloves and protective clothing should be worn, and care taken to avoid inhaling the solvent vapours.

The following formulation illustrates the general type of composition of a non-flammable water-rinsible paint remover:

	% by weight
Methylene chloride	88·8
Cellulose acetate	1·2
IMS 64 O.P.	4·9
Paraffin wax	0·8
Acetone	0·4
Toluene	2·8
Triethanolamine	1·1

Grease paints (also known as 'non-oxiding' paints)

The traditional type was based on green petroleum jelly, drying oil and pigment. Manufacture consisted in melting the petroleum jelly, adding the pigment ground in oil, and stirring until the mass had set sufficiently to avoid pigment settlement.

Thick films are applied by brush, but although the films remain soft and flexible, they are sufficiently rigid to resist the movement of water and rain.

This type of grease paint has been applied to some types of structural steelwork and steelwork difficult of access, such as bilge and ballast tanks on ships. Recent applications have included use as an anti-vandal or anti-thief treatment for walls and down pipes.

A more modern type is based on the calcium soaps of acids produced by the oxidation of higher petroleum hydrocarbons and waxes. These dry sufficiently hard to withstand occasional foot traffic. They are used for coating the interiors of box sections of bridges and for areas of steel structures which are difficult or impossible to clean sufficiently for conventional paint systems.

Trichlorethylene paints [9]

These paints derive their name from the solvent employed which confers several advantages on the paints. Among these are (a) rapid evaporation – the articles dry in 60 to 120 seconds and so allow fast operation of production lines; (b) rapid set-up of the film eliminates sagging in dipping operations; (c) good cover of edges; (d) no flash point and therefore no fire risk. In view of the toxicity of trichlorethylene, special attention is paid to plant design.

Bacteriostatic and fungistatic paints

Liquid paints Bacterial attack is generally confined to aqueous systems containing suitable nutrients such as cellulose ethers, starches, animal and vegetable glues and casein. Of these, the cellulose ethers are the most widely used, generally as thickeners for emulsion paints. The bacteria secrete enzymes which act as catalysts and break down the cellulose molecules into simple substances which are then assimilated by the bacteria. Biocides are effective against bacteria but not against enzymes which can be present in the paint raw materials [10].

A number of materials have been proposed and evaluated for 'in-can' preservation of emulsion paints. The earlier types included phenolic derivatives such as orthophenylphenol, *p*-chloro-*m*-cresol (PCMC) and sodium pentachlorphenate. Organic mercurials and organotin compounds have both proved very effective, but the former type is no longer used in view of its toxic nature. A range of effective organic bacteriostats is available [11] including fluorinated sulphonamides, organic amides and fatty-acid modified amides. 1:2-benzisothiazolin-3-one [12] is widely used and is highly effective at a concentration of 0·01 to 0·05 percent on the paint.

Fungistatic paints

Fungi and moulds of many types will grow on most surfaces under the appropriate conditions; fungistats should, therefore, be effective against a wide spectrum of growths.

The results of studies of the structure and growth processes of one of the most widespread of mildew-producing fungi (*Aureobasidium pullulans*) have been published by several investigators in the USA [13–16]. The object of these studies was elucidation of the mechanisms of growth and the food requirements of fungi as the primary stage in their control.

In interior conditions of high humidity or damp walls, both emulsion and solvent-type coatings can be affected. Emulsion paint films, by virtue of the presence of water-soluble cellulose derivatives and other ingredients, offer a favourable nutrient and are most affected. In an investigation into the preservation of emulsion paint systems, Carter [11] has stated that, in addition to the necessary fungicidal activity, biocides must show low solubility in water, low vapour pressure, and stability both to ultraviolet radiation and to sulphur staining. Fungicidal activity must also be sustained in the dry film. 2-(4: thiazolyl) benzimidazole is claimed to give protection against a wide variety of fungi, but it has no 'in-can' preservative properties.

The protection of paint films against fungi and moulds under exterior conditions has been discussed by Pauli [17] who emphasizes the importance of evaluating fungistats in the specific coating and under specific conditions. He lists a number of materials possessing the necessary sustained activity in addition to the other essential properties. These include modified barium metaborate [18] (which, however, has to be used at 15 to 20 percent by weight on the solids content), dithiocarbamates (used mainly in interior emulsion paints), phthalimide derivatives, substituted sulphonamides and sulphones.

Fungistatic paints generally contain sufficient biocide to prevent fungus growth on the film, but they are not fungicidal. When it is desired to paint surfaces which are infected with mould, steps must be taken to remove the growth and to sterilize the surface. The time-honoured method has consisted in scraping and/or hard brushing followed by treatment with a fungicidal wash.

Fungicidal wash

A variety of fungicides are employed, some in solution in water, others in solvent such as white spirit or aromatic solvent. The aqueous type is preferred for interior use where solvent vapour can be objectionable, whilst the water-insoluble solvent type is used on exterior work. The surface is thoroughly soaked with fungicide solution and is then allowed to dry out before being painted.

Methods for cleaning infected walls have recently been investigated by Whiteley and Bravery [19] who consider that fungicidal washes are less effective in preventing further growth than very thorough cleaning by water jet, preferably containing abrasive.

The preservation of stonework

The exterior architecture of many of our older cathedrals, churches and other buildings is a valuable heritage of which we are temporary custodians with a responsibility to future generations. Unfortunately neglect over many years has allowed progressive decay to take place, in consequence of which structures have become unsound and the intricate architectural features severely eroded. The problems and dangers are now recognized and remedial measures are being undertaken as rapidly as finances permit, cost being the rate-controlling factor.

Causes of decay Much of the stonework consists of either limestone (calcium carbonate) or magnesium limestone (calcium and magnesium carbonates), both of which are highly susceptible to attack by the following.

Frost Damage from frost occurs only on parts of structures which are frequently frozen while very wet. In this country this seldom occurs between the eaves and the damp-course [20].

Crystallization of salts resulting from the action of atmospheric sulphur dioxide producing calcium sulphate in the pores of limestone and calcium sulphate with magnesium sulphate in the magnesium limestone. Repeated crystallization of these salts results in disintegration of the surface of the stone and can produce deep cavities. This reaction, involving sulphur dioxide, is the most important cause of breakdown in the United Kingdom.

Dissolution of carbonates is caused by carbon dioxide in rainwater, forming water-soluble bicarbonates of calcium and magnesium.

Protective treatment The earlier treatments for preservation of the stone consisted in application of water-repellents such as emulsions of resins, waxes and other materials. These were not successful since the applied films were not sufficiently thick. Water penetration was not prevented and differential expansion effects accelerated the breakdown [21].

More modern approaches entail deep impregnation by resins or precipitation of other insoluble inorganic salts in the pores of the stone and are described by Price [20]. Resin impregnation is achieved by application of suitable monomers to give deep penetration followed by *in situ* polymerization. Monomers of interest include acrylics, epoxies and alkoxysilanes.

The second process involves precipitation of barium salts within the pores of limestones. Evaluation of the effectiveness of any process of stone preservation is slow and requires a number of years before conclusions can be drawn.

Paints for plastics

The extensive use of plastics has created a demand for coatings which are both decorative and protective. Many different types of plastic resin are in use with differing surface characteristics and, in some cases, properties vary between members of the same group.

It is generally accepted that the main problem in painting plastics is that of adhesion. Good adhesion requires a thoroughly cleaned surface but, as Risberg [22] has pointed out, other factors in the surface such as traces of mould release agent or migrated additive can play an important part.

It was formerly thought that a strong solvent which would 'bite' into the plastic surface was necessary for good adhesion. However, in many cases this led to swelling and subsequent crazing of the surface. It is now accepted that good adhesion is associated with the presence of polar groups at the paint/plastic interface but with non-polar plastics, such as the polyolefins, special surface treatments are required. In order that the polar groups can function, not only must the surface be clean but the paint must 'wet' the surface effectively. The topic of wetting is discussed by Arnoldous [23] in relation to the use of aqueous dispersions of acrylic and polyurethane resins on plastics. A knowledge of the critical surface tension of the plastic is useful in formulating a suitable coating. The surface tension of an aqueous paint can be adjusted to the same order by means of water-miscible solvents. The acrylic and urethane dispersions are now being used widely.

A number of solvent-borne coatings are in use among which are the two-pack polyurethanes, thermoplastic acrylics, PVA/PVC and modified alkyds. As mentioned above, the coating is usually matched to the plastic and many coatings are in use reflecting the ingenuity of the paint technologist.

Coating of pipelines.

The principal materials of construction are mild steel, concrete, uPVC and polyethylene. Problems of protection are most difficult with mild steel and to a lesser extent with concrete. The majority of concrete pipelines are buried and are treated with bitumen or tar compositions before being in-filled.

Steel pipelines are used both below and above ground. A number of protective systems is in use generally preceded by blast cleaning. The following schemes are in common use.

Buried lines (a) Wrapping at works with glass fibre impregnated with bitumen or tar. A special adhesive primer is used. The joints then require protection at site. (b) Electrostatic spray application of a heavy coating of epoxy powder [24]. The pipes are preheated to fuse and cure the resin.

Exposed (above ground) The pipes are primed and are commonly wrapped with adhesive PVC or polyethylene tapes which often are coloured as an aid to pipe identification. When black is used the polyethylene tape is stated to be more durable than the PVC material [25]. Glass fibre tapes impregnated with bitumen followed by a bituminous paint are also in use.

References to Chapter 24

[1] Scott, J. A., Aircraft finishes. *JOCCA*, 1969, **52**, 593.
[2] Hoey, C. E., Paints for high speed aircraft. *JOCCA*, 1970, **53**, 1026.
[3] Hill, R. R., Water repellent preservative finishes. *JOCCA*, 1973, **56**, 251.

[4] Scott Bader & Co. Ltd (Wollaston, Northants), 'Polidene for paint'. Tech. leaflet No. 649, 1974.
[5] Hurst, A. E. and Goodier, J. H., *Painting & Decorating*, 9th edn, p. 204. Charles Griffin, 1980.
[6] *Ibid.*, p. 205.
[7] Taylor, J. R. and Foster, H., Reflow thermosetting acrylic resins - some aspects of their performance and uses. *JOCCA*, 1968, **51**, 975.
[8] As ref. [5], p. 209.
[9] Geisgen, H. R., *Ind. Lack-Betrieb.*, 1973, **41**, 32.
[10] Miller, W. G., Incidence of microbial contamination of emulsion paint during the manufacturing process. *JOCCA*, 1973, **56**, 307.
[11] Carter, G., Some aspects of the prevention of microbiological attack on emulsion paint systems. *JOCCA*, 1973, **56**, 302.
[12] '*Proxel*' (ICI Ltd, London).
[13] Crang, R. E. and Pechak, D. G., *Aureobasidium pullulans*: firm structure and development. *JCT*, 1978, **50**, No. 639, 36.
[14] Zabel, R. A. and Terracina, F., Nutrition of saprobic fungi and control strategies for paint mildew caused by *Aureobasidium pullulans. Ibid.*, p. 43.
[15] Winters, H., Enzymology as it relates to mildew defacement of organic coatings. *Ibid.*, p. 48.
[16] Schmitt, J. A., Growth patterns of paint mildew and *Aureobasidium pullulans. Ibid.*, p. 51.
[17] Pauli, O., Anti-mildew coatings. *JOCCA*, 1973, **56**, 285.
[18] 'Busan 11-M1' (Buckman Laboratories Inc., Memphis, Tenn.).
[19] Whiteley, P. and Bravery, A. F., Masonry paints and cleaning methods for walls affected by organic growth. *JOCCA*, 1982, **65**, 25.
[20] Price, C. A., Stone-decay and preservation. *Chem. Brit.*, Oct. 1975, p. 350
[21] Clarke, B. L. and Ashurst, J., Stone preservation experiments. Building Research Establishment, 1972.
[22] Risberg, M., Plastics as a painting substrate. *JOCCA*, 1985, **68**, 197.
[23] Arnoldus, R., Acrylic and polyurethane dispersions in industrial coatings for plastics. *JOCCA*, 1985, **68**, 263.
[24] Tate, D. H., Pipeline coatings (report of lecture). *JOCCA*, 1986, **69**, 106.
[25] BS. 5493: 1977. Protective coating of iron and steel structures against corrosion. British Standards Institution, London.

25

Testing and evaluation

An exhaustive treatment of the testing and evaluation of liquid paints and paint films in the compass of one chapter is impossible, and the treatment is therefore restricted, in the main, to tests normally carried out in the course of paint production. In some cases, alternative methods or instruments will be described.

It is convenient to divide the tests into those involving (a) liquid paints, (b) application and wet films, and (c) dry paint films.

TESTS ON LIQUID PAINTS

Production control

Tests applied in production control are designed to ensure that the product conforms to a predetermined standard or set of values. Where quantitative measurements are involved, the maximum deviation from mean values will also have been established. Generally, samples are drawn at various stages of manufacture, and the factory mixing sheets are usually designed to show the results on intermediate and final tests on a batch of paint.

The following tests are carried out in normal production, but specialized products may require other tests.

Degree of dispersion

Often known as 'fineness of grind' degree of dispersion is determined on samples drawn from bead mill, ball mill, etc., before the final thinning down. The material is applied to a 'Hegman' or fineness-of-grind gauge, the use of which is described on p. 251. Using the 0 to 8 scale, an acceptable limit for undercoats is 6 and for the finishes 7½ or preferably 'off scale'.

Viscosity

Viscosity measurement is not only a useful check on make-up, but uniformity of viscosity is essential for modern production lines and other applications involving continuity of supply.

A number of instruments are in use for its determination. The most widely used are the following.

The ICI cone and plate viscometer [1] The instrument is illustrated in Fig. 25.1. The thermostatically controlled plate can be set at the desired temperature (usually 20° or 25°C) (68° or 77°F), while the rotating cone, which has an apical angle of 178°, gives a fixed rate of shear of 10 000 sec^{-1}.

A drop of the paint is placed in the centre of the plate and the cone lowered into position. The scale reading, in poise, is noted when it reaches a steady value.

A high proportion of modern paints, particularly the high-build types designed for airless spray, exhibit thixotropy to a greater or lesser degree, and the viscosity of such paints cannot be measured in the efflux type of viscometer. In the cone and plate, thixotropy is eliminated by shear before the viscosity reading is obtained. Other advantages are speed of measurement, small quantity of material required, and ease of cleaning.

The Ferranti–Shirley viscometer is a more elaborate and sophisticated cone and plate instrument which can be used to study flow properties over a range of shear rates and temperatures. A recording instrument can be attached to furnish results in the form of flow curves (p. 325). This viscometer is more suited to research or project work than to routine quality control.

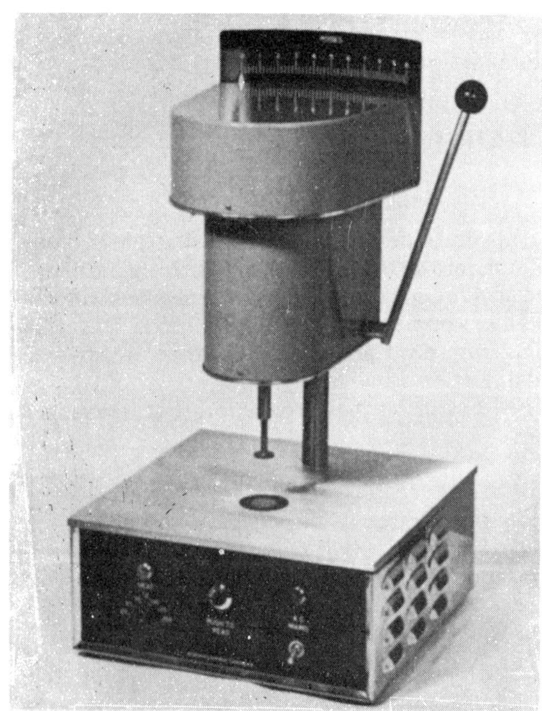

Fig 25.1 ICI cone and plate viscometer

The Ferranti portable viscometer is a co-axial cylinder instrument, the outer cylinder being rotated. The inner cylinder is connected through a spring to a dial, and this shows the torque transmitted through the liquid by the rotation of the outer cylinder. Three speeds are used and sets of cylinders of different sizes are supplied to cover a range of viscosities.

This instrument is very useful not only for assessing the degree of thixotropy in a paint, but, what is more important in practice, the rate of recovery of gel structure.

The Rotothinner [2] is a dual-purpose instrument used both as a thinning mixer and as a viscometer. It is illustrated in Fig. 25.2.

A frequent task in the control laboratory is the determination of the amount of thinner to be added to a batch to attain a specified viscosity, and with the conventional efflux cups this can be a time-consuming operation.

Using the Rotothinner, a given amount of paint is placed in a 250 cm^3 tin on the turntable and the stirrer lowered. The torque transmitted through the liquid

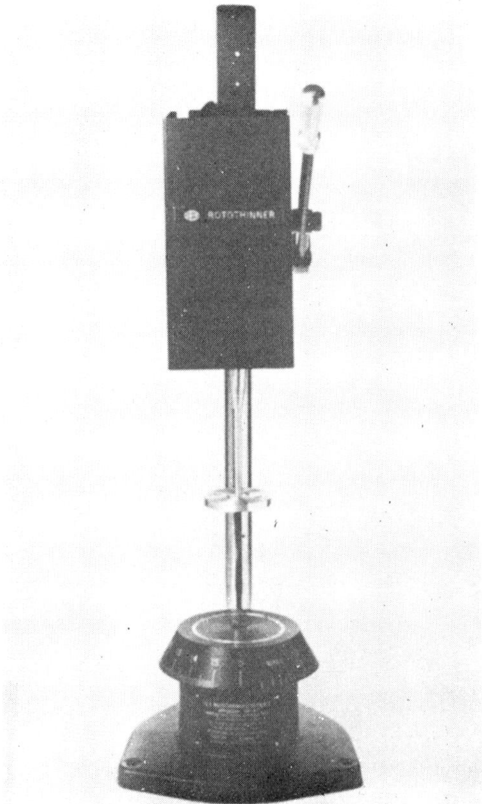

Fig 25.2 Rotothinner

exerts a twist on the turntable, the base of which is marked off in poises, and a fixed pointer indicates the viscosity. Thinners can be added from a graduated cylinder until the desired viscosity is reached.

The Brookfield viscometer is a portable instrument which indicates viscosity directly in centipoises by the torque on a rotating spindle immersed in the paint. A number of spindle sizes are available to cover a wide viscosity range.

The Stormer viscometer (Fig. 25.3) can be fitted either with a paddle or with a temperature-controlled rotating cylinder. In the former type the paddle is immersed in a sample of the paint and rotated by string and falling weights. The weights are adjusted to give a uniform speed of rotation as indicated by a stroboscope, and by means of tables the weights are converted to Krebs units.

Specific gravity

Specific gravity is determined using a cylindrical cup of 100 cubic centimeters

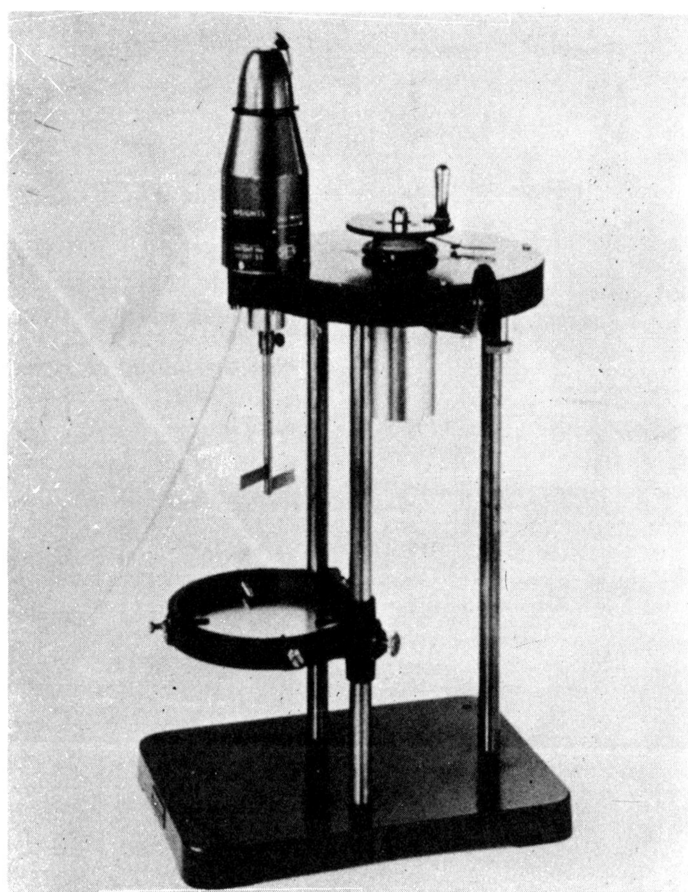

Fig 25.3 Stormer viscometer

furnished with a close fitting lid pierced with a small hole. It is used in a similar manner to a specific gravity bottle.

Specific gravity is a useful check on composition and is, of course, necessary for transport and shipping.

Volatile content

The volatile content is a useful check on composition and is often required in specification work. It is determined by weighing out 2 to 3 grams of paint into a flat-bottomed dish or tin lid (this is disposable) and heating for about 3 hours in an oven at 105°C (221°F). It is cooled, weighed, and reheated for a further ½ hour to check that the weight is constant. The volatile content is expressed as a percentage of the original paint.

Flash point

This is required for transport and shipping regulations. The majority of paints flash below 120°F (49°C), and the Abel (closed cup) method is used. The use of this apparatus is described in detail in British Standard 2000, Part 33 1982 [3].

Opacity

Opacity or hiding power is the ability of a paint to hide or obliterate the surface. It depends on the light-scattering and absorption properties of the dispersed pigments and on the thickness of the film. Opacity measurements are usually made on dry films of known thickness, but occasionally the need arises for the measurement on the liquid paint. This is best carried out by the use of the Pfund Cryptometer which is described in Chapter 2, p. 26.

APPLICATION AND WET FILMS

Brushability

Although the viscosity figure can afford a guide to the ease of brushing or suitability for spraying, other factors are involved which appear only during the application process. Brushability tests of decorative paints are made by application to areas of about 2 feet (0·6 metre) square to one side of which a piece of moulding is attached. The latter will show the behaviour of the paint on sharp edges and in hollows. On the panel, properties such as brush drag and ease of spreading, tendency to sag, levelling and wet-edge time can be assessed. For the latter, one half of the panel is coated, left for a short time, then the second half coated so that the two halves can be joined at the specified time. It should be possible to join up without leaving any visible sign (BS 3900) [4].

Sag resistance

This test is used for high-build paints to be applied by airless spray or brush. The paint is applied at a predetermined thickness to a steel or glass panel. A sag

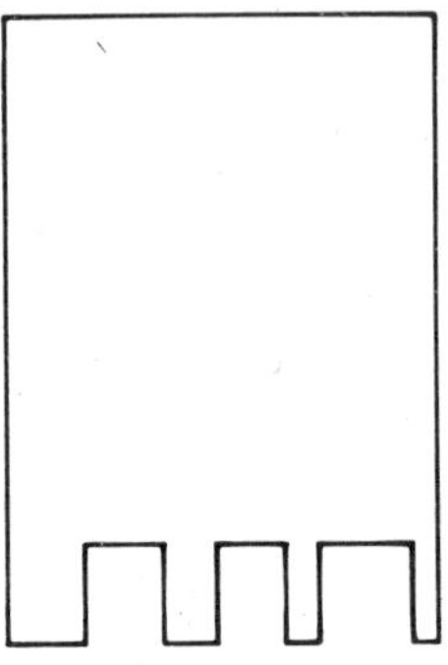

Fig 25.4 Sag comb

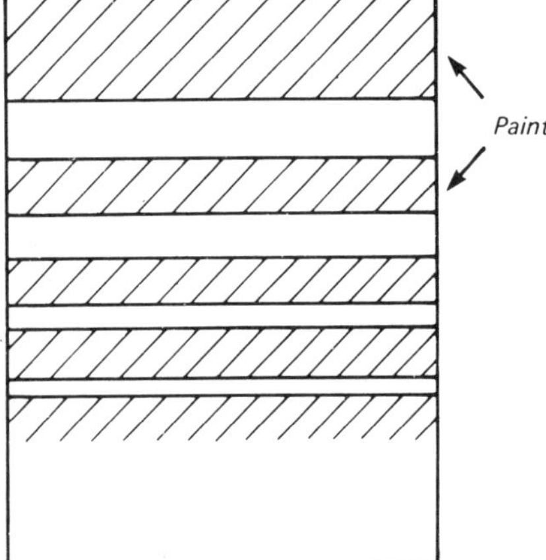

Fig 25.5 Sag test

comb (Fig. 25.4) is drawn across the film so that the substrate is based in strips of graded width (Fig. 25.5).

The panel is immediately placed in a vertical position. Any sagging becomes apparent by flow across the bared strips.

Measurement of wet-film thickness

A knowledge of the wet-film thickness is required in the sag resistance test (*above*). It is an invaluable aid in on-site painting. From a knowledge of the wet-film thickness and the solids content of the paint by volume, the dry-film thickness can be calculated (Chapter 1).

The most successful instrument for this purpose is the Wet-Film Thickness Gauge and is illustrated in Fig. 25.6. It consists of two solid metal disks and between them is mounted a slightly smaller third disk which is eccentric to the

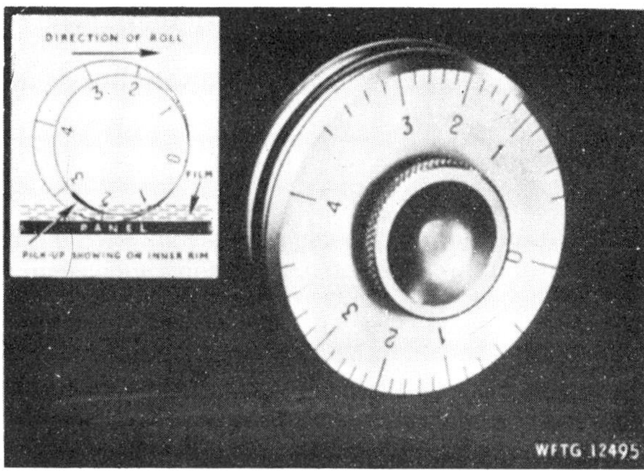

Fig 25.6 Wet film thickness gauge

other two, and at one point the three circumferences are in the same plane. This point is marked O and the point opposite is marked – the difference in diameter (in mils) between the centre and outside disks.

To use the gauge, it is placed on the film with point 4 downwards and then rolled along. The point at which paint is picked up on the centre disk is noted and the corresponding marking on the outer disk gives the thickness of the film.

A second method uses a comb in which the teeth, with the exception of those at the two ends, are stepped. The differences in length are usually 1 mil, so the instrument can be used only for approximate measurements.

In use the comb is held vertically on the paint film and a note made of the tooth which just touches the film.

Spraying properties

A high proportion of industrial air-drying and stoving paints are applied by one or other of the spraying methods (Chapter 20). Application of paint by the appropriate method will reveal any tendency to poor atomization, spray, orange peel, etc. These faults are discussed in Chapter 27.

Covering power

Covering power, often referred to as spreading rate, is expressed in square metres per litre or per 5 litres and represents the area which can be covered when the paint is applied at a given thickness by the specified method.

A theoretical figure for covering power can be calculated from the wet (or dry) film thickness and the volume solids, but 'practical' figures can be appreciably lower and depend on the nature of the substrate, the application technique, prevailing conditions and sundry losses. In fact, practical figures can vary from 60 to 90 percent of the theoretical values.

An approximate figure for brushing paints can be obtained by application of sufficient paint to cover a known area (by using, for example, a Morest chart, as described on p. 27) and then weighing the amount used.

Covering power or spreading rate should not be confused with opacity or hiding power.

Drying time

The drying time of a paint is the time required for the film to change from the liquid condition, as applied, to a solid film. Three types of drying or hardening are encountered.

1 Loss of volatile thinner, leaving a hard film. Drying times are short and of the order of 10 to 30 minutes. This type of drying occurs with lacquers and other non-convertible coatings generally.
2 Loss of volatile thinner followed by oxidation and polymerization of unsaturated fatty chains, as in oleoresinous varnishes and drying-oil alkyds. Drying times vary from 1 to 20 hours.
3 Chemical reaction between film components, where solvent may or may not be present. Examples are amine- and polyamide-cured epoxies, polyurethanes, and acid-cured amino/alkyd coatings. In each case the rate of change is increased by a rise in temperature.

Type **2** materials will be considered in the present discussion, but many of the methods can be applied to type **3** and, with limitations, to type **1**. The drying times of type **2** are influenced by humidity, air circulation, light intensity, and film thickness, in addition to temperature.

The drying process is characterized by two periods known as *surface dry* and *hard dry*. A film is surface dry when it is free from tackiness; it is also known as *tack-free*.

Assessment of drying time

Preparation of films Measurements of drying time (and of other properties which will be described later) require films of uniform and known thickness, particularly if used for comparison purposes. Both hand and automatic applications are in use.

Hand application includes brush and spray methods, as well as 'spreaders'. Among the latter are bar applicators (usually embodying an adjustable gap), steel shims with doctor blade, and spiral rod applicators. An interesting 'block spreader', which eliminates the personal factors associated with hand instruments, has been described by Monk [5]. The ICI Spinner [6] is an automatic spreader in which the paint is applied to the centre of a panel which is then spun at a uniform speed in its own plane. The thickness of the resulting film depends on the speed, duration of rotation, and the film properties of the paint.

Surface dry This condition, also known as 'touch dry', requires no definition. A yes/no method is described in BS 3900, Part C2 [7] of which the following are the salient points.

The coated panel, at the end of the specified period, is laid horizontally. 0·5 gram ballotini is then sprinkled on to the surface from a height of 15 cm (6 inches) and allowed to remain on the surface for one minute. If the paint is surface dry, the ballotini should be capable of removal by means of a soft camel-hair brush without damage to the film. Automatic instruments are available [6] to determine surface drying time when the yes/no method is not applicable.

Hard dry For many years decorators have used thumb pressure and rotation as a test for hardness of dry of a paint film. The principle has been embodied in an instrument known as the Hard Dry Apparatus, or 'mechanical thumb', which is described in detail in BS 3900, Part C3 [8]. The following is a very brief summary of the method.

A vertical steel plunger is tipped with rubber at the lower end. This rubber tip is capped with white drill which is renewed for each test. The plunger assembly is loaded to a total weight of 4 lb. By an automatic device, the plunger is lowered gently on to the paint surface, rotated through 270° and then raised. A film is hard dry if the substrate or previous coat is nowhere exposed.

The BK drying recorder [9] provides information on the whole of the drying process (Fig. 25.7). Films of known thickness are cast on long glass strips and, by means of a moving carriage, a vertical round-ended needle is drawn slowly through the paint film. The speed can be chosen so that the glass strip is traversed in 6, 12 or 24 hours. In the early period of drying, the paint flows round the needle and no record remains. As the paint sets, the needle cuts a groove and the appearance of this groove gives information on the relative speeds of surface- and through-drying. Rapid surface drying is shown by tearing of the skin. Too-slow surface dry is accompanied by an extended groove. When the paint film is hard dry, the needle rides on the surface and leaves little or no record.

Machines are available to hold six or ten samples and are used to provide an overnight record of the drying of production samples as well as for investigational work.

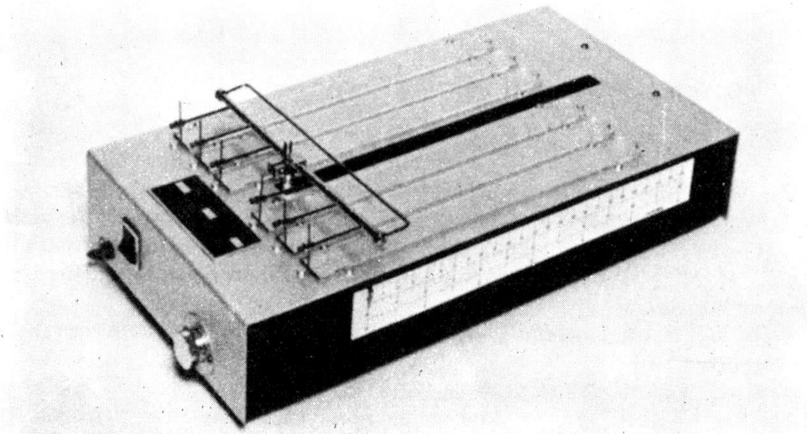

Fig 25.7 BK drying recorder

Overcoating time

With brushing paints, this corresponds to the British Standard definition of 'hard dry' (*above*). With spray paints, however, overcoating times do not necessarily require the film to be hard dry; they depend on both the nature of the film and of the paint to be used for overcoating.

Non-convertible films, such as cellulose nitrate, chlorinated rubber, vinyl and acrylic lacquers can be overcoated by spray with similar materials at any time, provided the bulk of solvent has evaporated. Fusion of coats takes place and no difficulties are encountered in respect of intercoat adhesion. Attempts to apply a second coat of these materials by brush normally leads to softening and working up of the first coat.

Oxidizable films such as alkyds and epoxy esters require atmospheric oxygen for curing and should be hard dry for recoating. For gloss on gloss the interval should not exceed 48 hours. After this time the film will be hard enough to resist penetration or 'bite' by solvents from the next coat, and adhesion is likely to be poor. The first coat, in this case, should be rubbed down. Alkyd undercoats and primers can, if necessary, be overcoated up to one week after application, but this time should not be exceeded.

Overcoating times of two-pack epoxy and epoxy/coal-tar pitch systems are restricted. Generally the solvents have largely evaporated and the film has set sufficiently at 16 hours to be overcoated. This can be carried out up to 48 hours, after which the first coat will be very hard, and if glossy, will require abrasion to promote adhesion.

Metal primers based on two-pack epoxy resins dry rapidly to a matt or semi-matt film. They can be overcoated at much shorter intervals than finishes. One-pack materials can be spray applied after 1 to 2 hours and two-pack materials after 3 to 4 hours. If the overcoating materials are brushed, double these times should be allowed. Since they dry to matt finishes, the maximum overcoating times are not as critical as for the gloss finishes, and when used as 'holding primers' two-pack epoxy types have been overcoated with undercoat after some months.

TESTS ON DRY FILMS

Measurement of film thickness

For most of the tests involving mechanical properties described below, a knowledge of the thickness of the dry film is essential for the correct interpretation of results. Dry film thickness can be calculated from wet film thickness and volume of solids but often it has to be checked as, for example, in specification work on structures.

According to circumstances, either destructive or non-destructive methods are employed [10].

Destructive methods are of three types. (a) Micrometer: used for plates or panels where a small area of paint can be removed, if necessary, to obtain the panel thickness. (b) Microscope method: a small area of the paint film is removed and mounted vertically on the microscope stage. (c) Dial thickness gauge: this has two legs which are planted on the paint film, and the probe, which is between them and attached to a dial, is set to rest firmly on the film. The dial reading is noted. The instrument is moved, the area of paint on which the probe had rested is removed, and the instrument replaced in the same position as before. The probe is now brought into contact with the substrate and the new reading noted.

Non-destructive methods A number of instruments are available using magnetic or electrical effects for ferrous and non-ferrous metals respectively.

With ferrous metals it is possible to use a permanent magnet and to measure the pull-off force required when it is placed on the surface. The presence of a paint film between magnet and steel reduces the force by an amount proportional to the thickness. This principle is employed in a widely used portable instrument [11] in which the pull-off force is applied to the magnet by turning a calibrated circular dial until the magnet detaches from the painted surface. Other instruments, for use on non-magnetic metals, employ an eddy-current principle [11].

Optical properties

Gloss Paint films can exhibit any degree of gloss from high gloss (98 to 100 percent reflectance) to dead matt (no reflectance) but the following five categories are often used: high gloss, semi-gloss, eggshell gloss, eggshell flat, and matt. There is no universally agreed standard of gloss or sheen for any of these, and manufacturers, painters and public bodies all differ in their interpretation of the terms. It is not feasible, therefore, to quote gloss values for any except the high gloss.

The measurement of gloss values is usually carried out by the method set out in BS 3900, Part D5 [12]. The following is a brief summary.

A film of uniform thickness is cast on a plane glass surface by means of 2-mil steel shims and a doctor blade. The film is allowed to harden for 24 hours. The gloss determination is carried out using a gloss head connected to a galvanometer.

Three gloss heads are available in which the angle of reflectance (relative to the normal to the surface) is 20°, 60° or 80°. The 60° instrument is used for high gloss finishes, the 20° for low sheen finishes, and the 80° for special effects. (For many years a 45° gloss head has been in use for gloss finishes, but the three instruments mentioned above provide a much wider range of measurements.) The equipment is standardized with a glossy black tile which is assumed to give a reflectance of 100 percent. The gloss head is placed on the sample and several readings taken at right angles to each other, the mean of the figures being taken as the gloss reading.

With high-gloss finishes the gloss retention is as important as the initial gloss, and the panel is therefore examined at weekly intervals over a period. The loss of gloss should not exceed 10 percent in the first 4 weeks.

Opacity of dry films

This is required for most specification work and is measured by the method described in BS 3900, Part D4 [13]. The following is a brief summary of the method.

Contrast ratio method A film of uniform thickness is cast on a thin sheet of cellulose acetate by fixing this on a plane surface under two steel shims of 2 mils thickness. The paint is spread over the surface with a doctor blade and allowed to dry.

The opacity or contrast ratio is obtained by measuring the total reflectance from the film when placed over a white and then over a black substrate. The difference is a measure of the amount of light which penetrates the film and is absorbed by the black.

The substrate consists of blocks of white and black glass fused together. The reflectance characteristics of these are first measured by reference to a magnesium carbonate standard. The paint film is placed on the white surface and the reflectance measured by means of the opacity reflectometer (Fig. 25.8) which comprises a lamp and photoelectric cell. It is connected to a galvanometer and the latter reading, with the opacity head and film on the white substrate, is set at 100. The film and opacity head are then placed on the black and the galvanometer reading noted. After making the correction for surface reflectances the opacity or contrast ratio of the paint is expressed as a percentage.

Opacity can also be measured by means of Morest charts in the same manner as for pigments.

Colour comparison

Colour matching and instrumental colour measurement have been discussed in Chapter 2. In production work a great deal of colour assessment is done visually, using either dry or wet standards. Dry standards are panels or patterns

Fig 25.8 Opacity reflectometer

which are compared with the sample after the latter has been applied to a suitable panel and allowed to dry. The comparison should be made in a north light. With a wet standard, this should be applied to a panel alongside the sample and the panel allowed to dry before comparison in a north light.

When a number of batches of the same colour are produced, colour comparisons should be made with the previous batch as well as with the standard. This ensures continuity of shade and prevents 'drift' from the standard colour.

Changes in the standards can be guarded against by recording the colour by means of a spectrophotometer or colorimeter as described in Chapter 2.

Mechanical properties

The mechanical properties of a paint film are so interdependent that it is difficult to measure a single property without involving at least one other to some degree. Many ingenious methods designed to minimize the involvement of other properties have been evolved, and the most widely used are described below.

Hardness

It is probable that the oldest method for assessing hardness was the use of the thumb-nail. This was very much a personal assessment.

Scratch test This method is widely used and is described in detail in BS 3900: Part E2 [14]. The following is a brief summary.

The paint is applied to clean abraded tin-plate, allowed to dry for the appropriate time, and fixed in a horizontal position in a slide in the apparatus (Fig. 25.9). Fixed over the slide is a pivoted counterbalanced arm to which is

Fig 25.9 Automatic scratch test unit

attached a needle with a hemispherical hardened steel point, 1 mm in diameter. This arm can be lowered so that the needle, which can be weighted, is vertical when in contact with the paint film.

In use the weighted needle is lowered gently on to the paint surface and the slide pulled out, causing the needle to move along the paint film. This may produce a scratch through to the metal, cut a groove in the paint, or the paint may be unaffected. The result recorded is the maximum load which can be applied to the needle before bare metal is visible through the scratch. Good adhesion between paint and metal is essential for consistent results. Manual and automatic forms of the apparatus are available.

Pendulum methods These are 'rocking' methods which assess hardness by a damping effect. In the Persoz instrument [15] the pendulum is attached to a horizontal bar to the underside of which two 8 mm diameter polished steel balls are attached. For standardization the pendulum is arranged with the steel balls in contact with plate glass, the pendulum displaced through 12° and the time noted for the amplitude to decrease to 4°. This is converted to 100. The test is then repeated on the paint film.

The Koenig pendulum works on a similar principle, but 5 mm polished steel balls are used. The opinion has been expressed [16] that the Persoz pendulum is to be preferred for softer films and the Koenig instrument for the harder types.

Pencil hardness test This is a very simple and useful test for assessing hardness. Pencils of decreasing hardness (6H to 4B) are drawn across the surface until one is found which leaves a black line. The method involves a number of variables but is very useful when experience in its use has been acquired.

Indentation methods These are based on measurement of the depth of penetration of a film by a ball-ended needle under an applied constant load. The penetration is recorded as a function of time. Film recovery after removal of the load can also be recorded so that resilience as well as hardness can be assessed. Several forms of apparatus are available, the best known being the ICI Micro-indentation Recorder [17]. In this apparatus measurements can be carried out over a wide range of temperatures. It can be used also for assessing changes in film properties with age and exposure.

Flexibility and extensibility

To some degree these terms are synonymous. The flexibility of any material is usually assessed by some kind of bending or flexing operation (often repeated several times), but on a paint film, to be meaningful, it has to be assessed on the film attached to a substrate. By their nature, bending operations involve extending both the film and metal panel surface but to different degrees. In the film the internal cohesive forces are put under strain as is also the adhesion between paint film and metal. Both extensibility and adhesion are therefore involved. Methods used for assessing extensibility with adhesion include bending (Bend test) and forming a 'dome' in the coated metal by a force applied either slowly (Erichsen or 'cupping' test) or rapidly (Impact test). These tests are summarized below.

Bend test This is described in detail in BS 3900, Part E1 [18]. The paint is applied to a strip, 4 in. by 2 in. (10 cm × 5 cm) of 30 s.w.g. chromate-dipped soft aluminium and allowed to harden for the specified time. The panel is then placed, paint side outermost, under a mandrel of prescribed diameter embodied in a hinge. The panel is bent through 180° in one second, removed, and the bend examined for cracks and loss of adhesion. Mandrels are in use ranging from 1 in. (25 mm) to $\frac{1}{8}$ in. (3·2 mm) in diameter. A conical mandrel is available which provides the information given by a series of hinge types. In this case a steel panel can be used with the appropriate bending aids.

Erichsen (cupping) test This measures the extensibility of a paint film when a portion of the substrate is extended into a 'dome' or 'cup'. The paint is applied to a prepared deep-drawing steel panel, and when ready for test is mounted against a rigid steel plate containing an orifice. A hemispherical hardened steel tool is placed against the back of the panel in line with the aperture. Pressure is applied to the tool so that a dome is gradually formed. It can be continued until the film ruptures or for a specified distance. The paint on the dome is then examined for cracking and loss of adhesion. This method is described in more detail in BS 3900, Part E4 [19].

Impact test The following is a brief summary of the method described in BS 3900, Part E3 [20]. The paint is applied to an 18 B. G. steel panel, 10 cm by 5 cm and allowed to dry for the specified time. It is then fixed, painted side down, between two horizontal die-blocks which have holes in the centres. The steel indentor has a hemispherical end and forms part of a steel block, the total weight of which is 10½ lb (4·77 kg). The block slides smoothly on two vertical steel supports and is allowed to fall on the test panel by a release mechanism. The dome formed in the panel is then examined for cracking and loss of adhesion of the paint. The depth of penetration, i.e. the height of the dome, can be controlled by spacing washers between the die-blocks.

Adhesion

The adhesion of a paint to a surface is influenced by a number of factors, chief among which are the nature of the binder and of the surface.

Generally, adhesion to wood and plaster surfaces presents few problems, provided the surfaces are clean and dry. On metals the van der Waals forces of attraction are not augmented, as on wood and plaster, by absorbency of the surface.

Adhesion to a metal surface is improved by increasing the effective area by chemical pretreatment or abrasion (Chapter 16), but the nature of the binder and cleanliness of the surface are of paramount importance.

Nature of the binder It is generally agreed that binders possessing polar groups (e.g. -OH, -COOH) show better adhesion than do those containing no such groups. For this reason, epoxy resins, epoxy esters and alkyds show good adhesion, whereas certain vinyls and polyethylene (applied by the powder coating process) do not adhere well to metals and require an appropriate primer.

Cleanliness of the surface The adhesion of a paint to a surface depends on good wetting and therefore on the absence of contamination. In practice it is extremely difficult, and in some conditions impossible, to remove all impurities. Most paints must therefore be able to tolerate a degree of contamination, but difficulties often arise in determining an acceptable degree of cleanliness.

Both chemical and physical methods are employed in preparing metal surfaces for painting, and these have been described in Chapter 16.

The role played by adhesion in paint performance is discussed in Chapter 13, but some of the methods used in its assessment are here described briefly.

Direct pull-off methods The most simple of these is the 'snatch' method using adhesive tape. This is pressed on to the paint surface, often over a cross-hatch (*below*), and then pulled off sharply.

Cross-hatch test Using a sharp knife, a number of parallel cuts are made through the film to the substrate at 1 mm intervals. These are crossed by a second series, making a number of 1 mm sided squares, and it is convenient to make 100 squares. If the adhesion is poor, some of these squares will fall out and the number can be expressed as a percentage. More often, however, the severity of the test needs to be increased by pressing a strip of adhesive tape across the squares followed by a quick pull-off.

The Arco knife is a mechanical cutting instrument which uses a cutting tool tipped with a 60° conical diamond. Successive parallel cuts are made in the film at decreasing distances apart until the film is removed by the lateral force of the tool. The result is expressed as the greatest distance between the two cuts when the coating begins to be detached.

The Houndsfield tensometer is capable of measuring adhesion in more precise terms by a direct pull-off test. The paint is applied to the thoroughly cleaned ends of two cylinders of the appropriate substrate and these are then glued together, end to end. The composite specimen is placed in the tensometer and the two cylinders drawn part. The force required to pull the paint from the surface is noted. The precise significance of the results provides an interesting speculation [21].

Some of the causes of poor adhesion with suggested remedies are set out in Chapter 27.

Other tests related to specific aspects of performance are described in Chapter 26.

References to Chapter 25

[1] Monk, C. J. H., Routine measurement of the viscosity of paint samples: a cone and plate viscometer with thermostatic control. *JOCCA*, 1966, **49**, 543.
[2] Sheen Instruments Ltd, Richmond, Surrey.
[3] BS 3442: 1974. Method for the determination of flash point by the Abel apparatus. British Standards Institution, 2 Park Street, London, W1A 2BS.
[4] BS 3900: Part A5. Large scale brushing test. British Standards Institution.
[5] Monk, C. J. H., *JOCCA*, 1963, **46**, 519.
[6] Sheen Instruments Ltd.
[7] BS 3900: Part C2. Surface drying test (ballotini method). British Standards Institution.
[8] BS 3900: Part C3. Hard drying time. British Standards Institution.
[9] Mickle Laboratory Engineering Co. Ltd, Gomshall, Surrey.
[10] BS 3900: Part C5. Determination of film thickness. British Standards Institution.
[11] Elcometer Instruments Ltd, Manchester.
[12] BS 3900: Part D5. Measurement of specular gloss of non-metallic paint films at 20°, 60° and 80°. British Standards Institution.

[13] BS 3900: Part D4. Comparison of contrast ratio (hiding power) of paints of same type and colour. British Standards Institution.
[14] BS 3900: Part E2. Scratch test. British Standards Institution.
[15] Persoz, B., The Hardness Pendulum. *Peint. Pig. Vernis.* 1945, **21**, 194.
[16] Fish, R. A and Morris, R. L. J., The Koenig and Persoz pendulum damping tests. *JOCCA*, 1972, **55**, 189.
[17] Monk, C. J. H. and Wright, T. A., A pneumatic micro-indentation apparatus for measuring the hardness of paint coatings. *JOCCA*, 1965, **48**, 520.
[18] BS 3900: Part E1. Bend test (cylindrical mandrel). British Standards Institution.
[19] BS 3900: Part E4. Cupping test. British Standards Institution.
[20] BS 3900: Part E3. Impact (falling weight) resistance. British Standards Institution.
[21] *See also* Funke, W., Adhesion and the gap between theory and practice. *JOCCA*, 1983, **66**, 120.

26

Performance and weathering

The majority of the test methods described in Chapter 25 are used to check paint properties during production, although some, such as adhesion and elasticity, have a bearing on performance. In this chapter we shall consider the performance of paint systems and discuss tests which can possibly be of assistance in predicting the performance or life of a paint system when exposed to various environments.

Performance, as applied to paint systems, can be defined broadly as resistance to their environment, the latter covering all types of exposure conditions. These can vary from weather resistance (or durability) to resistance to specific reagents or circumstances. The latter could include resistance to chemicals, oils, solvents and heat, all of which are fairly readily assessed. Information on these properties is available from most resin and pigment manufacturers.

Resistance to weather (weather-resistance or durability) is probably the aspect of performance which has received most attention. Maximum durability is demanded from paints applied to the exteriors of houses and buildings, ships, chemical plant, agricultural implements, and cars. The crucial test is, of course, performance of a complete system (primer, undercoat and finish) on the appropriate substrate, the correct preparation of which is of paramount importance.

PERFORMANCE OF PRIMERS

Primers for iron and steel

The formulation and properties of these primers have been discussed in Chapter 17. They are used under a wide variety of finishing systems, and to evaluate primers in complete systems, as suggested above, would be an interminable task in respect of both time and number of tests. In order to obtain results in a reasonable time, the primers are subjected to accelerated tests as one-coat systems on prepared steel panels. Whilst the results for these tests may not, in some cases, correlate as closely as one would wish with exterior exposure under a complete system, they do nevertheless provide a means of relative assessment and some guide to practical performance.

Two tests are commonly applied, involving exposure to (a) conditions of high humidity and condensation, testing the resistance of the film to water absorption and permeability, swelling and blistering, and (b) continuous salt spray which tests the resistance of the film to permeation by moisture and corrosion-promoting chloride ions.

Resistance to high humidity Details of this test are set out in BS 3900: Part F2 [1], of which the following is a summary.

The primers are applied to mild steel panels prepared in accordance with BS 3900: Part A3 [2]. The films are either stoved, or dried for seven days as appropriate and then placed in a humidity chest. This consists of a metal chamber containing a water bath, fan and thermostats. Humidity is maintained at 100 percent and there is an hourly temperature cycle between 42° and 48°C (107° and 118°F). By this means the panels are subjected to copious condensation.

Moisture absorption takes place and can lead to swelling of the film; if this is severe, blisters can form at the primer-metal interface. Highly pigmented films exhibit less swelling, but the greater permeability to both air and moisture can lead to severe rusting.

The test is usually run for 10 days after which the panels are removed and the films examined. Half the film is then removed with solvent type stripper and the metal examined for signs of rusting. If the panels are to be preserved, the cleaned surface is coated with a clear cellulose nitrate or polymethacrylate lacquer.

Resistance to continuous salt spray The following is a résumé of the method described in BS 3900, Part F4 [3]. The primer is applied to mild steel panels prepared in accordance with BS 3900 Part A3. The films are either stoved or dried for seven days as appropriate and then placed in the salt spray cabinet (Fig. 26.1). This is made of transparent plastic and contains racks for the panels and a reservoir of artificial sea-water. By means of compressed air and suitably arranged jets this is atomized into a salt mist which settles on the panels and then runs to waste. It is not re-circulated. The test is continued for ten days and then the panels are removed, washed and dried. Any deterioration in the film is noted. Half the film is then removed with solvent type stripper and the metal examined for signs of corrosion. If it is desired to preserve the panels, the cleaned area is protected with clear cellulose nitrate or polymethacrylate lacquer.

This test is carried out at ambient temperature which, in the United Kingdom, is below the glass transition temperature-range of most primers. If the test is carried out at a higher temperature, as in the ASTM method and in the humidity test (*above*), then it becomes more severe, but correlation with practical performance (again in the UK) is not improved. Timmins [4] has described a salt spray cabinet consisting of a weatherometer drum with the panels arranged round the inner periphery. The solution employed contained 0·4 percent ammonium sulphate and 0·05 percent sodium chloride, a more dilute form of a mixture suggested by Harrison [5]. The salt spray is interrupted at intervals by an air blow. Results with this apparatus, particularly with high-build coatings, are claimed to correlate more closely with actual practical performance than does the BS 3900/F3 method.

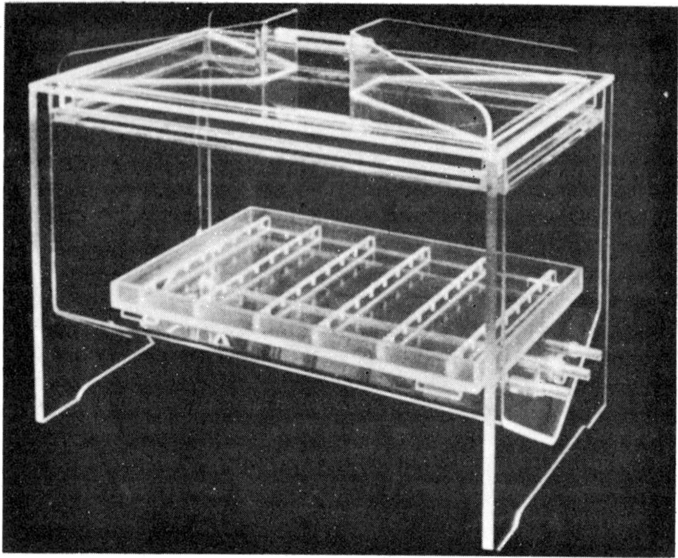

Fig 26.1 Salt spray cabinet

Primers for woodwork

These have been described in Chapter 18 where it is pointed out that two basic requirements for satisfactory performance are good adhesion and elasticity. Good adhesion to woodwork requires the wood to be dry, i.e. conditioned to the point where the moisture content is in equilibrium with the moisture in the air. Absorption and loss of water by the wood result in appreciable dimensional changes, the major proportion of the water passing along the grain from the ends. To minimize this movement it is essential to seal the ends of timber sections effectively before assembly into fabricated units. The consequences of the inefficient priming of wood are discussed in Chapter 27.

Accelerated weathering tests on wood primers can be carried out in the Marr weatherometer, but special attention must be given to sealing ends and edges. Difficulties arise in selecting identical wooden panels for standard and samples, without which it is difficult to draw any valid conclusions from the results. More often the primed panels or complete systems are exposed on the test fence at 45° (113°F) facing south. This method is slower but more meaningful results are obtained.

Problems which arise when primers are applied to timber which has been treated with certain types of wood preservative are mentioned in Chapter 18, but the following additional comments are apposite. Primer should not be applied to recently creosoted wood since bleeding, discoloration and lack of adhesion will result. If such timber has been exposed to weather for one to two years it can be sealed with a spirit type of clear sealer such as 'knotting' (p. 237) or a spirit sealer pigmented with aluminium ('white knotting'). Other wood primers vary in their effects. Copper and zinc naphthenates do not affect the adhesion of paint, but the former can cause slight discoloration through bleeding. The problem of painting over preservative-treated wood has received considerable attention in the USA where timber-clad houses are more popular. The use of trialkyl lead acetates on Southern pine has been investigated by Barnes [6] who showed that prevention of attack by micro-organisms resulted in increased durability of conventional pigmented systems. Little advantage accrued with clear systems which failed by UV degradation before the effects of micro-organisms became serious. The durability of both paint and varnish coatings on wood has been discussed by van Loon [7].

NATURAL WEATHERING

The crucial test for any paint used on an exterior surface is the durability or resistance to an environment comprising weather and location. Weather consists of several factors – sunshine, rain, temperature and, to a lesser extent, wind. The location (of a building, for example) determines the intensity and duration of sunshine and the rainfall reaching the building. In respect of the paint systems, the two most destructive influences are sunlight and moisture.

Sunlight

The most active wavelengths in causing breakdown of paint coatings are in the ultraviolet band, particularly wavelengths between 280 and 310 nm, but wavelengths up to 400 nm are harmful. The percentage of these harmful wavelengths in sunlight varies with the time of year and the geographical location.

It is a well-known fact that paint coatings on the south and south-west sides of buildings in the UK usually fail earlier than those on the north and east sides. Although sunlight is the major factor in failure, rain, carried on the prevailing south-westerly wind, also plays a big part.

In natural weathering tests, where painted panels are exposed in a position to receive the maximum sunshine and rain, it is usual, in the UK, to expose the panels on a test fence at an angle of 45° and facing south. Two internationally-known weathering stations are in the USA at Miami, Florida, where a high sunshine record is accompanied by high humidity and at Phoenix, Arizona, where dry desert conditions prevail. These are described in more detail later.

Effect of sunlight on paint films

Both pigment and binder can be affected by the harmful UV wavelengths.

Effect on pigments The most noticeable effect is fading of coloured pigments in pale shades. It results, of course, from the use of pigments with insufficient lightfastness. Apparent fading may be due, however, to chalking (*below*), but it is not difficult to differentiate between the two.

Colour changes which are sometimes experienced with green colours are worthy of mention. Greens made from a mixture of phthalocyanine blue and arylamide yellow often become bluer on exposure as a result of photochemical degradation of the yellow. Chrome greens, which continue to be used, albeit to a limited extent, in exterior paints, can become bluer in industrial atmospheres. This is attributed to the reduction of the chromate radical in the lead chrome by atmospheric sulphur dioxide.

Effect on binder It is an accepted fact that films of clear unpigmented binder fail as a result of UV degradation much sooner than the same binder when pigmented. The limited life of clear finishes on wood was confirmed by Barnes [6]. The extent of UV absorption and the consequent degradation depends on the nature of the polymer. Figures for the UV absorption of various polymers have been published by Marvuglio *et al* [8].

Clear films can be protected to some degree from the destructive effects of UV radiation by the introduction of 'UV absorbers'. Some of the more widely used are organic compounds of the substituted benzophenones or the benzotriazole type. They are usually marketed under trade names [9].

Alkyd resins containing linseed oil fatty acids, when exposed to light as films, yellow much more than those containing fatty acids from soya bean or sunflower oil. The yellowing is more apparent in pigmented white films. The yellowing of the linseed type is attributed to the linolenic acid present, an acid which is practically absent from the soya bean and sunflower oils. In these the predominant acid is linoleic which has little tendency to yellow.

Yellowing is not a problem when saturated fatty acids (such as those derived from the non-drying castor or coconut oils) are used to produce the non-drying alkyds. These are combined with amino resins in stoving finishes which present no serious yellowing problem.

It appears that yellowing of alkyd finishes is associated with certain types of unsaturation in the fatty acid chain and could be due to the formation of yellow oxidation products.

Effect on pigmented systems

With certain pigments the UV radiation is absorbed and converted into heat. These pigments act as efficient screens in protecting the binder from the UV radiation and so confer good durability on the paint film. Carbon black, red oxide of iron and phthalocyanine blue are the best-known examples. It has been shown [8] that oxide of iron, even in its transparent form and at low concentration (0·03%), will act as an efficient screen.

Small changes in visual appearance are more apparent in white and light colours than in the darker colours.

Black paints are usually pigmented with carbon black which is a very strong UV absorber. Consequently black gloss finishes retain their gloss and film integrity for longer periods than most other colours.

Dark colours In these also the effects of exposure to sunlight depend on the type or types of pigment used. Resistance is improved if a proportion of carbon black can be included. As mentioned earlier, oxides of iron are strong UV absorbers and films containing these pigments show good durability. Greens based on mixtures of phthalocyanine blue with either lead chrome or arylamide yellow can become progressively more blue on exposure. The causes have been described above.

Whites and pale shades are based on rutile titanium dioxide and can show very slow 'Chalking'. This is a condition of the paint surface when white pigment, with the appearance of 'chalk', can be removed by rubbing the surface with the tips of the fingers. Chalking is preceded by loss of gloss and is caused by photochemical breakdown of the surface layer of binder resulting in release of pigment particles. Rutile titanium dioxide absorbs UV radiation strongly. Much of this is dissipated as heat but some excitation of the crystals results in catalytic activity at the surface. Free radicals ($OH^{\cdot}$ and $OH_2^{\cdot}$) are produced [10] and these set up progressive oxidation of the polymer molecules. (With the anatase grade of titanium dioxide the excitation is much more intense with the result that, on exposure, paint films based on this pigment fail rapidly by severe chalking.)

The surface activity of the rutile crystals is reduced by coating the particles with oxides of aluminium, zirconium and silica either singly or in combinations. As a result paint films pigmented with the coated rutile grades exhibit a very low tendency to chalk and can function satisfactorily in 'long life' coatings.

Pale shades are produced by the addition of coloured pigment dispersions (or 'stainers') to white paint and these colours, many of which are organic, can vary considerably in lightfastness of tints from one class to another. The demand for very pale clean shades necessitates the use of organic pigments in which, as was mentioned in Chapter 7, stability is related to molecular complexity. Among the most stable (and lightfast) groups are the phthalocyanines, anthraquinones, quinacridones, dioxazines and heterocyclics. The high cost of many of these pigments is largely offset by their high tinting strength.

The onset and degree of chalking depend on the intensity and duration of the UV radiation. In the UK these effects tend to become noticeable in the summer months, but where more uniform climatic conditions exist, as in Florida, the onset of chalking can occur throughout the year.

Some modern types of building, such as high rise structures, have created a demand for 'long-life' paints which will, possibly, last for 20 to 25 years. The present day 'long-life' coatings based on specially formulated alkyds, silicone alkyds, acrylated rubber or chlorinated or fluorinated vinyls are considered to give protection for 15 years or more according to the geographical situation, the efficiency of surface preparation and conditions at the time of application.

Effect of moisture on paint films

The degree to which paint films absorb or are permeable to moisture is influenced by the nature of the polymer and depends, to a lesser extent, on the nature and degree of pigmentation (PVC). In gloss finishes the PVC is low but is much higher in undercoats. In exterior situations these should be protected by finish at the expiry of the overcoating period and should not be left exposed to the weather.

A few pigments can exert a deleterious effect if used in finishes. Examples are lithopone in which zinc sulphide can be oxidized to water-soluble zinc sulphate which can be leached out of the film, and Venetian red which frequently contains the slightly water-soluble calcium sulphate. In both cases the film is weakened.

Absorption of water by some types of binder can be accompanied by swelling of the film. This begins with the surface layers, but on prolonged contact with water the absorption and swelling can penetrate more deeply, ultimately reaching the substrate. This can result in blistering and a decrease in adhesion between film and substrate. On drying out, the film contracts, but recovery of adhesion where blisters had formed is only partial and often absent. Films of the conventional alkyd type absorb water in this way and therefore are not used in conditions of prolonged immersion in water. Rainfall affects only the surface layer of the film, but continual expansion and contraction contributes to ultimate failure. Furthermore, during winter in the UK the exposed film is damp or wet for considerable periods and, according to Whiteley *et al.* [11], long-term contact with water causes deterioration in alkyd (and one-pack polyurethane) films. It has been stated by Walker [12] that the adhesion of exposed films to substrates can fluctuate on exposure, possibly due to variations in the moisture content of the air.

The loss of adhesion (or 'water disbondment') of water-soaked films on metals has been discussed by Leidheiser and Funke [13]. They review the available literature and consider the loss of adhesion to be due to water permeation through the film and the formation of a film of water, several molecules thick, at the metal/paint interface. On a smooth metal surface water will 'creep' along the surface between metal and paint but this movement is impaired if the surface is roughened by, for example, blast cleaning. Again, if the movement of water is checked, local blistering can result.

With films of the non-oxidizing type such as vinyls, chlorinated rubber, two-pack polyurethanes and two-pack epoxies, water has little effect. The films exhibit a slight permeability to moisture, but the amount passing through, which decreases with increasing film thickness, is sufficiently low for these materials to be classified as 'moisture barriers'. They are used under conditions of prolonged immersion and do not exhibit the swelling which is characteristic of alkyds and oleoresinous systems. Chlorinated rubber, for example, is used on concrete swimming pools and is also sufficiently durable to be used in marine work and on exposed steel structures.

To sum up, the two major factors responsible for the deterioration of paint films in the UK are moisture and UV radiation in sunlight.

WEATHERING TESTS

Exposure to natural weathering

The ultimate test of the durability of a paint is exposure to natural weathering as a complete paint system on the appropriate substrate. The traditional procedure, practised by most paint manufacturers, is laboratory preparation and drying of panels for 7 days (this period is unnecessary with stoving finishes) followed by exposure on a test fence. This consists of a wooden or metal framework to which the panels are attached at an angel of 45° and facing south. At the latitude of the UK in the northern hemisphere this ensures maximum exposure to solar radiation. It will be realized, of course, that in actual practice, few paints are exposed at 45°. Most exposures are either vertical or horizontal. Furthermore, paints are exposed to the weather immediately after application and are not dried under ideal conditions for seven days before exposure. Paint films of the alkyd type are very sensitive to moisture during the drying and early hardening period, and the durability of the system (particularly the time to onset of chalking) is influenced by weather conditions immediately after exposure and by the time of year. The seven-day drying period does, however, reduce the effect of these variables in testing procedures.

The panels are examined at intervals of one or three months and the following properties are assessed as the film ages: gloss, chalking, fading (or colour change), checking, cracking, blistering, adhesion, and flaking. In some locations mould growth may be present. An additional panel of a paint of proven durability and of similar composition to those under test is included as a reference standard.

The results of exposure are influenced by the location of the test fence, and exposures from one fence are of limited value. Usable information can be obtained only if exposures are carried out at several different types of location, e.g. industrial, rural, and marine environments and, if possible, in selected overseas locations (*see below*). In industrial atmospheres and large cities generally, the paint films are subjected to attack by sulphur dioxide in the air. This is offset to some degree by solid impurities which settle and form a protective film on the surface. In rural and marine conditions the cleaner atmosphere allows more UV radiation to penetrate, and chalking takes place earlier and more intensely than in industrial locations.

Natural exposure tests are very slow and in the UK require at least three years, and more often four to five years, for a first assessment. (Inevitably many accelerated tests have been devised in an attempt to reproduce the types of failure experienced with natural weathering. Some of the instruments and methods which have been most successful are described below.)

Results for natural weathering can be obtained in a much shorter time by submission of panels for exposure at two stations in the southern USA, situated in Florida [14] and Arizona [15].

The Florida station enjoys high solar radiation and high humidity; one year's exposure is claimed to be equivalent to two to three years in the UK. In fact, one year's exposure in Florida, without substantial change in appearance, has long been a standard requirement for many finishes [16].

The Arizona station lies in a very dry area where the solar radiation is very high and contains a high proportion of UV wavelengths. This station uses an ingenious machine known as Emmaqua (Equatorial Mount with Mirrors for Acceleration plus water spray) for accelerating natural weathering. A number of highly polished aluminium mirrors reflect the sun's rays on to the test panels which are sprayed automatically with water for a short period during each hour. The water spray can be omitted if desired. The machine is mounted on an axis which is geared so that the sun's rays strike the mirrors at a constant angle during the hours of sunshine. According to Ellinger [16], about five weeks exposure in Emmaqua gives results comparable with those from six months normal exposure in Florida.

Accelerated weathering

A number of accelerated tests have been devised in attempts to reproduce, in a short time, the types and degree of failure experience with natural weathering. Among the accelerated weatherometers in use are the Marr Artificial Weathering Apparatus [17], described briefly below and in detail in BS 3900 Part F3 [18]; the Atlas Weather-o-Meter [19]; and the Xenotest [20].

The Marr artificial weathering apparatus is illustrated in Fig. 26.2. Briefly, it consists of a vertical steel drum with open top and arranged to make three revolutions per hour on its axis. The painted panels are placed in galleries round the periphery of the drum and are exposed to UV radiation from a carbon arc lamp (1600 watts, 130–145 volts) and to a spray of distilled water. A loose metal cover is suspended by wires so that it can be raised and lowered by a winch

Fig 26.2 Accelerated weathering equipment. Courtesy J.B. Marr

mechanism; it also contains apertures for the lamp, water, and compressed-air connections.

The apparatus operates on a programmed 24-hour cycle. During this period the panels receive constant UV radiation, but for certain intervals, the water spray is cut off and air is blown through to enable the panels to dry out. The panels are examined at intervals of seven days for loss of gloss, chalking, blistering, and general film integrity.

Accelerated weathering tests should generally be carried out on complete systems, but alkyd systems on steel panels show premature breakdown. Such finishes are therefore tested as single coats on aluminium panels. These should survive for 1000 hours in the Marr apparatus without appreciable change, but 'long life' coatings survive for much longer periods. In making any assessment of coatings it is usual to include in the test series a reference standard comprising a panel of paint of known performance and of similar composition to the paints under test. Hoey and Hipwood [21] consider that the Marr cycle, as described in BS 3900, Part F3 [18], correlates reasonably well with exterior behaviour in temperate climates.

In the Xenotest apparatus, the UV radiation is provided by a xenon arc lamp, but according to Capron and Crowder [22] it tends to accelerate fading, compared with sunlight, as the lamp ages. Nevertheless, the Xenotest gives fairly good correlation with natural exposure in many cases and its use is increasing.

References to Chapter 26

[1] BS 3900, Part F2. Determination of resistance to humidity under condensation conditions. British Standards Institution, 2 Park Street, London, W1A 2BS.
[2] BS 3900, Part A3. Standard panels for paint testing.
[3] BS 3900, Part F4. Resistance to continuous salt spray.
[4] Timmins, F. D., Avoiding paint failures by prohesion. *JOCCA*, 1979, **62**, 131.
[5] Harrison, J. B., The realistic prediction of primer performance - a case history. *JOCCA*, 1979, **62**, 18.
[6] Barnes, H. M., Coating durability of organolead-treated southern pine on exterior exposure. *JPT*, 1979, **51**, 43.
[7] Van Loon, J., The interaction between paint and substrate. *JOCCA*, 1966, **49**, 844.
[8] Marvuglio, P., Sharrock, R. F. and Kennedy, R. J., UV screening by pigments. *JOCCA*, 1978, **61**, 79.
[9] 'Aduvex' (Ward, Blenkinsop Ltd, Wembley, Middx).
'Cyasorb' (Cyanamid of Great Britain Ltd, Gosport, Hants).
'Tinuvin' (Ciba-Geigy Plastics & Additives Co. Ltd, Trafford Park, Manchester).
[10] Simpson, L. A., Influence of titanium dioxide pigment on the surface roughness of organic coatings. *JOCCA*, 1986, **69**, 232.
[11] Whiteley, P., Rothwell, G. W. and Kennedy, J., Deterioration of paint films by water. *JOCCA*, 1975, **58**, 200.
[12] Walker, P., Long term adhesion studies on multicoat systems exposed to natural weathering. *JOCCA*, 1969, **52**, 516.
[13] Leidheiser, H. and Funke, W., Water disbondment and wet adhesion of organic coatings on metals : A review and interpretation. *JOCCA*, 1987, **70**, 121.
[14] Sub-Tropical Testing Service Inc., Miami, Florida.

[15] DSET Laboratories Inc., Phoenix, Arizona.
[16] Ellinger, M. L., Accelerated Weathering Tests. *JOCCA*, 1979, **62**, 136.
[17] J. B. Marr & Co. Ltd, Kingston upon Thames, Surrey.
[18] BS 3900, Part F3. Resistance to accelerated weathering (enclosed carbon arc). British Standards Institution, London.
[19] Atlas Electric Devices Company, Chicago, Illinois.
[20] Heraeus Equipment Ltd, Brentwood, Essex. *see also* Hibbert, M., *JOCCA*, 1984, **67**, 264.
[21] Hoey, C. E. and Hipwood, H. A., An appraisal of artificial weathering methods for assessment of the durability of paint films. *JOCCA*, 1974, **57**, 151.
[22] Capron, E. and Crowder, J. R., The measurement of ultraviolet radiation in accelerated weathering machines. *JOCCA*, 1975, **58**, 9.

27

Common faults and remedies

This chapter contains a selection of the more common defects encountered in liquid paints and paint films. Some of these have been mentioned or have been given a brief description in earlier chapters.

When premature breakdown occurs in an exposed paint system the first reaction, usually, is condemnation of the paint, but this is seldom justified. A high proportion of breakdowns results from lack of adequate surface preparation, but other factors also can be involved. Some of these are set out below, but the list is not claimed to be exhaustive.

1 Inadequate surface preparation. The sophisticated coatings of today are less tolerant of surface contamination than were the older oil and oleoresinous paints.
2 Poor application technique or incorrect method of application.
3 Application under unsuitable conditions, e.g. low temperature, humidity.
4 Insufficient coating thickness.
5 Paint may be unsuited to the substrate or to the exposure conditions.
6 An inadequate painting specification (this should cover the points mentioned above).

Most premature failures can be avoided, provided that (a) a sound specification is prepared before the work is commenced, and (b) the specification is adhered to in all respects.

When failures occur, the defects are seldom encountered in isolation and, as will be evident in the following pages, many are inter-related.

ADHESION FAILURES

General The adhesion of a paint to a surface depends on good wetting and therefore on the cleanliness of the surface. Whilst, under controlled conditions, it is possible to produce a reasonably clean surface, situations are encountered, in maintenance work for example, when it is impossible to clean the surface thoroughly and the paints employed must tolerate some degree of surface contamination. The extent to which they do so varies considerably, the modern types being less tolerant than were the older oleoresinous or oil-based paints.

The role played by adhesion in paint performance has been discussed in Chapter 26 and methods for its measurement in Chapter 25. Here some of the causes of adhesion failure are discussed, together with some suggested remedies.

Embrittlement of the film This can lead to adhesion failure, accompanied by cracking and flaking [1]. Among the causes are the following.

1 Weathering of oxidizable type paints. Progressive hardening takes place through oxidation, accompanied by leaching out of water-soluble by-products.
2 Plasticizer loss, by evaporation or migration, from non-convertible films.
3 Overstoving, particularly of a primer.
4 Absorption of medium by an unsealed surface, resulting in an underbound and therefore brittle film.

Causes of adhesion failure of metals

1 Contamination of the surface by rust, scale, oil, grease or dirt which were not removed before painting. In cases of severe contamination, defects such as cissing or crawling can appear during application. The only remedy is complete removal of paint film and contamination, followed by repainting.

The degree of contamination of metal surfaces is, in general, difficult to assess, particularly when it is slight. New galvanized iron is an exception. When this is treated with T-wash (a dilute solution of phosphoric acid containing copper phosphate), blackening of the surface occurs only on clean metal, and any non-blackened areas require further cleaning. The T-wash produces an etched surface to which primers will adhere well.

2 Chemical reaction at the paint/metal interface, resulting in rupture of the adhesive bonds (van der Waals forces). This occurs when new galvanized iron is primed with an oxidizing type primer such as an alkyd or oleoresinous type. Water-soluble acidic scission by-products produced in the oxidation reactions are carried to the metal surface and react with the zinc to produce soluble zinc salts. The adhesion is thereby destroyed.

Remedies are as follows. (a) Prime with calcium plumbate primer. The alkaline nature of the pigment serves to neutralize the acidic scission products and reaction with the zinc is prevented. (b) Convert the zinc surface to phosphate (with simultaneous etching) by treatment with T-wash. (c) Treat the clean galvanized surface with etch primer. (d) Allow the metal to weather for at least six months before painting.

The use of cathodic protection, particularly in marine conditions, can lead to adhesion failure if a saponifiable medium is employed in the primer. Oleoresinous and alkyd type binders can be saponified by the alkali produced at the cathodes and the adhesion is destroyed. The obvious remedy consists in the use of non-saponifiable binders in primers when cathodic protection is employed. The two-pack epoxy has been found to be the most suitable binder.

3 Separation of the film due to underfilm rusting. This can result from the use of an insufficiently moisture-impermeable paint or from failure to provide sufficient thickness of film over rogue peaks produced in blast cleaning. In this case the general underfilm rusting is preceded by rust spots when the peaks corrode.

The remedy is complete removal by blast cleaning, followed by repainting with suitable materials to an adequate thickness.

4 **Highly polished metals**, even when clean, provide little key for the applied coating. Smooth aluminium, chromium, copper, brass and cadmium require to be abraded or etched in some way to provide a rougher surface. In addition to the condition of the metal surface, adhesion is influenced by the presence or absence of polar groups in the polymer binder of the paint. This topic has been examined by Sathyanarayana *et al.* [2] who have concluded that, in the case of alkyd paints pigmented with titanium dioxide, adhesion is affected by variations in the acid value of the alkyd resin, but variation in hydroxyl value has little effect. It is an accepted fact that whilst alkyds, epoxy esters and two-pack epoxies generally show good adhesion to metal surfaces, some other polymers, particularly certain vinyls, show poor adhesion. For this reason plastisols and organosols are usually applied over a two-pack epoxy primer to which they will adhere satisfactorily.

5 **Prolonged contact with water** of alkyd and oleoresinous films can result in swelling, often accompanied by blistering. When in this condition, adhesion of the film is at a very low value but is largely recovered on drying out. Areas of film which formed blisters are likely to crack and flake off.

If the substrate is mild steel, surface rusting also is likely to take place, but the degree will depend on the nature of the primer (the type and amount of inhibitive pigment) and the pH of the water. If the surface rusting is appreciable then, on drying out, recovery of adhesion will be partial and temporary.

Poor intercoat adhesion, resulting in peeling, can be due to any of the following.

1 Undercoat is too glossy and/or has been allowed to dry too hard before the finish is applied. The remedy consists in rubbing down to provide a key for the finish.

2 Moisture on the surface when painting was carried out. Most painting specifications include a ban on painting when the relative humidity is above 90 percent or the temperature below 4°C (39°F).

The addition of suitable cationic surfactants [3] to certain types of primer enables metal to be brush-painted when moist with water. The film of moisture is displaced by brushing shear and the water evaporates with the solvent. This process is not recommended for woodwork where the moisture will not be confined to the surface.

3 The use of incompatible binders in juxtaposition, for example, chlorinated rubber on some one-pack etch primers.

Causes of adhesion failure on wood

1 **The ends of joinery** were not adequately primed before insertion in brickwork, ingress of moisture causing swelling of the grain and dimensional changes in the wood. If the paint system on the sides of members is not sufficiently elastic it will crack – usually along the grain – allowing moisture to penetrate between wood and paint. The paint will then flake or peel.

This defect is commonly seen on wooden window sills and the lower horizontal bars. Since it is not practicable to remove windows and sills for further priming, frequent repainting is the sole expedient.

2 **The wood surface** was insufficiently dry when painted, so that the film of moisture prevented the establishment of adhesive bonds before the primer had dried. Trapped moisture can lead to blistering (q.v.). Moisture on the surface is not detrimental to water-based (acrylic emulsion) primers.

3 **An incorrect type of primer** has been used. 'Oily' wood such as teak affords indifferent adhesion for oleoresinous or alkyd type primers. A wash of white spirit has been recommended [4] as an adequate surface preparation. To ensure good adhesion, primers containing non-leafing aluminium pigment in a medium of tung oil/phenolic, tung oil/coumarone or urethane oil should be employed.

4 **Prolonged contact with water** will cause softening and swelling, often accompanied by blistering, in alkyd and oleoresinous paint systems. In the wet state adhesion is low but, on drying out, is substantially recovered except where blisters formed. Here the paint will crack and flake.

5 Certain types of wood preservative can affect adhesion (see Chapter 26, p. 442).

Adhesion failure on plastered walls

The most widely used decorative finish for plastered walls is vinyl emulsion paint in matt, 'vinyl silk' or semi-gloss finish. Failure in adhesion can be caused by a number of factors [1] of which the following are probably the most common.

1 **Highly impervious and/or smooth surfaces** Certain types of gypsum plaster, such as Keene's cement, can be trowelled to a hard glossy surface. This, and old gloss paint, provides little key for emulsion paints, and moisture absorption from condensation can destroy the adhesion. Roughening the surface with fine glass paper will improve the adhesion, but where conditions of heavy condensation exist, emulsion paints should be avoided. Solvent type anti-condensation paints are more suitable.

2 **Soft or friable surface**, possible an old type of distemper. Emulsion paints develop considerable tension in the film on drying, and unless the surface is hard and firm, cracking and peeling can result.

Old surfaces should be examined closely before repainting. If any doubt exists concerning the condition of the old coating it should be removed, or if this proves difficult, treated with a thin penetrating sealer to consolidate the surface and bind it to the substrate.

3 **Sound but highly porous** plaster surfaces will absorb the medium from the first coat applied and leave an underbound film. This absorption will be repeated if further coats are applied. Highly porous surfaces should be sealed thoroughly before painting; more than one application of sealer may be necessary. If the first coat of emulsion paint has been applied and found to be absorbed, a coat of sealer should be applied before any further coats of emulsion paint.

4 Damp walls A warm room will tend to draw the moisture inwards. If the coating is a flat emulsion paint it will be sufficiently porous to allow the moisture to pass through and evaporate. Nevertheless, stains may be formed and any soluble salts are likely to crystallize out on the surface (see Efflorescence). A semi-gloss coating is likely to provide a moisture barrier, and so water, often containing soluble salts, will accumulate at the plaster/paint interface. This will tend to destroy the adhesion and disrupt the film.

Treatment of the interior wall surfaces with waterproofing solutions is of little value. The source of dampness, whether it is rain penetration or rising damp, must be located and corrected before a satisfactory finish can be applied to the walls.

AERATION IN FILMS

This can be introduced in several ways.

1 By high-speed mixers, particularly with high-build or thixotropic paints. Air is drawn in by the vortex and dispersed. The film thickness and structure of these paints delays the release of air bubbles which can be a source of craters and pinholes.

When dry pigment is charged into a stirred liquid the speed of stirring should be as low as practicable and the pigment added uniformly to minimize the amount of air introduced from this source. This applies particularly to emulsion paints, in which air bubbles can be retained very firmly. The use of a defoamer at the pigment mixing stage can be very beneficial.

2 In roller application of solvent-type paint or emulsion paint, less aeration is produced in the film by the use of a lamb's-wool roller than by the plastic foam type.

3 Aeration can also be troublesome in dipping tanks, particularly with water-thinnable materials [5]. The use of defoamers and non-aerating circulating pumps are suggested as counter measures.

ALLIGATORING (or Crocodiling)

A severe form of checking (q.v.) in which the surface resembles the skin of an alligator. It occurs when a hard-drying finish is applied over a softer more elastic undercoat.

The remedy is removal of the complete system and, when repainting, to ensure that the finish is more elastic than the undercoat. It is wise, therefore, to purchase undercoat and finish from the same manufacturer.

BLISTERING

This is one of the most common defects and, in the majority of cases, is caused by liquids or gases under the film. According to Whiteley *et al.* [6], water is the most common cause of blistering and related failures such as cracking and flaking.

The paints most prone to blistering on prolonged contact with water are the decorative types based on alkyds, epoxy esters, urethane oils, and oleoresinous varnishes. It is significant that alkyds contain a much higher water-extractable fraction than paints with only little tendency to blister such as vinyls or two-pack epoxies.

Blistering involves an increase in the surface area of the paint film and, if the thickness is not reduced, an increase in volume of the film. The liquid in the blisters does not appear to exert any hydrostatic pressure; in fact, an alkyd or similar paint film, when soaked in water for a prolonged period, loses its adhesion over the entire surface (much as a postage stamp does). On drying out, the adhesion is largely recovered, except for the areas of blisters under which there may be little or no recovery.

Osmosis is considered to be the most general mechanism of blister formation [7], but on steel substrates electro-osmosis can play a large part. Mayne [8] considers that when blistering occurs on painted steel, most of the water is transferred by an electro-osmotic mechanism, the remainder by osmosis. Osmosis can result not only from water-soluble oxidation products in the film but also from water-soluble contamination on or in the surface. With metals it is essential to remove all traces of salts arising from pretreatment processes or, in the case of zinc-rich primers, from prolonged exposure.

On woodwork, blistering can arise from moisture trapped under the film, particularly if the painted woodwork is exposed to the sun soon after painting. Exterior wood surfaces can be contaminated not only with airborne dirt but also with other atmospheric contaminants which can penetrate the wood surface. Thorough washing down and drying before painting are essential.

A study of the relationship between blistering and wood structure has been made by Ulfvarson and Pattyramie [9]. They find that blistering occurs more easily and more intensely on panels of heartwood than on those of softwood. The direction of sawing is also important. Radially sawn panels showed more intense blistering then tangentially sawn panels. Although moisture is normally regarded as being responsible for blistering, the authors are of the opinion that blisters are produced by constituents which are extractable by water or by solvents.

Prevention of blistering

1 **Metals** (principally steel)

(a) Ensure that metal is clean. New steel must be free from rust and scale; exposed steel must be free from rust and corrosion products. Ammonium sulphate is particularly difficult to remove from pitted rusty steel and is best removed by water blast. When pretreatment is carried out, all traces of soluble salts must be removed.

(b) The metal must be thoroughly dry.

(c) Painting should not be carried out in temperatures below 4°C (39°F), at relative humidities above 90 percent, or in direct hot sunshine.

(d) When several coats are applied, the minimum overcoating intervals should be observed.

(e) For metals likely to be in contact with water for prolonged periods, paint systems with low moisture permeability should be chosen. Among these are chlorinated rubber, epoxy/pitch, two-pack polyurethane, polyurethane/pitch and vinyls.

2 Woodwork
(a) Wood must be dry and well seasoned.
(b) The ends of joinery must be primed with at least two coats of good quality primer.
(c) Priming should not be carried out under conditions of high humidity.
(d) Knots should be given two coats of knotting before woodwork is primed.

BLOOM

A deposit on glossy paint films resembling the 'bloom' on a grape. In most cases it can be removed with a damp cloth. The term 'haze' is used to indicate a permanent type of bloom.

Crystalline bloom has been discussed on p. 333. It is formed in urban and industrial atmospheres containing sulphur dioxide and ammonia.

Lacquer bloom or 'haze' is produced by condensation of moisture consequent upon the cooling effect of the rapid loss of volatile solvent. The moisture causes precipitation of cellulose nitrate or resin in a very fine state of division.

Stoving bloom sometimes appears on stoved films containing some pigment dyestuffs, usually of the arylamide type. At the temperature of stoving, a certain amount of dyestuff dissolves in the binder, and much of this separates on cooling as a bloom on the surface [10]. The bloom can be removed with a damp cloth.

BRUSH DRAG

Difficulty in spreading paint satisfactorily may be due to one of the following causes.

1 The viscosity of the paint is too high. The remedy is careful thinning with the correct thinners.
2 The undercoat may not be sufficiently dry, resulting in 'pick-up' and brushmarking (q.v.).
3 The paint may be the non-convertible type such as chlorinated rubber. Solvent from the top coat of chlorinated rubber will always penetrate and soften an undercoat on the same base, although this can be mitigated to some degree by including a proportion of white spirit in the thinners. If a second coat of chlorinated rubber has to be applied by brush, it should be laid on and no attempt should be made to brush out or lay off.
4 Brush drag, accompanied by brushmarking, will be experienced on highly absorbent surfaces. The remedy lies in efficient sealing of the surface.

BRUSHMARKING

This is associated with poor flow and is a common fault with highly pigmented materials such as decorative undercoats in which the pigment/extender content is high, or where there is intense pigment flocculation. The poor flow is a feature of pseudo-plastic consistency and this, in turn, is controlled by the type and relative amounts of the extenders (see p. 325).

Improved flow and reduced brushmarking can sometimes be achieved by the addition of flow-promoting or wetting agents. Pale boiled oil, lecithin and some oil-soluble non-ionics are also employed. If these expedients fail, the paint must be re-formulated.

Brushmarking can result also if a gloss finish is applied over an undercoat which is not sufficiently dry (see Brush Drag, above).

CHALKING

The chalking of white paints pigmented with titanium dioxide has been discussed on p. 55 and p. 445. This is a photochemical reaction resulting from the action of ultraviolet wavelengths in the 290–350 nanometre range.

Chalking can result also from other causes. Some exterior emulsion paints will chalk as a result of the leaching out of water-soluble colloids and additives. Acrylic emulsions are superior to vinyls in resistance to chalking, but if extenders are used, care must be taken in their selection. China clays, barytes and calcium carbonate are satisfactory, but talcs tend to chalk faster than carbonates [11].

Lithopone can cause rapid chalking in paints through oxidation of zinc sulphide to water-soluble zinc sulphate. It is not, therefore, used in exterior finishes.

CHECKING

The term used for short, shallow cracks in an exposed film, usually only in gloss finishes and affecting only the top coat. There are three general types.

1 **Line checking**, where the cracks are usually parallel, often following the lines of brushmarks.
2 **Crow's-foot checking** The shallow cracks are joined thus ⅄ with angles of approximately 120° between the cracks. Their appearance suggests a possible association with the boundaries of Bénard cells (see Flotation and Flooding, p. 96).
3 **Irregular checking** is a mixture of the line and crow's-foot types. It is the common type shown on exposed films.

It has been stated [12] that slight checking is not a serious defect. It indicates a release of stresses due to shrinkage in a film which might otherwise crack.

Severe checking, when the cracks are wider and have joined, is known as 'alligatoring' (q.v.) This often occurs if an insufficiently elastic finish is applied over a more elastic undercoat. The undercoat becomes visible through the cracks. By using contrasting colours, this defect has been used to produce decorative effects ('crackle' finish).

The use of resin-bonded plywoods for exterior cladding is increasing, but the veneers tend to develop line checking along the grain on exposure to the weather. The most suitable coatings are stated [13] to be textured types based on either oleoresinous media or emulsions, both vinyl and acrylic.

CISSING

This is retraction of a paint film from certain areas of a substrate or from an existing glossy film. The areas may be small and isolated, but if they are appreciable the defect is known as 'crawling'.

The basic cause of cissing is poor wetting, resulting from a high contact angle. On a metal substrate this may be due to contamination. The only remedy is cleaning off and ensuring that the metal is clean and dry.

Application of a gloss finish to an old glossy paint can result in cissing. The remedy is thorough rubbing down of the old glossy finish.

COBWEBBING

A defect encountered in conventional spraying of some types of medium, particularly chlorinated rubber. It is associated with the solvent blend and occurs when there is a deficiency of high-boiling solvent in the mixture.

COLOUR CHANGES

Colour changes on storage

Prolonged storage of paints containing coloured pigments can result in gravitational separation of pigments with dissimilar physical characteristics, and it is vital, therefore, that paints should be mixed very thoroughly before use. Hand stirring is the least efficient method. Mechanical stirrers or a paint reconditioner such as the 'Vibro shaker' [14] are much quicker and more effective. Many complaints of 'incorrect colour' originate in failure to mix the paint thoroughly before use.

A more troublesome defect is often encountered with pale shades of blue, green and grey. When containers are opened, the paint surface can exhibit a mottled appearance due to flocculation of stainers, and the bulk of the paint is therefore a different colour from that expected. In some cases the flocculates are readily re-dispersed by hand or mechanical stirring, but there are occasions when they are more resistant. These resistant flocculates are, however, usually disrupted by the shearing action of a brush (when coloured streaks are produced) and by a spray-gun giving a mottled appearance to the film. The addition of surfactants will often be of assistance, but which individual material is best in a given medium has to be determined by experiment. Remilling of the paint in the presence of the surfactant should also be tried. In very difficult cases, a change in the source or type of stainer may be necessary.

Prussian blue, when used in small quantity as a stainer in oxidizing paints, sometimes loses its colour completely on storage. It appears to be reduced to a colourless form, but the colour is recovered when the film is exposed to the air.

Colour changes on exposure

Many types of colour change can occur for a variety of reasons. The following list includes those which are most commonly encountered.

1 **Fading** is most apparent in pale shades and can occur, to different degrees, in both exterior and interior situations. The most frequent cause is the use, as stainers, of organic pigments of poor lightfastness. Many organic pigments possess good lightfastness in the pure state but lose it to varying degrees when diluted with white pigment. This fading, which is caused by the action of ultraviolet radiation, occurs more rapidly in the relatively pure air of country locations than in the more polluted atmospheres of towns and cities where much ultraviolet radiation is filtered out. Figures for the lightfastness of their pigments, both in the pure state and reduced with white pigment, are published by the manufacturers.

Apparent fading may, in effect, be chalking. This can be ascertained by washing with water. Any white 'chalk' would be removed.

2 **Action of atmospheric contaminants** In atmospheres containing sulphur dioxide, lead chromes appear to fade, but this is due to reduction of the chromate radicle by moist sulphur dioxide. For a similar reason, chrome greens gradually turn blue in industrial and urban atmospheres.

Ultramarine blue has been regarded as relatively fugitive when diluted with white and exposed to weather. This is erroneous. The pigment is highly sensitive to acids, and when used as a stainer in acidic urban atmospheres the colour is destroyed. If the paint is protected by a film of varnish, the colour is not destroyed.

3 **Alkaline substrates** New concrete, cement render and asbestos boards are highly alkaline and require paints in which both binder and pigments are alkali-resisting. Among the binders, chlorinated rubber, vinyls, two-pack epoxies and epoxy/pitch have been found satisfactory. A wide selection of pigments is available, but Prussian blue, which is attacked by alkali, should be avoided. For the same reason, chrome greens should not be used. Lead chromes are reddened by alkali.

4 **Yellowing** of alkyd paints is discussed later (p. 469).

5 **Discoloration**, due to mould growth, is common in damp locations, particularly if shielded from sunshine. The discoloration can take various forms and often looks like dirt on the surface. It can be differentiated from dirt by the action of dilute sodium hypochlorite solution. Mould stains on the surface are bleached, but dirt is unaffected. When infected surfaces have to be repainted, they should be stripped off, the substrate treated with a fungicidal wash (p. 419) and a fungicide incorporated in the new paint system.

CRACKING

This is a form of film breakdown in which the cracks extend to the substrate. In this respect it differs from alligatoring and checking, although it sometimes starts with the latter. The two principal types of cracking are as follows.

1 **Random cracking** in which the breaks in the film are of no regular pattern. This type occurs when a short-oil system is applied to a substrate in which appreciable movement can take place, e.g. steelwork in which considerable expansion and contraction can occur. It can also be caused by film embrittlement (q.v.).

The remedy is complete removal of the system and replacement with one which is more elastic.

2 **Line cracking**, in which the cracks form parallel lines, is commonly seen on window sills and the lower horizontal members of window frames. It arises from ingress of moisture, mainly at the ends of the woodwork, followed by expansion of the cells forming the grain. The paint will crack at the junction of summer and spring wood if it is insufficiently elastic to accommodate the movement. Cracking allows moisture to penetrate between paint and wood, resulting in flaking. It is not possible to correct this fault on windows without removing them from the masonry and priming the ends adequately. Failing this, the cracked paint must be removed and the woodwork repainted at the required intervals.

CRATERING ('fish eyes')

These are bowl-shaped depressions in a paint film caused by air bubbles breaking through the film when it has partially set (cf. Aeration). They are formed also in some stoving operations when the coated article has been given insufficient flash-off time.

'Rain spotting' is a form of cratering caused by raindrops on a partially set film.

CROCODILING, see Alligatoring

DRYING PROBLEMS

Slow drying may be due to one or more of the following causes.

1 Cold and/or damp conditions. The rate of solvent loss and of chemical reactions in the film are slowed down. It is recommended in BS 5493 [15] that no exterior painting be carried out when the temperature falls to 4°C (39°F) or below and the relative humidity rises to 90 percent. Alkyds, in particular, are very sensitive to adverse weather conditions (see Loss of Gloss).

2 Oil or grease on the surface is likely to be dissolved into the film and not only retard drying but also keep the film soft and liable to 'pick-up' if further coats are applied by brush.

3 The film is excessively thick, possibly due to high viscosity. Not only is solvent retained within the film for an extended period, but surface skinning may occur before the underlying paint has set and rivelling can result.
4 The drier content may be low or the drier combination may be incorrect.
5 On storage of the paint, adsorption of drier metals by certain pigments can extend the drying time. The principal offenders are carbon black and, to a lesser degree, rutile titanium dioxide.

Remedies consist in (a) addition of further driers to the paint; (b) remilling the paint when much, but not all, of the original drying property will be recovered; a smaller addition of drier is then required, (c) in the manufacture of carbon black paints which may be stored for some time, it is advisable to add a material such as 'Nuact' [16] at the dispersing stage. This material is a lead compound which satisfies the adsorption requirements of the carbon black.

EFFLORESCENCE

A 'growth' of white crystals which appears on wall surfaces as the walls dry out or as moisture permeates through to the warmer inside surface. It can, however, appear on exterior surfaces if the conditions are favourable. The crystals usually consist of sodium sulphate. If the interior wall is coated with a moisture-permeable film, such as a flat emulsion paint, the solution of sodium sulphate will pass through and the salt will crystallize on the surface. Usually it can be brushed off, but it will reappear until the wall dries out or the source of permeating moisture has been removed.

Should the wall surface be coated with a relatively impermeable film such as a semi-gloss emulsion or solvent-type wall paint, the sodium sulphate will accumulate behind the paint film, causing loss of adhesion and disruption of the film. In this case the paint should be removed and the wall be allowed to dry out, or the source of penetrating moisture removed before redecoration.

EMBRITTLEMENT

Two of the common causes are progressive oxidation of air-drying paints, and over-stoving or deficiency of plasticizing alkyd in alkyd/amino systems and plasticizer loss (by migration or evaporation) from films of non-convertible paints. The defects can be mitigated, if not completely remedied, by (a) using a longer oil-length medium in the air-drying oxidizable paints, or (b) choosing a less volatile plasticizer for non-convertible coatings and ensuring that surfaces are completely sealed to reduce plasticizer migration.

FLAKING

This is usually preceded by cracking (q.v.) and adhesion failures (q.v.).

FLASHING

Term applied to the uneven gloss sometimes observed on wall surfaces. It is generally caused by variation in the absorption of the wall surface. The remedy is application of a sealer over the entire surface before repainting.

Flashing can also be caused by uneven spreading of low sheen paints, such as eggshell finishes. This is often described as 'sheariness' (q.v.).

FLOTATION AND FLOODING

Both defects occur as a consequence of flocculation of (usually) one of the components in a mixture of dispersed pigments. Whether flotation or flooding takes place is influenced largely by the type of solvent evaporation, and this aspect has been discussed on p. 95. Other factors, however, are also involved (*below*). Briefly, in the early stages of drying, solvent evaporation is accompanied by convection currents in the paint film, and these operate until the viscosity of the film becomes too high. If the solvent evaporation takes place freely (i.e. ebullioscopic) from a large number of points in the surface, Bénard cells are produced, giving the surface a mottled appearance. This is *flotation.* When flotation occurs on films applied by dipping, the Bénard cells become elongated as a result of drainage and the effect is known as 'silking' (q.v.). If, however, the solvent evaporation is diffusive, i.e. takes place in still air under equilibrium conditions, the surface colour is uniform and the effect is known as *flooding*.

These defects are commonly encountered with some chrome greens and also in pale shades produced by tinting titanium dioxide white with phthalocyanine blue, carbon black, or ferrite yellow stainers. Generally it is the stainer which floats or floods, but under certain conditions, flotation of the titanium dioxide can occur.

The view has been advanced by Daniel [17] and by Patton [18] that the defects could be avoided if 'co-flocculation' of white and stainer could be induced.

The fundamental mechanisms of flocculation have been investigated by a number of workers [19]–[23] as part of the wider study of the stabilization of pigment dispersions. Among the factors involved in flocculation are pigment mobility, particle size and shape, electron drift and demand, the nature of the solvent and, in flotation and flooding, the humidity of the air.

Remedies for flotation and flooding

Addition of small quantities (1 part to 40 000 parts of paint) of certain silicone oils will convert flotation to the more uniform flooding.

In view of the many factors involved it is impossible to name a universal cure or palliative. Generally the addition of surface-active agent is the time-honoured approach, but the quantities are often critical. This applies to two widely used expedients, soya lecithin and Troykyd Anti-Float [24]. Other materials which have de-flocculated some organic pigments are fatty acids – the linoleic-rich types for air-drying alkyd media or saturated types (such as coconut-oil fatty acids) for stoving finishes.

GASSING

Evolution of gas, usually hydrogen, can take place in containers of paint pigmented with metallic aluminium or zinc. In some cases the pressure developed is sufficient to distort the container or force off the lid.

Aluminium pigments, in spite of the surface treatment, are very sensitive to acids. Media for use in aluminium paints must be of very low acidity with a maximum acid value of 4. Onset of attack is evidenced by a dulling or 'greying' of the aluminium particles. This is followed by release of gas, and ultimately gelling of the medium. Water in very small quantities can cause gassing in aluminium paints, but this can be prevented by the incorporation of molecular sieves [25] as moisture absorbers.

The safest method of avoiding gassing, particularly if aluminium paints are being sent to tropical climates, is the two-pack system in which medium and pigment are packed in separate containers, and mixed immediately before use.

Zinc, in the form of zinc-rich paints, is also reactive to acidity and moisture, resulting in gassing. Molecular sieves can be employed, but the safest remedy is the two-pack system.

Gassing takes place occasionally in emulsion paints on storage. It is a result of bacterial attack on the cellulosic thickener and indicates a deficiency of preservatives.

GLOSS, LOSS OF

Loss of gloss in finishes can occur within a few hours of application or after some months. It can take the form of 'haziness' or it can be a decrease in surface smoothness with an increase in diffuse at the expense of specular reflection. The main causes are the following.

1 Unsuitable solvent mixtures in which the slowest evaporating component is not the best solvent for the resin. This may induce a tendency to 'throw out' and to give a discontinuous film.

2 Two resins which appear to be compatible in a solvent mixture may show incompatibility on drying out. Matting or hazing can result. This defect is shown by certain alkyd/chlorinated rubber and alkyd/vinyl mixtures.

3 **Alkyd gloss paints** pass through a sensitive period in the course of drying. If during this period there is a drop in temperature and condensation is produced (as can happen at night), matting of the film takes place. This has been shown to be micro-rivelling. In most cases it is the horizontal surfaces which are affected, but a cold moist wind can affect vertical surfaces also. Pentaerythritol type alkyds are more prone to the defect than those based on glycerol. The remedy is to allow the paint to dry hard, to rub down, and then apply a further coat of finish. To avoid the trouble, exterior painting should not be carried out late in the day when temperatures are likely to fall appreciably.

On exterior exposure, loss of gloss can be a precursor to chalking. This indicates a breakdown in the surface layer of lightly pigmented medium (see Chalking).

If a gloss finish is applied over an undercoat which is underbound or not thoroughly dry, 'sinkage' can occur, resulting in loss of gloss.

Low sheen paints, which are applied usually to interior walls and woodwork, can exhibit a poor gloss or 'sheen' when dry, for the following reasons.

1 The wall surface has not been sealed adequately, resulting in sheariness or uneven gloss. This occurs with both solvent-type and emulsion paints. The remedy is application of a coat of sealer over the entire surface, followed, when dry, by a further coat of finish.

2 The previous coat may not be sufficiently dry or is underbound. In the latter case a coat of sealer is required, as in **1**. The former requires a further period of drying before application of an additional finishing coat.

3 The presence of moisture in pigments or extenders can cause these paints to dry to a lower degree of gloss. However, any appreciable amount of moisture usually affects the flow properties and is therefore evident before application.

LIFTING

This is usually caused by the strong solvent in a paint attacking the film below, causing swelling and destroying the adhesion to the substrate (the action is similar to, but less severe than that of paint remover). Lifting occurs when a chlorinated rubber paint is applied over an alkyd type which has not aged sufficiently.

Softening of an undercoat, which is not sufficiently hard, by a finish does not necessarily result in lifting. A brush-applied finish will show pick-up, resulting in brush drag and serious brushmarking of the finish.

OPACITY DEFECTS

Poor opacity in a film can result in 'grinning through' when the film fails to obliterate the substrate or coating beneath. The cause of poor opacity may be one or more of the following.

1 The film is too thin. There is a tendency in decorative work, particularly when using thixotropic paints which spread easily, to brush out the paint too far and so produce an excessively thin film.

2 The paint may have been overthinned. This is likely to result, also, in sagging and runs on vertical surfaces.

3 The paint may contain insufficient opaque pigment.

4 Some bright-coloured finishes, such as reds, yellows and blues, may require the use of organic pigments and these, in the main, possess poor opacity. Increasing the pigment content has little effect, and if a proportion of inorganic pigment cannot be incorporated, it is necessary to formulate the undercoat shade to be as close as possible to that of the finish.

The superior opacity of inorganic pigments can be utilized in many undercoats when their lack of brilliance is not a disadvantage.

5 Flocculation of the pigment Instances of the flocculation of titanium dioxide during storage and during drying of paints are frequently reported. Flocculation of stainers (which has little effect on opacity) is discussed under Colour Changes on Storage (q.v.). Flocculation of titanium dioxide can occur in emulsion paints on storage. The inclusion of a small quantity of finely divided silica is often effective in keeping the pigment in a dispersed condition.

Loss of opacity due to flocculation of titanium dioxide has been encountered during the cure of some acid catalysed alkyd/amino systems. The defect appears to be associated with the type of acid catalyst employed and is more likely to occur with hydrochloric acid than with other acid catalysts.

ORANGE PEEL

This is an uneven finish, resembling the peel of an orange, which sometimes appears on sprayed finishes. Possible causes are the following.

1 The viscosity of the paint is too high.

2 The solvent mixture is ill-balanced. If the proportion of highly volatile solvents is too high, the deposited film will be insufficiently fluid to flow out. (Dry spray also results from an excess of low-boiling solvent.) The remedy is an increase in the proportion of high-boiling solvent.

3 The air pressure must be correct for the type of material. 'Orange peel' can result from pressures which are too high or too low.

4 The gun is held too far from the surface. Much solvent is lost before the atomized paint reaches the surface. If the loss is excessive, dry spary results, and a discontinuous bitty film is produced.

Films exhibiting orange peel or bittiness should be rubbed down and re-sprayed.

PEELING

This is a consequence of breakdown of adhesion between paint and substrate or between coats of paint – see Adhesion Failures.

PIGMENT SETTLEMENT IN STORAGE

The factors involved in producing stable pigment dispersions have been discussed on p. 258.

With modern paints, many of which may be stored for considerable periods (six months to a year), it is unusual to encounter hard settlement. Except in the truly thixotropic paints there is a general downward movement of pigment particles under gravity (the finest particles are probably kept in suspension by Brownian movement) and it is essential, therefore, to mix all paints thoroughly before use. The practice of inverting closed tins at intervals serves to minimize settlement.

Pigment settlement is accelerated if paint is stored under warm conditions when the viscosity of the medium is reduced.

PINHOLING

This can be a mild form of cissing (q.v.). In thin films it can be due also to any of the following.

1 Aeration of the paint, small air bubbles being trapped in the film and released after the film has set. If the film sets rapidly, the bubbles fail to break and can be mistaken for bittiness. In thick films the release bubbles produce craters.
2 Particles of gelled resin can leave pinholes when they dry out. This type of pinholing occurs with some high molecular-weight polymers such as epoxies. It is claimed that epoxy resins produced by the 'advancement' process are less prone to this defect [26]. Particles of resin can be present in many types of finish as a consequence of the use of unsuitable thinners.
3 Too short a flash-off period before stoving. This applies particularly to water-thinned coatings.

RIVELLING (or 'Wrinkling')

A surface defect which can occur during the drying of some oxidizing paints or during the stoving of finishes based on tung oil. A surface skin is formed before the remainder of the film has set or cured.

Rivelling can occur under the following conditions.

Air drying paints (a) The film is too thick. The surface skin reduces the rate at which oxygen can reach the material below. (b) An excess of cobalt drier has been used. Rivelling is more pronounced if the medium contains tung oil. In this case the addition of small quantities of zinc naphthenate or calcium naphthenate can be beneficial [27].

Stoving paints A form of rivelling known as 'gas-checking' occurs with tung oil media (particularly if the tung oil is not properly bodied) in gas ovens if the products of combustion come into contact with the paint film. Other causes are an extended flash-off time and a too rapid increase in temperature.

Wrinkle finishes (p. 399) can be produced by addition of a proportion of unbodied tung oil to a short-oil alkyd medium.

SAGGING (Curtaining, Runs)

Sagging is the downward movement of a wet paint film on a vertical surface. If it occurs over a wide area, the term 'curtaining' is often used. 'Runs' denote sagging from restricted areas such as cracks, etc.

Sagging may be due to one or more of the following.

1 Excessive film thickness. In spraying, the gun is probably being held too close to the surface. Brushing paints should be applied uniformly and brushed out well.
2 High-build coatings, applied by airless spray, will sag if insufficient structuring agent is present.
3 The film is slow setting. This may be due to the type of binder, the use of too much slow-evaporating solvent in the solvent mixture, or too low temperature.
4 The paint has been thinned excessively.

SAPONIFICATION

This is due to the chemical decomposition of some types of binder by alkali present in new concrete, cement render, asbestos boards, or lime plaster. The signs are discoloration and stickiness; in severe cases the film may become semi-liquid and sag. Saponification can occur with alkyds, epoxy esters, oleoresinous and other binders based on esterified compounds.

Remedy: the paint film must be removed completely and the surface scrubbed clean and allowed to dry. Paints used on alkaline surfaces should be based on non-saponifiable resins such as chlorinated rubber, synthetic rubbers or vinyl resins. Alternatively the binders listed in para. 1 above can be used, provided the surface is sealed with at least two coats of a non-saponifiable sealer.

SHEARINESS

This is a defect usually associated with low sheen and emulsion paints. It appears as a variation in sheen or gloss, often accompanied by brushmarking.

The common causes are as follows.

1 An absorbent substrate, especially plaster. The remedy consists in application of one or two coats of sealer followed by a further coat of finish.
2 A short wet-edge time, so that the joining up of adjacent areas produces a 'fat edge' with a different degree of gloss from the main area. The wet-edge time of solvent-based finishes (both low sheen and gloss) can be extended by the addition of a small quantity of methylethylketoxime. With emulsion paints it is usually a sign of an absorbent substrate, and this requires treatment with sealer.

SILKING

A defect appearing on dip applied coatings in which flotation is present. The Bénard cells have become elongated as a result of drainage. For remedies, see Flotation and flooding.

THICKENING ON STORAGE

Possible causes are as follows.

Solvent-based paints

1 Reaction between a basic pigment and an acidic binder. The consistency can sometimes be reduced by addition of small quantities of benzoic, salicylic or citric acids.
2 Pigment flocculation, in highly pigmented systems such as undercoats. Addition of cationic or short chain non-ionic surfactant will often assist in de-flocculating the system.
3 Incompatibility, where two or more resins are used. Reformulation is necessary.
4 Incorrect solvent blend. Reformulation is necessary.
5 Delayed swelling of thickeners. Reformulation is necessary.
6 Adsorption of low molecular-weight polymer fractions by certain pigments, notably carbon black. In the early stages, thinning is possible, but this results in loss of opacity. Reformulation is necessary. Incorporation of a small quantity of Nuact [16] at the dispersing stage is beneficial.

Emulsion paints

1 Pigment flocculation. Polyphosphates used as sequestering agents for polyvalent ions can, over a period of time, change into orthophosphates which can cause pigment flocculation [28].
2 Coagulation of dispersed polymer through exposure to frost.
3 Alginates, used as protective colloids, can react with polyvalent ions if insufficient sequestering agent is present. A fall in pH also will cause an increase in the viscosity of alginate solutions.

THINNING (of emulsion paints) on storage

This is usually due to bacterial infection which may have been present in the raw materials or to insufficient preservative.

YELLOWING

The yellowing of white paint films has been a problem for many years, particularly with films based on linseed and tung oils (oleoresinous or oil paints) or alkyds containing linseed-oil fatty acids. The yellowing of linseed oil appears to be associated with the presence of the highly unsaturated linolenic acid (see p. 445).

Yellowing is accentuated by darkness and occurs to a greater degree in interior situations. Certain atmosphere contaminants, notably ammonia, can cause rapid yellowing.

Natural oils such as soya bean, sunflower, safflower and tall oil contain high percentages of linoleic acid (they are 'linoleic rich' oils) and are free, or essentially free from linolenic acid. White paints based on alkyds containing the fatty chains of these oils show a much reduced tendency to yellowing. These alkyds form the basis of present-day decorative finishes.

In an investigation into the effect on yellowing of the grade of titanium dioxide employed, Simpson [29] has shown that minimum yellowing results from the use of the rutile form as either a chloride grade or a fine sulphate grade. The same author reports a retarded rate of yellowing by the incorporation of either colloidal zinc oxide (which unfortunately can have side effects such as chemical reactivity and poor gloss retention) or ethyl-3-oxybutanoate (aceto-acetic ester).

Yellowing also occurs in films of oleoresinous and spirit varnishes based on phenolic and cresylic resins. The latter produces the more intense yellowing.

References to Chapter 27

[1] *Hess's Paint Film Defects*, 3rd edn, pp. 256 ff. Chapman & Hall, London, 1979.
[2] Sathyanarayana, M. K. *et al.*, Studies on adhesion of alkyd based titanium dioxide paints; role of acid and hydroxyl values of the medium. *JCT*, 1979, **51**, No. 657, 75.
[3] 'Duomeen TDO' (Akzo Chemie UK Ltd, Wandsworth, London).
'Colorol' (Croxton & Garry Ltd, Dorking, Surrey).
[4] Hurst, A. E, and Goodier, J. H., *Painting and Decorating*, 9th edn, p. 246. Charles Griffin & Co. Ltd, 1980.
[5] *Hess's Paint Film Defects*, p. 32.
[6] Whiteley, P. *et al.*, Deterioration of paint films by water. *JOCCA*, 1975, **58**, 200.
[7] Prosser, J. L. and Bullett, T. R., *JOCCA*, 1962, **45**, 836.
[8] Mayne, J. E. O., *JOCCA*, 1950, **33**, 538.
[9] Ulfvarson, U. and Pattyramie, C., The blistering of paints on wood induced by moisture. **I**: The influence of some structural variables of pinewood on blistering induced by moisture of an alkyd paint system. **II**: The influence of various treatments of pinewood on blistering induced by moisture. *JOCCA*, 1972, **55**, 207–26.
[10] Geissler, G., *Defazet*, 1973, **10**, 462.
[11] Allyn, G. in *Treatise on Coatings* (eds R. R. Myers and J. S. Long), Vol. 1, Part 1, p. 14. Edward Arnold, London, 1967.
[12] *Hess's Paint Film Defects*, p. 262.
[13] Gunn, D. J., The coating of external plywood cladding panels. *JOCCA*, 1980, **63**, 28.
[14] Westlairds Ltd, Datchet, Slough.
[15] BS 5493: 1977. Code of practice for protective coating of iron and steel structures against corrosion. British Standards Institution, 2 Park Street, London, W1A 2BS.
[16] Durham Chemicals Ltd, Durham.
[17] Daniel, F. K., Flocculation and related pigment interactions in paints. *Paint Tech.*, 1965, **29**, No. 2, 14.
[18] Patton, T. C., *Paint Flow and Pigment Dispersion*, 2nd edn. Wiley-Interscience, 1979.
[19] Brownlie, G. and Furans, G. C., Some factors related to pigment dispersion. *JPT*, 1966, **38**, 113.
[20] Carr, W., Optical properties of films. *JOCCA*, 1976, **59** 443.

[21] Crowl, V. T., Flocculation, flotation and flooding in phthalocyanine/titanium dioxide pigmented paints. *JOCCA*, 1967, **50**, 1023.
[22] *Idem*, The theory and practice of dispersion. *JOCCA*, 1972, **55**, 388.
[23] Parfitt, G. D., Dispersion. *JOCCA*, 1967, **50**, 822.
[24] Troy Chemical Corporation (Cole Chemicals Ltd, Purley, Surrey).
[25] 'Zeoliths' (Bayer).
[26] *Hess's Paint Film Defects*, p. 56.
[27] *Ibid.*, p. 81.
[28] *Ibid.*, p. 13.
[29] Simpson, L. A., Yellowing of alkyd paint films, *JOCCA*, 1981, **64**, 490.

Appendix A Bibliography

Application and cure

Bailey, A. G., *Electrostatic Spraying of Liquids—Fundamentals and Applications*. Research Studies Press, 1988.

Goodier, J. H., *Dictionary of Painting and Decorating*. Charles Griffin, 1974.

Hurst, A. E. and Goodier, J. H., *Painting and Decorating*, 9th edn. Charles Griffin, 1980.

OCCA. *Ultraviolet Curing*, Parts 1 & 2. Oil & Colour Chemists Association, 1976/78.

Philips, R., *Sources and Applications of UV Radiation*. Academic Press, 1983.

Roffey, C. G., *Photopolymerisation of Surface Coatings*. John Wiley, Chichester, 1982.

Tatton, W. H., and Drew, E. N., *Industrial Paint Application*, 2nd edn. Newnes Butterworth, 1981.

Colour and Pigments

Allen, R. L. M., *Colour Chemistry*. Nelson, London, 1971.

Allen, T., *Particle Size Measurement*, 3rd edn. Chapman & Hall, 1981.

Clulow, F. W., *Colour: its Principles and their Application*. Fountain Press, 1972.

Grigg, S. J. and Sing, K. S. W., *Adsorption, Surface Area and Porosity*, 2nd edn. Academic Press, 1982.

Ishihara, S., *Tests for Colour Blindness*. Lewis, London, 1959.

Lewis, P. A. and Patton, T. C. (Eds), *Pigment Handbook*, 3 vols. John Wiley, Chichester, 1988.

Lowell, S. and Shields, J. E., *Powder Surface Area and Porosity*, 2nd edn. Chapman & Hall, 1984.

McLaren, K., *The Colour Science of Dyes and Pigments*, 2nd edn. Adam Hilger: Bristol, 1983.

Saunders, K. H., and Allen, R. L. M., *Aromatic Diazo Compounds*, 3rd edn. Edward Arnold, 1985.

Stockham, J. D. and Fechtman, E. G. (Eds), *Particle Size Analysis*. Ann Arbor/Wiley: Chichester, 1977.

Corrosion

Evans, U. R., An *Introduction to Metallic Corrosion*, 3rd edn. Edward Arnold, 1979. *idem The corrosion and Oxidation of Metals*. Edward Arnold, 1960. *1st Supplementary Volume*, 1968. *2nd Supplementary Volume*, 1976.

Shrier, L. L. (Ed), *Corrosion*, 2nd edn, Vol. 1, *Corrosion of Metals and Alloys*, Vol. 2, *Corrosion Control*. Newnes Butterworth, 1976.

Defects

Hamburg, H. R. and Morgans, W. M. (Eds), *Hess's Paint Film Defects*, 3rd edn. Chapman & Hall, 1979.

Federation of Societies for Coatings Technology, *Pictorial Standards of Coatings Defects*. Philadelphia, Pa, 1979.

General

Dimwoodie, J. M., *Timber: its Nature and Behavior*. van Nostrand, 1981.

Harris, S. T., *Technology of Powder Coatings*. Portcullis Press, Redhill, 1976.

OCCA Manuals, Part 7, *Works Practice*. Oil & Colour Chemists Association, 1974.

OCCA Monographs: No. 1, T. A. Banfield, *Marine Coatings*; No. 2, J. W. Nickolson, *Water-borne Coatings*; No. 3, E. V. Schmid, *Painting on Zinc Surfaces and Zinc-containing Anti-corrosive Primers*; No. 4, E. G. Elphinstone, *The Determination of the Solid Content of Paints by Volume*. Oil & Colour Chemists Association.

OCCA, *The Surface Coating and Raw Materials Directory*, 1986.

Simpson, L. A., *Measuring Opacity*. TiINFO System 3.3.1 Tioxide UK Ltd, Stockton on Tees.

Waddams, A. L., *Chemicals from Petroleum*, 4th edn. John Murray, London, 1978.

Watt, I. M., *The Principles and Practice of Electron Microscopy*. Cambridge University Press, 1985.

Resins

Doi, M. and Edwards, S. F., *The Theory of Polymer Dynamics*. Oxford University Press, 1986.

Oldring, P. and Hayward G. (Eds), *Resins for Surface Coatings*.

Pierman, I. (Ed.), *Emulsion Polymerization*. Academic Press, N. Y, 1982.

Tess, A. W. and Fochlein, G. W., *Applied Polymer Science*. American Chemical Society, 1986.

Solvents

Durrans, T. H., *Solvents*, 8th edn revised by E. H. Davies. Chapman & Hall, 1971.

Flick, E. W. (Ed.), *Industrial Solvents Handbook*, 3rd edn. Noyes Publications, 1985.

Riddick, J. A., Bunger, W. B. and Sakano, T. K., *Organic Solvents. Physical Properties and Methods of Purification*, 4th edn. John Wiley: Chichester, 1986.

Surfactants, wetting and dispersion

Karsa, D. (Ed.), *Industrial Applications of Surfactants*. Special publication No. 69, Royal Society of Chemistry, London, 1986.

Parfitt, G. D. (Ed.), *Dispersion of Powders in Liquids with Special Reference to Pigments*, 3rd edn. Applied Science Publishers, 1981.

Patton, T. C., *Paint Flow and Pigment Dispersion*, 2nd edn. Wiley Interscience, 1979.

Tadros, T. F., *Surfactants*. Academic Press, London, 1984.

Appendix B Composition of inorganic pigments and extenders

(where composition is not apparent from the name)

Material	Composition
Antimony vermilion	Antimony sulphide
Asbestine	Natural calcium magnesium silicate
Barytes	Natural barium sulphate
Bauxite residue	Mainly oxide of iron
Black, Bone	Carbon mixed with calcium phosphate
Black, Carbon	Almost pure carbon
Black, Drop	The best quality Bone Black
Black, Gas	Identical with Carbon Black
Black, Lamp	Carbon (identical with Vegetable Black)
Blacklead	Identical with graphite
Black, Vegetable	Carbon (identical with Lamp Black)
Blanc Fixe	Precipitated barium sulphate
Blue, Bronze	Ferric ammonium (or sodium) ferrocyanide
Blue, Brunswick	Prussian Blue with extender
Blue, Celestial	Prussian Blue with extender
Blue, Cerulean	Mixed oxides of cobalt and tin
Blue, Chinese	Term formerly applied to best Prussian Blue
Blue, Cobalt	Mixed oxides of cobalt and aluminium
Blue, French	Identical with Ultramarine Blue
Blue, Milori	A variety of Prussian Blue
Blue, Prussian	Ferric potassium ferrocyanide
Blue, Thénard's	Identical with Cobalt Blue
Blue, Ultramarine	Complex sulphosilicate of aluminium and sodium
Brown, Purple	Iron oxide
Bronze powders	Alloys of copper
Cadmopone	Co-precipitated cadmium sulphide and barium sulphate

Calcite	Natural crystalline calcium carbonate
China clay	Natural hydrated aluminium silicate
Chrome, Lemon	Lead chromate and lead sulphate
Chrome, Mid	Lead chromate
Chrome, Orange	Basic lead chromate
Chrome, Primrose	Lead chromate, lead sulphate and alumina
Chrome, Scarlet	Lead chromate, lead molybdate and lead sulphate
Graphite	Semi-crystalline carbon
Green, Bronze	A chrome green modified with other pigments, often oxides of iron
Green, Brunswick Green, Chrome	Mixtures of lead chrome and Prussian Blue
Green, Cobalt	Mixed oxides of cobalt and zinc
Green, Guignet's	Hydrated oxide of chromium
Gypsum	Natural hydrated calcium sulphate
Kaolin	Identical with china clay
Kieselguhr	A natural diatomaceous silica
Lapis lazuli	Natural form of Ultramarine Blue (q.v.)
Lead Orange Lead, Red	Oxide of lead (Pb_3O_4)
Lead, White	Basic lead carbonate
Lithopone	Co-precipitated zinc sulphide and barium sulphate
Magnesite	Natural magnesium carbonate
Maroon, Cadmium	Co-precipitated cadmium sulphide and selenide
Mica	Natural hydrated aluminium potassium silicate
Micaceous oxide	Natural black iron oxide
Ochres	Hydrated oxides of iron and siliceous matter
Red, Cadmium	Co-precipitated cadmium sulphide and selenide
Red, Indian	Synthetic iron oxide
Red, Turkey	Synthetic iron oxide
Red, Venetian	Synthetic iron oxide with calcium sulphate
Sienna, Burnt	Iron oxide containing manganese oxide and siliceous matter
Sienna, Raw	Hydrated iron oxide, manganese oxide and siliceous matter
Talc	Natural hydrated magnesium silicate
Umber, Burnt	Iron oxide, manganese oxide and siliceous matter
Umber, Raw	Hydrated iron oxide, manganese oxide and siliceous matter
Veridian	Hydrated chromium oxide
Vermilion	Mercuric sulphide
Violet, Cobalt	Cobalt pyrophosphate
White, Antimony	Antimony oxide
White, Chinese	Zinc oxide
White, Flake	Basic lead carbonate (identical with white lead)

White, Mineral	Identical with Gypsum
White, Paris	Calcium carbonate
White, Satin	Calcium sulphate and aluminate
White, Strontium	Strontium sulphate
White, Zinc	Identical with Lithopone
Whiting	Calcium carbonate
Witherite	Natural barium carbonate
Yellow, Cadmium	Cadmium sulphide

Appendix C Some colouring-matters and resins of historic interest

Aureolin A double nitrite of cobalt and potassium. A bright transparent yellow, very expensive but used occasionally as an artists' water colour.

Arsenic Yellow or King's Yellow Arsenic sulphide (As_2S_3). Originally the natural ore 'orpiment' was used but later the pigment was manufactured. It was used as a yellow pigment before the advent of lead chromes.

Carmine Lake Produced from carminic acid (see Cochineal, below) by precipitation with calcium and aluminium salts. It possesses a brilliant red colour but poor lightfastness.

Cochineal A dyestuff extracted from the dried bodies of the female cochineal beetle. The principal component is carminic acid, an anthraquinone derivative.

Crimson Lake A lake precipitated from carminic acid by aluminium and tin salts. A bright colour and less fugitive than Carmine Lake.

Dutch Pink A yellow lake prepared from the extract of the bark of a North American oak. The colouring principle is quercitrin, a tetrahydroxyflavonol.

Egyptian Blue A pigment introduced in Egypt in about 3000 B.C. It remained the principal blue pigment in general use for many centuries; a crystalline silicate of copper.

Indigo (natural) A pigment obtained from various species of *Indigofera*. This has now been replaced by the synthetic material.

Logwood Prepared from the dye extracted from the wood of *Haematoxylin campechianum* (indigenous to Mexico and Central America). Brown to blue-black lakes are precipitated with various mordants.

Madder Lake The dyestuff from the root of *Rubia tinctorum*; contains alizarin and some purpurin. The lake is formed by treatment with aluminium salt and alkali. It shows good lightfastness and was formerly used as a standard for this property. The synthetic product is used as an artists' colour.

Persian Berries Lake A yellow lake prepared from an extract of the unripe berries of a species of buckthorn. The colouring principle is rhamnetin, a flavonol.

Purple Lake A lake produced in a manner similar to that for Carmine Lake (*above*) but with the addition of lime to produce the deep purple tone. It is used as an artists' colour.

Sepia A dark-brown pigment obtained from marine animals of the family Sepiidae (cuttlefish). Its principal use was in artists' colours.

Tyrian Purple A bright pigment much used by some ancient peoples. It was obtained from a variety of snail, *Murex brandaris*. The colouring-matter is 6:6-dibromoindigotin.

Vandyke Brown A natural brown earth containing ferric oxide. It continues to be used as an artists' colour.

Weld A yellow lake prepared from an extract of *Reseda luteola*. The colouring principle is luteolin, a tetrahydroxyflavanone.

Resins

The following list contains some resins of historic interest only; others are used on a small and diminishing scale.

Amber A fossil type and the hardest known resin. Used in some forms of jewellery.

Copals This term covers both fossil and 'modern' types. The latter continue to be used, but the former, which are soluble in oils only after thermal treatment ('gum running'), have been largely replaced by the hard synthetic resins.

Dragon's blood A red-coloured resin exuded from a species of plant native to eastern Asia. At one time it was used for colouring lacquers.

Gamboge A yellow gum resin obtained from certain trees indigenous to Thailand. It has poor chemical resistance and lightfastness.

Kauri A hard fossil copal obtained from New Zealand. It is now very expensive and used only for special purposes.

Bibliography

Barry, T. Hedley, *Natural Varnish Resins*. Ernest Benn, London, 1932.

Appendix D Electrode potentials

Galvanic series	**Electrochemical series**	
Cathodic end	*Volts*	
Gold	Au	+ 1·50
Graphite		
Silver	Ag	+ 0·80
Stainless steel 18/8/3 (passive)		
Stainless steel 18/8 (passive)		
Inconel (passive)		
Nickel (passive)		
Monel		
Copper/nickel alloys		
Bronzes		
Copper	Cu	+ 0·34
Brasses	H	0·00 (reference standard)
Inconel (active)		
Nickel (active)		
Tin		
Lead	Pb	− 0·126
Lead/tin solders	Sn	− 0·136
Stainless steel 18/8/3 (active)	Ni	− 0·250
Stainless steel 18/8 (active)	Cd	− 0·402
Cast iron		
Steel & iron	Fe	− 0·440
Cadmium		
Zinc	Zn	− 0·762
Aluminium	Al	− 1·670
Magnesium alloys		
Magnesium	Mg	− 2·380
Anodic end		

Appendix E Temperature conversion tables

°C	°F	°C	°F	°C	°F	°C	°F
− 17·8	0	22	71·6	70	158·0	128	262·4
− 15·0	5	23	73·4	72	161·6	130	266·0
− 12·2	10	24	75·2	74	165·2	132	269·6
− 9·4	15	25	77·0	76	168·8	134	273·2
− 6·6	20	26	78·8	78	172·4	136	276·8
− 4·0	25	27	80·6	80	176·0	138	280·4
− 1·1	30	28	82·4	82	179·6	140	284·0
0	32	29	84·2	84	183·2	142	287·6
1	33·8	30	86·0	86	186·8	144	291·2
2	35·6	31	87·8	88	190·4	146	294·8
3	37·4	32	89·6	90	194·0	148	298·4
4	39·2	34	93·2	92	197·6	150	302·0
5	41·0	36	96·8	94	201·2	160	320·0
6	42·8	38	100·4	96	204·8	170	338·0
7	44·6	40	104·0	98	208·4	180	356·0
8	46·4	42	107·6	100	212	190	374·0
9	48·2	44	111·2	102	215·6	200	392·0
10	50·0	46	114·8	104	219·2	210	410·0
11	51·8	48	118·4	106	222·8	220	428·0
12	53·6	50	122·0	108	226·4	230	446·0
13	55·4	52	125·6	110	230·0	240	464·0
14	57·2	54	129·2	112	233·6	250	482·0
15	59·0	56	132·8	114	237·2	260	500·0
16	60·8	58	136·4	116	240·8	270	518·0
17	62·6	60	140·0	118	244·4	280	536·0
18	64·4	62	143·6	120	248·0	290	554·0
19	66·2	64	147·2	122	251·6	300	572·0
20	68·0	66	150·8	124	255·2	310	590·0
21	69·8	68	154·4	126	258·8	315	600·0

Conversion factors

$$°C = (°F - 32) \times \frac{5}{9} \qquad °F = \frac{°C \times 9}{5} + 32$$

Appendix F Conversion factors (× = multiply by)

Length	
'Thou' or 'mils' to micrometres (μm)	× 25.4
Inches to centimetres	× 2.54
Yards to metres	× 0.914
Miles to kilometres	× 1.609
Area	
Square inches to square centimetres	× 6.4516
Square yards to square metres	× 0.836
Volume	
Pints to litres	× 0.568
Gallons (Imp.) to litres	× 4.54
Gallons (US) to litres	× 3.79
Cubic feet to litres	× 28.4
Cubic yards to cubic metres	× 0.764
(1 cubic metre = 1000 litres = 220.3 Imp. gallons = 35.2 cubic feet)	
Mass	
Ounces to grams	× 28.35
Pounds to kilograms	× 0.454
Cwts to kilograms	× 50.85
Density	
Lbs/cubic foot to kg/cubic metre	× 16.0
Lbs/Imp. gallon to kg/litre	× 0.1
Miscellaneous	
Gallons (Imp.) per minute to litres per second	× 0.0758
Psi to N/m^2	× 6895
Psi to bars	× 0.0689
1 bar = 14.5038 lb/in^2 = $10^5 N/m^2$ = 0.987 Atm.	
1 Atm. = 14.6959 lb/in^2 = 101.325 KN/m^2 = 1.013 bar	

Appendix G Specific gravities

Common pigments		Extenders	
Aluminium paste, Standard	1.47	Barytes, micronised	4.4
Antimony oxide	5.5	Barytes, white	4.5
Black oxide of iron	4.17	Blanc fixe	4.3
Cadmium colours	4.2–4.7	Calcite	2.7
Calcium plumbate	5.7	'Celites'	2.24
Carbon black	1.7–2.0	China clay	2.2–2.6
Chrome green, light	5.0	Dolomite	2.8–2.9
Chrome green; mid	4.3	Gypsum	2.3
Chrome green; deep	3.4	Magnesite	3.0
Chromium oxide	5.2	Mica, water ground	2.8
Ferrite yellow	4.0	'Microcal'	2.8
Graphite, natural	2.5	Pearlite	2.4
Graphite, manufactured	2.25	Silica, natural	2.6–2.8
Lead chromes	5.9–6.2	Silica, manufactured	2.0–2.2
Lead sulphate, basic	6.4	Talc	2.65
Micaceous iron oxide	4.7	Whiting	2.8
Pigment dyestuffs	1.4–1.7		

Prussian blue	1.97
Red lead	8.8
Red oxides	4.2–4·4
Ultramarine blue	2.35
Vegetable (lamp) black	1.8–1.9
White lead	6.8
Yellow ochre	3.0
Zinc chrome, Type 2	3.4
Zinc chrome, Type 5	3.7
Zinc oxide	5.65
Zinc phosphate	3.30

Metals

Aluminium	2.6
Bronzes	8.5–8.7
Cast iron	7.0–7.2
Copper	8.85
Lead	11.34
Magnesium	1.75
Steel	7.7–7.8
Wrought iron	7.6–7.9
Zinc	7.1

Oils

Castor, blown	0.98
Castor, first pressings	0.958–0.969
Castor, dehydrated	0.964
Linseed, acid refined	0.930–0.934
Linseed, pale boiled	0.945–0.955
Linseed, raw	0.930–0.936
Linseed, stand oils	0.95–0.97
Soya bean, refined	0.924–0.928
Sunflower, refined	0.922–0.927
Tung, A. Fordii	0.939–0.943
Tung, A. Montana	0.934–0.943

Solvents

Acetone	0.792
'Aromasol H'	0.88
isoButanol	0.805
n-Butanol	0.812
isoButyl acetate	0.87
n-Butyl acetate	0.88
Cyclohexanol	0.947
Cyclohexanone	0.946
Diacetone alcohol (tech.)	0.942
Dipentine	0.864
Ethanol	0.830
Ethyl acetate	0.903
Ethyl alcohol (abs.)	0.79
isoPropanol	0.785
Propylene glycol	1.06
Propylene glycol, ether/esters	0.873–0.970
'Shellsols'	0.859–0.892
'Solvesso 100'	0.872
'Solvesso 150'	0.895
SBP 1	0.681
SBP 2	0.695
SBP 3	0.724
SBP 4	0.700
SBP 5	0.715
SBP 6	0.767
SBP 7	0.701

IMS 64 OP	0.82	SBP 8	0.729
Methyl ethyl ketone	0.806	SBP 11	0.776
Methyl isobutyl ketone	0.80	Toluene	0.87
Methylene chloride	1.336	White spirit	0.78
Naphtha, 90/160	0.854	White spirit 115	0.79
Naphtha, 90/190	0.880	Xylene	0.87
Pine oil	0.925–0.945		

Solid resins

Chlorinated rubber	1.60	Modified phenolics	1.1
Ethyl cellulose	1.14	Rosin	1.06
Gilsonite	1.05–1.10	Vinyls	1.35–1.39
Manila copal	1.07		

Miscellaneous materials

Acrylic plastics	1.18–1.20	Pitch	1.10–1.15
Asphalt	1.0–1.4	Polythene	0.93–0.96
'Bentones'	1.7–1.8	PVC	1.34–1.37
Brick, hard	1.18–1.22	Sand	2.3–2.5
Brick, soft	1.5–1.9	Sea water (non-polluted)	1.025
Cork	0.21–0.25	Timbers	0.5–0.8
Concrete	2.0–2.5	Triethanolamine	1.122
Creosote	1.06–1.08		
Glass, soda	2.6		
Ice	0.92		
Methyl ethyl ketoxime	0.923		
Phosphoric acid, ortho	1.70		

Index